World Geomorphological Landscapes

Series editor

Piotr Migoń, Wroclaw, Poland

For further volumes:
http://www.springer.com/series/10852

Francisco Gutiérrez • Mateo Gutiérrez
Editors

Landscapes and Landforms of Spain

Springer

Editors
Francisco Gutiérrez
Mateo Gutiérrez
Earth Sciences
University of Zaragoza
Zaragoza
Spain

Every effort has been made to contact the copyright holders of the figures and tables which have been reproduced from other sources. Anyone who has not been properly credited is requested to contact the publishers, so that due acknowledgment may be made in subsequent editions.

ISSN 2213-2090 ISSN 2213-2104 (electronic)
ISBN 978-94-017-8627-0 ISBN 978-94-017-8628-7 (eBook)
DOI 10.1007/978-94-017-8628-7
Springer Dordrecht Heidelberg New York London

Library of Congress Control Number: 2014934839

Printed on acid-free paper

Springer is part of Springer Science+Business Media (www.springer.com)

Series Editor Preface

Landforms and landscapes vary enormously across the Earth, from high mountains to endless plains. At a smaller scale, Nature often surprises us creating shapes which look improbable. Many physical landscapes are so immensely beautiful that they received the highest possible recognition—they hold the status of World Heritage properties. Apart from often being immensely scenic, landscapes tell stories which not uncommonly can be traced back in time for tens of million years and include unique events. In addition, many landscapes owe their appearance and harmony not solely to the natural forces. Since centuries, or even millennia, they have been shaped by humans who modified hillslopes, river courses, and coastlines, and erected structures which often blend with the natural landforms to form inseparable entities.

These landscapes are studied by Geomorphology—'the Science of Scenery'—a part of Earth Sciences that focuses on landforms, their assemblages, surface and subsurface processes that moulded them in the past and that change them today. Shapes of landforms and regularities of their spatial distribution, their origin, evolution and ages are the subject of research. Geomorphology is also a science of considerable practical importance since many geomorphic processes occur so suddenly and unexpectedly, and with such a force, that they pose significant hazards to human populations and not uncommonly result in considerable damage or even casualties.

To show the importance of geomorphology in understanding the landscape, and to present the beauty and diversity of the geomorphological sceneries across the world, we have launched a new book series *World Geomorphological Landscapes*. It aims to be a scientific library of monographs that present and explain physical landscapes, focusing on both representative and uniquely spectacular examples. Each book will contain details on geomorphology of a particular country or a geographically coherent region. This volume, the second in the series, introduces the geomorphology of Spain—a country with highly diverse landscapes, from coastal flats and deltas to very high mountains of different origin. More than 20 selected examples from mainland Spain and its islands are presented, along with fascinating stories behind the marvellous sceneries. Thus, the book is not only suitable for scientists and students of Geography and Earth Science, but can also provide guidance to holidaymaking geoscientists as to where to go to enjoy the very best scenery.

The World Geomorphological Landscapes series is produced under the scientific patronage of the International Association of Geomorphologists—a society that brings together geomorphologists from all around the world. The IAG was established in 1989 and is an independent scientific association affiliated with the International Geographical Union and the International Union of Geological Sciences. Among its main aims are to promote geomorphology and to foster dissemination of geomorphological knowledge. I believe that this lavishly illustrated series, which sticks to the scientific rigour, is the most appropriate means to fulfil these aims and to serve the geoscientific community. To this end, my great thanks go to Prof. Francisco and Prof. Mateo Gutiérrez for coordinating the efforts of Spanish geomorphological community and expertly editing the book, as well as to all individual contributors who worked together to show us the Spanish landscape at its best.

Piotr Migoń

Contents

Short Biodata of Authors

Roy W. Alexander graduated with B.Sc. Geography from the University of Newcastle, M.Sc. Ecology from the University of Durham and Ph.D. Botany from Chelsea College, London. He has conducted research into vegetation-erosion relationships in northern England and SE Spain, where much of his work has focussed on the distribution and dynamics of biological soils crusts at El Cautivo. He is Professor of Environmental Sustainability at the University of Chester.

Juan Luis Arsuaga specialises in palaeobiological, palaeodemographical and palaeopathological aspects. He works at the Complutense University of Madrid, and is the Head of the UCM-ISCIII Joint Centre for Research into Human Evolution and Behaviour.

Daniel Ballesteros is carrying out his Ph.D. in the Department of Geology of the University of Oviedo. His thesis is focused on speleogenesis in the Picos de Europa combining caving, geomorphological, geological and geochronological techniques. He has explored and surveyed around 26 km of cave passages.

Teresa Bardají obtained her Ph.D. in Geology from the Complutense University of Madrid (Spain) and Reader at the University of Alcalá since 1988. Her main research lines include Quaternary sea-level changes in Mediterranean and Atlantic settings, as well as Neotectonics and Palaeoseismology. She has conducted geological and geomorphological mapping in different areas of Spain. Former President of the Spanish Quaternary Association (2001–2009) is currently President of the INQUA National Committee.

Santiago Beguería is Tenured Scientist at the Aula Dei Experimental Station (CSIC), where he uses models to study the evolution of soil erosion and stream flow.

José María Bermúdez de Castro conducts research on Palaeobiology and dental anthropology, and has worked in the National Natural Sciences Museum (Higher Council of Scientific Research, CSIC), and at the CENIEH, where he currently leads the Palaeobiology of Hominids Program.

Jaime Bonachea obtained his Ph.D. at the Universidad de Cantabria (Spain) with the European mention. He joined the Universidad de Zaragoza (Spain) in 2007 and 2010. He is currently a lecturer at the Universidad de Cantabria. His main research interests are landslide and sinkhole risk modelling, as well as global geomorphic change. He has more than 100 contributions in international journals and congresses and has participated in several international and national research projects.

Viola M. Bruschi obtained the Degree in Geological Sciences at the Università degli Studi di Modena (Italy). In 2007 she obtained her PhD at the Universidad de Cantabria (Spain). Nowadays, she is a researcher at the Universidad de Cantabria. Her research is mainly focused on geomorphological processes, geological heritage, geological risk, and global geomorphic change. She has more than 80 contributions in international journals and congresses and has been involved in national and international research projects. Since 2006, she is member of the "Working Group on Geomorphosites" of the "International Association of Geomorphologists".

José María Calaforra is Professor of Geodynamics in the Department of Hydrogeology and Analytical Chemistry at the University of Almería. After completing his doctorate in Geological Sciences, he began his research work in the Water Resources and Environmental Geology Research Group, focussing on geomorphological and hydrogeological topics. He is author of the book "Gypsum Karstology" and of more than 200 scientific papers dealing with karst and cave speleogenesis.

Alfonso Benito-Calvo defended his Ph.D. at the Complutense University of Madrid, and has received the María Jesús Ibáñez National Award for Spanish Geomorphological and Quaternary Studies. He works on geomorphological mapping, Late Cenozoic landscape evolution and GIS modelling.

Adolfo Calvo-Cases graduated with B.Sc. and Ph.D. in Geography at the University of Valencia, Spain, where he is employed as Professor of Physical Geography. He has conducted research on landform evolution and hillslope processes in different areas of the Mediterranean and northern Britain.

Yolanda Cantón graduated with a B.Sc. in Biology at the University of Granada and a Ph.D. in Biological Sciences in the University of Almería. She began her research at the EEZA (CSIC) on surface hydrology and erosion of semiarid ecosystems in 1994. Since 2000 she works at the University of Almería as Lecturer on soil erosion and GIS. Her research is currently focused on the ecohydrology of semiarid ecosystems, especially on the role of BSCs on water and C balance of these systems.

Eudald Carbonell focused his expertise on the study of technological, cultural and archaeological aspects. He is Professor at the Universitat Rovira i Virgili and Director of the Institute of Human Palaeo-Ecology and Social Evolution (IPHES).

Luis Carcavilla graduated B.Sc. in Geology, M.Sc. in Applied Hydrology and Ph.D. in Geoconservation. He is Full Scientific Researcher in the Geological Survey of Spain (Instituto Geológico y Minero de España, IGME). His research themes include geoconservation, geological heritage management and popularisation of science and has published eight books and a large number of scientific contributions.

Juan Carlos Carracedo graduated and received his Ph.D. in Geology at the Universidad Complutense, Madrid. Since 1980 he worked at the Spanish Research Council (CSIC) as a Research Professor and Director of the Estación Volcanológica de Canarias (EVC-CSIC). After retirement, he has continued his scientific activity as Emeritus Researcher associated to the University of Las Palmas de Gran Canaria (Canary Islands). Prof. Carracedo has worked in the Canary Islands over 40 years and published more than 200 scientific articles on the geology, palaeomagnetism and geomorphology of the Canaries, in addition to two dozen books, e.g. *Canarian Volcanoes: I. Tenerife, II. Gran Canaria, III. Lanzarote and Fuerteventura; IV. La Palma-La Gomera-El Hierro* (Ed. Rueda, 1978–1980 and 2008); *La Erupción de Lanzarote de 1730* (Cabildo de Lanzarote 1991); *Geology and Volcanology of La Palma and El Hierro, Canary Islands* (Estudios Geológicos 2001); *Geological Guide of the Canary Islands, Classic Geology in Europe* (Terra Publishing, London 2002); *The Geology of Spain-The Canary Islands* (The Geological Society, London 2002); *Los Volcanes del Parque Nacional del Teide* (Serie Técnica, Parques Nacionales, Ministerio Medio Ambiente 2006); El Volcán Teide (Serv. Pub. Caja-Canarias 2006); *Encyclopedia of Islands: The Canary Islands* (University of California Press 2009); *Active Volcanoes of the World: Teide Volcano, Geology and Eruptions of a Highly Differentiated Oceanic Stratovolcano* (Springer-Verlag 2013).

Rosa M. Carrasco is Ph.D. in Geology and Lecturer in Geology and Geomorphology in the Universidad de Castilla-La Mancha. Her work and research topics have focused on tectonic geomorphology, paleoglacial geomorphology and mass movements. She is expert in geographical information systems (GIS) applied to geomorphology and environmental studies. She has published numerous peer-reviewed papers on Spanish and international scientific journals and also is author or co-author of several books: *Geomorfología de El Valle del Jerte* (1999),*Geomorfología. Principios, Métodos y Aplicaciones* (1996)*and Guía Metodológica para los Estudios de Medio Físico*(2006).

Antonio Cendrero is Fellow of the Royal Academy of Sciences of Spain, Emeritus Professor at Universidad de Cantabria (Spain) and Honorary Professor at the National Universities of La Plata and Mar del Plata (Argentina). His research or teaching activity are in Universidad Complutense de Madrid (Spain), Dartmouth College, University of Texas at Austin and California State University (USA) and University of Sulaimaniyah (Iraq). Invited lectures or short courses in over 20 countries. Member of national and international organisations, including the Steering Committee of The Global Terrestrial Observing System (UN) and Vice-Presidency of Cogeoenvironment (IUGS). His research interests include volcanic geology, environmental geology and land-use planning, geomorphic hazards, human activities-geomorphic processes interactions and global geomorphic change. He has author or co-authored over 260 scientific contributions.

Javier Chueca-Cía studied Geography at the University of Zaragoza (Spain) and at the International Institute for Geo-Information Science and Earth Observation (Enschede, Netherlands). He is currently Profesor Tit-ular at the Department of Geography in the University of Zaragoza. His main research topics are glacial, periglacial and snow processes (avalanche hazard analysis) and geomorphological mapping and he has published more than 50 peer-reviewed papers on these subjects. His field research experience includes stays at Iceland, Poland and Norway and at the Iberian Peninsula he has mainly worked in the Spanish and French Central Pyrenees.

Javier de Pedraza is Ph.D. in Geology and Senior Lecturer in Geomorphology in the Complutense University of Madrid. His main research topics include geomorphology of Variscan massifs and environmental geology, with emphasis in landscape evolution, granitic landforms, glacial and periglacial geomorphology, environmental evaluation/restoration and conservation. These topics have been published in numerous peer-reviewed articles on Spanish and international scientific journals. He is also co-author of several books; *Formas Graníticas de la Pedriza* (1989), *Geomorfología. Principios, Métodos y Aplicaciones* (1996) *and Guía Metodológica para los Estudios de Medio Físico* (2006).

José Ramón Díaz de Terán got his Ph.D. in Geology at Universidad de Oviedo (Spain). He is Senior lecturer in Geomorphology at Universidad de Cantabria (Spain). He has participated as researcher in both international and national projects. He is co-author of more than 100 scientific contributions. He has lectured in several national and international universities. He is member of the European Centre on Geomorphological Hazards.

Andrés Diez-Herrero graduated B.Sc. in Geology, M.Sc. in Applied Hydrology and Ph.D. Fluvial Geomorphology and Hydrology. He is Full Scientific Researcher in the Geological Survey of Spain (Instituto Geológico y Minero de España, IGME), Former lecturer on Environmental Geology and Water Resources in the University Complutense of Madrid, the European University of Madrid, the SEK University of Segovia and the University of Castilla-La Mancha. His main research topics include flood hazard and risk analysis using geological and geomorphological methodologies, palaeohydrology and dendrogeomorphology. He has published around 30 papers in peer-reviewed international journals and more than 90 chapters in scientific books. See detailed information in www.andresdiezherrero.es.

María José Domínguez-Cuesta is Lecturer at the Department of Geology of the University of Oviedo since 2012. Her research interest includes landscape evolution, slope instability processes and natural hazards.

David Domínguez-Villar is Ph.D. in Geology, Researcher in the Spanish national research centre for human evolution (CENIEH) and Honorary Research Fellow at the University of Birmingham. He is expert in granite geomorphology, Quaternary paleoclimate and speleothem science, including stable isotope geochemistry/geochronology of stalagmites and cave processes. His research has been published in numerous international and national scientific journals. He is author or co-author of the books *La Naturaleza de Torrelodones* (1999), and *Speleothem Science: from process to past environments* (2012).

Gema Fernández-Maroto holds a degree and Ph.D. in Geology from the Universidad de Oviedo (Spain) with a Certificate in Environmental Engineering. She has collaborated in several international and national research projects. Currently, her research is focused on rock falls and slope processes mechanism related to geotechnical properties. Nowadays, she is Lecturer at the Universidad de Cantabria.

Germán Flor graduated with BSc and PhD from the University of Oviedo, and has worked in the Faculty of Geology since 1970, carrying out research on sedimentology and geomorphology of beaches, aeolian dunes and estuaries in Spain. He is now Senior Lecturer of Marine Geology at University of Oviedo, Spain.

Germán Flor-Blanco completed his B.Sc. and Ph.D. in the University of Oviedo. His research has been focused mostly on sedimentology and geomorphology of coastal systems in Spain. He is now Reader in Marine Geology at the University of Oviedo, Spain.

José M. García-Ruiz is Research Professor at the Pyrenean Institute of Ecology (CSIC). His main fields of interest are soil erosion, environmental hydrology and landscape evolution.

Fernando Gázquez graduated in Environmental Sciences in 2007 at the University of Almería (Spain). Afterwards, he carried out his Ph.D. on the study of the geochemical characteristics of gypsum speleothems as palaeoclimatic proxies, including those of the gypsum karst of Sorbas. Since 2013 he has been involved in the study of underground environments and cave minerals as Martian analogues at the Department of Mineralogy and Petrology of the University of Valladolid (Spain).

Angel Ginés and Joaquín Ginés graduated and obtained their Ph.D. at the Universitat de les Illes Balears (Spain). Both have taught for more than 10 years as assistant lecturers in the Biology and Earth Science Departments of the same University. Their main research topics were, respectively, karren landform processes and coastal karst evolution. At the present time, they share as research interests cave geomorphology and speleogenesis, especially the study of cave sediments and speleothems related to Quaternary sea-level changes. **AG** is co-editor of the books *Karren Landforms* (1996) and *Karst Rock Features. Karren Sculpturing* (2009). **JG** is co-editor of the monographs *Geomorfología litoral: Migjorn y Llevant de Mallorca* (2007) and *El Carst: patrimoni natural de les Illes Balears* (2011). Furthermore, both were co-editors of the publications *El Carst i les Coves de Mallorca/Karst and Caves in Mallorca* (1995) and *Mallorca: a Mediterranean benchmark for Quaternary studies* (2012).

Alberto González-Díez is graduated in Geology and M.Sc. in Engineering Geology at Universidad Complutense de Madrid (Spain). He defended his Ph.D. in Geology at Universidad de Oviedo (Spain). Currently, he is Senior lecturer at Universidad de Cantabria. He is the President of the Spanish Society of Geomorphology (SEG) since 2012.

F. Javier Gracia graduated in Earth Sciences from the University of Zaragoza and has lectured for 23 years at the University of Cádiz, South Spain. His research covers different topics like coastal geomorphology and dynamics, coastal hazards, historical evolution of coastlines, karst and lacustrine geomorphology. He has published more than 50 peer-reviewed scientific papers and has edited several monographs, like Geomorfología litoral, procesos activos (2000) or *Las dunas en España* (2011). Between 2008 and 2010, he was president of the Spanish Society of Geomorphology, and has been co-editor of the journal *Cuaternario y Geomorfología* (2010–2012).

Jesús Guerrero studied Geology at the Universities of Zaragoza (Spain) and Oxford Brookes (UK). He carried out his Ph.D. in geomorphology in the University of Zaragoza with merits. His main research topics are evaporite karst, fluvial geomorphology, halokinetic processes and geological hazard analysis in Spain and the USA. He is currently teaching Geology and Hydrogeology at the University of Zaragoza.

Francisco Gutiérrez studied Geology at the Universities of Zaragoza (Spain) and Aberdeen (UK). He is currently Full Professor at the Department of Earth Sciences in the University of Zaragoza and has carried out research stays at the University Complutense of Madrid, the Faculty of Geographical Sciences in Utrecht University and the Colorado Geological Survey. His main research topics include evaporite karst, landslides, tectonic geomorphology and geomorphological mapping. He has published more than 60 peer-reviewed papers and has coedited eight special issues in international journals. Francisco is a member of the Executive Committee of the International Association of Geomorphologists (2005) and of the editorial board of the journals Geomorphology (2003) and International Journal of Earth Sciences (2007).

Mateo Gutiérrez studied Geology at the Universidad Complutense of Madrid. He has taught several geological subjects at Madrid University, the Colegio Universitario of Teruel, and the University of Zaragoza, where he became Full Professor in 1980. He has given postdoctorate courses on Geomorphology at Sao Paulo University (Brazil) and San Juan University (Argentina). His main research topics include regional geomorphology, karst, tectonic geomorphology, aeolian landforms, weathering features, periglacial geomorphology, and soil erosion. He has published more than 100 peer-reviewed papers, has edited the book Geomorfología de España and is author of the books Climatic Geomorphology (Elsevier), Geomorfología (Pearson) and Geomorphology (CRC Press-Balkema).

Adrian M. Harvey graduated with B.Sc. and Ph.D. from University College, London. He was employed for 40 years by the University of Liverpool, carrying out research in modern and Quaternary fluvial systems in northern Britain, Spain, the USA and the Middle East. He is now Emeritus Professor of Geomorphology in the University of Liverpool.

Montserrat Jiménez-Sánchez is Lecturer at the Department of Geology of the University of Oviedo since 2001. Her research interests include landscape evolution, with special reference to glacial, karstic and slope instability processes, geoarchaeology and geomorphological heritage.

Asunción Julián-Andrés studied Geography at the University of Zaragoza (Spain) and at the International Institute for Geo-Information Science and Earth Observation

(Enschede, Netherlands). She is currently *Profesor Titular* at the Department of Geography in the University of Zaragoza. Her main research topics are glacial, periglacial and snow processes (avalanche hazard analysis) and geomorphological mapping. Her field research experience includes stays at Poland and Norway and at the Iberian Peninsula she has mainly worked in the Spanish and French Central Pyrenees. She has published more than 50 peer-reviewed papers.

Roberto Lázaro-Suau is B.sc. and Ph.D. in Biology at the Valencia University, Spain. After 4 years working in private companies, he worked for 20 years as upper technician at the Estación Experimental de Zonas Aridas (EEZA-CSIC) in Almería, Spain, in topics related to plant ecology and experimental field installations. Since 2002 he is a staff researcher in the same institute, where he mainly conducts research on plant ecology.

Juan I. López-Moreno is Tenured Scientist at the Pyrenean Institute of Ecology (CSIC). His main research topics are water resources evolution, snow accumulation and snow hydrology.

Carlos Martí-Bono is Tenured Scientist at the Pyrenean Institute of Ecology, where he studies glacial landforms and evolution.

Ángel Martín-Serrano studied Geology at the Universidad Complutense de Madrid. He is currently Researcher in the Instituto Geológico y Minero de España, where he has served as Head of the Geomorphology and Geological Mapping Section, and currently as a co-Director of the Research Department. His main investigation topics include stratigraphy, geomorphology, mapping, paleosoils and neotectonics, mainly focused on the Variscan basement and the Cenozoic basins of central Spain. He has also conducted research in República Dominicana, Andes and the Antarctic Peninsula. He has published numerous scientific papers, the reference methodological guidelines for the elaboration of geomorphological and geoenvironmental maps in Spain and more than 45 geological and geomorphological sheets of Spain, República Dominicana and Argentina. He is the main author of the Geomorphological Map of Spain at 1:1,000,000 scale.

Anne Mather completed her B.Sc. in the University of Hull and her Ph.D. at the University of Liverpool. She has worked on Neogene, Quaternary and Modern fluvial systems especially in Spain and Chile. She is now Reader in Physical Geography at the University of Plymouth.

Ana Isabel Ortega obtained her Ph.D. which merited the Extraordinary Doctorate Award at the University of Burgos. She is Vice President of the Edelweiss Speleological Group, and has wide experience in speleological surveying, karst geomorphology and archaeology.

María Ortuño-Candela studied Geology at the Universities of Granada (Spain) and Washington (Seattle, USA). She got her Ph.D. in Earth Sciences at the University of Barcelona, where she is currently working as a postdoc researcher and teacher, as part of the RISKNAT group and Geomodels Institute. The main topics of her research are neotectonics, paleoseismology and mass movements, with special focus on slow gravitational landslides. Her field experience includes stays in Kamchatcka, Italy, Canada, Nicaragua and Mexico. At the Iberian Peninsula, she has worked mainly at the Eastern Betics Shear Fault Zone and the Central Pyrenees.

Lluís Pallí Ph.D. in Geology (Autonomous University of Barcelona 1972), is Professor of Geology at the University of Girona. He is a specialist in geomorphology.

Alfredo Pérez-González carried out his Ph.D. Thesis at the Complutense University of Madrid. During his career, he has worked for the Spanish Geological Survey (IGME), at the University of Zaragoza and the Complutense University of Madrid, and is now the Director of the CENIEH. His research interests include Neogene and Quaternary landscape evolution and site formation processes.

Juan Puigdefabregas graduated in Biology and Ph.D. from de University Navarra, Pamplona, Spain. He was employed for 42 years by CSIC (Spain) carrying out research on geo-ecology in mountain and arid environments in Spain, Maghreb, Sahel, Patagonia and NW China. He is now Emeritus Professor in the CSIC at the Estación Experimental de Zonas Aridas, Almería, Spain.

Juan Remondo is Senior Lecturer at Universidad de Cantabria (Spain). He got his Ph.D. in Geology at Universidad de Oviedo (Spain). He has participated in Lecturing and research activities in several foreign centres in Italy, The Netherlands, Canada, Argentina, Brazil and Mexico. He has produced more than 150 scientific contributions, most of them in international journals and volumes. His main research interests are natural hazard assessment and mapping, GIS and SDA applications in Geomorphology, mass movement processes, relationships between human activities and earth surface processes and global geomorphic change.

Laura Rodríguez-Rodríguez is Ph.D. student at the Geology Department of the University of Oviedo. Her research is focused on the extent and chronology of former glaciations in the central part of the Cantabrian Mountains, including the Northwest portion of the Picos de Europa Regional Park.

Inmaculada Rodríguez-Santalla has a Degree in Geology from the Complutense University of Madrid (1993) and obtained her Ph.D. in Environmental Sciences (1999) from the Alcalá de Henares University. Her doctoral thesis dealt with the evolution of the Ebro Delta (1999). Her professional activity in the Community of Madrid began in 1989, using Remote Sensing and Geographic Information Systems (GIS) applied to different disciplines. Subsequently, she has held different positions in various companies and public agencies, participating in numerous technical reports related to coastal geomorphology management. Currently, she is a Professor at the Rey Juan Carlos University, teaching various subjects related with Geology and Environmental Sciences. Her research activity is focused on Coastal Geomorphology and Integrated Coastal Management. She has directed and participated in different national and international research projects and has published numerous articles and conference papers. Additionally, she has supervised several doctoral theses and graduate works dealing with the research topics mentioned above.

Joaquín Rodríguez-Vidal Ph.D. in Geology by the University of Zaragoza (Spain) is currently Full Professor at the Department of Geodynamics and Palaeontology in the University of Huelva since 1993, where he also serves as the Director of the Geomorphology and Quaternary Group. He has led projects in northern Africa and the Iberian Peninsula linked with their recent geological evolution. In addition to his 200 scientific papers, he has been Elected Member of the Spanish Geological Commission and President of AEQUA (Spanish Quaternary Association).

Carles Roqué Ph.D. in Geology (Autonomous University of Barcelona 1993), is Lecturer at the Department of Environmental Sciences, University of Girona. He works on diverse geomorphological topics (granite landforms, karst and active processes) and has broad experience in geomorphological and geological mapping.

Ángel Salazar graduated B.Sc. in Geology and M.Sc. in Engineering Geology. He is Senior Science Technician in the Geological Survey of Spain (Instituto Geológico y Minero de España, IGME). Formerly, Professional Geologist employed at consulting companies, foundations and as freelance. His main interests include surficial deposits and Cenozoic sediments, geomorphological and geological mapping and geological heritage assessment and conservation. He has published 24 geological and geomorphological maps from local (1:25,000–1:50,000) to national scales (1:1,000,000–1:2,000,000), as well as more than 25 contributions in journals and books. A significant part of the 25 contributions presented in scientific meetings deal with geological heritage and geomorphological mapping.

Yasmina Sanjuan is contracted at the Pyrenean Institute of Ecology (CSIC), and is undertaking her Ph.D. research on the geomorphic evolution of the subalpine belt in the Pyrenees.

Jorge Sanjurjo has Ph.D. in Biology and Researcher at the Institute of Geology of the University of Coruña, Spain, where he is responsible for the Geochronology Unit. His main research interests are the study of biodeterioration in granite rocks in urban and natural environments and the application of luminescence dating techniques to archaeological and historical monuments, as well as sediments from natural environments. He is author of 54 peer-reviewed papers in national and international journals. He has worked in Spain, Argentina, Portugal and Syria.

Rogelio Linares holds a Ph.D. in Geology (Autonomous University of Barcelona 1995) and is Lecturer at the Department of Geology, Autonomous University of Barcelona. He is a specialist in geomorphology and hydrogeology.

María Pilar Serrano-Muela is contracted at the Pyrenean Institute of Ecology. She conducts research on the hydrological behaviour of mid-mountain areas.

Pablo G. Silva graduated from the Complutense University of Madrid (1988) and presented his Ph.D. Thesis on Tectonic Geomorphology of the Guadalentín Depression (1994). Since 1992, he is Reader in the Department of Geology of Salamanca University, teaching graduate courses on Geomorphology and Geophysics, and master courses on Tectonic Geomorphology, Quaternary and Natural Hazards. His main research interest include tectonic geomorphology and paleoseismology, but also he has expertise in alluvial fan, fluvial and coastal geomorphology in a variety of regions within the Iberian Peninsula. He has published over 80 peer-reviewed papers, and is author of several Geological and Geomorphological Maps edited by the Spanish Geological Survey (IGME), covering different regions of the Betic Cordillera, the Cantabrian Mountains and the Neogene basins of central Spain. Currently, he is leading a research project on Tectonic Geomorphology and Paleoseismology in the Betic Cordillera QTECTBETICA CGL2012 37581 C02-01.

Albert Solé-Benet graduated in Geology from the University of Barcelona, Spain, and obtained a Ph.D. in Soil Science from Université Paris VI, France. He has mostly worked on the morphological, mineralogical and physical properties of soils. He is a tenured scientist from the Spanish National Research Council (CSIC) since 1990 and works in Almería, Spain, at the Experimental Station for Arid Zones (EEZA).

Luis Somoza is Ph.D. from the Complutense University of Madrid, Spain. His doctoral thesis dealt with Quaternary sea-level changes and neotectonics recorded by emerged and submarine marine terraces along the Mediterranean coast of Spain. He has worked for nearly 18 years at the Marine Geology Division of the Geological Survey of Spain (IGME). He was formerly on the research staff at the Oceanographic Institute of Spain in Malaga and Professor of Geomorphology at the University of Salamanca. He obtained a fellowship in the Department of Geography at the University of Durham (United Kingdom) for his studies on Holocene sea-level changes. His research interests include coastal geomorphology, sea-level changes, geodynamics of marine terraces in the Mediterranean Sea and Atlantic Ocean and ice-sheet shelf evolution in Antarctica. He has carried out seven field campaigns in Antarctica and numerous oceanographic surveys. He has published over 100 peer-reviewed papers.

Martin Stokes completed his B.Sc. in the University of Liverpool and his Ph.D. in the University of Plymouth. His research has been primarily in the sedimentology and geomorphology of Neogene and Quaternary fluvial systems in Spain, Morocco and the USA. He is now Senior Lecturer in Geological Sciences at the University of Plymouth.

Blas Valero-Garcés is Research Professor at the Pyrenean Institute of Ecology (CSIC). His main fields of interest are limnogeology and environmental changes during the Quaternary.

Marcos Vaqueiro is Industrial Engineer by the Universidad of Vigo, Spain, and member of the UIS Commission for Pseudokarst since 2008. He is currently Researcher at the Institute of Geology of the University of Coruña. His main research interests include pseudokarst in non-soluble rocks, speleogenesis of granite cavities and dissolution of siliceous rocks. He has published more than 30 papers and books on karst and pseudokarst (granite rocks) and has participated in more than 120 speleological expeditions in Spain, Germany, Sweden, Portugal and Poland.

Juan Ramón Vidal-Romaní is Professor of Geomorphology at the University of Coruña, Spain, and Member of the UIS Commission for Pseudokarst since 2010. His main research interests are the study of Quaternary geology in Galicia, karst, granite geomorphology and pseudokarst, marine paleo-levels and glacial and volcanic geomorphology. He has conducted research in Spain, Madagascar, Argentina, Australia, Sweden and Portugal. He has published 47 books or book chapters, 37 geomorphological maps and around 130 peer-reviewed papers in national and international journals.

Elizabeth Whitfield (nee Maher), completed her B.Sc. and Ph.D. degrees in the University of Liverpool. She has worked mostly on Quaternary fluvial systems in Spain, but also in other Mediterranean areas, Turkey and Greece. She is now Lecturer in Earth Sciences at John Moores University, Liverpool.

Mario Zarroca Ph.D. in Geology (Autonomous University of Barcelona 2012), is junior Lecturer at the Department of Geology, Autonomous University of Barcelona. He is a specialist in geophysics applied to the geomorphology and hydrogeology.

The Geology and Geomorphology of Spain: A Concise Introduction

1

Francisco Gutiérrez, Mateo Gutiérrez, and Ángel Martín-Serrano

Abstract

Spain has a remarkable geomorphological diversity largely due to its geological and climatic variety. From the geological perspective, the Iberian Peninsula may be divided in two broad geological domains; the Iberian Massif in the western sector, and the mountains belts and Cenozoic basins related to Alpine tectonics in the eastern sector. The Iberian Massif (Variscan Spain) mainly consists of Paleozoic metamorphosed sedimentary formations intruded by plutonic rocks. This region is characterised by extensive planation surfaces locally interrupted by inselbergs, and includes outstanding examples of granitic landscapes. The Alpine Mountain Belts, related to the convergence between Europe, the Iberian microplate, and Africa, contain excellent examples of landscapes controlled by active tectonics. In these Alpine orogens, extensive limestone outcrops have favoured the development of outstanding poljes, dolines and karren fields Glacial landscapes are best developed in the Pyrenees, which still contain a number of active cirque glaciers. The Cenozoic Basins include some of the finest areas to examine stunning conglomerate monoliths, dramatic badlands, dune fields, deflation basins associated with lunette dunes and yardangs, and a wide variety of features related to evaporite dissolution. The Canarian Archipelago is a late Cenozoic chain of hot-spot-related volcanic islands located in the Atlantic Ocean west of the Sahara coast. The evolution of the Canaries is characterised by the growth of large volcanic edifices, punctuated by the development of giant landslides. The Teide volcano (3,718 m a.s.l.) in Tenerife rises more than 7 km above the adjacent abyssal plain. A total of 18 eruptions have been documented over the last 500 years, some of them with great societal impact; the 1730-1736 Timanfaya eruption covered more than 20 % of Lanzarote island. The around 10,000 km-long coastline of the Spanish territory display a wide variety of coastal landscapes, including rías, estuaries sequences of raised beaches, deltas, lagoons and spit bars, and dune fields.

Keywords

Spain • Geomorphological diversity • Regional geomorphology • Geoheritage

1.1 General Physiographic Features

Spain is the southernmost country in Europe and covers an area of 505,990 km^2, of which 12,465 km^2 corresponds to the Balearic Islands in the Mediterranean Sea and the Canary Islands in the Atlantic Ocean (Fig. 1.1). The Spanish territory has a remarkable geomorphological diversity largely due to its geological and climatic variability (Gutiérrez 1994;

F. Gutiérrez (✉) · M. Gutiérrez
Department of Earth Sciences, University of Zaragoza, Zaragoza, Spain
e-mail: fgutier@unizar.es

Á. Martín-Serrano
Instituto Geológico y Minero de España, Madrid, Spain
e-mail: a.martinserrano@igme.es

F. Gutiérrez and M. Gutiérrez (eds.), *Landscapes and Landforms of Spain*, World Geomorphological Landscapes,
DOI: 10.1007/978-94-017-8628-7_1, © Springer Science+Business Media Dordrecht 2014

Fig. 1.1 Physical map of Spain. Numbers correspond to the chapters of the sites and areas covered by the book. *2* Guadalentín tectonic depressions; *3* Neogene sedimentary basins of Almería; *4* Granite landforms in Galicia; *5* La Pedriza granitic massif; *6* Conglomerate monoliths in the Ebro Basin (several sites); *7* Tramuntana Range; *8* Atapuerca karst and palaeoanthropological sites; *9* Evaporite karst of Calatayud; *10* Gypsum karst of Sorbas; *11* Gallocanta Lake; *12* Playa lakes of Bujaraloz-Sástago; *13* Picos de Europa National and Regional Parks; *14* Ordesa and Monte Perdido National Park; *15* Maladeta Massif; *16* Block streams in the Tremedal Massif; *17* Badlands in the Tabernas Basin; *18* Ebro River Delta; *19* Doñana National Park; *20* Raised beaches in the Cantabrian coast; *21* Olot volcanic field; *22* Teide Volcano; *23* Timanfaya and Montañas del Fuego in Lanzarote; *24* Structural collapses in the Canary Islands

Martín-Serrano 2005; Benito-Calvo et al. 2009). Moreover, the pressure on the environment caused by long-sustained human activity and the dynamics of a wide variety of hazardous surface processes, make this country an excellent natural laboratory to investigate anthropogenic impacts on geomorphic systems and geohazards (e.g. Vilaplana 2008; García-Ruiz and López-Bermúdez 2009; Bruschi et al. 2013; Chap. 26).

The wide climatic variability is related to several geographical factors (Font 1983; Instituto Geográfico Nacional 1995): (1) The territory covers a wide latitudinal range, from around 44°N in northern Spain to 28°N in the Canary Islands, coinciding with the latitude of the Sahara Desert (Fig. 1.1). The annual average precipitation in the eastern islands of the Canaries and in the southern leeward flank of the western islands may reach values below 100 mm (Fig. 1.2). (2) The Iberian Peninsula is located between the Atlantic Ocean and the Mediterranean Sea. A large proportion of the precipitation in Spain is related to fronts coming from the Atlantic Ocean that typically traverse the Peninsula from NW to SE. The annual precipitation in most of the northern sector of the Peninsula exceeds 1,200 mm, whereas there is an extensive sector in the south-east where the yearly rainfall is below 400 mm. The Gata Cape, Almería Province, has a mean

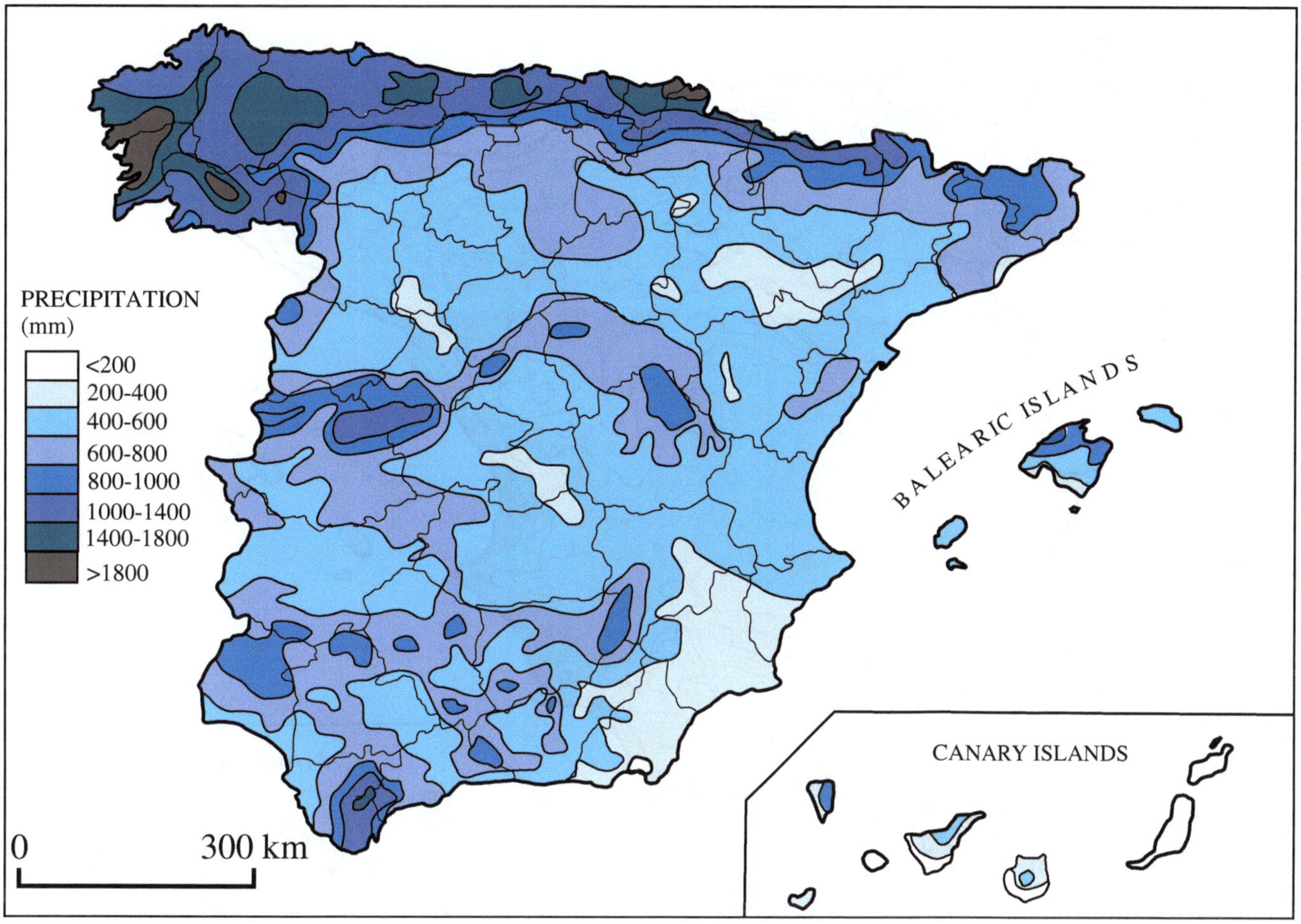

Fig. 1.2 Map of mean annual precipitation in Spain (modified from Instituto Geográfico Nacional 1995)

annual precipitation of ca. 130 mm (Fig. 1.2). Areas with annual precipitation below 400 mm also occur in the central sector of the depressions associated with the interior Cenozoic basins (Fig. 1.2). (3) The topography of the Iberian Peninsula is characterised by a mosaic of morphostructural depressions (Cenozoic basins) and mountain belts (mostly Alpine orogens), some of which are located next to the coast, acting as barriers for moist air currents (Fig. 1.1). The sharp topographic contrasts, together with the orientation of the slopes with respect to the atmospheric circulation, determine striking temperature and precipitation gradients with a decisive imprint on the geomorphology (Figs. 1.2, 1.3). For example, the distance between some active glaciers in the Pyrenees (Chaps. 14 and 15) and playa lakes with wind-fluted yardangs and evaporite deposition in the semi-arid Ebro Depression (Chap. 12) is just 150 km. (4) Spain has the second highest mean elevation in Europe (660 m), after Switzerland. This overall high altitude is related to the extensive area covered by mountain ranges and the presence of extensive elevated plateaus (*mesetas*) in central Spain (Gutiérrez 1994), largely corresponding to planation surfaces and structural surfaces (Fig. 1.1). These topographic characteristics have a significant influence on some climatic features with geomorphological significance. The average annual number of days with temperature below 0 °C exceeds 120 days/year in most of the mountain areas above 1,200 m a.s.l. in the northern half of Spain, and is typically higher than 60 days/year in the central *mesetas* (Fig. 1.4). On the other hand, the areas with the highest mean annual temperature are distributed in southern Spain and along the southern-central Mediterranean strip (Fig. 1.3).

Another important characteristics of the Spanish climate, particularly in the Mediterranean fringe and the mountain regions, are the frequent occurrence of severe rainfall events, which may have dramatic geomorphic effectiveness and are responsible for the natural disasters with the highest number of fatalities (e.g. White et al. 1997; Gutiérrez et al. 1998; White and García-Ruiz 1998; Ferrer et al. 2004; Ortega and Garzón-Heydt 2009). The maximum daily rainfall for a return period of 50 years exceeds 100 mm/day in most of the mountain areas and reaches values above 200 mm/day in some sectors on the Mediterranean strip (Fig. 1.5).

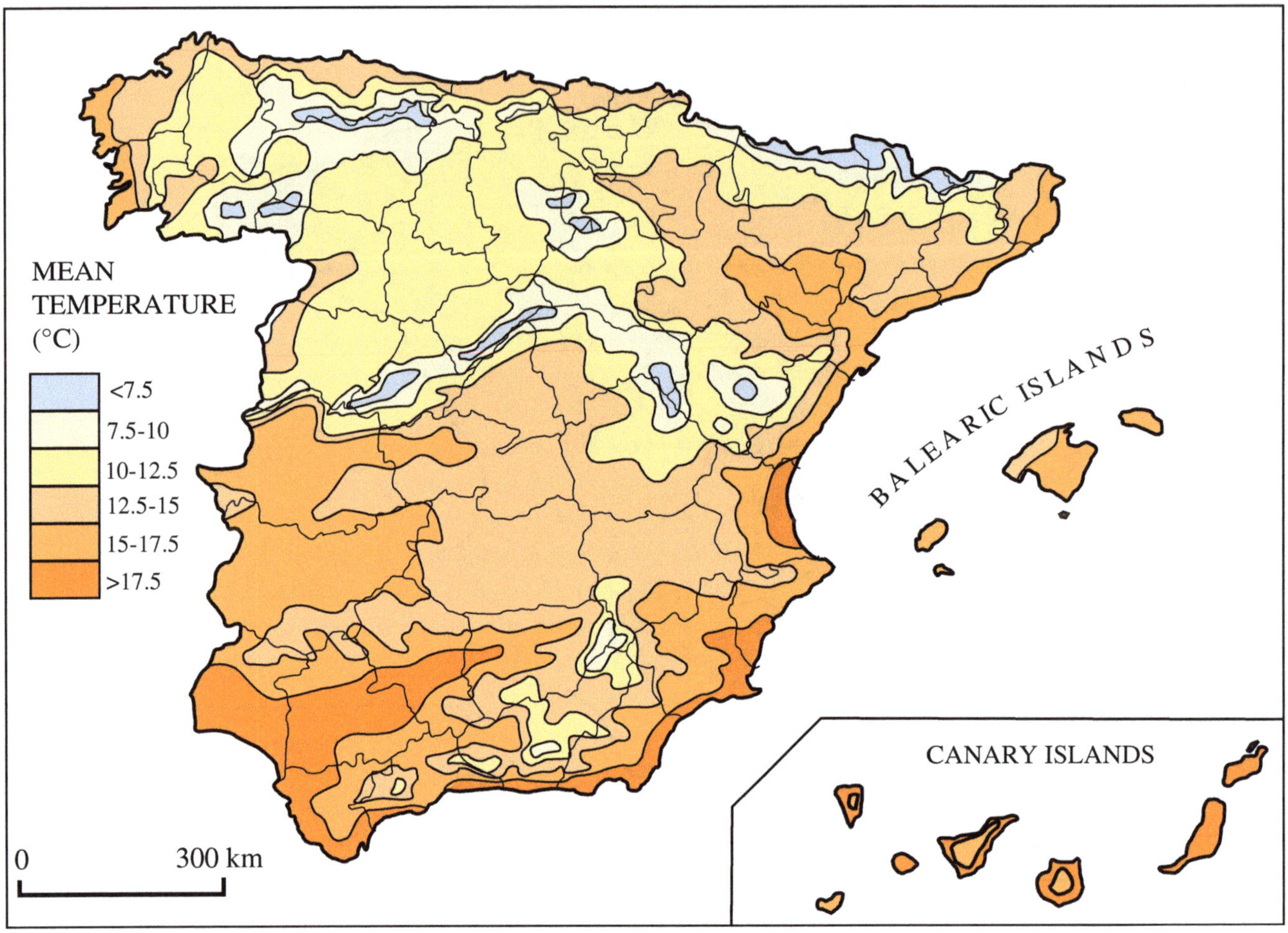

Fig. 1.3 Map of mean annual temperature in Spain (modified from Instituto Geográfico Nacional 1995)

The available records include a large number of rainfall events exceeding 400 mm/day, with top values higher than 800 mm/day (Martín-Vide 2002).

An additional underlying reason why Spain has a great potential for geomorphological investigations is its outstanding geological diversity (Pérez-González et al. 1989; Gutiérrez 1994; Gibbons and Moreno 2002; Vera 2004; Martín-Serrano 2005). The Iberian Peninsula is commonly divided into two broad geological areas (Fig. 1.6): (1) The Iberian (or Hesperian) Massif in the western sector, frequently regarded as Variscan (or Hercynian) Spain. (2) The Alpine mountain belts and Cenozoic basins of the eastern sector, related to the general N–S convergence and collision between Europe, the Iberian microplate and Africa since the late Mesozoic. The "Geomorphological Map of Spain and the Continental Margin at 1:1,000,000 scale" (Martín-Serrano 2005) constitutes an excellent companion for this publication (downloadable at http://www.igme.es/internet/cartografia/cartografia/tematica.asp?mapa=geomorfologico1000).

1.2 The Iberian Massif

The Iberian Massif is by far the least known area of Spain from the geomorphological perspective. It is the best exposure of the European Variscan orogen, generated by the collision between Laurasia and Gondwana in the late Palaeozoic. This structurally complex area mainly consists of Palaeozoic metamorphosed sedimentary formations intruded by plutonic rocks, chiefly granitoids. The Mesozoic was a period dominated by erosion, which led to the development of extensive planation surfaces. They are particularly frequent in the southern half of the Iberian Massif and are commonly preserved in elevated areas. Compressional Alpine tectonics in this portion of Iberia has been accommodated by the development of intraplate mountain systems (e.g. Central System, Montes Galaico-Leoneses), large peripheral Cenozoic basins (Tajo and Duero) and small internal Cenozoic basins controlled by reverse and strike-slip faults, locally showing evidence of recent activity (e.g.

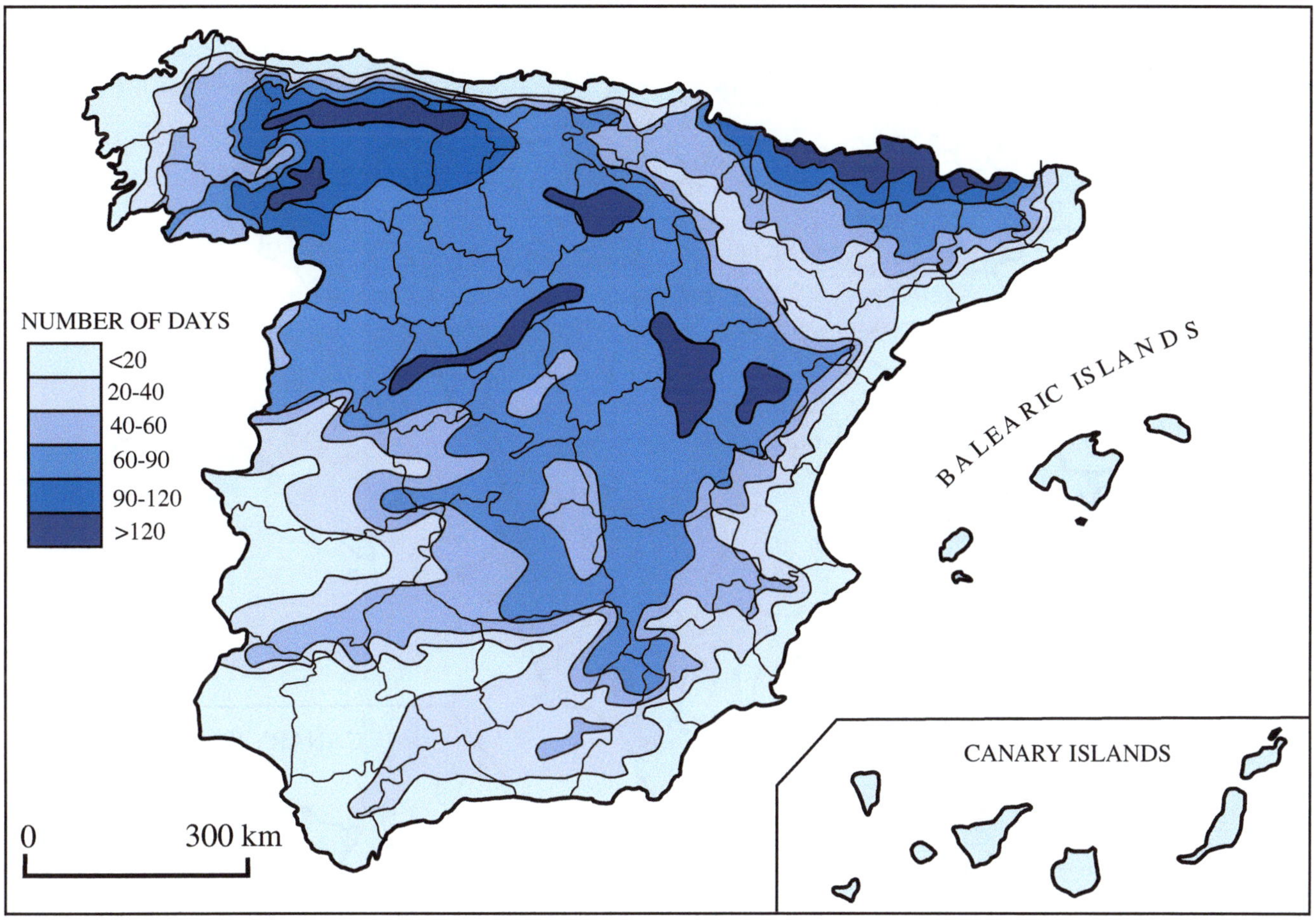

Fig. 1.4 Map depicting the average annual number of days with temperature below to 0 °C (modified from Instituto Geográfico Nacional 1995)

Martín-González 2009). The planation surfaces constitute valuable geomorphic markers to identify and assess Cenozoic tectonic deformation. In some sectors, they have been compartmentalised into uplifted and downthrown blocks, and in others they show gentle uplifts (Sierra Morena, Montes de Toledo) (Fig. 1.1). The 180-km-long Guadiana Cenozoic basin is the largest one, which has controlled the path of the Guadiana River. The 700-km-long and ENE–WSW trending Central System corresponds to an uplifted portion of the Variscan basement, bounded by double-verging reverse faults developed since the early Cenozoic (De Vicente et al. 2007). This Alpine pop-up morphostructure reaches 2,592 m in elevation and separates the Duero and Tajo Cenozoic basins, as well as the northern and southern mesetas in central Spain (Figs. 1.1, 1.6).

One of the most characteristic features of the landscape in the Iberian Massif is the presence of planation surfaces, which may form extensive plains locally interrupted by residual reliefs; inselbergs (Martín-Serrano and Molina 2005). The mature topography of this relatively stable area has favoured the development and preservation of palaeoweathering profiles that record past climate conditions and constitute a valuable correlation tool for regional geomorphology. The majority of the palaeoweathering profiles can be correlated with Mesozoic continental formations (e.g. Purbeck, Weald, Utrillas, Areniscas de Bucaço, Areniscas silíceas de Salamanca), constituting a highly useful source of information on the basin margins during those periods. The iron-rich composition of both the palaeoregoliths and the continental deposits indicates tropical conditions. The weathering profiles locally underlie the sedimentary fill of Cenozoic basins (e.g. Duero Basin; Molina et al. 1997; Martín-Serrano and Molina 2005).

Some of the most striking geomorphological features in the Iberian Massif are related to the underlying lithology and structure. In some areas, differential erosion of folded Palaeozoic rocks with contrasting resistance to erosion (e.g. quartzites and slates) has produced a distinctive Appalachian type of topography. The best example is found in the Montes de Toledo, due to the excellent preservation of the planation surface on the crest of the ranges. The development of the Appalachian topography has been favoured by intense weathering under tropical conditions during the Mesozoic and regional tectonic rejuvenation. In some

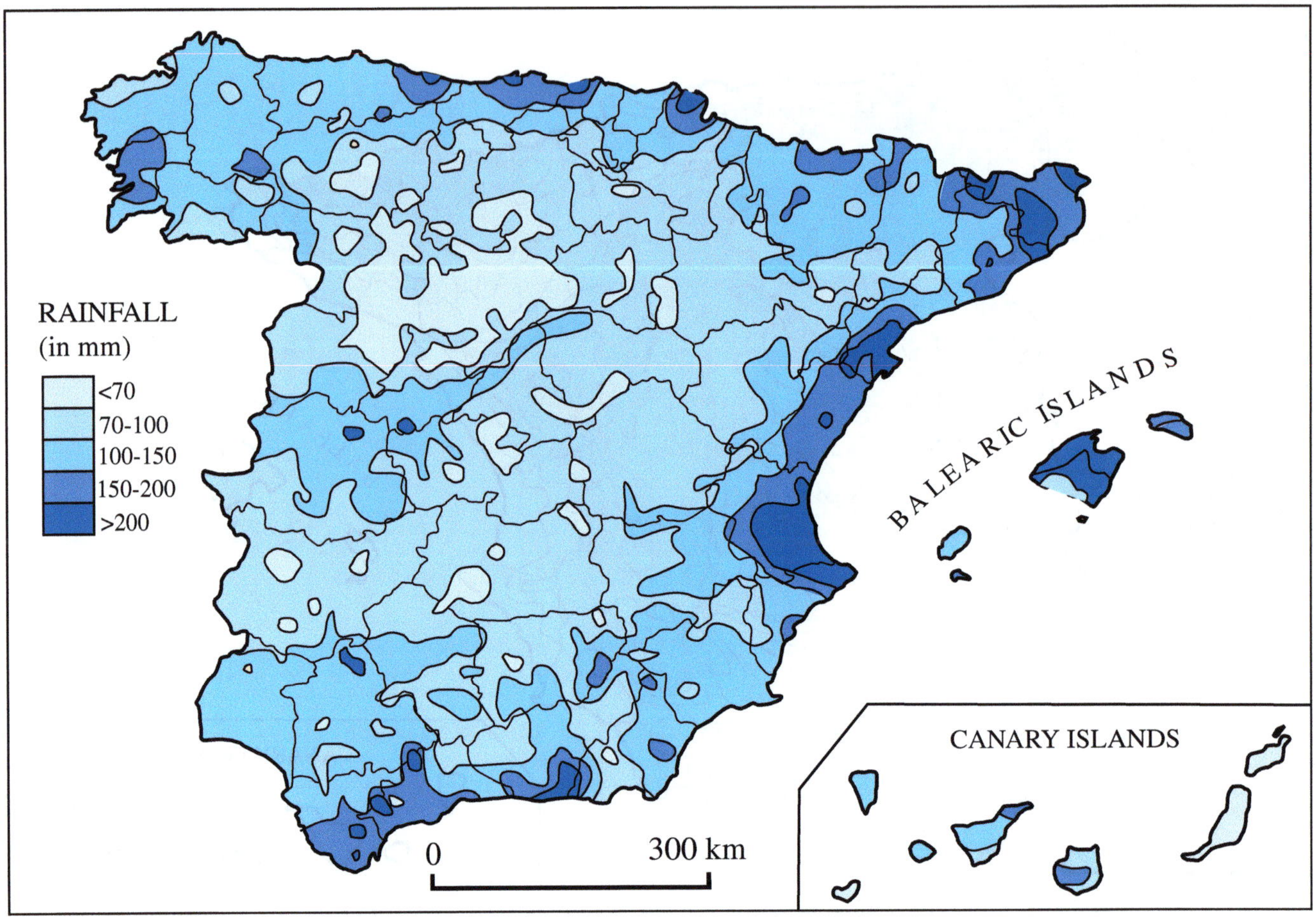

Fig. 1.5 Map of maximum daily rainfall for a return period of 50 years (modified from Instituto Geográfico Nacional 1995)

sectors, the Appalachian erosional features are buried by Mesozoic and/or Palaeogene formations, like in the western margin of the Duero Basin (Blanco et al. 1982; Molina et al. 1989) or in the SE edge of Sierra Morena, Relumbrar Range (Nozal 2002).

Some areas of Galicia and the Central System display spectacular examples of granite geomorphology with bornhardts, tors (Fig. 1.7), fields of corestones, flared slopes, etchsurfaces, pseudokarren and speleothems (Pedraza et al. 1989, Vidal-Romaní and Twidale 1998; Twidale and Vidal-Romaní 2005; Chaps. 4 and 5).

The best example of karst geomorphology in the Iberian Massif is found in the Picos de Europa Massif, with peculiar depressions of mixed karstic and glacial origin (Smart 1986), shaft-dominated caves more than 1.5-km deep, and gorges with impressive walls more than 1-km high (Miötke 1968; Chap. 13). Outstanding surface and underground karst features developed in Cambrian limestones are also found in the southern sector of the Iberian Massif, like the Gruta de las Maravillas in Aracena, Huelva, and Cerro del Hierro, Sevilla. The glaciated areas are restricted to the highest elevations in the northern and central sector of the Iberian Massif. A peculiar feature is the development in the late Pleistocene of ice caps on the planated summits of some mountain ranges, linked to radiating outlet glaciers (Cowton et al. 2009; Carrasco et al. 2013). The largest ice mass was developed in the Sanabria Lake area, Zamora, where the ice cap and outlet glaciers reached more than 400 km^2 (Cowton et al. 2009). The moraine-dammed Sanabria Lake, Zamora, which covers 319 ha, constitutes the largest glacial lake in Spain. In the Central System, the largest valley glaciers developed in the northern flanks of the Gredos and Bejar ranges (Palacios et al. 2012; Carrasco et al. 2013). Periglacial features including nivation hollows, rock glaciers, scree slopes, and patterned ground typically occur in mountainous areas above 700–800 m a.s.l. (Martín-Serrano and Molina 2005).

Another characteristic feature of the Iberian Massif is the presence of extensive piedmont alluvial deposits (*raña* surfaces). They consist of siliceous clast-supported conglomerates with fluvial sedimentological features and a limited thickness, which may overlie basement rocks or Tertiary sediments. These sediments are strongly overprinted by weathering processes, including (1) disintegration of clasts

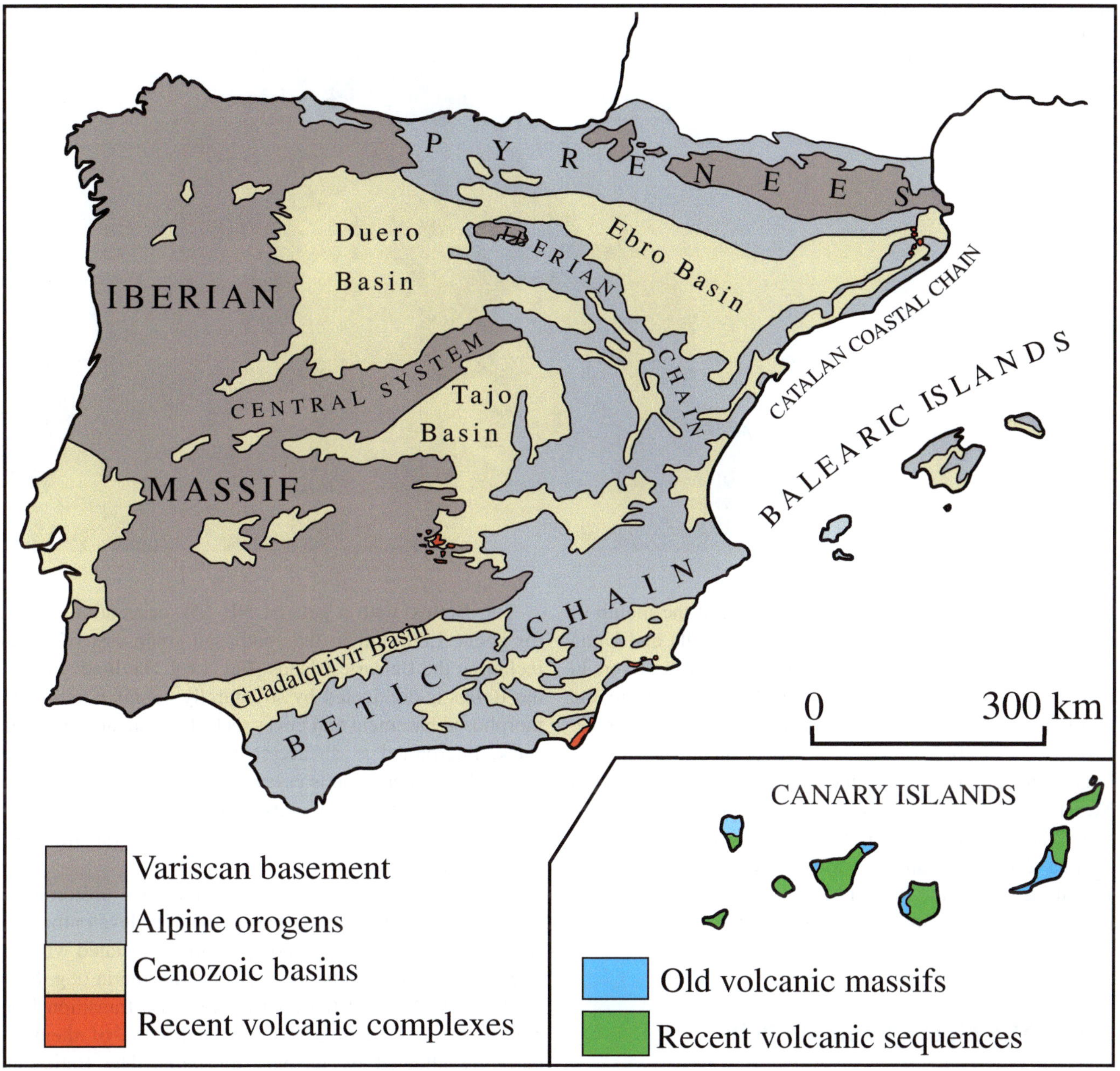

Fig. 1.6 Map illustrating the main geological units in Spain (modified from Martín-Serrano et al. 2005)

into finer-grained particles; (2) segregation of oxihydroxides, silica and other compounds that are mobilised under hydromorphic conditions and (3) transformation of clay minerals (Molina 1991). The result is a well-developed ultisol–oxisol (Espejo 1986). These thin alluvial mantles situated above the terrace sequences in water divide areas record the initial phases of development of the present-day drainage network (Martín-Serrano 1991; Pinilla 1993; Molina-Ballesteros and Cantano-Martín 2002). Fluvial incision and the development of the drainage network in the Iberian Massif progressed from the Atlantic margin towards the interior of the peninsula. The entrenchment of the drainage network and its headward expansion led to the rejuvenation of the relief, especially in the western sector of the Iberian Massif. Deep fluvial dissection affects the whole Portuguese basement, two-thirds of the Sil and Miño rivers, and the Portuguese section of the Duero and Guadiana rivers, whereas entrenchment in the Tajo Valley has reached Madrid Basin in the central sector of the Peninsula. Fluvial incision in this context has produced excellent examples of antecedent and/or superimposed drainages: (1) synorogenic gorges carved by the Cares, Deva and Sella rivers in Picos de

Fig. 1.7 Manqueospese Castle, Ávila, built on a granitic tor, northern sector of the Central System (*Photograph* C. Roqué)

Europa; (2) hanging valleys with waterfall at the mouth of numerous rivers in north-western Galicia (Xalla and Umia rivers); (3) canyons excavated by the Sil and Miño rivers in central Galicia (Birot and Solé 1954); (4) gorges in the marginal sector of the Central System (Voltoya, Alberche, Lozoya and Sorbe rivers); (5) deep discordant water gaps opened by the Tajo and Guadiana rivers in the southern Meseta across Variscan structures, most probably inherited from the Tertiary fluvial network (Martín-Serrano and Molina 2005). Perhaps, the most spectacular example of fluvial incision in the Iberian Massif is found in the Arribes del Duero, a 400-m-deep canyon carved in granitoids (Fig. 1.8).

1.3 The Alpine Mountain Belts

The Betic Chain and the Pyrenees are Alpine orogens resulting from the collision of the Iberian microplate with the European and African plates, respectively (Fig. 1.6). These are the mountain belts with highest peaks in mainland Spain. The Mulhacén in the Betics and the Aneto in the Pyrenees, have elevations of 3,482 m and 3,404 m a.s.l., respectively. The Pyrenees is essentially a "blocked" plate margin with negligible relative motion, whereas the Betics is currently affected by considerable convergence (4 mm/year) and tectonic activity (Zazo et al. 1998; Silva et al. 2003). One of the main geomorphic differences between these Alpine collision orogens and the rest of the intraplate mountain belts in Spain are the limited extent of planation surfaces, attributable to rapid deformation in a plate margin context.

The **Betics**, with a general NE–SW orientation, extend for about 1,000 km in the south and south-east of Spain, including the Balearic Islands (Fig. 1.6). The Inner Zone of the Betics is dominated by structurally complex and metamorphosed basement and cover rocks forming an antiformal stack, interpreted as an accreted terrane (Alborán microplate). The Outer Zone is essentially a suite of allocthonous south-verging structural units made up of Mesozoic and Cenozoic sedimentary sequences detached from the Variscan basement.

The Betic Chain also includes numerous postorogenic Late Miocene-Quaternary basins whose development is related to the general N–S compression associated with the ongoing convergence between Africa and Iberia (e.g. Silva et al. 1993). Most of these basins record a transition from marine to continental deposition, and some of them are currently affected by tectonic inversion. The Betics are clearly the best region in Spain to investigate the impact of active tectonics on landscape development (Chap. 2). The eastern Betics display excellent examples of fault-controlled mountain fronts and alluvial fan systems (e.g. Silva et al. 2003; Martínez-Díaz et al. 2012), whose development may be affected by multiple factors such as tectonic activity, climate variability and base level changes (Harvey 1996; Silva et al. 1992; Harvey et al. 1999; García-Meléndez et al. 2003). A number of studies conducted in this region illustrate the crucial role played by geomorphological studies in the identification of faults and the assessment of their seismogenic potential (e.g. Silva et al. 1997; García-Tortosa et al. 2011; Rodríguez-Pascua et al. 2012). In the Betic Chain, the evolution of the drainage network, largely guided

Fig. 1.8 Arribes del Duero, an impressive canyon carved by the Duero River into the granitic Variscan basement. Image taken next to the Aldeadavila de la Ribera Dam, Salamanca (*Photograph* José Bonilla)

by the postorogenic basins, has been the focus of pioneering studies addressing issues such as the impact of capture induced base-level changes (Harvey and Wells 1987; Goy et al. 1994; Mather 2000; Maher et al. 2007; Whitfiled and Harvey 2012), the transition of alluvial fan systems into fluvial systems (Silva et al. 2008), the impact of active faulting and folding on transverse drainages (Maher and Harvey 2008) or the morphostratigraphic record of incision waves (García et al. 2003). These investigations reveal the need for a regional approach for examining long-term changes in fluvial systems (Chap. 3).

Limestone karst is well-developed in numerous regions, mostly in the Outer Betics (Durán and López-Martínez 1999), with magnificent examples of poljes controlled by active faults (e.g. Zafarraya Polje; Lhenaff 1986; Reicherter et al. 2003; Fig. 1.9), doline fields (e.g. Grazalema and Mágina ranges), and karren fields such as those found in the Torcal de Antequera (Pezzi 1977; Fig. 1.10) and the Tramuntana Range (Chap. 7). There is also a number of remarkable cave systems, some of them with significant economic (e.g. Nerja, Drac and Artá show caves), engineering (Hundidero-Gato Cave and the failed Montejaque Dam project) and palaeoanthropological (Finlayson et al. 2006) implications. Landforms related to evaporite dissolution are mainly developed on halokinetic Triassic halite-bearing evaporites (Calaforra and Pulido-Bosch 1999) and in the Messinian gypsum of Sorbas Basin, with a peculiar stratigraphically controlled multilevel cave system, mainly carved in argillaceous units and unique speleothems (Calaforra and Pulido-Bosch 2003; Chap. 10).

Evidence of Quaternary glaciation is restricted to Sierra Nevada, which is the southernmost glaciated area in Europe. Here, valley glaciers reached ca. 3 km in length during their maximum extent (Gómez-Ortiz et al. 2012), which apparently occurred before the global Last Glacial Maximum. Rock glaciers have been reported at the foot of the headwalls of some cirques (Gómez-Ortiz et al. 2012). The spatial distribution of large landslides in the Betics is mainly controlled by litho-structural factors, active tectonics and fluvial incision (e.g. Gelabert et al. 2003; Azañón et al. 2005; Delgado et al. 2011) and, unlike the Pyrenees, debutressing related to deglaciation has a negligible impact.

The **Pyrenean** orogen, with a prevalent E–W structural and topographic grain, extends for around 650 km in northern Spain, including the eastern portion of the Cantabrian Cordillera, underlain by post-Variscan sequences affected by contractional structures (Fig. 1.6). In this collisional plate margin, the orogenic phase and the inversion of post-Variscan basins took place from late Cretaceous to Miocene times. The Spanish sector of this double-verging mountain belt can be divided into two main structural units. The Axial Pyrenees, in the core of the orogen, constitute an antiformal stack made up of Variscan basement. The Southern Pyrenean zone is an allochthonous system of south-verging thrusts, mostly affecting post-Variscan successions and locally including Palaeogene sequences deposited in former foreland basins incorporated into the orogen. The topography shows a general decrease in elevation from the axial zone, with peaks above 3,000 m, towards the southern margin of the orogenic wedge.

Fig. 1.9 Zafarraya Polje, Málaga, bounded on its southern margin by a rectilinear and scarcely dissected fault-controlled escarpment (foreground). This active fault is considered to be the source of the catastrophic 1884 Andalucía earthquake (*Photograph* F. Gutiérrez)

Fig. 1.10 Karren field in Torcal de Antequera, Málaga (*Photograph* F. Gutiérrez)

The regional geomorphology of the Pyrenees is dominated by differential erosion processes controlled by the E–W structure and N–S glacio-fluvial transverse valleys coherent with the general topographic trend. Differential erosion of erodible sediments, mostly Palaeogene and Triassic argillaceous and evaporitic formations, has generated broad E–W trending erosional depressions which display the best developed pediment and terrace sequences (e.g. Peña 1983, 1994). The transverse drainages have carved deep and narrow valleys with local widenings associated with less resistant lithologies.

In the great part of the Pyrenees, the headwaters of the catchments were occupied by valley glaciers in the late Pleistocene (Fig. 1.11), which reached the maximum extent well-before the global Last Glacial Maximum (García-Ruiz et al. 2003, Jiménez-Sánchez et al. 2013). In the central Pyrenees, the valley glaciers, in some cases more than 500-m thick and 30-km long, reached elevations below 900 m a.s.l.

Fig. 1.11 Terminal moraines and nested debris cones at the foot of steep glacial troughs carved into an erosional escarpment underlain by folded Mesozoic and Palaeogene rocks, Partacua Range, central Pyrenees (*Photograph* F. Gutiérrez)

In the more humid Cantabrian Mountains, the front of some palaeoglaciers has been situated below 500 m a.s.l. These alpine glaciers carved cirques, over-deepened basins and deep troughs with steep slopes. Locally, lateral moraines blocked tributary drainages generating marginal enclosed basins with lacustrine deposition. In some valleys it has been possible to establish chronological associations between frontal moraines and outwash terraces and identify older glacial phases on the basis of morphostratigraphical relationships and geochronological data (Lewis et al. 2009; Jiménez-Sánchez et al. 2013). At the present time, there are about 20 cirque glaciers restricted to massifs higher than 3,000 m a.s.l. in the central Pyreenes (Fig. 1.12). These glaciers expanded during the Little Ice Age, as revealed by historical data and fresh moraines, and are currently affected by rapid recession (Chueca-Cía et al. 2005; Chaps. 14 and 15). Periglacial activity is represented by both active and relict talus slopes, protalus ramparts, grêzes litées (Peña et al. 1998; García-Ruiz et al. 2001), rock glaciers (Serrano et al. 2010) and patterned ground.

Landslides in the Pyrenees constitute a major morphogenetic process and, together with flooding, pose the main geomorphological hazard. A number of villages have been destroyed or abandoned due to landslide activity (Inza, Salinas de Jaca, Puigcercós, Montclús, Pont de Bar; Fig. 1.13). In glaciated valleys with weak lithologies, deep-seated landslides related to the debutressing of oversteepened slopes may display very high spatial frequencies (Guerrero et al. 2012). The development of most large landslides is favoured by litho-structural factors, such as the presence of thick halite-bearing evaporites (Gutiérrez et al. 2012a), or the favourable attitude of the strata (Pinyol et al. 2012). In addition to glacial debutressing, fluvial erosion, severe precipitation (Corominas and Moya 1999) and seismic shaking (e.g. González-Díez et al. 1999; Gutiérrez et al. 2008a; Rosell et al. 2010; Zarroca et al. 2013) are the main natural triggering factors. The occurrence of shallow landslides and debris flows has been largely influenced by changes in land use and land cover (e.g. Martí et al. 1997; Remondo et al. 2005; Beguería 2006; García-Ruiz et al. 2010a). Changes in plant cover have a significant influence on the magnitude and frequency of floods, erosion processes and sediment transport. Plant colonisation after farmland abandonment resulted in a progressive decline in the number of floods and in the sediment yield at both small catchment (García-Ruiz et al. 2010b) and regional scales (Beguería et al. 2006). Floods have a particularly severe geomorphic and societal impact in relatively small and steep drainage basins, where catastrophic flash floods related to convective storms may develop in a very short period of time. The 1996 Arás flood caused 87 fatalities in a campground built in the active lobe of an alluvial fan, fed by a drainage basin around 18 km^2 in area and 1,200 m in relief (White et al. 1997; Gutiérrez et al. 1998) (Fig. 1.14).

The Outer Zone of the Pyrenees, with thick limestone sequences, includes some of the most remarkable karst massifs in Spain with doline and karren fields (e.g. Sierra de Boumort, Lérida; Llano de Cupierlo, Guara Ranges), poljes frequently with vague structural control (e.g. Saganta, Abeles), deep canyons, ponors and springs, as well as caves with large-vertical development (Arañonera Cave, 1,350 m) (e.g. Rodríguez-Vidal 1986; López-Martínez 1986).

Fig. 1.12 Cirques separated by sharp ridges (arêtes) in the granodioritic Maladeta Massif, central Pyrenees. The Aneto Glacier in the north-facing cirque (*right*) (*Photograph* F. Gutiérrez)

Fig. 1.13 Head scarp of the landslide that caused the partial destruction and abandonment of the Puigcercós village in the second half of the nineteenth Century, eastern Pyrenees (*Photograph* F. Gutiérrez)

Evaporite karst features mainly correspond to lake basins developed in collapse structures related to dissolution of subjacent Triassic and Eocene formations (Estaña, Montcortés, Bañolas; e.g. Canals et al. 2006; López-Vicente et al. 2009; Gutiérrez et al. 2012a). Moreover, numerous dam projects have been severely affected by karst-related water leakage problems (e.g. Belsué, Canelles, Camarasa; Milanovic 2000).

The **Iberian Chain** and the **Catalan Coastal Chain** in NE Spain are intraplate Alpine orogens resulting from the tectonic inversion of Mesozoic basins during the Paleogene (Fig. 1.6). During the Neogene, extensional tectonics generated horsts and graben morphostructures superimposed on the previous contractional structures. The Neogene grabens in the Iberian Chain are filled by alluvial and lacustrine sediments, whereas those of the Catalan Coastal Chain may

Fig. 1.14 The Arás alluvial fan developed at the mouth a steep mountain torrent incised into unconsolidated moraine deposits on the western margin of the Gállego River valley, central Pyrenees. A flash flood in August 1996 killed 87 people in Las Nieves campsite (*white arrow*). This highly vulnerable resort was built on the active lobe of the fan between a pre-existing artificial channel (*right black arrow*) and the natural fan channel, which was subsequently used to construct a channelization with a much larger section (*left black arrow*). After a long litigation, the administration was condemned to compensate the victims with an economic indemnity (*Photograph* F. Gutiérrez)

include marine sequences. The Mesozoic successions have a high proportion of limestone units that form extensive outcrops. The Iberian Chain, with a general NW–SE trend, is a broad-elevated area 400-km long and 250-km wide. The Catalan Coastal Chain, with a NE–SW orientation, extends obliquely for about 200-km along the Mediterranean coast. This mountain chain displays a conspicuous horst and graben topography consisting of two ranges (Littoral Cordillera and Pre-littoral Cordillera) separated by an axial graben system (Pre-littoral Depression). In the northern sector of the chain, there are quite extensive outcrops of Variscan plutonic rocks with a well-developed granitic landforms, including inselbergs, fields of corestones, caves, tafoni, pseudokarren and spelothems (Vilaplana 1987; Roqué et al. 2011, 2013). The Catalan Coastal Chain, forming a topographic barrier adjacent to the Mediterranean Sea, is one of the most prone areas in the Spain to the occurrence of flash floods and rainfall-triggered landslides (e.g. Corominas and Moya 1999; Vilaplana 2008; Llasat et al. 2010).

One of the most outstanding geomorphological characteristics of these orogens, especially the Iberian Chain, is the presence of extensive planation surfaces cut across deformed pre-Neogene rocks, chiefly Mesozoic carbonate rocks (Gutiérrez and Peña 1994). This general plateau-like topography is locally interrupted by residual reliefs, frequently underlain by more resistant Palaeozoic rocks, neotectonic grabens, erosional depressions and fluvial valleys. The flat topography developed on carbonate rocks has favoured the development of doline fields (e.g. Villar del Cobo and Pozondón in Teruel and Los Palancares and Cañada del Hoyo in Cuenca) (Gutiérrez and Peña 1979, 1994), karren (Ciudad Encantada, Cuenca) and poljes (e.g. Gutiérrez and Valverde 1994; Gracia et al. 2003 and references therein; Chap. 11). Pleistocene glaciers were restricted to cirques carved in the hard-rock massifs higher than 2,000 m located in the northern sector of the Iberian Chain (Cebollera, Urbión and Moncayo). In the high country, active and relict periglacial features are relatively abundant, including nivation hollows, protalus ramparts, rock glaciers, patterned ground, grèzes litées and remarkable block streams (e.g. Gutiérrez and Peña 1977; Chap. 16).

The drainage network in the central sector of the Iberian Chain is largely controlled by the post-orogenic grabens and records the successive capture of different tectonic depressions by headward expansion (Gutiérrez et al. 2008b). The horst and graben topography of the Catalan Coastal Range is crossed perpendicularly by major transverse drainages (Arche et al. 2010). In the outcrops of limestone-rich Mesozoic successions, streams typically flow deeply entrenched in canyons (Fig. 1.15) with frequent tufa accumulations (e.g. Vázquez-Urbez et al. 2011). Both the Iberian Chain and the Catalan Coastal Chain have good examples of landforms associated with active tectonics and gravitational normal faults, such as mountain fronts, triangular facets and disrupted drainages (Perea et al. 2012; Gutiérrez et al. 2012b; Carbonel et al. 2013a).

1.4 The Cenozoic Basins

Spain has four large Cenozoic basins that cover around one-third of the country area (Fig. 1.6). These morphostructural depressions control the path of the main fluvial systems, from which they receive their names; Ebro, Duero, Tajo and

Fig. 1.15 Albarracín village (Iberian Chain, Teruel), built in the inner side of a tight meander in the Guadalaviar canyon, carved into Jurassic carbonate rocks. The village was declared Heritage of the Humanity by the UNESCO in 1998 (*Photograph* F. Gutiérrez)

Guadalquivir basins (Gutiérrez-Elorza et al. 2005a). The formation and development of these sedimentary basins have been mainly controlled by the tectonic evolution of the surrounding Alpine orogens and, in the case of the Ebro, Duero and Tajo basins, by the capture of the depressions and the progressive change from endorheic-aggradational to exorheic-incisional conditions. The Ebro and Guadalquivir depressions are foreland basins of the Pyrenees and Betic Chain, respectively. The Duero and the Tajo depressions are essentially intracratonic structures bounded by Alpine contractional structures.

The ENE-WSW Guadalquivir Basin has been opened to the sea during its entire evolution and the Miocene-Quaternary fill mainly consists of marine sediments. The southern half of the basin is dominated by olistostromes made up of chaotic Mesozoic and Cenozoic rocks, whereas the northern half is mainly underlain by autochthonous soft marly sediments. The upper and middle reach of the Guadalquivir valley, currently abutting the rectilinear northern margin of the basin, displays an extensive terrace sequence on the southern margin (Díaz del Olmo et al. 1989; Baena and Díaz del Olmo 1997). In the lower reach, the river splits into several anastomosed channels flowing through an extensive marshland separated from the sea by a long spit bar with a large superimposed dune field (Doñana National Park) (Rodríguez-Ramírez et al. 1996; Chap. 19). The growth of the spit bar has induced the rapid siltation of the tidal estuary-marshland after the Flandrian transgression (ca. 6.5 ka) and the shifting of the river channel to the SE (Zazo et al. 1999).

The Ebro, Duero and Tajo basins have relatively similar sedimentary and geomorphic evolution. Most of the outcropping sediments in these depressions are Palaeogene–Miocene continental formations with subhorizontal structure deposited under endorheic conditions. These sediments typically display a roughly concentric facies distribution, with conglomerates at the margins that grade distally to fine-grained alluvial fan facies, and lacustrine evaporites and carbonates in the depocentral sectors. The end of the endorheic fill is commonly recorded by Miocene limestone units which may connect, physically or altitudinally, with planation surfaces at the basin margins (e.g. Gutiérrez et al. 1982; Benito-Calvo and Pérez-González 2007). These endorheic basins were captured in the Miocene by the external drainage network. The new exorheic conditions lead to the development of the present-day drainage network responsible for the dissection of the sedimentary fill and the development of stepped pediment and terrace sequences. The oldest exorheic morphosedimentary units typically correspond to extensive and prominent mantled pediments (e.g. *rañas*) that record unconfined alluvial–fluvial systems developed before the entrenchment of the fluvial systems and the formation of staircased terraces (e.g. Martín-Serrano 1991; Pinilla 1993; Lucha et al. 2012).

The landscape within these Cenozoic basins is largely influenced by the distribution of lithofacies. The marginal conglomerates form elevated areas and locally stunning monoliths more than 300-m high with precipitous cliffs controlled by vertical fractures (e.g. Riglos, Montserrat; Chap. 6). Erosional depressions with extensive mantled

Fig. 1.16 Badlands landscape in Bardenas Reales Natural Park, Ebro Cenozoic Basin (*Photograph* F. Gutiérrez)

Fig. 1.17 Argillaceous sediments capped by a resistant gypsum unit in El Planerón, central sector of the Ebro Basin (*Photograph* F. Gutiérrez)

pediments and badlands landscapes (Fig. 1.16) occur in the areas dominated by argillaceous facies (e.g. Desir et al. 2005). The limestone, and locally gypsum units cap buttes, mesas and structural platforms (Fig. 1.17), which may display fields of shallow solution sinkholes. The reliefs formed by relatively thin limestone caprocks or cemented gravels underlain by erodible sediments are frequently surrounded by sequences of talus flatirons (Gutiérrez-Elorza et al. 2010). The Ebro and Tajo basins have thick halite- and glauberite-bearing evaporitic units. In these areas, the Quaternary alluvial deposits are locally thickened recording dissolution-induced synsedimentary subsidence (Benito et al. 2010; Guerrero et al. 2013; Silva et al. 2013). In these areas, sinkholes may show a high activity in the densely populated lower alluvial levels, resulting in high risk scenarios (Galve et al. 2009; Gutiérrez et al. 2009; Carbonel et al. 2013b; Fig. 1.18).

Aeolian activity has a significant geomorphic imprint in the southern sector of the Duero basin, where the bedrock is dominated by friable arkosic sandstones and the fluvial

Fig. 1.18 Colapse sinkhole related to suballuvial dissolution of evaporites developed in the N-232 highway, next to Zaragoza city, on March 3, 2013 (*Photograph* F. Gutiérrez)

systems are mainly nourished by sands. Here, there are extensive sand sheets and dune fields, as well as blowouts, deflation basins and lunette dunes (Bateman and Díez 1999; García-Hidalgo et al. 2007; Gutiérrez-Elorza et al. 2005b; Bernat-Rebollal and Pérez-González 2008). In the Ebro Basin, where there is very limited availability of sands, aeolian accumulations are very scarce, but the strong wind has carved yardangs in gypsiferous rocks and unconsolidated lake deposits in the leeward margin of playa lakes (Gutiérrez-Elorza et al. 2002; Chap. 12). Another common feature of the Cenozoic basins is the occurrence of endorheic areas that may host ephemeral saline lakes. Subsurface evaporite dissolution and aeolian deflation are generally the main processes involved in their genesis (Gutiérrez et al. 2013 and references therein).

1.5 The Canarian Archipelago and other Neogene Volcanic Provinces

An additional major geological unit in Spain is the Canarian Archipelago, a chain of hot-spot-related volcanic islands that extends for around 500 km across the eastern Atlantic, with its eastern edge about 100 km from the Sahara coast. These islands have grown upon the slow moving Jurassic oceanic lithosphere next to the passive margin of the African plate. The Cenozoic volcanic sequences, dominated by basaltic rocks, record a long period (>20 Ma) of eruptive activity, from the early building up of submarine sea mounts to the polyphase development of volcanic edifices, eventually affected by giant landslides (Carracedo et al. 2002). The islands constitute steep piles of volcanic rocks rising several kilometres above the abyssal plain, with less than 10 % of the volume emerged above the sea. The Teide stratovolcano in Tenerife is the highest peak in Spain (3.718 m a.s.l.), with more than 7 km of relief with respect to the adjacent abyssal plain (Carracedo 2008; Chap. 22).

Long-sustained subaerial volcanism has produced large volcanoes with subcircular base or elongated ridges where the emission of magma is controlled by persistent rift systems and active volcanic underplating. The age of these hot-spot-related islands show a general decrease towards the west. All the islands have significant Miocene volcanic sequences, whereas La Palma and El Hierro developed in Plio-Quaternary times. A total of 18 historical eruptions have been documented over the last 500 years. The penultimate event was the 1971 Teneguía volcano eruption in La Palma, and the most recent event the 2011 submarine eruption of El Hierro (Pérez-Torrado et al. 2012). The exceptionally prolonged 1730–1736 eruption of Timanfaya, Lanzarote, is the second largest basaltic fissure eruption documented in historical time. It lasted for more than 2000 days, and eruptive activity from a 15-km-long fissure produced more than 30 cones and lava flows covering around 225 km^2; over 20 % of the island (Carracedo et al. 1992).

The areas affected by recent volcanic activity are dominated by poorly dissected lava fields and cones (Rodríguez-González et al. 2012). Lava tubes and sinkholes resulting from the collapse of their roofs, locally designated as *jameos*, are relatively frequent in some sectors. The 17-km-long Viento-Sobrado Cave, Icod de los Vinos, Tenerife, is the largest lava tube in the European Union. The old massifs are characterised by a deeply entrenched drainage network and isostatic uplift induced by

Fig. 1.19 Las Cañadas del Teide, Tenerife Island. The escarpment in the background corresponds to the head scarp of a giant landslide that created a large erosional depression in which the Teide volcano (3,718 m a.s.l.) has built up. The image illustrates the phonolitic lava dome of Montaña Rajada (*middle left*), the pumice-covered Montaña Blanca (*middle right*) and the lateral margins of two blocky basaltic flows (*lower part*). View from the path to Altavista Mountain Hut (*Photograph* F. Gutiérrez)

erosional unloading (Menéndez et al. 2008) and landforms related to differential erosion like volcanic necks and protruding dykes. In these areas, wave erosion and mass wasting processes have generated spectacular coastal cliffs (e.g. Romero and Dóniz 2005). A striking feature in the Canaries are the giant landslides (up to thousands of km^3) related to the gravitational collapse of volcanoes, favoured by the continuous growth of the edifices, the presence and injection of dykes in rift zones, and pressurised fluids (Cendrero and Dramis 1996; Masson et al. 2002; Hürlimann et al. 2004; Carracedo et al. 2011; Chap. 24). These mass movements, with an estimated frequency of 125–170 ka, are expressed as huge arcuate escarpments and chaotic landslide deposits, mostly accumulated in the sea floor, which may exceed 1,000 km in length (Urgelés et al. 1997) (Fig. 1.19). The development of these structural collapses may cause dramatic changes on the magma-feeding system and influence the subsequent volcanic and geomorphic evolution of eruptive complexes. Emplacement of magma at shallower depths after the collapses may lead to compositional changes, with frequent emission of more viscous felsic lavas (Carracedo et al. 2011; Boulesteix et al. 2012).

There are three main Neogene-Quaternary volcanic districts in mainland Spain; the Almería-Murcia region, Campo de Calatrava and the Olot Volcanic field (Fig. 1.6). The Almería-Murcia region covers a large area of the eastern Betics, from the Gata Cape to El Mar Menor, and has the most heterogeneous and complex volcanic products, mostly of Miocene age (López-Ruiz et al. 2004). Here, the old volcanic landforms (cones, calderas, lava flows) are highly degraded. The Campo de Calatrava area is situated on the south-eastern edge of the Iberian Massif, between La Mancha, the Toledo Mountains and Sierra Morena. It includes around 200 emission points scattered over ca. 5,000 km^2, which record Late Miocene-Quaternary (4.7–1.75 Ma) strombolian and hydromagmatic activity (Ancochea 2004). In addition to cones, volcanic necks and lava flows up to 7-km long, the area displays more than a hundred of maars that may be classified into two types depending on the rocks affected by the hydromagmatic explosion: well-preserved and defined craters developed in resistant basement rocks, and larger craters but largely subdued by erosion associated with more erodible Neogene sediments. Some of the latter have evolved through a process of differential erosion and relief inversion into peculiar elliptic buttes capped by lacustrine limestone, once deposited in the bottom of a crater lake (Martín-Serrano et al. 2009). The Olot volcanic field in the Pyrenees, which forms part of the Catalan Volcanic Zone, is the area with youngest activity (0.5–0.1 Ma) and the best preserved volcanic landforms. Alkali basalts and basanites reached the surface through post-orogenic normal faults during strombolian and phreatomagmatic eruptions, generating around 50 monogenic volcanoes and lava flows up to 10-km-long routed along valleys, that locally dammed fluvial systems (Martí et al. 2011; Chap. 21).

1.6 The Coastal Regions

An outstanding asset of the Spanish geomorphology is the extent and diversity of the coastal environments, with ca. 8,000 km of coastline surrounding mainland Spain, as well as the islands of the Canarian and Balearic archipelagos

Fig. 1.20 Macarella Cove at the mouth of a valley incised into Miocene limestones of the Migjorn area, Menorca Island (*Photograph* F. Gutiérrez)

Fig. 1.21 Dune system developed on the downwind edge of Bolonia Beach, Cádiz Province. Sand transport was significantly reduced in the mid-twentieth Century by pine reforestation (*Photograph* F. Gutiérrez)

(Fig. 1.20). The main factors that control the geomorphology and Quaternary geology of the Spanish coasts include (Goy and Zazo 2005): (1) Eustatic changes; in the last glacial cycle, at ca. 18 ka, the sea level dropped more than 100 m below the present-day position, favouring the development of extensive aeolian accumulations (Sanjaume and Gracia 2011), (2) Lithostructural factors, including substantial differential uplift and neotectonic deformation in numerous sectors, especially in the Betic Cordillera (Zazo et al. 1993, 1999), (3) Geographical location between the European and African continents and the Atlantic and Mediterranean basins, both connected from the Early Pliocene through the Strait of Gibraltar, (4) Tidal range, (5) Relative orientation of prevailing winds and the coastline and (6) Human activity, chiefly the construction of coastal structures and dams in the drainage basins that have modified sedimentation and erosion patterns. A remarkable feature of the Spanish coasts is the presence of good sequences of raised marine terraces and alluvial fans associated with areas affected by neotectonic uplift, which constitute excellent records of sea level change and valuable markers to identify and assess recent tectonic deformation (e.g. Zazo et al. 2003, 2013; Rodríguez-Vidal et al. 2004). Caves and speleothems have been also used to constrain the timing of sea level changes (Tuccimei et al. 2010).

The geomorphology of the coasts in the Mediterranean, including the Balearic Archipelago, is largely determined by the distribution of areas affected by positive and negative vertical tectonics and the microtidal regime (range below 50 cm). The subsiding areas are characterised by the development of lakes and lagoons associated with spit bars and dune systems related to the Flandrian transgression (ca. 6.5 ka), as well as beach-ridge progradational complexes (Goy et al. 1994, 2003). The uplifting sectors are dominated by Plio-Quaternary sequences of alluvial fans and marine terraces, as well as rock cliffs. These coasts also include deltas (e.g. Somoza et al. 1998) related to the progradation of fluvial systems after the Flandrian transgression, largely influenced by human activity in the last centuries (e.g. Lario et al. 1995). The Ebro Delta, with an emerged surface of approximately 325 km^2, is the third largest delta in the Mediterranean. It has a triangular shape with two large spits generated by littoral drift on each flank of the delta. The drastic reduction in sediment supply and the cancellation of major floods due to dam construction, has favoured significant changes, mainly in the mouth sector, where a maximum retreat of 2 km has been recorded over the last 50 years (Sánchez-Arcilla et al. 1998; Chap. 18).

The geomorphic features of the Atlantic coast in the Gulf of Cadiz are largely controlled by the distribution of the different geological units and the presence of active faults with different orientations with respect to the coastline. In the Cenozoic Guadalquivir Basin, the low relief coast has extensive estuaries and marshes associated with the lower reaches of the main rivers, partially closed by spit bars up to 30-km long and large dunes fields (Borja et al. 1999; Zazo et al. 1999; Chap. 19). The Atlantic coast developed on the Gibraltar Strait area has a strong structural control and displays good sequences of raised marine terraces, cliffs interrupted by small bays and coves, and coastal dunes (Fig. 1. 21). The most relevant landforms in the Atlantic coast of northern Spain are the Galician *rías* (deep fluvial valleys partially submerged by the sea; Pagés-Valcarlos 2000), estuaries (Flor and Flor-Blanco 2006), rock cliffs and stepped sequences of *rasas* (Álvarez-Marrón et al. 2008; Chap. 20). The latter, interpreted as wave-cut platforms, correspond to flat surfaces cut across bedrock, perched up to several hundred metres above the sea level and locally covered by a thin veneer of deposits. In Galicia, geomorphic, stratigraphic and geochronological data indicate that shore platforms cut across hard rocks correspond to inherited features developed during the previous interglacials, when the sea level was similar to today (Blanco-Chao et al. 2003).

Acknowledgments This work has been supported by the Spanish national project CGL2010-16775 (Ministerio de Ciencia e Innovación and FEDER) and the Regional project 2012/GA-LC-021 (DGA-La Caixa).

References

Álvarez-Marrón J, Hetzel R, Niedermann S, Menéndez R, Marquínez J (2008) Origin, structure and exposure history of a wave-cut platform more than 1 Ma in age at the coast of Northern Spain: a multiple cosmogenic nuclide approach. Geomorphology 93:316–334

Ancochea E (2004) La región volcánica del Campo de Calatrava. In: Vera JA (ed) Geología de España. Sociedad Geológica de España. Instituto Geológico y Minero de España, Madrid, pp 676–677

Arche A, Evans G, Clavell E (2010) Some considerations on the initiation of the present SE Ebro river drainage system: post- or pre-Messinian? J Iberian Geol 36:73–85

Azañón JM, Azor A, Pérez-Peña JV, Carrillo JM (2005) Late Quaternary large-scale rotational slides induced by river incision: the Arroyo de Gor area (Guadix basin, SE Spain). Geomorphology 69:152–168

Baena R, Díaz del Olmo F (1997) Resultados paleomagnéticos de la Raña del Hespérico Meridional (Montoro, Cordoba). Geogaceta 21:31–34

Bateman MD, Díez A (1999) Thermoluminiscence dates and palaeoenvironmental information of the Quaternary sand deposits, Tierra de Pinares. Catena 34:277–291

Beguería S (2006) Changes in land cover and shallow landslide activity: a case study in the Spanish Pyrenees. Geomorphology 74:196–206

Beguería S, López-Moreno JI, Gómez-Villar A, Rubio V, Lana-Renault N, García-Ruiz JM (2006) Fluvial adjustments to soil erosion and plant cover changes in the Central Spanish Pyrenees. Geogr Ann 88A:177–186

Benito-Calvo A, Pérez-González A (2007) Quantitative reconstruction of Late Cenozoic landscapes: a case study in the Sierra de Atapuerca (Burgos, Spain). Earth Surf Proc Land 33:196–208

Benito-Calvo A, Pérez-González A, Magri O, Meza P (2009) Assessing regional geodiversity: the Iberian Peninsula. Earth Surf Proc Land 34:1433–1445

Benito G, Sancho C, Peña JL, Machado MJ, Rhodes EJ (2010) Large-scale karst subsidence and accelerated fluvial aggradation during MIS6 in the NE Spain: climatic and paleohydrological implications. Quatern Sci Rev 29:2694–2704

Bernat-Rebollal M, Pérez-González A (2008) Inland aeolian deposits of the Iberian Peninsula: sand dunes and clay dunes of the Duero Basin and the Manchega Plain. Palaeoclimatic considerations. Geomorphology 102:207–220

Birot P, Solé L (1954) Recherches morphologiques dans le Nord-Ouest de la Península Ibérique. Mém. et Doc. du CNRS 4:9–61

Blanco JA, Corrochano A, Montigny R, Thuizat R (1982) Sur l'age du debut de la sedimentation dans le bassin tertiaire du Duero (Espagne). Attribution au Paleocene por datation isotopique des alunites de l'unité inferieure. C R Acad Sci 295:259–262

Blanco-Chao R, Costa-Casais M, Martínez-Cortizas A, Pérez-Alberti A, Trenhaile AS (2003) Evolution and inheritance of a rock coast: western Galicia, northwestern Spain. Earth Surf Proc Land 28:757–775

Borja F, Zazo C, Dabrio CJ, Díaz del Olmo F, Goy JL, Lario J (1999) Holocene eolian phases and human settlements along the Atlantic coast of southern Spain. Holocene 93:333–339

Boulesteix T, Hildenbrand A, Gillot PY, Soler V (2012) Eruptive response of oceanic islands to giant landslides: new insights from the geomorphologic evolution of the Teide–Pico Viejo volcanic complex (Tenerife, Canary). Geomorphology 138:61–73

Bruschi VM, Bonachea J, Remondo J, Gómez-Arozamena J, Rivas V, Méndez G, Naredo JM, Cendrero A (2013) Analysis of geomorphic systems' response to natural and human drivers in northern Spain; implications for global geomorphic change. Geomorphology 196:267–279

Calaforra JM, Pulido-Bosch A (1999) Gypsum karst features as evidence of diapiric processes in the Betic Cordillera, Southern Spain. Geomorphology 29:251–264

Calaforra JM, Pulido-Bosch A (2003) Evolution of the gypsum karst of Sorbas (SE Spain). Geomorphology 50:173–180

Canals M, Got H, Juliá R, Serra J (2006) Solution-collapse depressions and suspensates in the limnocrenic lake of Banyoles (NE Spain). Earth Surf Proc Land 15:243–254

Carbonel D, Gutiérrez F, Linares R, Roqué C, Zarroca M, McCalpin J, Guerrero J, Rodriguez V (2013a) Differentiating between gravitational and tectonic faults by means of geomorphological zapping, trenching and geophysical surveys. The case of Zenzano Fault (Iberian Chain, N Spian). Geomorphology 189:93–108

Carbonel D, Rodríguez V, Gutiérrez F, McCalpin JP, Linares R, Roqué C, Zarroca M, Guerrero J (2013b) Sinkhole characterisation combining trenching, ground penetrating radar (GPR) and electrical resistivity tomography (ERT). Earth Surf Proc Land DOI:10.1002/esp.3440

Carracedo JC (2008) (ed) El Volcán Teide. Caja Canarias, Tenerife, 3 vols

Carracedo JC, Rodríguez-Badiola E, Soler V (1992) The 1730–1736 eruption of Lanzarote, Canary Islands: a long, high-magnitude basaltic fissure eruption. J Volcanol Geoth Res 53:239–250

Carracedo JC, Pérez-Torrado FJ, Ancochea E, Meco J, Hernán F, Cubas CR, Casillas R, Rodríguez-Badiola E, Ahijado A (2002) Cenozoic volcanism II: the Canary Islands. In: Gibbons W, Moreno T (eds) The Geology of Spain. The Geological Society, London, pp 439–472

Carracedo JC, Guillou H, Nomade S, Rodríguez-Badiola E, Pérez-Torrado FJ, Rodríguez-González A, Paris R, Troll VR, Weismaier S, Delcamp A, Fernández-Turial JL (2011) Evolution of ocean-island rifts: the northeast rift zone of Tenerife, Canary Islands. Geol Soc Am Bull 123:562–584

Carrasco RM, Pedraza J, Domínguez-Villar D, Villa J, Willenbring JK (2013) The plateau glacier in the Sierra de Béjar (Iberian Central System) during its maximum extent. Reconstruction and chronology. Geomorphology 196:83–93

Cendrero A, Dramis F (1996) The contribution of landslides to landscape evolution in Europe. Geomorphology 15:191–211

Chueca-Cía J, Julián-Andrés A, Saz-Sánchez MA, Creus-Novau J, López-Moreno JI (2005) Responses to climatic changes since the Little Ice Age on Maladeta Glacier (Central Pyrenees). Geomorphology 68:167–182

Corominas J, Moya J (1999) Reconstructing recent landslide activity in relation to rainfall in the Llobregat River basin, Eastern Pyrenees, Spain. Geomorphology 30:79–93

Cowton T, Hughes PD, Gibbard PL (2009) Palaeoglaciation of Parque Natural Lago de Sanabria, northwest Spain. Geomorphology 108:282–291

Delgado J, Vicente F, García-Tortosa F, Alfaro P, Estévez A, López-Sánchez JM, Tomás R, Mallorquí JJ (2011) A deep seated compound rotational rock slide and rock spread in SE Spain: structural control and DInSAR monitoring. Geomorphology 129:252–262

Desir G, Marín C, Guerrero J (2005) Badlands and talus flatirons in the Bardenas Reales region. Field-trip guide B-3. In: Desir G, Gutiérrez F, Gutiérrez M (eds) Sixth international conference on geomorphology, Zaragoza, vol 2, pp 57–93

De Vicente G, Vegas R, Muñoz-Martín A, Silva PG, Andriessen P, Cloetingh S (2007) Cenozoic thick-skinned deformation and topography evolution of the Spanish Central System. Global Planet Change 58:335–381

Díaz del Olmo F, Vallespí E, Baena R, Recio JM (1989) Terrazas pleistocenas del Guadalquivir occidental. Geomorfología, suelos, paleosuelos y secuencia cultural. In: Díaz del Olmo F, Rodríguez-Vidal J (eds). El Cuaternario en Andalucía occidental. Asociación Española para el Estudio del Cuaternario, pp 67–86

Durán JJ, López-Martínez G (1999) (eds) Karst en Andalucía. Instituto Tecnológico y Geominero de España. Madrid, 199 p

Espejo R (1986) Procesos edafogenéticos y edad de las formaciones tipo raña relacionadas con las estribaciones meridionales de los Montes de Toledo. Anales Edafol Agrob 45(5–6):655–680

Ferrer M, González de Vallejo LI, García JC, Rodríguez JA (2004) Pérdidas por terremotos e inundaciones en España durante el periodo 1987–2001 y su estimación para los próximos 30 años (2004–2033). Consorcio de Compensación de Seguros, Madrid 126 p

Finlayson C, Giles-Pacheco F, Rodríguez-Vidal J, Fa DA, Gutiérrez-López JA, Santiago A, Finlayson G, Allue E, Baena-Preysler J, Cáceres I, Carrión JS, Fernández-Jalvo Y, Gleed-Owen CP, Jiménez-Espejo FJ, López P, López-Sáez JA, Riquelme JA, Sánchez-Marco JA, Giles-Guzman F, Brown K, Fuentes N, Valarino CA, Villalpando A, Stringer CB, Martínez-Ruiz F, Sakamoto T (2006) Late survival of Neanderthals at the southernmost extreme of Europe. Nature 443:850–853

Flor G, Flor-Blanco G (2006) Morphodynamic processes in the estuarine barrier of Guernica (Northern of Spain): problems and solutions for their management. J Coastal Res 48:41–48

Font I (1983) Climatología de España y Portugal. Instituto Nacional de Meteorología, Madrid 286 p

Galve JP, Gutiérrez F, Lucha P, Bonachea J, Remondo J, Cendrero A, Gutiérrez M, Gimeno MJ, Pardo G, Sánchez JA (2009) Sinkholes in the salt-bearing evaporite karst of the Ebro River valley upstream of Zaragoza city (NE Spain). Geomorphological mapping and analysis as a basis for risk management. Geomorphology 108:145–158

García AF, Zhu Z, Ku TL, Sanz de Galdeano C, Chadwick OA, Chacón-Montero J (2003) Tectonically driven landscape development within the eastern Alpujarran Corridor, Betic Cordillera, SE Spain (Almería). Geomorphology 50:83–110

García-Hidalgo JF, Temiño J, Segura M (2007) Holocene aeolian development in central Spain; chronology, regional correlations and causal processes. Quatern Sci Rev 26:2661–2673

García-Meléndez E, Goy JL, Zazo C (2003) Neotectonics and Quaternary landscape development within the Eastern Huércal-Overa Basin (Betic Cordillera, southeast Spain). Geomorphology 50:111–133

García-Ruiz JM, López-Bermúdez F (2009) La Erosión del Suelo en España. Sociedad Española de Geomorfología, Zaragoza 441 p

García-Ruiz JM, Valero-Garcés B, González-Sampériz P, Lorente A, Martí-Bono C, Beguería S, Edwards L (2001) Stratified scree in the Central Spanish Pyrenees: paleoenvironmental implications. Permafrost Periglac Process 12:233–242

García-Ruiz JM, Valero-Garcés BL, Martí-Bono C, González-Sampériz P (2003) Asynchroneity of maximum glacier advances in the central Spanish Pyrenees. J Quat Sci 18:61–72

García-Ruiz JM, Beguería S, Alatorre LC, Puigdefábregas J (2010a) Land cover changes and shallow landsliding in the flysch sector of the Spanish Pyrenees. Geomorphology 124:250–259

García-Ruiz JM, Lana-Renault N, Beguería S, Lasanta T, Regüés D, Nadal-Romero E, Serrano-Muela P, López-Moreno JI, Alvera B, Martí-Bono C, Alatorre LC (2010b) From plot to regional scales: Interactions of slope and catchment hydrological and geomorphic processes in the Spanish Pyrenees. Geomorphology 120:248–257

García-Tortosa FJ, Alfaro P, Sanz de Galdeano C, Galindo-Zaldívar J (2011) Glacis geometry as a geomorphic marker of recent tectonics: the Guadix-Baza basin (South Spain). Geomorphology 125:517–529

Gelabert B, Fornós JJ, Gómez-Pujol L (2003) Geomorphological characteristics and slope processes associated with different basins: Mallorca (Western Mediterranean). Geomorphology 52:253–267

Gibbons W, Moreno T (2002) (eds) The geology of Spain. Geological Society, London, 649 p

Gómez-Ortiz A, Palacios C, Palade B, Vázquez-Selem L, Salvador-Franch F (2012) The deglaciation of the Sierra Nevada (Southern Spain). Geomorphology 159–160:93–105

González-Díez A, Remondo J, Díaz de Terán JR, Cendrero A (1999) A methodological approach for the analysis of the temporal occurrence and triggering factors of landslides. Geomorphology 30:95–113

Goy JL, Zazo C (2005) El Litoral. In: Martín-Serrano A (ed) Mapa Geomorfológico de España y del Margen Continental a escala 1:1,000,000. Instituto Geológico y Minero de España, Madrid, pp 137–156

Goy JL, Zazo C, Rodríguez-Vidal J (1994) Cordilleras Béticas-Islas Baleares. In: Gutiérrez-Elorza M (ed) Geomorfología de España. Rueda, Madrid, pp 123–157

Goy JL, Zazo C, Dabrio CJ (2003) A beach-ridge progradation complex reflecting periodical sea-level and climate variability during the Holocene (Gulf of Almería, Western Mediterranean). Geomorphology 50:251–268

Gracia FJ, Gutiérrez F, Gutiérrez M (2003) The Jiloca karst polje-tectonic graben (Iberian Range, NE Spain). Geomorphology 52:215–231

Guerrero J, Gutiérrez F, García-Ruiz JM, Galve JP, Lucha P, Carbonel D, Bonachea J (2012) Landslide map of the upper Gallego Valley (central Spanish Pyrenees). J Maps 8:484–491

Guerrero J, Gutiérrez F, Galve JP (2013) Large depressions, thickened terraces, and gravitational deformation in the Ebro River valley (Zaragoza area, NE Spain): Evidence of glauberite and halite interstratal karstification. Geomorphology 196:162–176

Gutiérrez F, Gutiérrez M, Sancho C (1998) Geomorphological and sedimentological analysis of a catastrophic flash flood in the Arás drainage basin (Central Pyrenees, Spain). Geomorphology 22:265–283

Gutiérrez F, Ortuño M, Lucha P, Guerrero J, Acosta E, Coratza P, Piacentini D, Soldati M (2008a) Late Quaternary episodic displacement on a sackung scarp in the central Spanish Pyrenees. Secondary paleoseismic evidence? Geodinámica Acta 21:187–202

Gutiérrez F, Gutiérrez M, Gracia FJ, McCalpin JP, Lucha P, Guerrero J (2008b) Plio-Quaternary extensional seismotectonics and drainage network development in the central sector of the Iberian Range (NE Spain). Geomorphology 102:21–42

Gutiérrez F, Galve JP, Lucha P, Bonachea J, Jordá L, Jordá R (2009) Investigation of a large collapse sinkhole affecting a multi-storey building by means of geophysics and the trenching technique (Zaragoza city, NE Spain). Environ Geol 58:1107–1122

Gutiérrez F, Linares R, Roqué C, Zarroca M, Rosell J, Galve JP, Carbonel D (2012a) Investigating gravitational grabens related to lateral spreading and evaporite dissolution subsidente by means of detailed mapping, trenching, and electrical resistivity tomography (Spanish Pyrenees). Lithosphere 4:331–353

Gutiérrez F, Carbonel D, Guerrero J, McCalpin JP, Linares R, Roque C, Zarroca C (2012b) Late Holocene episodic displacement on fault scarps related to interstratal dissolution of evaporites (Teruel Neogene Graben, NE Spain). J Struct Geol 34:2–19

Gutiérrez F, Valero-Garcés B, Desir G, González-Sampériz P, Gutiérrez M, Linares R, Zarroca M, Moreno A, Guerrero J, Roqué C, Arnold LJ, Demuro M (2013) Late Holocene evolution of playa lakes in the central sector of the Ebro depression based on geophysical surveys and morpho-stratigraphic analysis of lacustrine terraces. Geomorphology 169:177–197

Gutiérrez M (1994) Introducción a la geomorfología española. In: Gutiérrez M (ed) Geomorfología de España. Editorial Rueda, Madrid, pp 1–24

Gutiérrez M, Peña JL (1977) Las acumulaciones periglaciares del Macizo del Tremedal (Sierra de Albarracín). Boletín Geológico y Minero 92:101–110

Gutiérrez M, Peña JL (1979) El karst de Villar del Cobo (Sierra de Albarracín). Estud Geol 35:651–654

Gutiérrez M, Peña JL (1994) Cordillera Ibérica. In: Gutiérrez M (ed) Geomorfología de España. Editorial Rueda, Madrid, pp 251–286

Gutiérrez M, Valverde M (1994) El sistema de poljes del río Guadazaón (Cordillera Ibérica, Provincia de Cuenca). Cuaternario y Geomorfología 8:87–95

Gutiérrez M, Meléndez A, Soriano MA (1982) Las series terminales neógenas en el sector centromeridional de la depresión del Ebro y su relación con la morfogénesis. Boletín Geológico y Minero de España 93:465–469

Gutiérrez-Elorza M, Desir G, Gutiérrez-Santolalla F (2002) Yardangs in the semiarid central sector of the Ebro Depression (NE Spain). Geomorphology 44:155–170

Gutiérrez-Elorza M, Gutiérrez-Santolalla F, Nozal F, Pérez-González A, Salazar A (2005a) Las Cuencas Cenozoicas. In: Martín-Serrano A (ed) Mapa Geomorfológico de España y del Margen Continental 1:1.000.000. Instituto Geológico y Minero de España. Madrid, pp 107–134

Gutiérrez-Elorza M, Desir G, Gutiérrez-Santolalla F, Marín C (2005b) Origin and evolution of playas and blowouts in the semiarid zone of Tierra de Pinares (Duero Basin, Spain). Geomorphology 72:177–192

Gutiérrez-Elorza M, Lucha P, Gutiérrez F, Moreno A, Guerrero J, Martín-Serrano A, Nozal F, Desir G, Marín C (2010) Are talus flatirons sequences in Spain climate-controlled landforms? Zeitschrift für Geomorphologie 54:243–252

Harvey AM (1996) The role of alluvial fans in the mountain fluvial systems of southeast Spain: implications of climatic change. Earth Surf Proc Land 21:543–553

Harvey AM, Wells SG (1987) Response of Quaternary fluvial systems to differential epeirogenic uplift: Aguas and Feos River systems, southeast Spain. Geology 15:689–693

Harvey AM, Silva PG, Mather AE, Goy JL, Stokes M, Zazo C (1999) The impact of Quaternary sea-level and climatic change on coastal alluvial fans in the Cabo de Gata ranges, southeast Spain. Geomorphology 28:1–22

Hürlimann M, Martí J, Ledesma A (2004) Morphological and geological aspects related to large slope failures on oceanic islands. The huge La Orotova landslides on Tenerife. Canary Islands. Geomorphology 62:143–158

Instituto Geográfico Nacional (1995) Atlas Nacional de España. Ministerio de Obras Públicas, Transportes y Medio Ambiente, Madrid

Jiménez-Sánchez M, Rodríguez-Rodríguez L, García-Ruiz JM, Domínguez-Cuesta MJ, Farias P, Valero-Garcés B, Moreno A, Rico M, Valcárcel M (2013) A review of glacial geomorphology and chronology in northern Spain: timing and regional variability during the last glacial cycle. Geomorphology 196:50–64

Lario J, Zazo C, Dabrio CJ, Somoza L, Goy JL, Bardají T, Silva PG (1995) Record of recent Holocene sediment input on spit bars and deltas of South Spain. J Coastal Res (Special Issue) 17:241–245

Lewis CJ, McDonald EV, Sancho C, Peña JL, Rhodes EJ (2009) Climatic implications of correlated Upper Pleistocene glacial and fluvial deposits on the Cinca and Gállego Rivers (NE Spain) based on OSL dating and soil stratigraphy. Global Planet Change 67:141–152

Lhenaff R (1986) Les grands poljés des Cordilleres Bétiques andalouses et leurs rapports avec l'organisation endokarstique. Karstologia Memoires 1:101–112

Llasat MC, Llasat-Botija M, Rodríguez A, Lindbergh S (2010) Flash floods in Catalonia: a recurrent situation. Adv Geosci 26:105–111

López-Martínez J (1986). Geomorfología del Macizo Kárstico de la Piedra de San Martín (Pirineo Occidental). PhD Thesis. University of Zaragoza, 529 p

López-Ruiz J, Cebriá JM, Doblas M, Benito R (2004) La región volcánica de Almería-Murcia. In: Vera JA (ed) Geología de España. Sociedad Geológica de España. Instituto Geológico y Minero de España, Madrid, pp 678–682
López-Vicente M, Navas A, Machín J (2009) Geomorphic mapping in endorheic catchments in the Spanish Pyrenees: an integrated GIS analysis of karstic features. Geomorphology 111:38–47
Lucha P, Gutiérrez F, Galve JP, Guerrero J (2012) Geomorphic and stratigraphic evidence of incision-induced halokinetic uplift and dissolution subsidence in transverse drainages crossing the evaporite-cored Barbastro-Balaguer Anticline (Ebro Basin, NE Spain). Geomorphology 171–172:154–172
Maher E, Harvey AM (2008) Fluvial system response to tectonically induced base-level change during the late-Quaternary: the Río Alias, southeast Spain. Geomorphology 100:180–192
Maher E, Harvey AM, France D (2007) The impact of a major Quaternary river capture on the alluvial sediments of a beheaded river system, the Río Alias, SE Spain. Geomorphology 84:344–356
Martí C, Valero B, García-Ruiz JM (1997) Large, historical debris flows in the central spanish Pyrenees. Phys Chem Earth 22:381–385
Martí J, Planagumà L, Adelina G, Canal E, Pedrazzi D (2011) Complex interaction between strombolian and phreatomagmatic eruptions in the Quaternary monogenetic volcanism of the Catalan Volcanic Zone (NE of Spain). J Volcanol Geoth Res 201:178–193
Martín-González F (2009) Cenozoic tectonic activity in a Variscan basement: Evidence from geomorphological markers and structural mapping (NW Iberian Massif). Geomorphology 107:210–225
Martínez-Díaz JJ, Masana E, Rodríguez-Pascua MA (2012) (eds) Active faults in Iberia. J Iberian Geol 38, 302 p
Martín-Serrano A (1991) La definición y el encajamiento de la red fluvial actual sobre el Macizo Hespérico en el marco de su geodinámica alpina. Revista de la Sociedad Geológica de España 4:337–351
Martín-Serrano A (2005) (ed) Mapa Geomorfológico de España y del Margen Continental 1:1,000,000. Instituto Geológico y Minero de España, Madrid, 232 p
Martín-Serrano A, Molina E (2005) El Macizo Ibérico. In: Martín-Serrano A (ed) Mapa Geomorfológico de España y del Margen Continental 1:1,000,000. Instituto Geológico y Minero de España, Madrid, pp 65–85
Martín-Serrano A, Vegas J, García-Cortés A, Galán L, Gallardo-Millán JL, Martín-Alfageme S, Rubio FM, Ibarra PI, Granda A, Pérez-González A, García-Lobón JL (2009) Morphotectonic setting of maar lakes in the Campo de Calatrava Volcanic Field (Central Spain, SW Europe). Sed Geol 222:52–63
Martín-Vide J (2002) Las lluvias máximas diarias. In: Ayala-Carcedo F, Olcina J (eds) Riesgos naturales. Ariel Ciencia, Capellades, pp 913–920
Masson DG, Watts AB, Gee MJR, Urgeles R, Mitchell NC, Le Bas TC, Canals M (2002) Slope failures on the flanks of the western Canary Islands. Earth-Sci Rev 57:1–35
Mather AE (2000) Adjustment of a drainage network to capture induced base-level change: an example from the Sorbas Basin, SE Spain. Geomorphology 34:271–289
Menéndez I, Silva PG, Martín-Betancor M, Pérez-Torrado FJ, Guillou H, Scaillet S (2008) Fluvial dissection, isostatic uplift, and geomorphological evolution of volcanic islands (Gran Canaria, Canary Islands, Spain). Geomorphology 102:189–203
Miötke FD (1968) Karstmorphologische studien in the Glacial-überformten Höhenstufe der "Picos de Europa", Nordspanien. Jaher. der Geographischen Gesselschaft zu Hannover, Sonderheft 4, 161 p
Milanovic P (2000) Geological engineering in Karst. Zebra, Beograd 347 p
Molina E (1991) Geomorfología y Geoquímica del paisaje. Dos ejemplos en el interior de la Meseta ibérica. Acta Salmanticensia 72, Univ. Salamanca, 156 p
Molina E, Vicente A, Cantano M, Martín-Serrano A (1989) Importancia e implicaciones de las paleoalteraciones y de los sedimentos siderolíticos del paso Mesozoico-Terciario en el borde suroeste de la Cuenca del Duero y Macizo Hercínico Ibérico. Stvd Geol Salmanticensia 5:177–186
Molina E, García J, Vicente MA (1997) Paleoweathering profiles developed on the Iberian Hercynian Basement and their relationship to the oldest Tertiary surface in central and western Spain. In: Widdowson W (ed) Paleosurfaces: recognition, reconstruction and paleoenvironmental interpretation. Geological Society, London, pp 175–186
Molina-Ballesteros E, Cantano-Martín M (2002) Study of weathering processes developed on old piedmont surfaces in Western Spain: new contributions to the interpretation of the "Raña" profiles. Geomorphology 42:179–292
Nozal F (2002) Cartografía Geológica y Memoria (Geomorfología) de la Hoja de Siles (865). Mapa Geológico de España a escala 1:50.000 (MAGNA) IGME
Ortega JA, Garzón-Heydt G (2009) Geomorphological and sedimentological analysis of flash-flood deposits. The case of the 1997 Rivillas flood (Spain). Geomorphology 112:1–14
Pagés-Valcarlos JL (2000) Origen y evolución geomorfológica de las rías atlánticas de Galicia. Revista de la Sociedad Geológica de España 13:393–403
Palacios D, Andrés N, Marcos J, Vázquez-Selem L (2012) Maximum glacial advance and deglaciation of the Pinar Valley (Sierra de Gredos, Central Spain) and its significance in the Mediterranean context. Geomorphology 177–178:51–61
Pedraza J, Sanz MA, Martín A (1989) Formas Graníticas de la Pedriza. Agencia del Medio Ambiente de la Comunidad de Madrid, Madrid 205 p
Peña JL (1983) La Conca de Tremp y las Sierras Prepirenaicas Comprendidas entre los Ríos Segre y Noguera-Ribagorzana. Estudio Geomorfológico, Instituto de Estudios Ilerdenses 373 p
Peña JL (1994) Cordillera Pirenaica. In: Gutiérrez M (ed) Geomorfología de España. Editorial Rueda, Madrid, pp 159–225
Peña JL, Chueca J, Julián A (1998) Los derrubios estratificados del sector central pirenaico: cronología y límites altitudinales. In: Salvador F, Schulte T, García Navarro A (eds) Procesos Biofísicos Actuales en Medios Fríos. Publicaciones de la Universidad de Barcelona. Barcelona, pp 205–216
Perea H, Masana E, Santanach P (2012) An active zone characterized by slow normal faults, the northwestern margin of the Valencia trough (NE Iberia): a review. J Iberian Geol 38:31–52
Pérez-González A, Cabra P, Martín-Serrano A (1989) (eds) Mapa del Cuaternario de España 1:1,000,000. Instituto Tecnológico y Geominero de España, Madrid, 279 p
Pérez-Torrado FJ, Carracedo JC, Rodríguez-González A, Soler V, Troll VR, Wiesmaier S (2012) La erupción submarina de La Restinga en la isla de El Hierro, Canarias: Octubre 2011–Marzo 2012. Estudios Geológicos 68, doi:10.3989/egeol.40918.179
Pezzi M (1977) Morfologías Kársticas del Sector Central de la Cordillera Subbética. Cuadernos Geográficos de la Universidad de Granada. Serie Monográfica 2, 209 p
Pinilla A (1993) (ed) La Raña en España y Portugal. Centro de Ciencias Medioambientales. Consejo Superior de Investigaciones Científicas. Madrid, 396 p
Pinyol NM, Alonso EE, Corominas J, Moya J (2012) Canelles landslide: modelling rapid drawdown and fast potential sliding. Landslides 9:33–51
Reicherter KR, Jabaloy A, Galindo-Zaldíbar J, Ruano P, Becker-Heidmann P, Morales J, Reiss S, González-Lodeiro F (2003) Repeated paleoseismic activity of the Ventas de Zafarraya fault (S Spain) and its relation with the 1884 Andalusian earthquake. Int J Earth Sci 92:912–922

Remondo J, Bonachea J, Cendrero A (2005) A statistical approach to landslide risk modelling at basin scale: from landslide susceptibility to quantitative risk assessment. Landslides 2:321–328

Rodríguez-González A, Fernández-Turiel JL, Pérez-Torrado FJ, Paris R, Gimeno D, Carracedo JC, Aulinas M (2012) Factors controlling the morphology of monogenetic basaltic volcanoes: The Holocene volcanism of Gran Canaria (Canary Islands, Spain). Geomorphology 136:31–44

Rodríguez-Pascua MA, Pérez-López R, Garduño-Monroy VH, Giner-Robles JL, Silva PG, Perucha-Atienza MA, Hernández-Madrigal VM, Bischoff J (2012) Paleoseismic and geomorphologic evidence of recent tectonic activity of the Pozohondo Fault (Betic Cordillera, SE Spain). J Iberian Geol 38:239–251

Rodríguez-Ramírez A, Rodríguez-Vidal J, Cáceres L, Clemente L, Belloumini G, Manfra L, Improta S, Andrés JR (1996) Recent coastal evolution of the Doñana National Park (SW Spain). Quatern Sci Rev 15:803–809

Rodríguez-Vidal J (1986) Geomorfología de las Sierras Exteriores Oscenses y su Piedemonte. Instituto de Estudios Altoaragoneses, Huesca 172 p

Rodríguez-Vidal J, Cáceres LM, Finlayson JC, Gracia FJ, Martínez-Aguirre A (2004) Neotectonics and shoreline history of the Rock of Gibraltar, Southern Iberia. Quatern Sci Rev 23:2017–2019

Romero P, Dóniz FJ (2005) Los relieves volcánicos españoles. In: Martín-Serrano A (ed) Mapa Geomorfológico de España y del Margen Continental a Escala 1:1.000.000. Instituto Geológico y Minero de España. Madrid, pp 159–172

Roqué C, Linares R, Rodríguez R, Zarroca M (2011) Granite caves in the north-east of the Iberian Peninsula: artificial hypogea versus tafoni. Zeitschrift für Geomorphologie 55:341–364

Roqué C, Zarroca M, Linares R (2013) Subsurface initiation of tafoni in granite terrains; geophysical evidence from NE Spain: geomorphological implications. Geomorphology 196:94–105

Rosell J, Linares R, Fité F, Roqué C, Llompart C, Losantos M (2010) Montclús. Misteriós Naixement i Desaparició d'un Poble. Institut Geològic de Catalunya, Barcelona 111 p

Sánchez-Arcilla A, Jiménez JA, Valdemoro H (1998) The Ebro Delta: morphodynamics and vulnerability. J Coastal Res 14:134–152

Sanjaume E, Gracia FJ (2011) Las Dunas de España. Sociedad Española de Geomorfología, Madrid 747 p

Serrano E, de Sanjosé JJ, González-Trueba JJ (2010) Rock glacier dynamics in marginal periglacial environments. Earth Surf Proc Land 35:1302–1314

Silva PG, Harvey AM, Zazo C, Goy JL (1992) Geomorphology, depositional style and morphometric relationships of Quaternary alluvial fans in the Guadalentín Depression (Murcia, SE Spain). Zeitschrift für Geomorphologie 36:325–341

Silva PG, Goy JL, Somoza L, Zazo C, Bardají T (1993) Landscape response to strike-slip faulting linked to collisional settings: Quaternary tectonics and basin formation in the Eastern Betics (SE Spain). Tectonophysics 224:289–303

Silva PG, Goy JL, Zazo C, Bardají T, Lario J (1997) Paleoseismic indications along aseismic segments of Quaternary faults in the Guadalentín Depression (SE Spain). J Geodyn 24:105–115

Silva PG, Goy JL, Zazo C, Bardají T (2003) Fault-generated mountain fronts in southeast Spain: geomorphologic assessment of tectonic and seismic activity. Geomorphology 50:203–225

Silva PG, Bardají T, Calmel-Ávila M, Goy JL, Zazo C (2008) Transition from alluvial to fluvial systems in the Guadalentín Depression (SE Spain) during the Holocene: Lorca Fan versus Guadalentín River. Geomorphology 100:140–153

Silva PG, López-Recio M, Tapias F, Roquero E, Morín J, Rus I, Carrasco García P, Rodriguez-Pascua MA, Pérez-López R (2013) Stratigraphy of the Arriaga Palaeolithic sites. Implications for the geomorphological evolution recorded by thickened fluvial sequences within the Manzanares River valley (Madrid Neogene Basin, Central Spain). Geomorphology 196:138–161

Smart PL (1986) Origin and development of glacio-karst closed depressions in the Picos de Europa, Spain. Zeitschrift für Geomorphologie 30:423–443

Somoza L, Barnolas A, Arasa A, Maestro A, Rees JF, Hernández-Molina FJ (1998) Architectural stacking patterns of the Ebro Delta controlled by Holocene high-frequency eustatic fluctuations, delta-lobe switching and subsidence processes. Sed Geol 117:11–32

Tuccimei P, Soligo M, Ginés J, Ginés A, Fornós J, Kramers J, Villa IM (2010) Constraining Holocene sea levels using U-Th ages of phreatic overgrowths on speleothems from coastal caves in Mallorca (Western Mediterranean). Earth Surf Proc Land 35:782–790

Twidale CR, Vidal-Romaní JR (2005) Landforms and Geology of Granite Terrains. Balkema, Leiden 352 p

Urgelés R, Canals M, Baraza J, Alonso B, Masson DG (1997) The most recent megaslides of the Canary Islands: the El Golfo Debris Avalanche and the Canary Debris Flow, west El Hierro Island. J Geophys Res 102:20305–20323

Vázquez-Urbez M, Pardo G, Arenas C, Sancho C (2011) Fluvial diffluence episodes reflected in the Pleistocene tufa deposits of the River Piedra (Iberian Range, NE Spain). Geomorphology 125:1–10

Vera JA (2004) (ed) Geología de España. Sociedad Geológica de España. Instituto Geológico y Minero de España, Madrid, 884 p

Vidal-Romaní JR, Twidale CR (1998) Formas y Paisajes Graníticos. Universidade da Coruña, A Coruña 411 p

Vilaplana JM (1987) Guía dels Paitsatges Granitics dels Països Catalans. Kapel, Barcelona 182 p

Vilaplana JM (2008) Los riesgos naturales en Cataluña. Generalitat de Catalunya, Barcelona 226 p

White S, García-Ruiz JM (1998) Extreme erosional events and their role in mountain areas of Northern Spain. Ambio 27:300–305

White S, García-Ruiz JM, Martí C, Valero B, Errea MP, Gómez-Villar A (1997) The 1996 Biescas campsite disaster in the central Spanish Pyrenees and its temporal and spatial context. Hydrol Process 11:1797–1812

Whitfiled E, Harvey AM (2012) Interaction between the controls on fluvial system development: tectonics, climate, base level and river capture. Río Alias, Southeast Spain. Earth Surf Proc Land 37:1387–1397

Zarroca M, Linares R, Roqué C, Rosell J, Gutiérrez F (2013) Integrated geophysical and morphostratigraphic approach to investigate a coseismic? Translational slide responsible for the destruction of Montclús village (Spanish Pyrenees). Landslides, in press

Zazo C, Goy JL, Dabrio CJ, Bardají T, Somoza L, Silva PG (1993) The last interglacial in the Mediterranean as a model for the present interglacial. Global Planet Change 7:109–117

Zazo C, Silva PG, Goy JL, Hillaire-Marcel C, Lario J, González A (1998) Coastal uplift in continental collision plate boundaries: data from the last interglacial marine terraces of the Gibraltar strait area (South Spain). Tectonophysics 301:95–109

Zazo C, Dabrio CJ, Borja F, Goy JL, Lezine AM, Lario J, Polo MD, Hoyos M, Boersma JR (1999) Pleistocene and Holocene eolian facies along the Huelva coast (southern Spain): climatic and neotectonic implication. Geol Mijnbouw 77:209–224

Zazo C, Goy JL, Dabrio CJ, Bardají T, Hillaire-Marcel C, Ghaleb B, González-Delgado JA, Soler V (2003) Pleistocene marine terraces of the Spanish Mediterranean and Atlantic coast: record of coastal uplift, sea level highstands and climate changes. Mar Geol 194:103–133

Zazo C, Goy JL, Dabrio CJ, González-Delgado J, Bardají T, Hillaire-Marcel C, Cabero A, Ghaleb B, Borja F, Silva PG, Roquero E, Soler V (2013) Retracing the Quaternary history of sea level on Spanish Mediterranean-Atlantic coasts: a geomorphological and sedimentological approach. Geomorphology 196:36–39

The Guadalentín Tectonic Depression, Betic Cordillera, Murcia

2

Pablo G. Silva

Abstract

The tectonic Guadalentín Depression is an elongated Quaternary sedimentary basin generated by a system of left-lateral strike-slip faults on the eastern Betic Cordillera. The Lorca-Alhama de Murcia fault (LAF) is the most relevant structure, controlling the 100-km-long western margin of the depression with a prominent mountain front. Climatic conditions are semiarid, and sedimentation is dominated by alluvial fans. The interaction between tectonics, alluvial sedimentation, and climate results in the development of widespread tectonic landforms and alluvial fan surfaces with different degree of calcrete development. The variable morpho-sedimentary arrangements record different styles of faulting and uplift history on the range front faults. Large fans occur associated with the main gaps and step-overs between the different fault segments bounding the Quaternary basin (e.g. Lorca Fan). Telescopic fan sedimentation under limited distal aggradation eventually turned into distal trenching throughout the late Holocene. This geomorphic evolution constitutes a good example of the transformation of the ancient main fan-feeding channel into a true fluvial channel (present-day Guadalentín River) linked to a well-preserved Late Bronze geoarcheological record.

Keywords

Tectonic geomorphology • Mountain fronts • Alluvial fans • Drainage development • SE Spain

2.1 Introduction

The Guadalentín Depression (Murcia, SE Spain) constitutes one of the Quaternary sedimentary basins associated with the NE-SW active strike-slip fault system of the Eastern Betic Shear Zone (EBSZ; Larouzière et al. 1988; Silva et al. 1993). This tectonic depression, more than 100 km in length, is one of the most outstanding examples of recent tectonic landscapes in the Betic Cordillera, and the area of well documented historical and modern seismic activity (Mw 5.1, 2011 Lorca Earthquake; Alfaro et al. 2011). The depression is bounded by fault-controlled mountain fronts that provide evidence of significant Late Quaternary tectonic activity (Fig. 2.1) recorded by a large variety of tectonic landforms, mostly related to strike-slip faults (e.g. micro pull-apart basins, linear sag ponds, small pressure and shutter ridges, and offset drainages; Silva et al. 1992a, 1997, 2003; Silva 1994, 1996; Martínez-Díaz et al. 2012). Mountain front tectonic activity is well recorded by proximally trenched and distally aggrading alluvial fan sequences, whose stratigraphic and geomorphic relationships provide evidence of (a) their uplift history (Harvey 1984; Silva et al. 1992b);

P. G. Silva (✉)
Dpto. Geología, Escuela Politécnica Superior de Ávila,
Universidad de Salamanca, Salamanca, Spain
e-mail: pgsilva@usal.es

F. Gutiérrez and M. Gutiérrez (eds.), *Landscapes and Landforms of Spain*, World Geomorphological Landscapes,
DOI: 10.1007/978-94-017-8628-7_2, © Springer Science+Business Media Dordrecht 2014

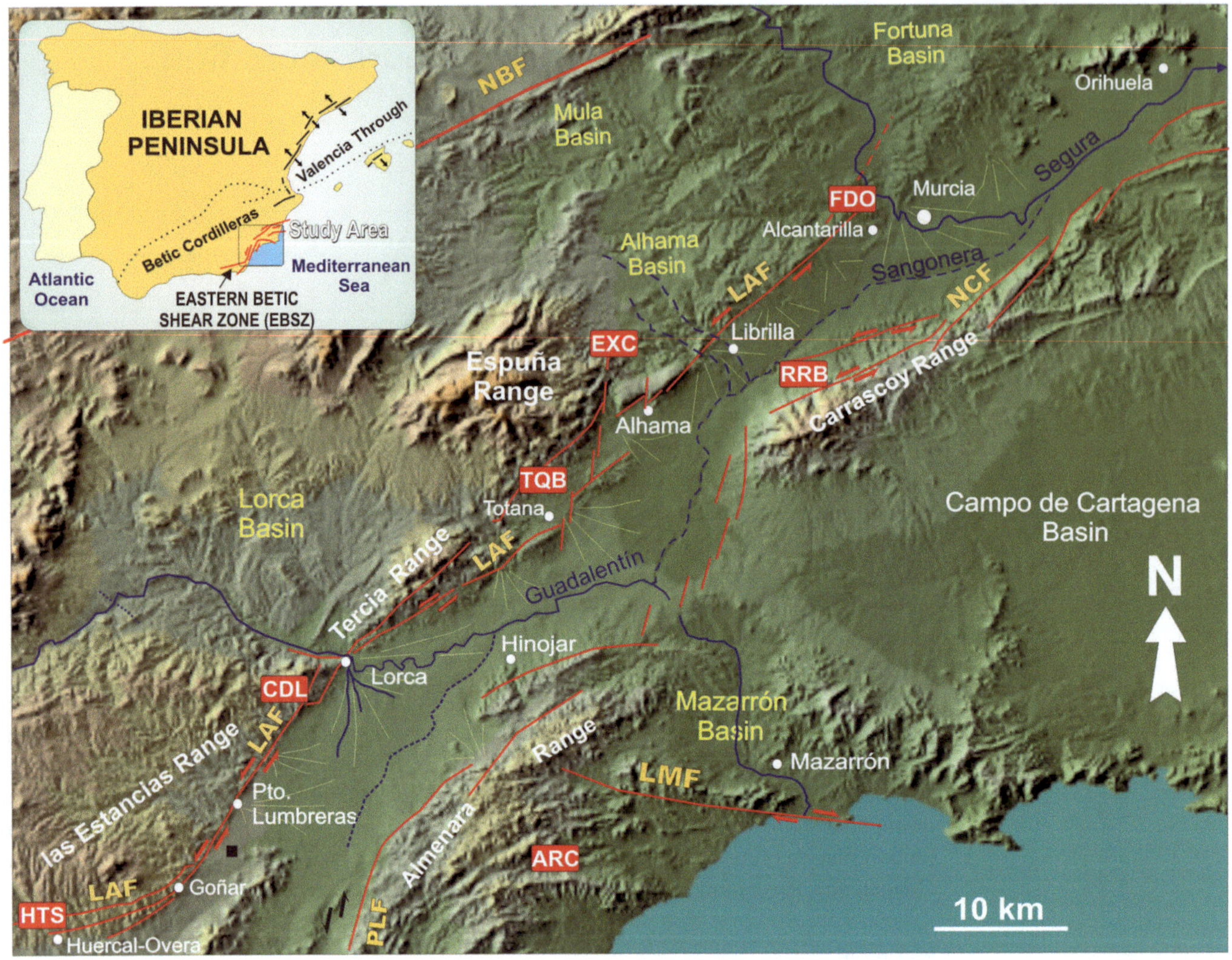

Fig. 2.1 Geomorphological and tectonic setting of the Guadalentín Depression. Blue: main drainage systems related to the Guadalentín River. *Red* main strike-slip fault systems controlling major range fronts. *Yellow* location of the main Late Neogene sedimentary basins. *White* main ranges and localities of the region. *LAF* Lorca-Alhama de Murcia Fault; *PLF* Palomares Fault; *NCF* North Carrascoy Fault; *LMF* Las Moreras Fault; *NBF* North Betic Fault; *HTS* horse-tail splay termination of the LAF; *CDL* Contractional duplex of Lorca; *TQB* triangular Quaternary pull-apart basins; *EXC* extensional system of Canacarix; *RRB* El Romeral Rock-Bar fault; *FDO* fault die-out at surface

(b) the progressive development of the drainage network during the Late Quaternary (Harvey 1990, 1997); (c) and the transition from alluvial to fluvial systems in the semi-arid SE Spain during historical times (e.g. Bronze Age; Silva et al. 1996, 2008; Calmel-Avila 2002). Additionally, the depression constitutes a remarkable example of a semi-endorheic environment drained artificially in very recent historical times and affected by severe flooding events (López-Bermúdez et al. 2002). The preserved Late Holocene alluvial landforms and sedimentary sequences within the depression provide a high-quality geoarchaeological record illustrating the relationships between human populations and drainage changes since the early Bronze to Medieval times (Silva et al. 2008).

2.2 Geological and Geographical Setting

The Guadalentín Depression is located in the central sector of the EBSZ. This is a crustal-scale structure defined by a set of post-orogenic NE-SW left-lateral strike-slip faults in the eastern Betic Cordillera, such as the Lorca-Alhama de Murcia (LAF), Palomares (PLF), and North-Carrascoy (NCF) faults (Fig. 2.1). These faults have been affected by successive transtensional and transpressional activity from the Late Neogene to the present time (Bousquet 1979; Larouzière et al. 1988). Tectonic activity generated the elongated Guadalentín Depression bounded by prominent fault-controlled mountain fronts from the Middle Pleistocene (Silva et al. 1993, 2003). The main range fronts are

Fig. 2.2 **a** Panoramic view of the North Carrascoy mountain front and related alluvial fan systems. **b** Panoramic view of a beheaded channel along the left-lateral Puerto Lumbreras-Lorca fault segment (*LAF*) south of Lorca city. Note *red car* in the deflected channel bed for scale (Burruezo fan, location in Fig. 2.5)

developed along the LAF reaching elevations above 900 m a.s.l. This fault comprises three large segments expressed as distinctive range fronts (Fig. 2.1), which also correspond to the main structural and seismic segments (Silva et al. 1992a, 1996, 1997, 2003; Martínez-Díaz et al. 2003, 2012). The Puerto Lumbreras (Estancias Range) and Lorca-Totana (La Tercia and Espuña ranges) fronts are mainly developed on Palaeozoic metamorphic rocks (dominantly schists and marbles) of the Alpujárride and Maláguide Betic complexes, but also on the terrigenous and marly formations of the ancient Late Neogene Lorca Basin (La Tercia Range). The Alcantarilla front (Alhama Range), with lower elevations, is entirely developed in these erodible sediments. To the east, the Almenara (PLF) and NCF range fronts are mainly underlain by more resistant metamorphic rocks of the Nevado-Filábride and Almágride Betic complexes (Fig. 2.2). Deformed marly Neogene sediments locally occur along the fault traces. Only the northern portion of the Almenara front (Hinojar Range) is entirely developed on sedimentary Neogene terrains and is the only fault displaying a landscape dominated by erosional features, indicating the lower degree of tectonic activity (Silva et al. 2003).

The depression is currently drained by a ca. 80-km-long axial fluvial system, the Guadalentín River, which flows into the Segura River in the vicinity of Murcia city (Fig. 2.1). The Guadalentín River is an ephemeral and flashy fluvial system slightly incised (7–17 m) into the Holocene fill of the depression (Fig. 2.3; Silva et al. 2008). This torrential rambla has produced major damaging floods in historical and recent times (i.e. López-Bermúdez et al. 2002), related to high-intensity convective storm events typical of the Mediterranean environments (López-Gómez and López-Gómez 1987). However, the most significant drainage systems in this zone are the numerous channels and large gullies (ramblas) that feed the small and steep alluvial fans developed at the foot of the mountain fronts (Harvey 1990; Silva et al. 1992b). Most of the fan systems are characterised by proximal trenching and distal aggradation from at least the Late Pleistocene, but also currently in relation to storms events. In most cases, the fan channels are disconnected from the axial drainage, which constitutes the base level (Guadalentín River). Endorheic conditions prevail in the central and southern sector of the depression, where only some inter-fan channels are connected to the axial drainage causing incision on distal fan surfaces related

Fig. 2.3 Upper part of the Holocene fill within the northern sector of the Guadalentín Depression (Rambla de Algeciras). Note the big boulders recording torrential activity. The *yellow cap* indicates the position of the Calcolithic soil horizon, a characteristic stratigraphic marker in this sector of the depression. Photo courtesy of Mary Ivonne Calmel-Avila (2002)

to headward erosion. Only in the northern sector of the depression, along the Alcantarilla Range front, through-fan dissection occurs and ancient fan and inter-fan channels are properly integrated in the regional drainage network as tributaries of the trunk Guadalentín River (Fig. 2.1). In this sector, aggressive headward erosion favours the development of badland landscapes within the depression and on its margins.

2.3 Geomorphology

2.3.1 Tectonic Landforms

The largest tectonic landforms developed in the Guadalentín Depression correspond to the mountain fronts associated with the main faults bounding the depression. The 80-km-long LAF is the main fault on the NW margin of the Guadalentín Depression (Fig. 2.1). This oblique-slip fault has accommodated significant left-lateral (8–20 km) and vertical displacement from the Messinian to the present day (Weijermars 1987; Silva et al. 1997). The Quaternary vertical slip rate has been estimated at 0.8–0.4 mm/yr (Silva et al. 2003; Masana et al. 2004). The fault is subdivided into several morpho-structural segments separated by significant erosional gaps through which the main drainages enter into the Guadalentín Depression (Silva 1994), forming large fans such as the Lorca fan, whose apex is located in the gap between the Lorca and Totana fault segments (Fig. 2.4). To the north, the Segura River fan at Murcia City is developed at the surface termination of the LAF (Fig. 2.1). However, Quaternary tectonic activity on the fault-controlled range fronts promoted the development of many small- to medium-size alluvial fans fed by minor rambla systems (Fig. 2.5). The main morphological features associated with the range front faults include the following (Silva 1994, 1996; Silva et al. 1997, 2003):

(a) the development of a terminal horsetail splay (HTS) in the southern termination of the LAF south of the locality of Puerto Lumbreras. Pleistocene tectonics generated a staircased topography in the range front, affecting the oldest fan surfaces and previous Plio-Pleistocene sedimentary sequences.
(b) the splitting of the basin-bounding fault into branches between Lorca and Alhama de Murcia, giving place to a set of intervening contractional duplexes (e.g. Lorca CDL), triangular strike-slip basins (TQB), and small pull-apart zones in which the Pleistocene fan surfaces are deformed and faulted. Strike-slip activity of the fault generates kilometric pressure ridges, linear tectonic ridges, beheaded fans and channels, as well as a small-scale horst and graben topography (Figs. 2.2, 2.5, 2.6).
(c) the development of linear mountain fronts with low sinuosity (Smf < 1.5) and low valley-floor width-to-valley height ratios (Vf < 0.6), in which the occurrence of metre- to decametre-scale channel offsets and deflections are common. Other minor tectonic landforms are also frequent, such as shutter ridges, beheaded fans, and flexures on fan surfaces, illustrating the dominant Late Pleistocene strike-slip activity on the fault (Fig. 2.5). Only the northern fault segments of the LAF (Librilla-Alcantarilla) and PLF (Hinojar) display high sinuosity values (Smf > 2.0) and valley-floor width ratios (Vf > 0.8), indicating the prevalence of erosion over tectonics in this sector of the depression.
(d) the development of younger N–S trending normal faults intersecting the main fault zones has a major role in the

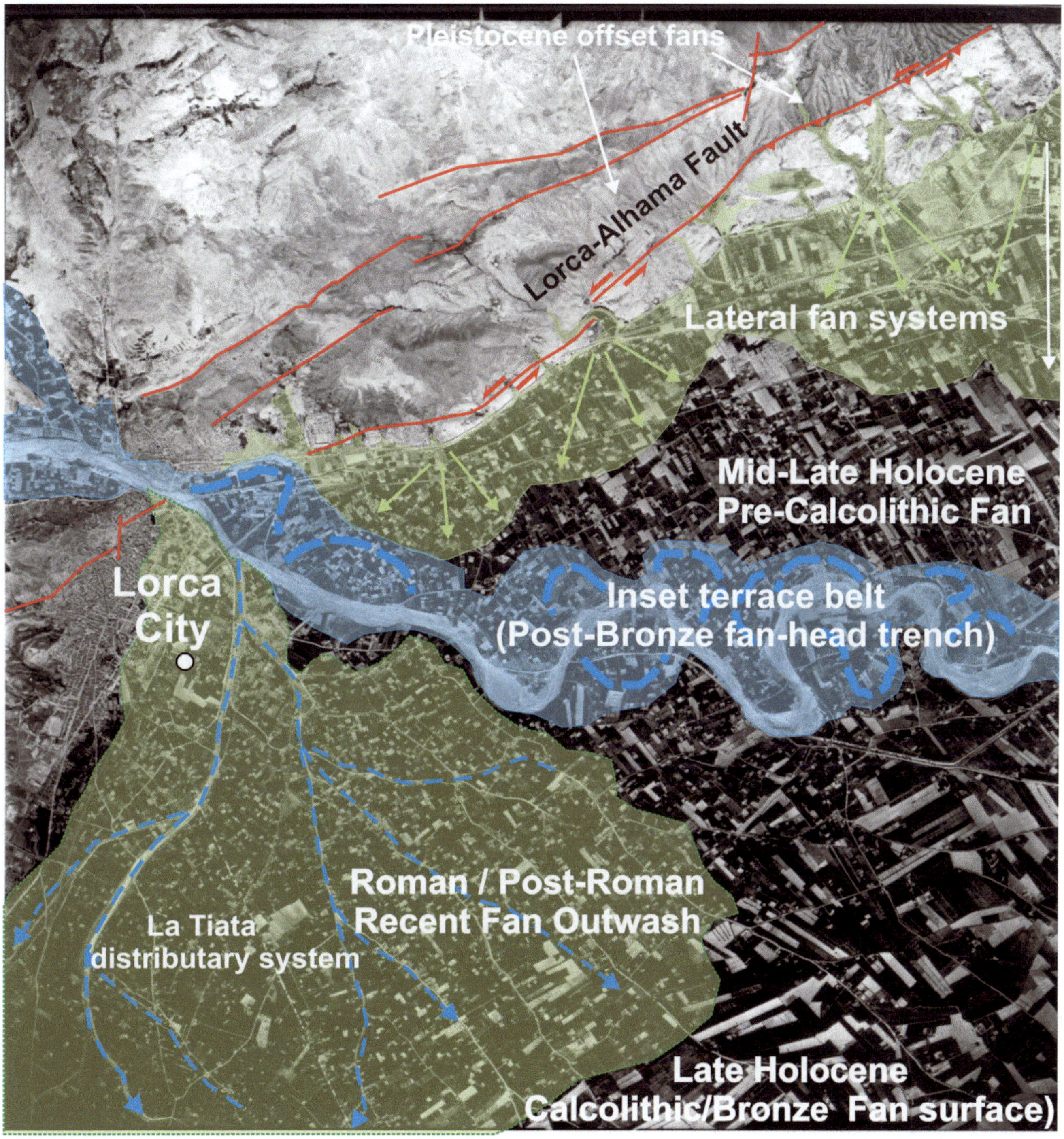

Fig. 2.4 Aerial photograph taken in 1953 (1:30,000 scale) showing the modern and ancient surfaces of the Lorca fan in the proximal sector. The terraces associated with the present-day rambla dissecting the fan surface are indicated in blue and holds an overall post-Bronze Age working as the primary fanhead trench incised in the Late Holocene fan surface. Green-coloured zones depict modern (Roman and post-Roman) fan lobes. Note fault branching and offset fan surfaces in the NE upper quadrant [modified from Silva et al. (2008)]

landscape development from Late Pleistocene to Holocene times, leading to the generation of a transverse micro-horst and graben topography and controlling in some cases the geometrical arrangement of the most recent fan sequences (Fig. 2.7).

The LAF has been active throughout the entire Quaternary period, but early transpressive tectonics favoured the development of mountain fronts from the Lower Pleistocene (Silva 1994). Conversely, tectonic activity since the Late Pleistocene seems to be controlled by pure strike-slip

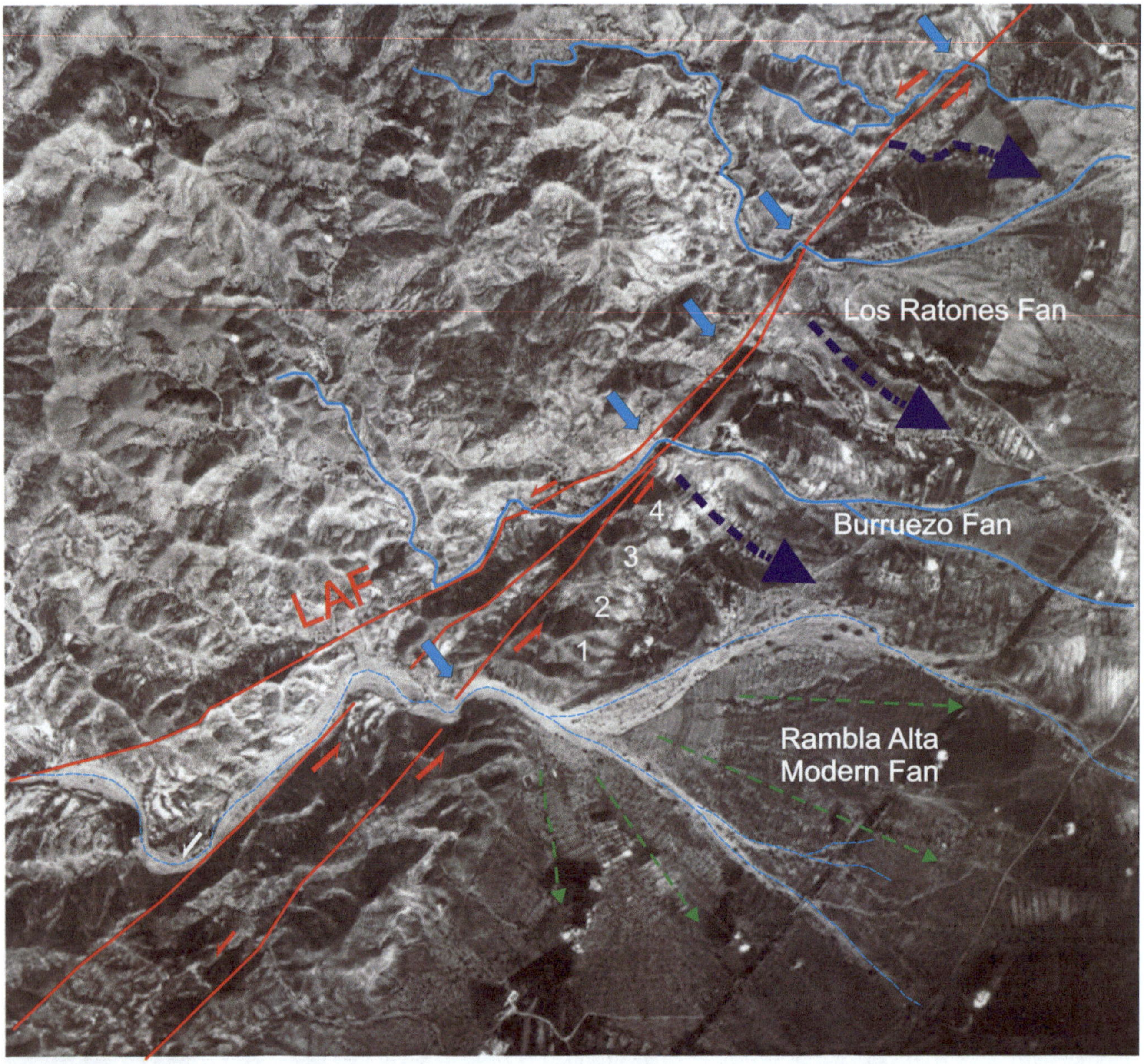

Fig. 2.5 Aerial photograph taken in 1953 (1:30,000 scale) showing left-lateral channel offsets and deflections along the LAF south of Lorca City. Note the Burruezo fan and beheaded channel shown in Fig. 2.2. *Blue arrows* indicate major channel deflections and offsets with bayonete-like pattern. Numbers (1–4) indicate the position of Lower–Middle Pleistocene fan lobes incorporated into the marginal relief due to fault activity

faulting and local transtensive tectonics (Silva et al. 1997). Paleoseismological investigations carried out via trenching north and south of Lorca city reveal two recent faulting events with loose bracketing ages (830–2130 BC and 1760 BC–1650 AD) and ascribed to large earthquakes with estimated magnitudes of 6.5–7.0 Mw (Masana et al. 2004; Ortuño et al. 2012), but with apparently minor geomorphic impact. In contrast, in the Librilla zone, the entire pre-Bronze Age Holocene sedimentary sequence is tilted more than 30° at its contact with the El Romeral Rock-Bar fault, a transverse fault zone probably belonging to the southern branch of the LAF, but buried by Late Holecene deposits. (Calmel-Avila 2002). The geoarcheological record around the El Romeral Rock-Bar indicates fault activity during the Late Calcolithic–Early Bronze transition (2700–2400 BC) along the southern branch of the LAF (Silva et al. 2008). Some synsedimentary liquefaction features in sediments corresponding to the Roman period (200 BC–90 AD) around the Totana–Librilla sector also suggest the occurrence of seismic activity during this time (Silva et al. 2008;

Fig. 2.6 Some characteristic tectonic landforms in the Guadalentín Depression. **a** Pressure and shutter ridges developed along the LAF south of Lorca city. Note the interaction of the fault zone with the alluvial fan surfaces of the second depositional phase (backfilling). **b** View of the LAF zone south of the Lorca city (see location in Fig. 2.6a) with remains of backfilled fan units on the fault zone. **c** Recurrent displacement of Middle to Late Pleistocene alluvial paleosols in the Palomares fault (*PLF*) at the southern sector of the Guadalentín Depression

Calmel-Avila et al. 2009). In addition to this probable paleoseismic evidences, historical records indicate at least three main seismic events with intensities of VIII-VII MSK (1679, 1784, and 1818 AD) similar to that occurred on May 2011. No surface faulting has been reported for any of the historical events, but significant rockfall events are the most widespread secondary environmental effects (Alfaro et al. 2011).

2.3.2 Alluvial Fan Systems

Alluvial fans are the most extensive landforms in the Guadalentín Depression. Small to medium fan systems develop along the foot of the main mountain fronts, while large fans (i.e. Lorca, Nogalte, Lebor, Algeciras, and Librilla) are associated with drainages that take advantage of gaps between different fault segments and range fronts (Figs. 2.1, 2.4). Alluvial fans in this region record a complex Pleistocene history (Harvey 1990; Silva et al. 1992b) with (a) early periods dominated by aggradation, followed by (b) fan-surface stabilisation and calcrete formation (Alonso-Zarza et al. 1998), and eventually (c) fan-head trenching and distal aggradation (Silva et al. 1992b). Different alluvial fan sedimentary styles controlled by variables such as mountain front geomorphology, lithology, and degree of tectonic activity have been identified. Silva et al. (1992b) recognised three main depositional alluvial fan sequences along the entire Guadalentín Depression (Fig. 2.7).

The first depositional phase, dominated by cemented debris-flow conglomerates with thick calcrete profiles at the surface (Alonso-Zarza et al. 1998), can be assigned to the Middle Pleistocene (>330 ka) (Sohbati et al. 2011; Ortuño et al. 2012). This syntectonic sequence displays a cumulative wedge-out with a proximal offlap arrangement related to continuous uplift along a range front controlled by oblique-slip faults. These oldest fan surfaces have a

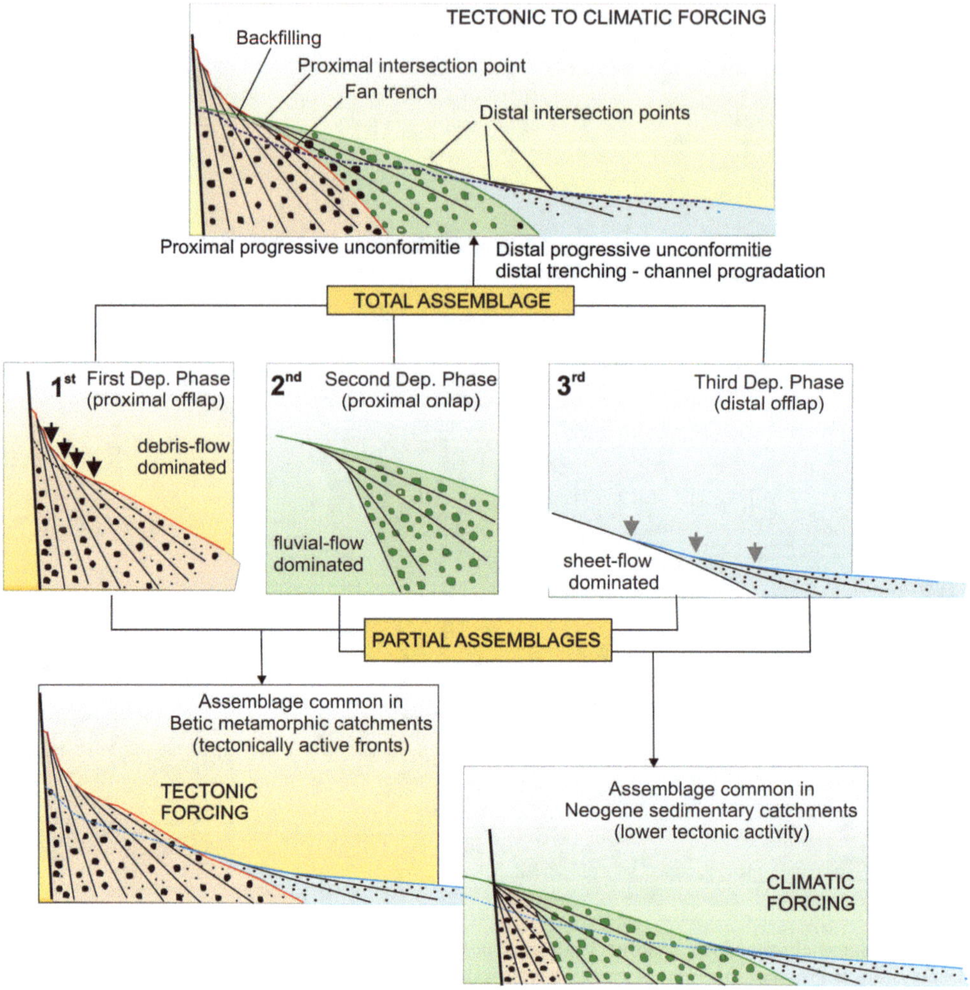

Fig. 2.7 Stratigraphic and morphological arrangement of alluvial fan bodies along the margins of the Guadalentín Depression recording tectonic and climatic forcing during deposition of the three main Middle Pleistocene–Holocene morpho-sedimentary units [modified from Silva et al. (1992b)]

dominant geomorphic expression in the southern horse-tail splay termination of the LAF (Goñar sector) and in the proximal zones of the intervening strike-slip triangular basins developed in the branched central sector of the LAF, between Lorca and Alhama de Murcia.

The second depositional phase comprises debris flow and alluvial gravel and sand facies with a proximal onlap arrangement recorded by proximal fan aggradation and backfilling in the mountain front catchments. Fan surfaces of this phase display some degree of cementation, but no true calcrete development (Alonso-Zarza et al. 1998). Recent numerical dates (Sohbati et al. 2011; Ortuño et al. 2012) allow assigning this phase to the Middle–Late Pleistocene, with ages from ca. 290 to 106–107 ka BP. The final development of these fan surfaces by means of proximal onlap aggradation and backfilling can be preliminary assigned to the last interglacial period (OIS 5). Fan surfaces of this second depositional phase dominate the mountain front piedmonts in the southern and central sectors of the depression. Proximal onlap aggradation can be interpreted as a progressive attenuation of tectonic uplift along the range fronts.

The third depositional phase mainly comprise sandy to gravely sheet flood deposits, with inset fluvial-like gravel channels generated by distributary systems. This phase is characterised by the development of proximal fanhead trenches and distal aggradation, with the progressive down-fan migration of intersection points and the formation of telescopic fan systems prograding onto the playa-lake and palustrine environments located in the centre of the depression. Available chronological data (OSL, Th/U, C14) from geoarchaeological (Calmel-Avila 2000, 2002; Silva et al. 2008) and paleoseismic research (Martínez-Díaz et al. 2003; Masana et al. 2004; Ortuño et al. 2012) allowed a finer subdivision of the third depositional fan sequence into several Late Pleistocene (<100 ka BP), Early Holocene, Bronze, Roman, Muslim, and historically recent phases of distal fan aggradation (Silva et al. 2008). Detailed analyses of the larger fans generated during the Holocene (e.g. Lorca Fan; Fig. 2.4) indicate that palustrine environments and fan aggradation prevailed until at least 2500–2300 BC (Early Bronze Age) in the southern and central sectors of the depression. Major intrabasinal fluvial incision started from the Late Bronze Age, when significant headward erosion reached the central sector of the Guadalentín Depression between Librilla and Totana (Fig. 2.3), achieving an entrenchment of up to 17 m (Calmel-Avila 2002). However, semi-endorheic conditions remained in the central zone of the depression upstream (Totana zone, Fig. 2.1) until the sixteenth to seventeenth centuries. In this zone, the ancient palustrine environments were fragmented by fluvial dissection evolving into smaller ephemeral playa-lake systems, occasionally flooded by the Guadalentín River (Silva et al. 2010a, b).

Although Middle Pleistocene fan development was favoured by tectonic activity on the basin-bounding faults, Late Pleistocene to Holocene sedimentation was mainly controlled by climate, and fans in the Murcia region evolved under very limited distal aggradation and proximal trenching (Harvey 1990, 1997). The most important factor in recent alluvial fan dynamics is the effectiveness of rainstorm events, controlling the production of sediment and run-off in the mountain catchments, as well as the generation of new distal fan lobes with telescopic arrangement (Silva et al. 2008). Data from significant flood events which occurred in the Guadalentín Depression indicate that precipitation events of 286 mm may produce peak-discharge values of 3,090 m^3/s, and sediment supply to individual large-sized fans (e.g. Nogalte Fan) can reach volumes of 813 m^3, ultimately accumulated in the distal fan segments (López-Bermúdez et al. 2002). More than 200 flood events have been documented since 1482 AD within the Guadalentín Depression, and some references to major floods during Roman and Muslim times are also available (Camarasa-Belmonte 2002).

2.4 Evolution

The Eastern Betic Shear zone was generated by an overall N–S compression related to the crustal-scale indentation processes associated with the development of the Aguilas Arc from the Middle Miocene until the Quaternary (Larouzière et al. 1988; Silva et al. 1993; Bardají et al. 2003). Late Neogene activity along this large transcurrent zone gave rise to intense magmatic phenomena and to the formation of transtensive and transpressive marine basins of various types (Larouzière et al. 1988). These Neogene basins (i.e. Lorca, Hinojar, Mazarrón, and Mula-Fortuna basins) developed on both sides of the present Guadalentín tectonic depression, where ancient Betic metamorphic paleomassifs were located (Fig. 2.1; Silva et al. 1993). Neogene basins were affected by progressive uplift from the Messinian giving place to the deposition of thick evaporite sequences in the western basins (Lorca and Mula-Fortuna) induced by the significant sea level drop that occurred at the end of the Messinian (Montenat et al. 1990). Important paleogeographical changes took place across the entire area from the Late Pliocene onwards. During this period, the stress field rotated to a NNW–SSE orientation, causing the dislocation, differential uplift, and a generalised inversion of the Late Neogene basins (Montenat et al. 1990; Silva et al. 1993).

The development of the current landscape started in the Late Pliocene, with the formation of a large sedimentary trough (Guadalentín Depression) in a zone previously occupied by the ancient Betic paleomassifs. The uplifted Neogene formations were incorporated into the marginal reliefs and mountain fronts, underlying the catchments that fed the Plio-Quaternary alluvial fan sequences. The landscape within these ancient sedimentary zones is dominated by erosional landforms including cuestas and mesas developed in the more resistant lithologies and extensive badlands in the marly and silty Late Neogene sequences (Silva 1994). The oldest late Pleistocene alluvial fan sequences are also incorporated into the mountain fronts and tectonic push-ups generated along the LAF (Fig. 2.6), recording continuous uplift and left-lateral slip during the early Quaternary (Silva et al. 2010a, b; Silva and Bardají 2012). Local rotation of the stress field to N170E associated with the development of complex tectonic structures (duplexes, step-overs, bends, and branching) within in the fault zone, especially in the complex central sector of the fault (Martínez-Díaz 2002), took place during the Lower–Middle Pleistocene transition (Silva et al. 2010a). This new tectonic scenario gave place to the rearrangement of the mountain front morphology, faulting of ancient fan sequences, and new sedimentary assemblages in Middle to Late Pleistocene alluvial fan sequences (Silva et al. 1992b, 2010b). From a geomorphological point of view, the development of the tectonic landforms associated with the alluvial fan sequences and strike-slip tectonics preserved in the piedmont areas dates from this period. Most of the mountain front uplift was achieved during the Pliocene and Early Pleistocene (>85 %), while the remaining vertical displacement (<15 %) can be assigned to the Middle Pleistocene. During this last phase, dominant strike-slip tectonics, and limited uplift, controlled the geomorphological development of mountain fronts and associated alluvial fan sequences. This interpretation is supported by offset key stratigraphic markers and the geomorphic and stratigraphic relationships of the alluvial fan sequences along the fault zones (Silva et al. 1992b, 2010a, b). From the Late Pleistocene, the geomorphological evolution has been mainly controlled by climate, inducing the generation of fanhead trenches and progressive distal fan progradation related to the aridification of the area during the Late Holocene (Calmel-Avila 2002). These climatic conditions favoured the progradation of the fan channels towards basin centre and extrabasinal capture processes that promoted the transformation of ancient alluvial systems into confined fluvial systems, leading to the present-day drainage network and the fragmentation of ancient palustrine systems in basin centre (Silva et al. 1996, 2008; Bardají et al. 2003).

2.5 Conclusions

The Guadalentín tectonic Depression is an elongated Quaternary sedimentary basin generated by left-lateral oblique faults on the eastern Betic Cordillera. The 100-km-long LAF is the main basin-bounding structure, comprising several segments and the associated mountain fronts. Climatic conditions are semiarid, and alluvial fans are the dominant sedimentary environment. The interaction among tectonics, climate, and alluvial sedimentation results in the development of a wide range of tectonic landforms and extensive alluvial fan surfaces with different degree of calcrete development and variable morpho-sedimentary assemblages. The latter record the different styles of faulting and uplift history on the range front faults. Large fans are associated with major gaps between fault segments at the basin margin (e.g. Lorca Fan). Their progressive development under limited distal aggradation throughout the Late Holocene constitutes a nice example of the transformation from alluvial to fluvial channel systems linked to a well-preserved Late Bronze geoarcheological record.

References

Alfaro P, Delgado J, García-Tortosa FJ, Lenti L, López JA, López-Casado C, Martino S (2011) Widespread landslides induced by the Mw 5.1 earthquake of 11 May 2011 in Lorca, SE Spain. Eng Geol 137–138:40–52

Alonso-Zarza AM, Silva PG, Goy JL, Zazo C (1998) Fan-surface dynamics, plant-activity and calcrete development: Interactions during ultimate phases of fan evolution in the semiarid SE Spain (Murcia). Geomorphology 24:147–167

Bardají T, Silva PG, Goy JL, Zazo C (2003) Evolución geomorfológica durante el Cuaternario de la Cuenca de Mazarrón (SE España). In: Flor G (ed) Actas de la XI Reunión Nacional de Cuaternario. Oviedo, Spain

Bousquet JC (1979) Quaternary strike-slip faults in Southern Spain. Tectonophysics 52:277–286

Calmel-Avila M (2000) Procesos hídricos holocenos en el Bajo Guadalentín (Murcia, España). Cuaternario y Geomorfología 14(1–2):65–78

Calmel-Avila M (2002) The Librilla "rambla", an example of morphogenetic crisis in the Holocene (Murcia, SE Spain). Quatern Int 93–94:101–108

Calmel-Avila M, Silva PG, Bardají T, Goy JL, Zazo C (2009) Drainage system inversion in the Guadalentín Depression during the Late Pleistocene-Holocene (Murcia, Spain). In: Romero C, Belmonte F, Alonso F, López-Bermudez F (eds) Advances in studies on desertification. Servicio de Publicaciones Universidad de Murcia, Murcia

Camarasa-Belmonte A (2002) Crecidas e inundaciones. In: Ayala F, Olcina J (coord) Riesgos Naturales. Editorial Ariel S.A. Barcelona, Spain

Harvey AM (1984) Debris flows and fluvial deposits in Spanish Quaternary alluvial fans: implications for fan morphology. Can Soc Petrol Geol Mem 10:123–132

Harvey AM (1990) Factors influencing Quaternary alluvial fan development in Southeast Spain. In: Rachocki AH, Church U (eds) Alluvial fans: a field approach. Willey, New York

Harvey AM (1997) The role of alluvial fans in arid zone fluvial systems. In: Thomas DGS (ed) Arid zone geomorphology, 2nd edn. Wiley, Chichester

Larouzière D, Bolze J, Bordet P, Hernández J, Montenat C, Ott D'Estevou P (1988) The Betic segment of the lithospheric transalboran shear zone during the Late Miocene. Tectonophysics 152:41–52

López-Bermúdez F, Conesa-Garcia C, Alonso-Sarriá F (2002) Floods: magnitude and frequency in ephemeral streams of the Spanish Mediterranean region. In: Bull LJ, Kirby MJ (eds) Dryland rivers: hydrology and geomorphology of semi-arid channels. Wiley, Chichester, pp 329–350

López-Gómez J, López-Gómez A (1987) Los Climas secos de España según el Sistema de Köppen. Papeles de Geografía Física 12:5–10

Martínez-Díaz JJ (2002) Stress field variety related to fault interaction in a reverse oblique-slip fault: the Alhama de Murcia Fault, Betic Cordillera, Spain. Tectonophysics 356:291–305

Martínez-Díaz JJ, Masana E, Hernández-Enrile JL, Santanach P (2003) Effects of repeated paleoearthquakes on the Alhama de Murcia fault (Betic Cordillera, Spain) on the Quaternary evolution of an alluvial fan system. Ann Geophys 46:775–792

Martínez-Díaz JJ, Masana E, Ortuño M (2012) Active tectonics of the Alhama de Murcia fault, Betic Cordillera, Spain. J Iberian Geol 38(1):170–181

Masana E, Martínez-Díaz JJ, Hernández-Enrile JL, Santanach P (2004) The Alhama de Murcia fault (SE Spain), a seismogenic fault in a diffuse plate boundary: Seismotectonic implications for the Ibero-Magrebian region. J Geophys Res 109:B01301

Montenat C, Ott D'Estevou P, Delort T (1990) Le Basin de Lorca. Doc et Trav IGAL 12–13:239–259

Ortuño M, Masana E, García-Meléndez E, Martínez-Díaz JJ, Canora C, Stepancikova P, Cunha P, Sohbati R, Buylaert JP, Murray AS (2012) Paleoseismic study of a slow-moving and silent fault termination: the Alhama de Murcia-Góñar system (Eastern Betics, Spain). Geol Soc Am Bull 124(9/10):1474–1494

Sohbati R, Murray AS, Buylaert JP, Ortuño M, Cunha P, Masana E (2011) Luminescence dating of Pleistocene alluvial sediments affected by the Alhama de Murcia fault (Eastern Betics, Spain)—a comparison between OSL, IRSL and post-IR IRSL ages. Boreas 10:2–13

Silva PG (1994) Evolución Geodinámica de la Depresión del Guadalentín desde el Mioceno Superior hasta la actualidad: Neotectónica y Geomorfología. Ph.D. Thesis, University Complutense of Madrid, Madrid, Spain, 642 pp

Silva PG (1996) Geometría fractal de la Falla de Lorca-Alhama de Murcia (SE España). Geogaceta 20(6):1385–1389

Silva PG, Goy JL, Zazo C (1992a) Características estructurales y geométricas de la Zona de Falla de Lorca-Alhama. Geogaceta 12:7–11

Silva PG, Harvey AM, Zazo C, Goy JL (1992b) Geomorphology, depositional style and morphometric relationships of Quaternary alluvial fans in the guadalentin depression (Murcia, SE Spain). Zeitschrift für Geomorphologie 36:661–673

Silva PG, Goy JL, Somoza L, Zazo C, Bardají T (1993) Landscape response to strike-slip faulting linked to collisional settings: Quaternary tectonics and basin formation in the Eastern Betics, Southeast Spain. Tectonophysics 224:289–303

Silva PG, Goy JL, Zazo C, Bardají T (1996) Evolución reciente del drenaje en la Depresión del Guadalentín (Murcia). Geogaceta 20(6):1385–1389

Silva PG, Goy JL, Zazo C, Lario J. Bardají T (1997) Palaeoseismic indications along "aseismic" fault segments in the Guadalentín Depression (SE Spain). J Geodyn 24:105–115

Silva PG, Goy JL, Zazo C, Bardají T (2003) Fault generated mountain fronts in Southeast Spain: geomorphologic assessment of tectonic and seismic activity. Geomorphology 250:203–226

Silva PG, Calmel-Avila M, Bardají T, Goy JL, Zazo C (2008) Transition from alluvial to fluvial systems in the Guadalentín Depression (SE Spain) during the Holocene: Lorca Fan versus Guadalentín River. Geomorphology 100:144–153

Silva PG, Bardají T, Goy JL, Zazo C, González-Hernández FM (2010a) Geomorphology and active processes. Geological map of Lorca (953). Mapa Geológico de España a Escala 1:50.000 (MAGNA) 3a Serie. Instituto Geológico y Minero de España (IGME). Digital Edition. Madrid, España. 2 Maps and Memoir

Silva PG, Bardají T, Goy JL, Zazo C, González-Hernández FM (2010b) Geomorphology and active processes. Geological map of Totana (954). Mapa Geológico de España a Escala 1:50.000 (MAGNA) 3a Serie. Instituto Geológico y Minero de España (IGME). Digital Edition. Madrid, España. 2 Maps and Memoir

Silva PG, Bardají T (2012) Geomorphology and active processes. Geological map of Puerto Lumbreras (974). Mapa Geológico de España a Escala 1:50.000 (MAGNA) 3a Serie. Instituto Geológico y Minero de España (IGME). Digital Edition. Madrid, España. 2 Maps and Memoir

Weijermars R (1987) The Palomares brittle-ductile shear zone of Southern Spain. J Struct Geol 9:139–157

The Late Neogene to Quaternary Drainage Evolution of the Uplifted Neogene Sedimentary Basins of Almería, Betic Chain

3

Adrian M. Harvey, Elizabeth Whitfield (nee Maher), Martin Stokes, and Anne Mather

Abstract

The evolution of an incising drainage network controls regional geomorphic development, but is in turn controlled by four sets of dynamic factors. These are as follows: tectonics, including both regional epeirogenic uplift and more local tectonic deformation; climatic change, affecting variations in flood power and sediment supply; base level; and local factors such as river capture, related to the development of the drainage network itself. The geomorphology of four uplifted Neogene sedimentary basins in the eastern Betic Cordillera of Almería, Spain, demonstrates how these factors interact and operate over a range of temporal and spatial scales. The basins were marine basins until the early Pliocene, when differential epeirogenic uplift caused emergence and the initiation of the drainage networks; first in the Tabernas, then in the Sorbas, and finally in the Vera and Almería basins. The last two became terrestrial in the early Pleistocene. The modern landscape reflects the influence of differential regional uplift rates on the long-term dissectional history, operating regionally over the whole period of landform development. The extremes are represented on the one hand by the deeply dissected Tabernas basin and on the other hand by the centre of the Almeria basin, which is dominated by coalescent aggrading alluvial fans. The Quaternary climatic signal is another regional signal, expressed by the sediment-led terrace sequence, with aggradation occurring primarily during Pleistocene global glacials and incision during the interglacials. These regional signals are modified locally by the other factors. Local neotectonic deformation is particularly important in the Tabernas and Almeria basins. Base-level change induced by tectonic activity and by river capture is important locally throughout the area, but the effects of base-level change induced by Quaternary sea-level change are

A. M. Harvey (✉)
School of Environmental Sciences/Geography, University of Liverpool, Roxby Building, PO Box 147Liverpool, L69 7ZT, UK
e-mail: amharvey@liverpool.ac.uk

E. Whitfield (nee Maher)
School of Natural Sciences and Psychology, Liverpool John Moores University, Byrom Street, Liverpool, L3 3AF, UK
e-mail: e.whitfield@ljmu.ac.uk

M. Stokes · A. Mather
School of Geography, Earth and Environmental Sciences, University of Plymouth, Drake Circus, Plymouth, PL4 8AA, UK
e-mail: M.stokes@plymouth.ac.uk

A. Mather
e-mail: A.mather@plymouth.ac.uk

F. Gutiérrez and M. Gutiérrez (eds.), *Landscapes and Landforms of Spain*, World Geomorphological Landscapes,
DOI: 10.1007/978-94-017-8628-7_3, © Springer Science+Business Media Dordrecht 2014

restricted to the coastal zone. River capture has had profound effects, modifying the drainage areas. The Vera basin has gained drainage area substantially, whereas that of the Almeria basin has decreased. The most important effects have been on base level and incision rates, especially in the Sorbas basin. The overriding long-term control on drainage development and therefore on landform dynamics has been the pattern of regional epeirogenic uplift, onto which the Quaternary climatically controlled aggradation/dissection sequence has been imposed. These are regional signals that have been modified locally by the more spatially and temporally restricted signals generated by base-level change and river capture.

Keywords
Drainage evolution • Tectonics • Climatic control • Base-level • River capture • Uplifted sedimentary basins • Almeria

3.1 Introduction

3.1.1 The Significance of Drainage Evolution

The evolution of the drainage network is of fundamental importance in controlling the rate and pattern of landform development. Incision of the channel network influences slope processes and therefore sediment supply to the fluvial system and hence the basin-wide sediment flux from source areas to the oceans. It is particularly important in the context of the evolution of uplifting sedimentary basins and their transformation from depositional systems to through-fluvial transport systems.

In addition to the essentially static factors of passive tectonics and gross relief and of underlying geology, including lithology, four sets of dynamic factors can be identified that control drainage evolution:

1. Tectonics: including regional epeirogenic uplift and more localised active tectonic deformation. These factors operate over Neogene to modern timescales.
2. Climate: controlling flood power, erosion rates, and therefore sediment flux, is especially important over Quaternary glacial/interglacial timescales. During the Holocene, human impact has modified the climate signal.
3. Base level: including regional base levels, related to Quaternary sea levels, modified by tectonic activity, and local base levels related to incision rates, local tectonics, and to river capture.
4. Local factors, related to the evolving drainage net itself, including effects in addition to those of base-level change, related to incision, river capture, and headwards erosion.

There have been previous studies of the relative importance of and the interplay between these sets of factors, and of the signals generated by each, particularly in relation to the geomorphology of alluvial fans (Frostick and Reid 1989; Ritter et al. 1995; Harvey 2002a; Pope and Wilkinson 2005), but fewer studies relating to the wider context of the evolution of whole drainage basins or how the interplay may vary over the timescales involved (Harvey 2006; Whitfield and Harvey 2012; Stokes et al. 2012a).

The uplifted Neogene sedimentary basins of Almería in southeast Spain provide an excellent opportunity to address these topics (Harvey 1987, 2006; Mather et al. 2001). The modern landforms exhibit an enormous range of geomorphic styles (Fig. 3.1), from areas where dissection is advanced, in incised valley systems (Fig. 3.1c) and canyons and in deeply dissected badland terrain (Fig. 3.1b), to parts of basins where aggradation in coalescent alluvial fan systems still dominates (Fig. 3.1a). These contrasts can be seen as a response to Quaternary tectonic activity, including both epeirogenic uplift and more localised ongoing neotectonic deformation, acting on the extraordinarily varied underlying Neogene geology, and modified by the development of the drainage network itself.

In this chapter, we examine these basins and address two related questions:

1. What are the relative importance of and the nature of the interplay between tectonics, climate, base-level, and local factors?
2. How do these relationships change during the evolution of the fluvial systems as the sedimentary basins undergo transformation from marine basins to basins dissected by an evolving fluvial network?

3.2 Almería: The Context for Drainage Evolution

3.2.1 Regional Geology and the Neogene–Quaternary Stratigraphic Sequence

The Almería region comprises several mountain blocks and intervening basins (Fig. 3.2), defined by a series of major left-lateral strike-slip faults with vertical component, part

Fig. 3.1 Representative terrain types within the uplifted sedimentary basins of Almeria. **a** Coalescent aggrading alluvial fans, near Nijar, Almería basin. **b** Deeply dissected badland terrain, Tabernas basin. **c** Deeply incisional terrain of major valleys, Sorbas basin. Here, the deep incision has been accentuated by river capture. The two terraces **B** and **C** on which the major part of Sorbas village is built pre-date the capture. The incision below the lower of these two terraces **C**, including deposition of the lowest terrace **D**, shown in the foreground, post-dates the capture event, at ca. 70 ka (see Stokes et al. 2002). **d** The site of the Aguas/Feos capture at Los Molinis, Sorbas basin, looking south into the abandoned Feos valley, former course of the Aguas/Feos river (Harvey and Wells 1987; Harvey et al. 1995). The col, floored by Terrace C Gravels (*left-hand side* of the photograph, marked by the *red soils*) was abandoned when the capture took place. The modern Rio Aguas now flows east (towards the *left* of the photograph and has incised c. 90 m below the floor of the col. Note the landslide (*centre* of the photograph) triggered by the rapid incision. **e** Coalescent alluvial fans at the head of the Tabernas basin, fed by catchments in the Sierra de los Filabres. **f** The incised valley of the Rio Jauro, head of the Vera basin, cut into the Plio-Pleistocene Salmeron formation of alluvial fan sediments, overlain by early Pleistocene braided river gravels, representing through fluvial drainage. The section is capped by a mature calcrete. Below, in the foreground is a younger fluvial terrace. Flow of the river is from *left* to *right*. **g** The knickpoint on the Rio Alias near Argamasson 4 km upstream of the Carboneras fault, held up by an outcrop of resistant sandstone. The knickpoint was generated by movement on the fault. This has been an effective knickpoint twice: note the cemented terrace gravels (*left-hand side* of the photograph). Flow of the river is towards the camera

of the trans-Alboran shear zone, that has developed essentially since the late Miocene (Bousquet 1979; De Larouziere et al. 1988; Sanz de Galdeano 1990; Weijermars 1991; Mather et al. 2001; Sanz de Galdeano et al. 2010). The main mountain groups themselves are composed dominantly of metamorphic rocks of the Nevado-Filábride series and the Alpujárride nappe of the Interior Zone of the Betic Cordillera, that resulted from the Cenozoic

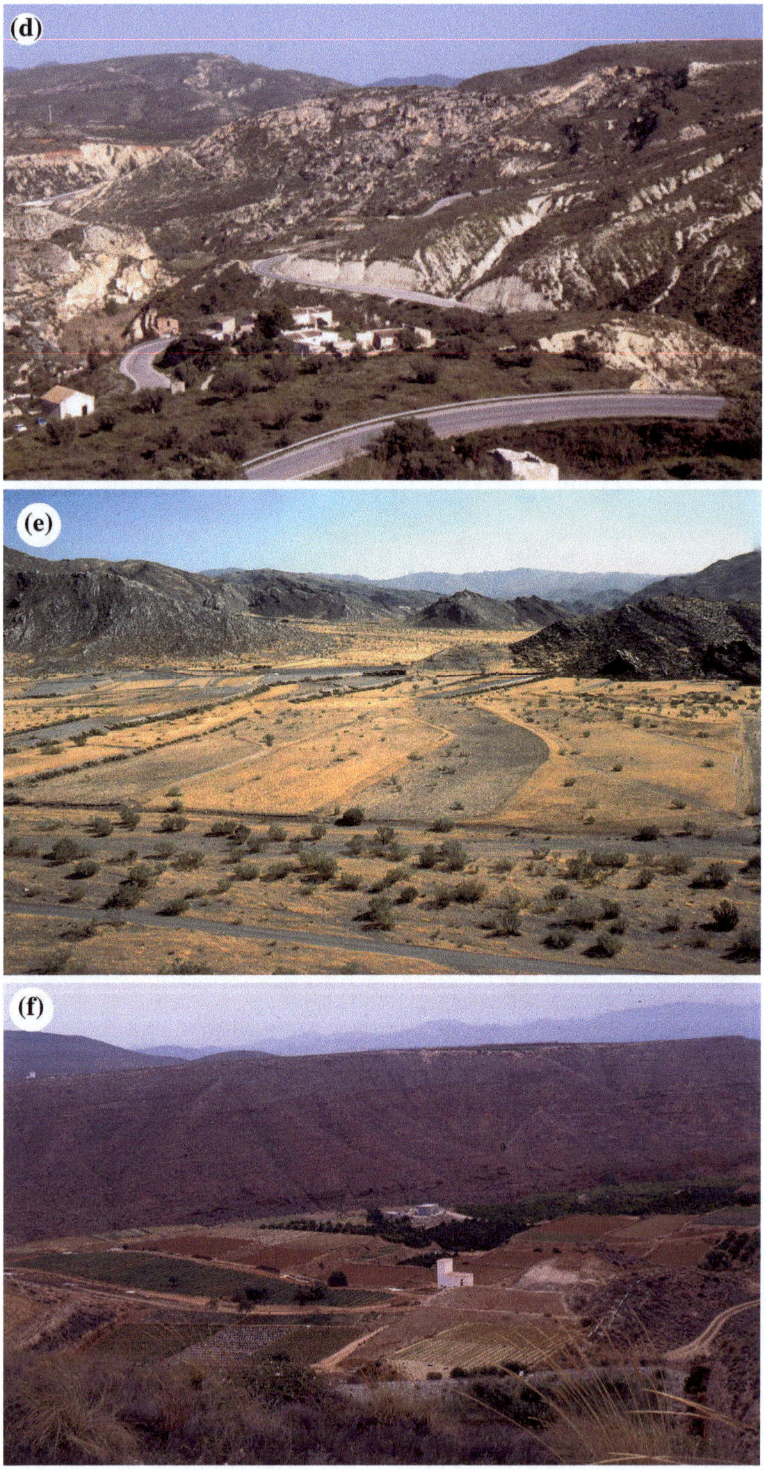

Fig. 3.1 (continued)

collision of the African and European plates (Weijermars 1991). To the south-east of the Carboneras fault zone, the low mountains of the Cabo de Gata ranges are of Late Miocene (Tortonian) volcanic rocks. The intervening intra-montane basins comprise Neogene to Quaternary sedimentary rocks, initially dominantly marine then of terrestrial origin.

By the Tortonian, the gross tectonic framework had been created by the uplift of the Sierra de los Filabres, south of which lay one large sedimentary basin. The early Messinian

Fig. 3.1 (continued)

saw the emergence of the Sierras Alhamilla/Cabrera separating the northern basins from the Almería basin to the south, and the deposition of shallow marine sediments in a sequence, more or less common to all the basins, involving sandstones, then reefs at the basin margins and marls in the basin centres (Weijermars 1991; Mather et al. 2001). The Messinian salinity crisis (Hsu et al. 1977) interrupted this sequence; sea levels fell, causing erosion, then on the recharge by seawater, evaporitic gypsum was deposited in marginal basins as they were isolated by oscillating sea levels. Gypsum formed in the Sorbas basin and extended west into the Tabernas basin and south into the northern part of the Almería basin. Finally at the end of the Messinian, marine conditions were re-established in all four basins with the possible exception of the Tabernas basin.

Marine withdrawal then began. By the early Pliocene, marine conditions were restricted to the Vera and Almería basins, with the exception of a short-lived marine phase that affected the Sorbas basin (Fig. 3.2b). By the early Pleistocene, the shoreline had retreated more or less to the position of the present coast except in the western part of the Almería basin.

3.2.2 Regional Uplift and Neotectonic Patterns

The post-Pliocene sequence of emergence reflects differential epeirogenic uplift of the mountain blocks and basins. The pattern of uplift (Fig. 3.2b), established by considering the modern elevations of early Pliocene shoreline sediments (Mather 1991; Braga et al. 2003; Harvey 2006), reveals maximum uplift of the mountain blocks, then of the basins, with maximum basin uplift in the Tabernas basin, followed by the Sorbas, Vera, and Almería basins. The differential uplift created tectonically induced gradients influencing the patterns of the developing consequent drainage and the incision rates of the river systems.

In addition to the spatial patterns of differential epeirogenic uplift, there has been ongoing neotectonic deformation during the Quaternary, localised along individual structures (Fig. 3.2b), involving sufficient deformation to directly influence the evolving geomorphology. These structures include the main regional fault systems, the Palomares, Carboneras, and Alhamilla fault systems (Keller et al. 1995; Mather and Stokes 2001; Maher and Harvey 2008; Giaconia et al. 2012), and secondary faults, especially in the western part of the Tabernas basin (Sanz de Galdeano et al. 2010). Also important are blind faults in the underlying basement causing the development of growth folds in the Neogene cover. These affect a number of NE–SW lineaments within the Sorbas basin (Mather and Westhead 1993; Maher 2005; Harvey 2007), including the Urra alignment, accentuated by gypsum-related deformation (Mather et al. 1991). Others occur within the Almería basin (Maher 2005; Whitfield and Harvey 2012) and both E–W and NW–SE alignments in the Tabernas basin (Harvey et al. 2003; Harvey 2006).

3.2.3 Other Factors Affecting Drainage Development

There are other sets of factors that have affected drainage development (Harvey 2001). Quaternary climatic cycles have modified run-off and sediment production, therefore affecting critical stream power (Bull 1979), hence

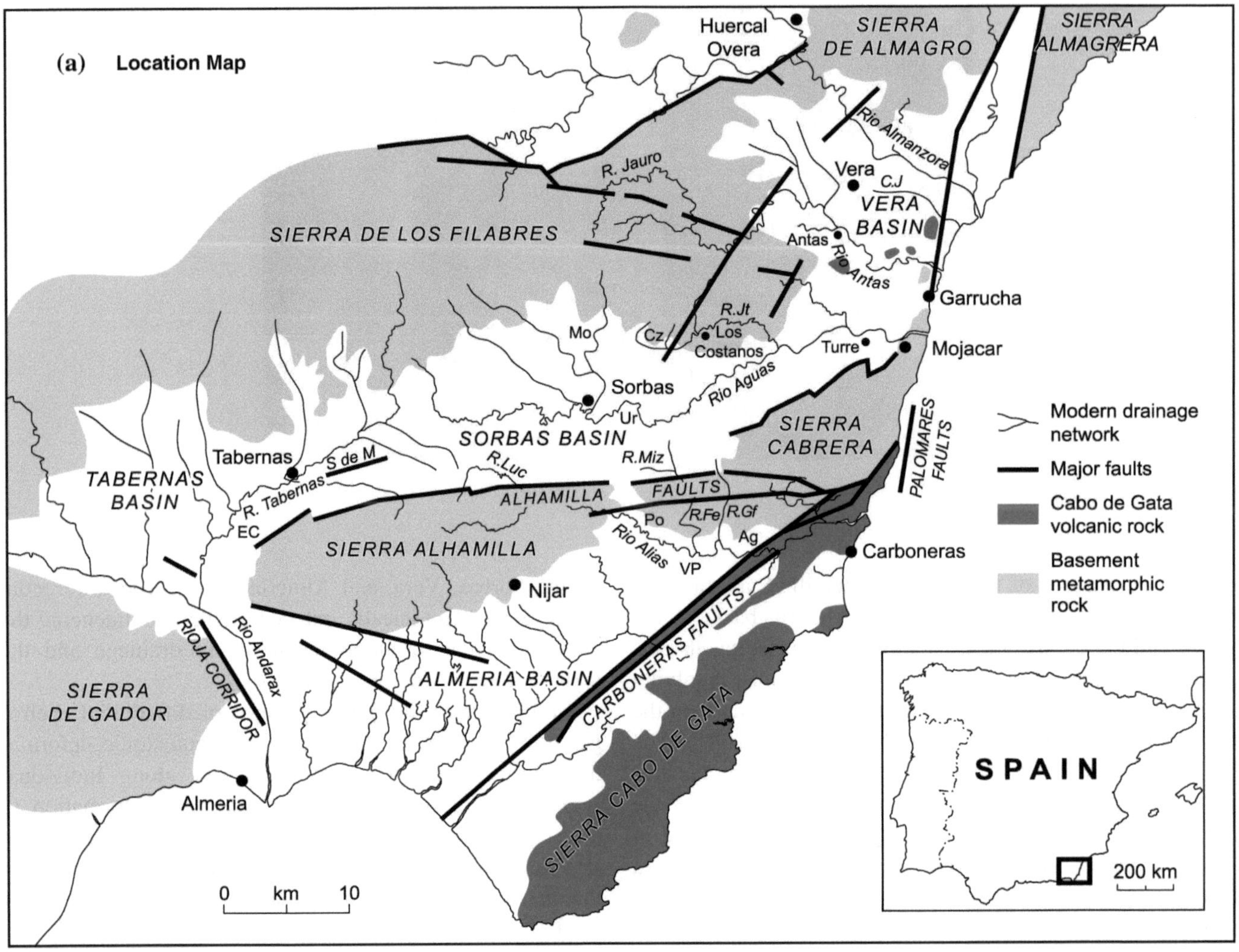

Fig. 3.2 Location, uplift, and tectonic framework. **a** Map showing all locations mentioned in the text. Abbreviations as follows: Ag Argmasson; C.J Cañada de Julián; Cz Carietiz; EC El Cautivo; Mo Moras; Po Polopos; R.Fe Rambla de los Feos; R.Gf Rambla Gafares; R.Jt Rio Jauto; R.Luc Rambla de Lucainena; R.Miz Rambla Mizala; S de M Serrata del Marchante; Ur Urra; VP Venta del Pobre. **b** Regional patterns of post-lower Pliocene epeirogenic uplift (from Braga et al. 2003; modified after Mather 1991; Harvey 2006), numbers indicate net uplift of lower Pliocene shoreline sediments: elevations in metre above modern sea level. Also shown is the approximate position of the lower Pliocene shoreline and the major basin-defining tectonic structures and patterns of ongoing Quaternary neotectonic deformation

aggradation or incision of the river systems. Quaternary glacial periods were generally associated with aggradation and interglacials with incision (Harvey 2006), though there is recent evidence of aggradation being especially important during glacial-to-interglacial transitions (see later). The results are that, as elsewhere (see Stokes et al. 2012b), the major valleys show climatically generated sequences of river terraces set into the overall epeirogenically driven incisional development of the valley systems themselves.

Quaternary climatic cycles also controlled regional base levels through their influence on eustatic sea levels of the Mediterranean (Thurber and Stearns 1965; Ovejero and Zazo 1971; Zazo et al. 1981, 2003; Goy and Zazo 1986) affecting the distal portions of the fluvial systems, and acting counter to the climatically driven sediment signal, with incision associated with glacial period low sea levels (Whitfield and Harvey 2012). Local base levels have also been modified by tectonic activity (Maher and Harvey 2008) and main stream incision caused by other factors, such as river capture (Mather et al. 2002; Stokes et al. 2002). River capture itself (Harvey and Wells 1987; Harvey et al. 1995; Mather and Harvey 1995; Mather 2000a, b; Mather et al. 2000, 2007), and other major changes to drainage organisation such as headwards erosion (Stokes and Mather 2003), have had fundamental effects on the composition of the evolving fluvial network.

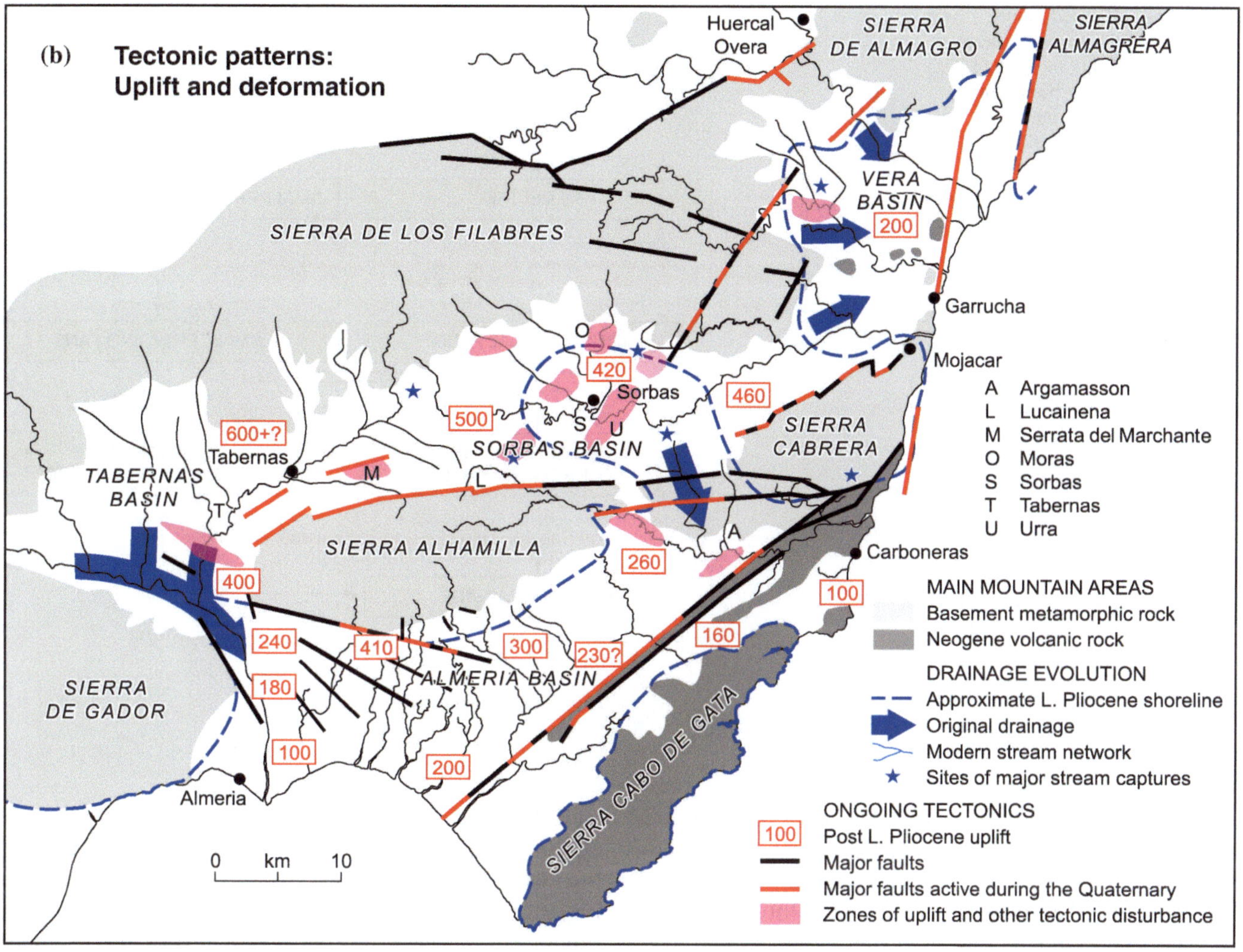

Fig. 3.2 (continued)

3.3 Basin-by-Basin Description

A simple model can be proposed (Fig. 3.3), summarising the major changes that could be expected to take place in the transformation of marine sedimentary basins into dissecting fluvial systems. As regional epeirogenic uplift takes place and sediment input to the marine basins increases, the last phases of marine sedimentation might involve fan-delta deposition. On emergence and marine retreat, these would be replaced by aggrading alluvial fans. This may be in the context of an enclosed basin with internal drainage or of an open externally drained basin. Dissection may be induced proximally by fan trenching or distally by headwards erosion from a neighbouring basin or in relation to lowered regional base levels. In both cases, aggradation is transformed into dissection by an incising through drainage. Local complications to this sequence may be brought about, for example, through tectonic activity or through drainage reorganisation by river capture. We consider the sequence in the Almería basins, basin-by-basin, initially using the Sorbas basin as a template.

3.3.1 The Sorbas Basin: The Template

Following the late Messinian post-salinity crisis recharge the "Sorbas" sea extended into the Sorbas basin, depositing the sediments of the shallow marine Sorbas Member. Later, the shallow marine environment gave way to the terrestrial sediments of the Cariatiz formation (Mather 1991), low energy floodplain or coastal plain sediments in the basin centre (Zorreras member) (Martín-Suárez et al. 2000), and fringing alluvial fans around the basin margins (Moras member). At that time, it appears that the drainage was internal as several intermittent lake phases are recorded within the basin sediments. The sequence culminated in a short-lived marine phase during the early Pliocene, which entered the basin from the south across a structural low

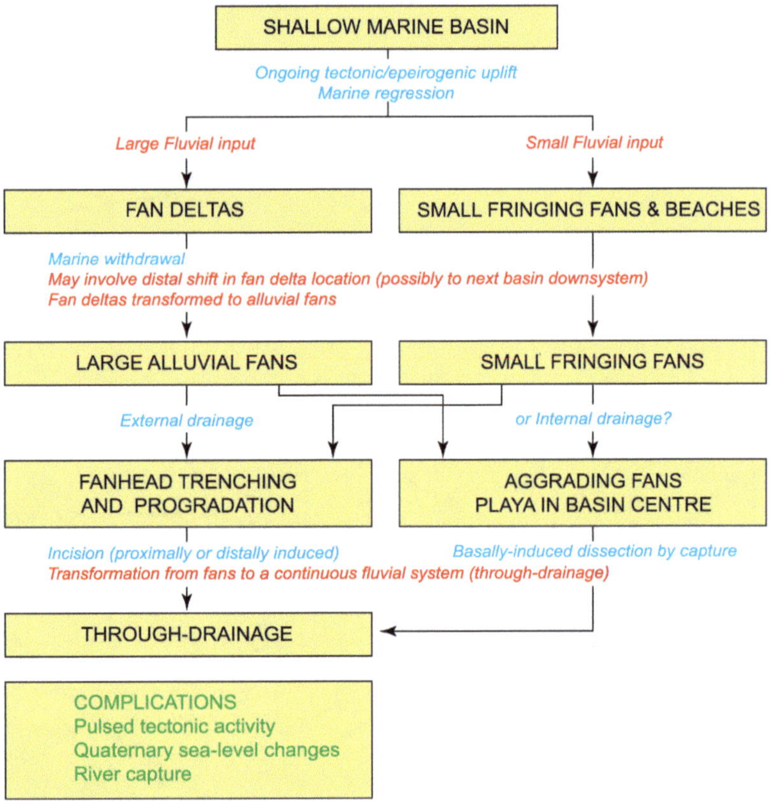

Fig. 3.3 A proposed model, summarising the possible stages in the development of uplifted marine sedimentary basins and their transformation into continuous, dissecting fluvial systems (based on Harvey et al. 2012)

between the Sierras Cabrera and Alhamilla and deposited the "Zorreras" marine band of Mather (1991). This was the last marine phase to affect the Sorbas basin.

Later during the Pliocene, a pulse of uplift especially in the Sierra de los Filabres released large quantities of gravels which form the conglomerates of the Gochar formation (Mather 1991; Mather and Stokes 2001) in the form of a series of large alluvial fans feeding an axial aggrading braided river system (Mather 1991; Mather and Harvey 1995). This system exited the basin to the south through the structural low between the Sierras Cabrera and Alhamilla (Fig. 3.4a). This was the origin of the "Aguas/Feos" river system. It drained the Sorbas basin and was initially superimposed across the Cabrera/Alhamilla zone (Fig. 3.4a) by the retreating "Zorreras" sea. The drainage became antecedent with the continued uplift of the Sierras (Harvey and Wells 1987). This river system fed a large fan delta south of the Sierras, on the northern margin of the still marine Almeria basin (see below), as the Polopos formation (Mather 1993a). Within the Sorbas basin itself, the final stages of terrestrial basin-filling culminated in the creation of the (early Pleistocene) end-Gochar formation depositional surface across the centre of the basin (Mather 1991, 1993b; Mather and Harvey 1995; Mather et al. 2002; Stokes et al. 2002).

Ongoing tectonic deformation during the Pleistocene, particularly around the margins of the basin at Cariatiz in the north (Mather 1991) and on the southern rim associated with the Lucainena fault system (Harvey and Wells 1987; Giaconia et al. 2012), caused some drainage adjustment (see Fig. 3.4; for all locations mentioned in the text see Fig. 3.2a). On the northern margin, basinal drainage was captured by the east-flowing headwaters of the Castanos/Jauto system and diverted into the lower Aguas system and so into the Vera basin (Mather 1991; Mather et al. 2000). On the southern margin headwards erosion by the subsequent Rambla de Lucainena headwater of the Alias, captured Sorbas drainage (Mather 2000a). On the western watershed Sorbas drainage was lost to the incising Tabernas drainage (Mather and Harvey 1995; Harvey 2006).

In the centre of the basin in the Sorbas area (Mather 1991) and in the Moras area (Harvey 2007) underlying basement faults creating growth folds in the Neogene–Quaternary cover influenced the incising drainage without causing any obvious captures or diversions. At Urra, tectonism in the late Pleistocene, associated with movement on a major underlying structure, locally modified by gypsum deformation (Mather et al. 1991; Harvey 2001, 2007), altered the incisional sequence by the creation of an extra terrace.

During the early–middle Pleistocene, a major change took place, the switch from basin-centre aggradation to dissection, presumably uplift-related, creating the incising river valley of the master Aguas/Feos river. The evolution of this river system was marked by three mid–late Pleistocene inset gravel terraces (Fig. 3.1c) (labelled A, B, C, oldest to

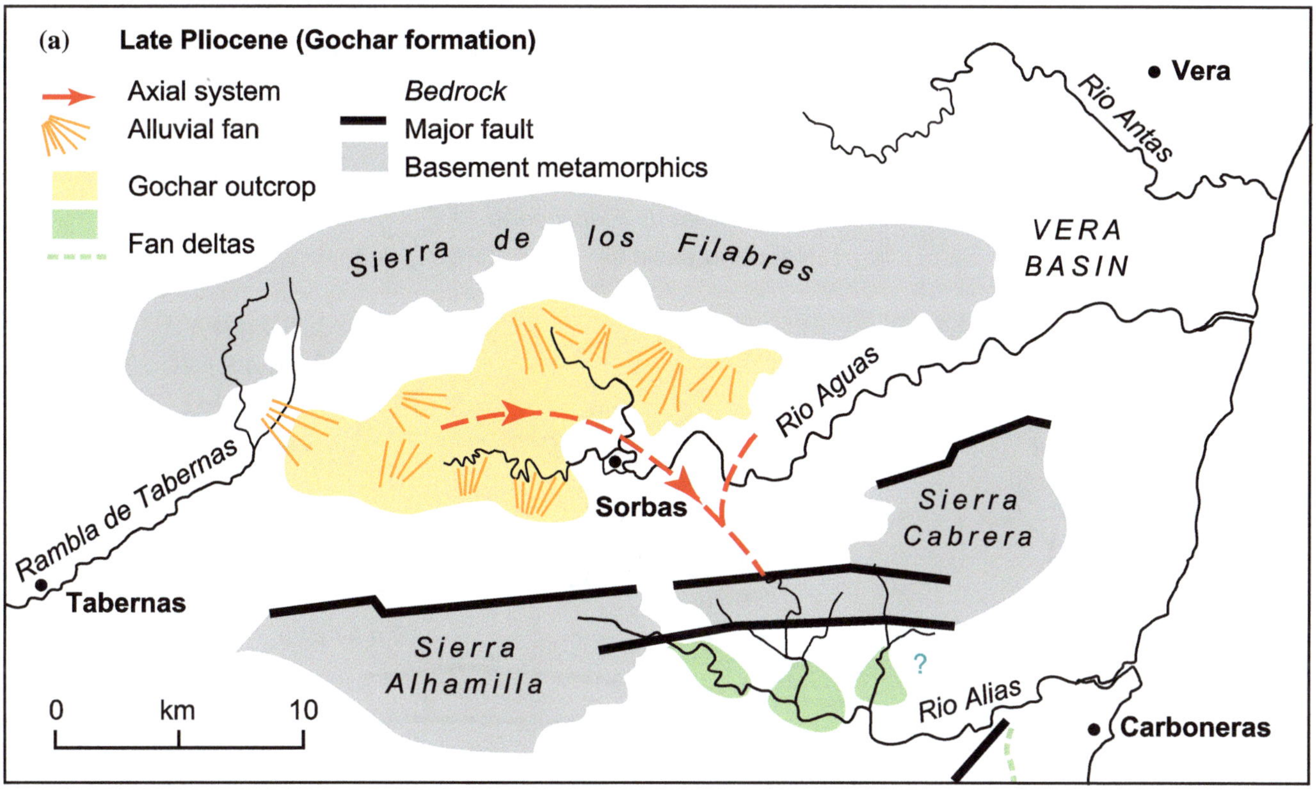

Fig. 3.4 The Sorbas basin. **a** Late Pliocene-early Pleistocene reconstruction of the palaeogeography of the Sorbas basin, as evidenced by the sedimentology of the Gochar and Polopos formations (see Mather 1991, 1993a; Mather and Stokes 2001). **b** Pleistocene to recent evolution of the fluvial system of the Sorbas basin, focussing on the effects of the Aguas/Feos capture event ca. 70 ka (see Harvey and Wells 1987; Harvey et al. 1995; Harvey 2001). **c** Reconstructed profile of the Aguas/Feos terraces that preceded the capture event (modified from Harvey 2001)

youngest by Harvey and Wells. 1987; Harvey et al. 1995). These can be traced throughout the Sorbas basin and through the gap across the Sierras Alhamilla/Cabrera into the northern end of the Almería basin (Figs. 3.4b, c). In that basin, the Aguas/Feos formed the main headstream of the Rio Alias (Harvey 2001; Maher et al. 2007).

The dates on these terrace sediments that have been obtained so far (Candy et al. 2003, 2004a, 2005; see below) indicate that, as in many other areas in the Mediterranean region (Santisteban and Schulte 2007; Mather 2009; Whitfield et al. in press), terrace aggradation took place during global glacials, perhaps especially during glacial-to-interglacial transitions. During glacials, presumably effective frost action in the mountain source areas increased sediment supply beyond the threshold of critical stream power (Bull 1979) resulting in sustained aggradation. Terrace C, the youngest terrace that can be traced through the sierras into the Almería basin, was aggrading until 60–68 ka, during MIS4. Terrace B appears to date from MIS 6 or 8 and Terrace A from some earlier unspecified date. During the intervening interglacials, incision took place.

During Terrace C aggradation, a major event took place that radically altered the fluvial systems of the Sorbas basin and the neighbouring Vera and Almería basins (Fig. 3.4b). The lower Aguas, a subsequent stream, draining towards the Vera basin, cut headwards along the outcrop of the weak Messinian Abad Marl to capture the upper Aguas (the main stem of the Aguas/Feos) near Los Molinos (Fig. 3.1d) (Harvey and Wells 1987; Harvey et al. 1995), thus beheading the Rambla de los Feos, which was the previous outlet for the drainage. The effects of this capture were not only to divert water and sediment from the Sorbas basin and its southward path towards the Almería basin, into an eastwards path towards the Vera basin, but were to have profound effects on the regional geomorphology.

Upstream of the capture site, the dramatic fall of ca. 90 m in the local base level caused an incision wave to work upstream, creating a deeply incised valley (see Fig. 3.1c) below the earlier terraced landscape (Mather et al. 2002; Stokes et al. 2002). The rapid incision generated incised bedrock meanders and incised meander cut-offs (Harvey 2007), a wave of tributary headwater captures (Mather 2000b), and triggered landslides on the oversteepened valley sides (Hart et al. 2000; Mather et al. 2003). The capture-related base level lowering and wave of incision has also had groundwater and subterranean drainage

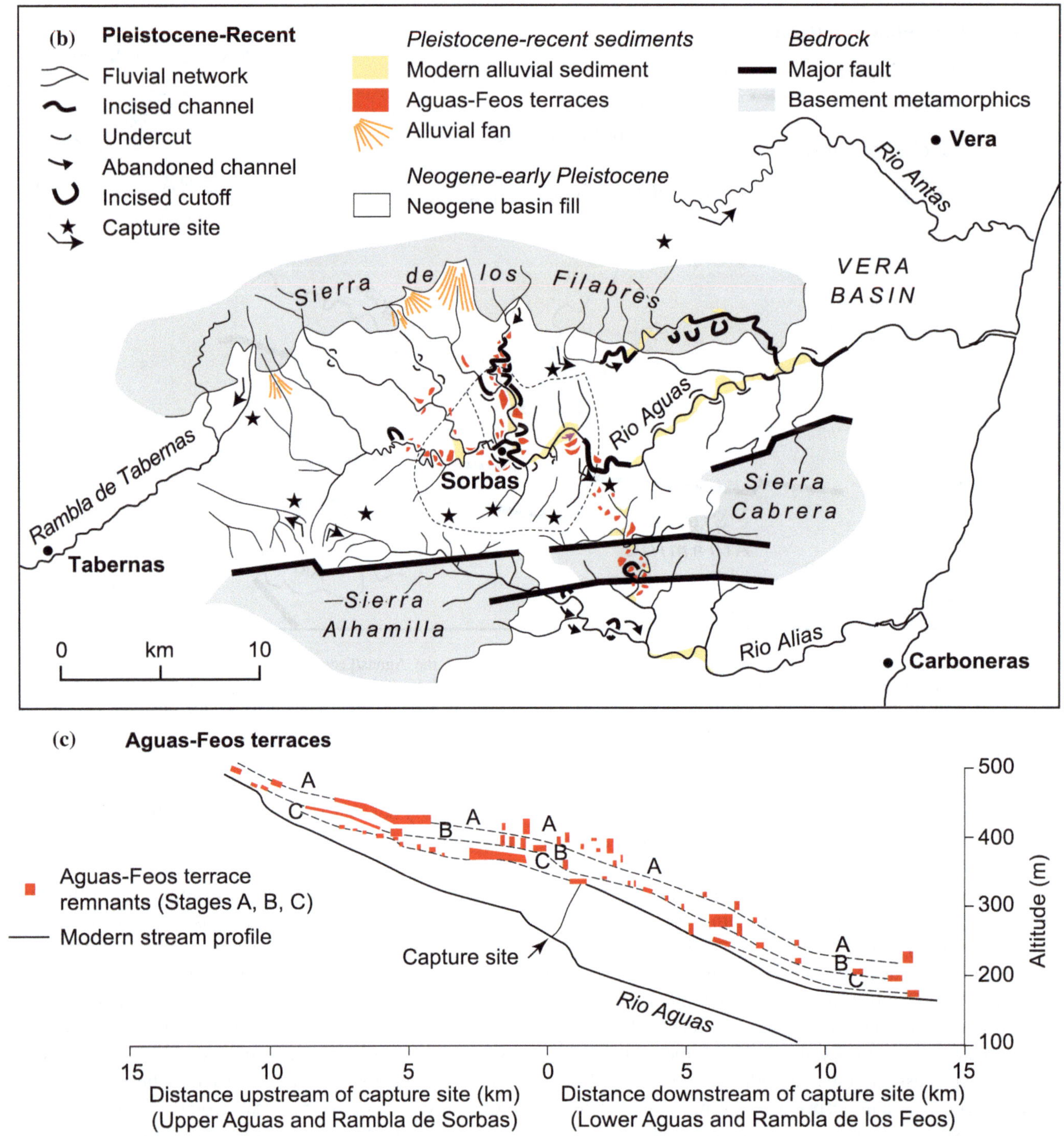

Fig. 3.4 (continued)

development implications where cave karst development within the Messinian gypsum is attributed to the base-level fall (Calaforra and Pulido-Bosch 2003). Downstream of the capture site, on the lower Aguas, the increased stream power had similar but less dramatic incisional effects, especially in the first several kilometres downstream from the capture site creating an erosional landscape dominated by badlands and landslides (Hart et al. 2000). The Feos, the former course of the master stream, was beheaded leaving only local drainage, incapable of removing the sediment supplied from the hillslopes and small tributaries. This accumulated at the valley margins as colluvium and small alluvial fans (Harvey et al. 1995; Harvey 2006). Further downstream, the Alias (see below), having lost the greater

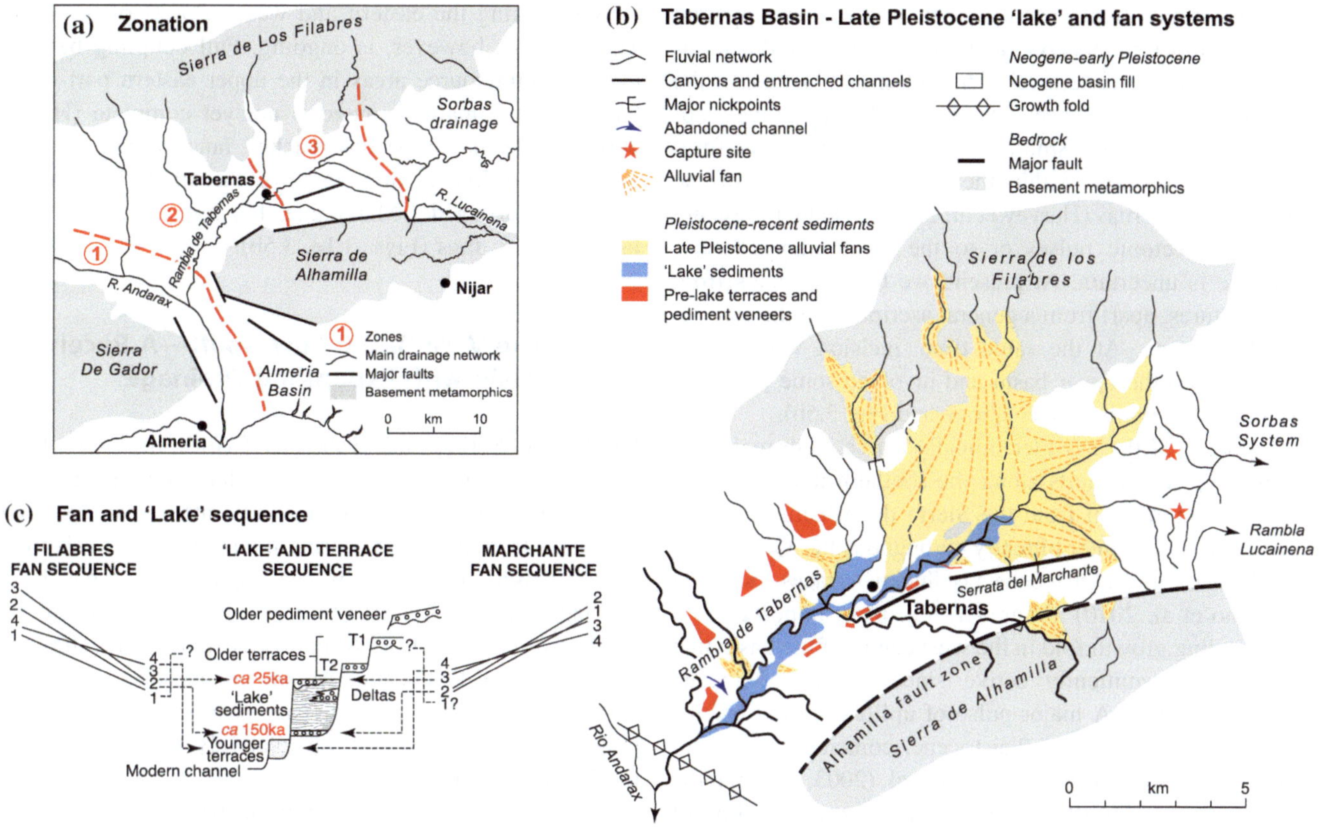

Fig. 3.5 The Tabernas basin **a** Zonation of the basin; *1* The Pliocene fluvial system of the Andarax fluvial system feeding the fan delta in the Rioja Corridor, the throat between the Tabernas and Almería basins (after Postma 1984); *2* The deeply dissected badland and canyon zone in the mid-basin; *3* The little dissected upper basin comprising coalescent late Quaternary alluvial fans. **b** The late Pleistocene 'lake' and fan system (modified from Harvey 2001; Harvey et al. 2003). **c** Correlation of 'lake', fan and terrace units within the Tabernas basin (modified from Harvey et al. 2003)

part of its headwaters, was transformed from a major river into a much smaller stream (Maher et al. 2007).

Late Pleistocene terraces that post-date the capture event can be traced along the Río Aguas and the Feos/Alias systems. They have been labelled D1, D2, D3, and E (Harvey and Wells 1987) with D1 probably pre-dating the Last Glacial Maximum, D3 from the very end of the Pleistocene (Kelly et al. 2000; Candy et al. 2003, 2004b, 2005), and E, probably human-induced, to the Holocene (Harvey and Wells 1987). Terrace D2, also of late Pleistocene age, is restricted to the Urra/Sorbas area and is related to the gypsum-related tectonic disturbance in the Urra area (Mather et al. 1991; Harvey 2007).

3.3.2 The Tabernas Basin: Maximum Uplift and Deformation

The Tabernas basin (Fig. 3.5; for locations see Fig. 3.2a) shows the greatest post-lower Pliocene uplift (Fig. 3.2b) and the steepest tectonically induced gradients. These run to the south through the basin outlet into the western end of the Almería basin. There is little evidence for the timing of the initiation of the drainage. There were clearly intermittent marine conditions following the Messinian salinity crisis, as gypsum is present both on the eastern and on northern margins of the basin, but there is little evidence of sustained end-Miocene marine conditions. The only equivalent of the Sorbas member of the Sorbas basin is restricted to the south-western part of the Tabernas basin. What is clear is that by the Pliocene the bulk of the basin was exposed, with a large depositional fluvial system in the western part of the basin, fed by the Rio Andarax to which the Tabernas system was a tributary. This system was incised into the uplifted Neogene basin-fill of Tortonian marls and turbidites. It fed a large fan delta extending from the south-west corner of the basin into the throat of the Rioja corridor (location: see Fig. 3.2a) linking the Tabernas with the Almería basin (Postma 1984).

The steep tectonically induced gradients and the high uplift rate, together with the erodible nature of the Tortonian basin-fill created a dominantly erosional landscape in the centre of the basin through the majority of the Pleistocene.

The dissectional trend was punctuated by the development of hillslope pediments (Alexander et al. 1994), and a series of fragmentary river terraces. The oldest and highest of the pediments and the terraces now form calcreted gravel-capped mesas (Nash and Smith 1998); the younger ones form fragmentary terraces within the developing valley of the Rambla de Tabernas (Harvey et al. 2003). Whether these are related to tectonic pulses or to the Quaternary climatic sequence is uncertain. At present, we have no dates from these features, apart from a general ascription to the early to mid-Pleistocene. At the same time, incision would have extended into the upper basin and involved some piracy of earlier Sorbas drainage (see above and Fig. 3.5b).

The basin is the most tectonically active, with major E–W trending thrust faults along the southern margin generated by compression from the Sierra de Alhamilla. These have been active intermittently throughout the Quaternary and displace mid-Pleistocene (?) fluvial terraces (Sanz de Galdeano et al. 2010) and travertines. Most important is a NW trending growth fold in the west centre of the basin that has been intermittently active since the late Miocene (Haughton 2001). A major pulse of uplift occurred on this structure during the mid-Pleistocene, coincident with tectonic activity recognised by García et al. (2003, 2004) in the Andarax valley. Uplift on this structure caused ponding of the drainage and the accumulation of a linear body of lacustrine and palustrine sediments (Fig. 3.5b) filling the ENE–WSW valley of the Rambla de Tabernas through the centre of the basin (Harvey et al. 2003).

A U/Th date from the base of this sequence (Delgado-Castilla and Pascual-Molina 1993) of ca. 150 ka and OSL dates from near the top of the sequence (Alexander et al. 2008) of ca. 20 ka, indicating that during virtually the whole of the late Pleistocene the drainage was impeded. The uplift along the growth fold appears to have caused back tilting of the calcreted gravel-capped mesas in the north of the basin and may have been partially responsible for eastward migration of the streams within the El Cautivo badlands (location: see Fig. 3.2a) (Alexander et al. 1994, 2008) and associated drainage reorganisation. The backtilting, together with the impeded drainage, contributed to the burial of the upper part of the basin by large coalescent alluvial fans (Fig. 3.1e) fed from the Sierra de los Filabres, the Serrata del Marchante, and the Sierra de Alhamilla (Harvey 1984; Delgado-Castilla 1993; Harvey et al. 2003).

Once the barrier caused by the growth fold had been breached after c 20 ka, the Rambla de Tabernas rapidly incised, triggering the development of the modern Tabernas badlands on the lower slopes of the western part of the Tabernas basin (see Fig. 3.1b). The high rates of erosion on these badland slopes may have been accentuated by human impact. By headwards erosion, the incision created a deep canyon linking the eastern and western parts of the basin. This process, however, is ongoing. Full coupling between the mountain source areas in the upper eastern part of the basin and the main drainage is not yet complete (Harvey 2002b). The dissection has reached fans issuing from the Sierra Alhamilla, is imminent for the small fans issuing from the Serrata del Marchante, but has yet to reach the large Filabres fans (Figs. 3.1e, 3.5b).

3.3.3 The Vera Basin: Low Uplift—A Receiving Basin with Expanding Drainage

Unlike the Sorbas and Tabernas basins, the Vera basin (Fig. 3.6; locations: see Fig. 3.2a) has been primarily a receiving basin, fed by larger rivers rather than by small drainages from the neighbouring mountain areas. This trend has increased over time by the capture of Sorbas basin drainage by the Aguas system (see above) and by the headwards erosion of the Almanzora to capture the drainage of the Huercal-Overa and Almanzora basins to the north (Stokes and Mather 2003). This basin style reflects the general low elevation and the relatively low uplift rate of the Vera basin (Fig. 3.2b).

Marine conditions persisted in this basin later than in the Tabernas and Sorbas basins, until the late Pliocene (Stokes 1997, 2008). The early–mid-Pliocene Cuevas formation of mudstones in the basin centre and sandstones with patch reefs around the basin margin was followed by the late Pliocene Espiritu Santo formation of large bodies of fan-delta deposits (Stokes 1997, 2008). This trend indicates an increased fluvial input to a shallowing basin. The fan deltas were fed by fluvial systems emanating (1) from the north of the basin (Sierra Almagro and adjacent areas), presumably from a forerunner of the lower Almanzora, and (2) from the east of the basin by a large fluvial system emanating from the Sierra Almagrera, which at the time lay further south than its present position, into which it has since moved along the Palomares fault (Völk 1967; Postma and Roep 1985; Stokes 1997, 2008).

During the late Pliocene, the sea withdrew and the Espíritu Santo formation was succeeded by the terrestrial Salmerón formation (Fig. 3.1f) (the equivalent of at least the younger part of the Gochar formation of the Sorbas basin). This formation comprises basin-marginal alluvial fans, fed from the north (by the proto-Almanzora and related systems) and from the west (by the Jauro system headstream of the Antas) into an area of internal drainage characterised by extensive bajada fringing a central playa lake (Stokes 2008) In the west, the alluvial fan sediments of the Salmerón formation are overlain by early-–mid-Pleistocene braided river deposits of the Rio Jauro/Antas (Fig. 3.1f) (Stokes and Mather 2000).

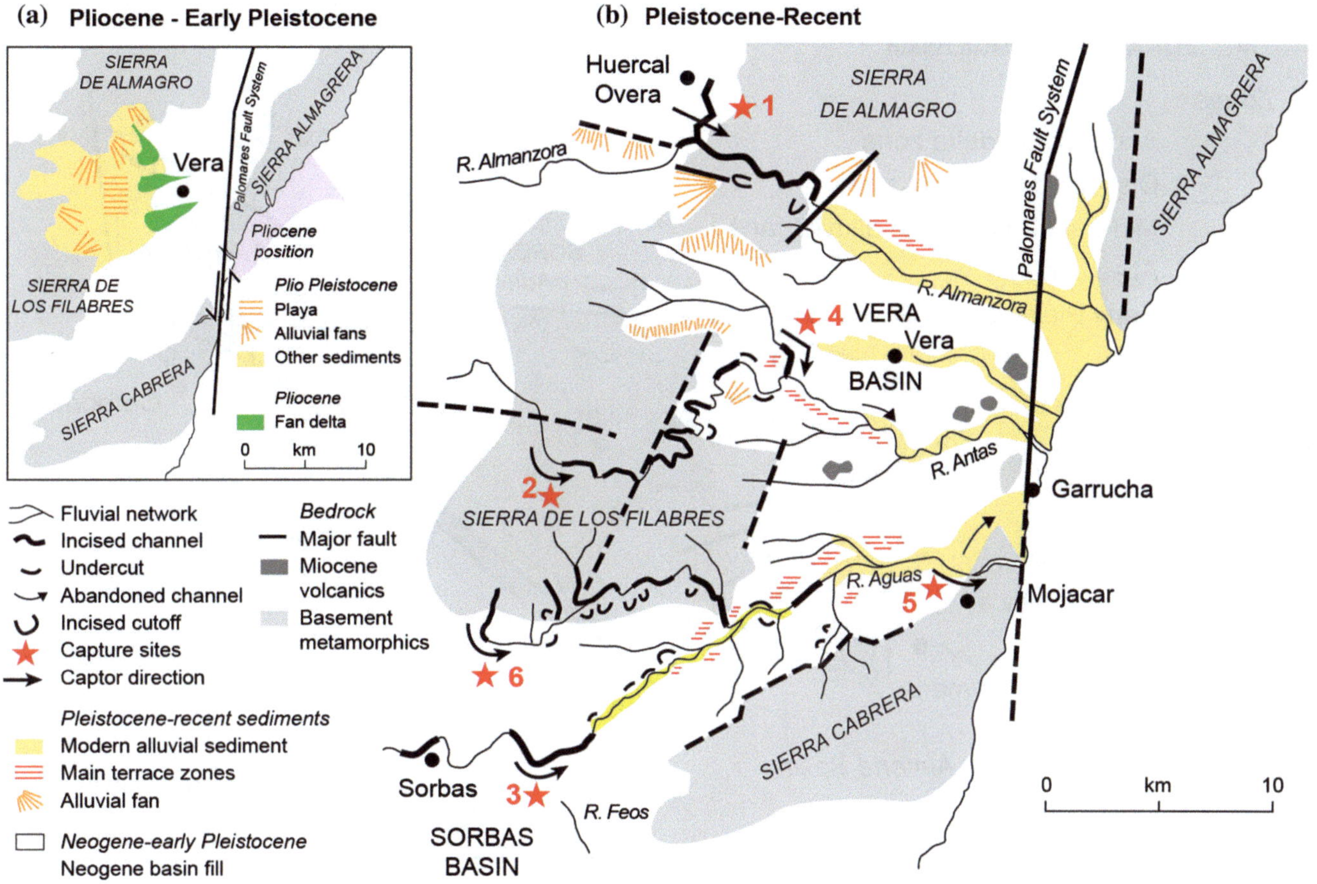

Fig. 3.6 The Vera basin. **a** Pliocene to early Pleistocene palaeogeography (based on Stokes 1997, 2008; Mather and Stokes 2001; Mather and Stokes 2001). **b** Pleistocene to recent evolution of the fluvial network (based on Stokes 1997, 2008; Mather and Stokes 2001; Mather and Stokes 2001). Major capture sites (*1–6*), as follows: *1* Headwards erosion/capture of the Huercal-Overa drainage by the lower Alanzora; *2* Capture of the *upper* Cariatiz systems by the *upper* Jauro/Antas system; *3* Capture of the *upper* Aguas (the Aguas/Feos system) by the *lower* Aguas; *4* Capture of the Cañada de Julián by the Rio Antas; *5* Capture/diversion of the lowermost Aguas from the Garrucha abandoned channel to its present course through Mojacar; *6* Capture of the Cariatiz system by the Rio Jauto

Pulsed epeirogenic uplift continued, though at lower rates than in the Sorbas basin. In addition, tectonic deformation occurred along the major structures. These include the NNE–SSW Palomares fault system, bounding the basin to the east, and similarly orientated fault zones in the western part of the basin, causing local reorganisation of the drainage (Stokes 2008), especially of the Jauro system (Stokes and Mather 2000).

During the mid-Pleistocene, a switch from aggrading braided rivers to dissection and the development of incised through-drainage took place (Figs. 3.1f, 3.6b), almost certainly triggered by a pulse of tectonic activity (Stokes and Mather 2000; Stokes 2008), though by now the basin was open to the Mediterranean and a contribution from distally induced dissection cannot be ruled out. The mid- to late Pleistocene incision, as elsewhere, was punctuated by a climate-related terrace sequence on the main rivers, the Almanzora, Jauro/Antas, and lower Aguas. There are some OSL and U-Series dates available on the terrace sequences, with terrace aggradation on the Almanzora dating from 140–120 ka and 28 ka (Meikle 2009; Meikle et al. 2010), and on the lower Aguas from 120 ka and 70–22 ka (Schulte et al. 2008), plus what are obviously Holocene low terraces.

During the Pleistocene, the Vera basin received major increases in its contributing drainage area through three major river captures, all three outside the basin itself. Headwards erosion by the Almanzora (Stokes and Mather 2003) in the early–mid-Pleistocene captured the hitherto internal drainage of the Huercal-Overa basin. Terrace sediments of the transverse reach of the Almanzora across the Sierra Almagro appear to record significant increases in palaeodischarge during incision. This is linked either to progressive incision and capture of the Huercal-Overa basin or to climate-change-related variability of precipitation and storminess (Stokes et al. 2012b).The lower Aguas captured drainage from the Sorbas basin, first in the early Pleistocene by the Rambla de Castaños (location Fig. 3.2a) capturing the Cariatiz systems in the northern part of that basin, then

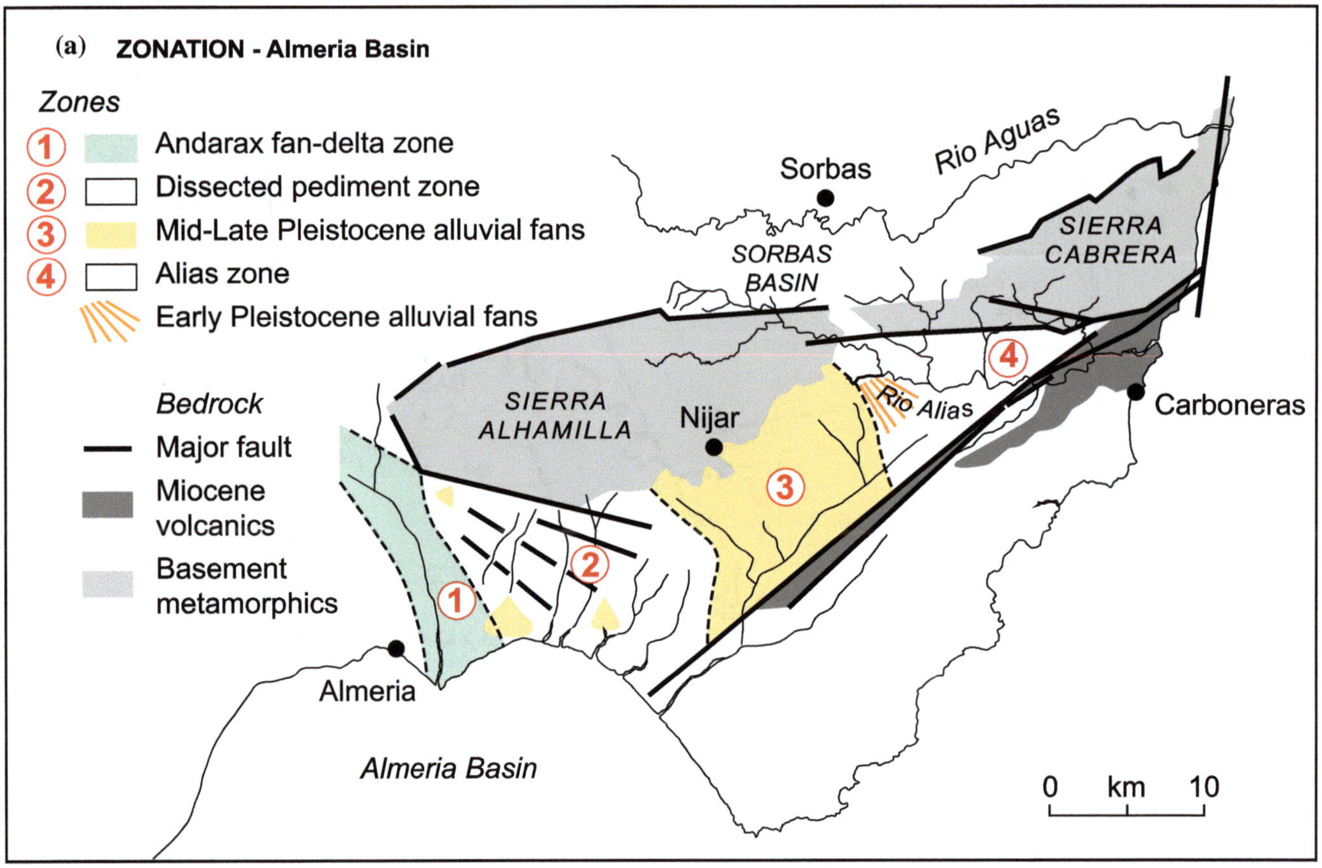

Fig. 3.7 The Almería basin: **a** Zonation: *1* The Andarax valley, dominated by the Pliocene fan delta of the Andarax and its Quaternary extension seawards. *2* Uplifted segment of the western part of the basin, dominated by offlap of successive Pleistocene coastal sediments, mantled by dissected pediment surfaces. *3* The downsagged basin centre dominated by coalescent mid–late Quaternary alluvial fans. *4* The northern part of the basin dominated by the dissecting Rio Alias system. **b** The Alias system, showing features associated with the incisional history, including zones of accelerated incision due to base-level change—*1* The coastal zone, affected by Quaternary sea-level changes; *2* the zone upstream of the Carboneras faults; 3 the Alhamilla zone (after Maher 2005; Maher et al. 2007; Maher and Harvey 2008; Whitfield (nee Maher) and Harvey 2012)

in the late Pleistocene by the lower Aguas itself capturing the Aguas/Feos (see above) and hence the bulk of the drainage of the Sorbas basin. Thirdly, the headwaters of the Jauro/Antas (Mather et al. 2000) had earlier substantially increased that drainage area, at the expense of the headstreams of the Cariatiz system.

In addition, there were minor captures within the basin, re-routing elements of the drainage (Fig. 3.6b). The Antas captured the Cañada Julián (location: see Fig. 3.2a) beheading the valley downstream through Vera. During the late Pleistocene (?), the coastal reach of the Aguas was diverted through Mojacar away from its former course towards Garrucha, presumably in response to incision induced by low sea levels during the last global glacial. The last marine lowstand caused incision of the seaward end of the Rio Aguas presumably causing the diversion through Mojacar, but also extending upstream as far as the canyon and nickpoint upstream of Turre (location: see Fig. 3.2a).

3.3.4 The Almería Basin: A Complex Receiving Basin with the Least Uplift

The Almería basin is a complex receiving basin, generally with the lowest elevations and the lowest uplift rates of all four basins. It is bounded to the northwest by the Sierra Alhamilla and to the northeast by a complex fault zone along the southern boundary of the Sierra Cabrera. It is bounded to the southeast by the Carboneras fault zone. The location of the Almeria basin in relation to other basins, i.e. as a receiving basin, and the geological structure within the basin itself results in there being four contrasted zones within the basin (Fig. 3.7a):

1. In the west, the Rioja corridor (locations: see Fig. 3.2a) serves as the outlet from the Tabernas basin, juxtaposing a zone of deep subsidence, occupied by successive fan deltas of the Río Andarax, with the outlet of the Tabernas basin, the basin with the maximum uplift rates.

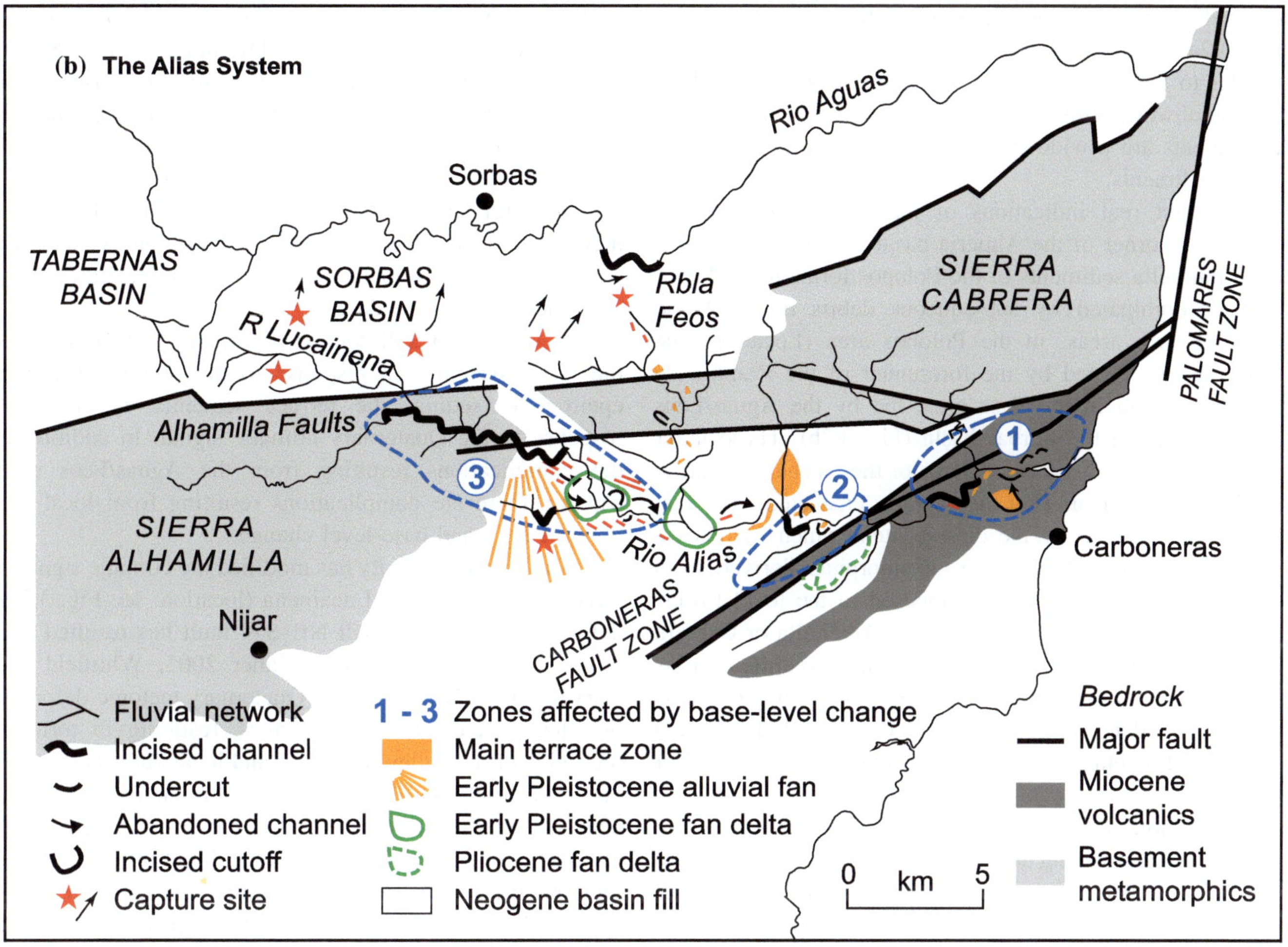

Fig. 3.7 (continued)

2. East from the Andarax valley to Nijar is a fault bounded and elevated zone (the triangular zone south of the Sierra Alhamilla on Fig. 3.7a) where the Neogene to Quaternary basin-fill sediments have been uplifted and dissected. The episodic uplift of the Sierra Alhamilla is expressed by the offlap of successive Pleistocene shorelines through this area.
3. The basin centre from Nijar to Venta del Pobre is a zone of tectonic sag where the Neogene sediments are buried by coalescent Quaternary alluvial fans (Fig. 3.1a). This was possibly a zone of internal drainage during the Quaternary. Even today, the ill-developed external drainage to the southwest is scarcely incised.
4. The northern margin of the basin is controlled by the Sierra Cabrera southern boundary fault zone, which merges eastwards with the Carboneras fault zone. The northern segment of the Almería basin has been a receiving area for the Lucainena/Alias river system and until the late Pleistocene also for the Aguas/Feos, the master drainage of the Sorbas basin (see above). Throughout the majority of the Pleistocene, the drainage has been channelled to the east in a general alignment parallel with the fault zone, rather than being directed into the basin centre.

It is the northern part of the Almería basin, where the drainage evolution is illustrated by the Río Alias, that we will be most concerned in this section (Fig. 3.7b).

The upper Messinian stratigraphy of this area resembles that of the Sorbas basin, with pockets of gypsum along the southern flank of the Sierra Cabrera, followed by marls that are the equivalent of the Sorbas member. Thence, however, the stratigraphy is more similar to that of the Vera basin. The area remained marine throughout the Pliocene, the lower Pliocene dominated by the generally massive shallow marine bioturbated sandstones of the Cuevas formation, and the upper Pliocene in the east of the area, by fan-delta sediments similar to the Espiritu Santo formation of the Vera basin. Other than the obvious, that these sediments originated from river systems draining the Sierra Cabrera, their specific origin is uncertain, though they may have been derived from a proto-Gafares system (location: see Fig. 3.2a), that early on lost its headwaters in a manner similar to the later Aguas/Feos capture (Blum 2007). Further southwest, within the rest of the Almería basin, their equivalents are a wide range of shallow marine and near-shore sediments.

The first real indications of marine retreat from the northwest corner of the Almeria basin are the Plio-Pleistocene fan-delta sediments of the Polopos formation (Mather 1993a), dominated by sub-aqueous debris flows. These occur in two areas, in the Polopos area (Location: see Fig. 3.2a) itself, fed by the forerunner of the Rambla de Lucainena, and in the Feos area, fed by the Aguas/Feos system draining the Sorbas basin (Fig. 3.7b). These latter appear to be the distal equivalents of the Gochar formation of the Sorbas basin. That they clearly emanated from that basin is evidenced by the clast content which includes a high proportion of high-grade metamorphics derived from the Sierra de los Filabres, as opposed to the local lower-grade metamorphics from the Sierras Alhamilla or Cabrera (Maher 2005). In the Polopos area, these sediments pass upwards into sub-aerially deposited alluvial fan sediments with an abundance of palaeosol horizons, the whole affected by syn-sedimentary and post-sedimentary faulting (Mather 1993a).

This whole sequence is capped by Stage A gravels of Maher (2005), forming an alluvial fan, fed by the Lucainena headwater of the Alias, and grading east into a high-level river terrace, fed from the Sorbas basin by the Aguas/Feos (Terrace A of Harvey and Wells 1987; Harvey et al. 1995). As far as can be ascertained (there is only fragmentary evidence of this terrace further east), this formed the mid-Pleistocene valley of the combined Ríos Alias and Aguas/Feos, eastwards towards the coast (Maher 2005; Whitfield and Harvey 2012).

As elsewhere the mid- and late Pleistocene were dominated by incision, punctuated by phases of terrace aggradation. Terraces B and C (after Maher 2005) can be correlated with the same terraces in the Sorbas basin (after Harvey and Wells 1987) through their continuity along the Aguas/Feos system. The terrace sequence represents a climatic signal, with aggradation phases associated with global glacials and incision phases with interglacials.

The major capture of the Aguas/Feos within the Sorbas basin (see above) during the aggradation of Terrace C had radical implications for the Rio Alias, which lost a high proportion of its drainage area through the beheading of the Rambla de los Feos. The Lucainena/Alias became the main stream with the Feos as a minor tributary. Downstream of the confluence the river was much reduced in size and power, expressed in the contrasts in sedimentology and palaeochannel morphology between Terrace C (pre-capture) and Terrace D (post-capture) deposits (Maher et al. 2007).

Terrace D, as in the Sorbas basin, can be subdivided into two phases (the equivalents of D1 and D3 in th Sorbas basin), but rather than exhibiting an inset relationship as in the Sorbas basin, the lack of deep incision between the two has resulted in the younger terrace sediments burying the older sediments and forming one topographic terrace. Preliminary OSL dates (Maher 2005) suggest that these sediments are of late Pleistocene and end-Pleistocene ages, respectively. The Holocene Terrace E is fragmentary along both the Feos and the Alias.

Whereas the overall context of drainage development of the Alias system is a response to the regional tectonic/epeirogenic setting, the terrace sequence is clearly a response to the Quaternary climatic signal. In addition to the complications resulting from the Aguas/Feos river capture, there are complications resulting from local tectonic activity and base-level change.

Local tectonic activity has modified the climatic signal in three main areas. Near Lucainena (location: see Fig. 3.2a), deformation along a small NE–SW fault has resulted in a minor within-basin capture (Maher 2005; Whitfield and Harvey 2012). Near Polopos, Quaternary tectonic deformation has accelerated incision rates, resulting in tortuous incised bedrock valley meanders and local cut-offs (Harvey 2007; Whitfield and Harvey 2012). Similar features occur near Argamasson where the Alias crosses the Carboneras fault zone, but more significant is the direct influence of the fault on the terrace sequence. During the aggradation of Terrace C, there was substantial movement of the fault, involving a downthrow to the east which triggered a local incision phase, separating two phases of Terrace C aggradation (Maher and Harvey 2008), and creating a local nick-point which propagated several km upstream (Fig. 3.1g).

At the seaward end of the system, low sea levels during global glacials triggered incision, at times when the climatic signal was producing aggradation. The results were a substantial phase of incision, together with the development of incised valley meanders, early during a regional climatically driven terrace aggradation phase. The incised valley was later buried by the terrace sediments. This interplay between base-level change and sediment supply appears to have affected all three Terraces B, C, and D, but the upstream effects were limited to a few kilometres from the modern coast (Maher 2005; Whitfield and Harvey 2012).

3.4 Discussion

3.4.1 Regional Sequence Summary

The regional sequence of geomorphic development can now be illustrated by considering the evolving palaeogeography in a series of four critical time slices (Fig. 3.8). Between the

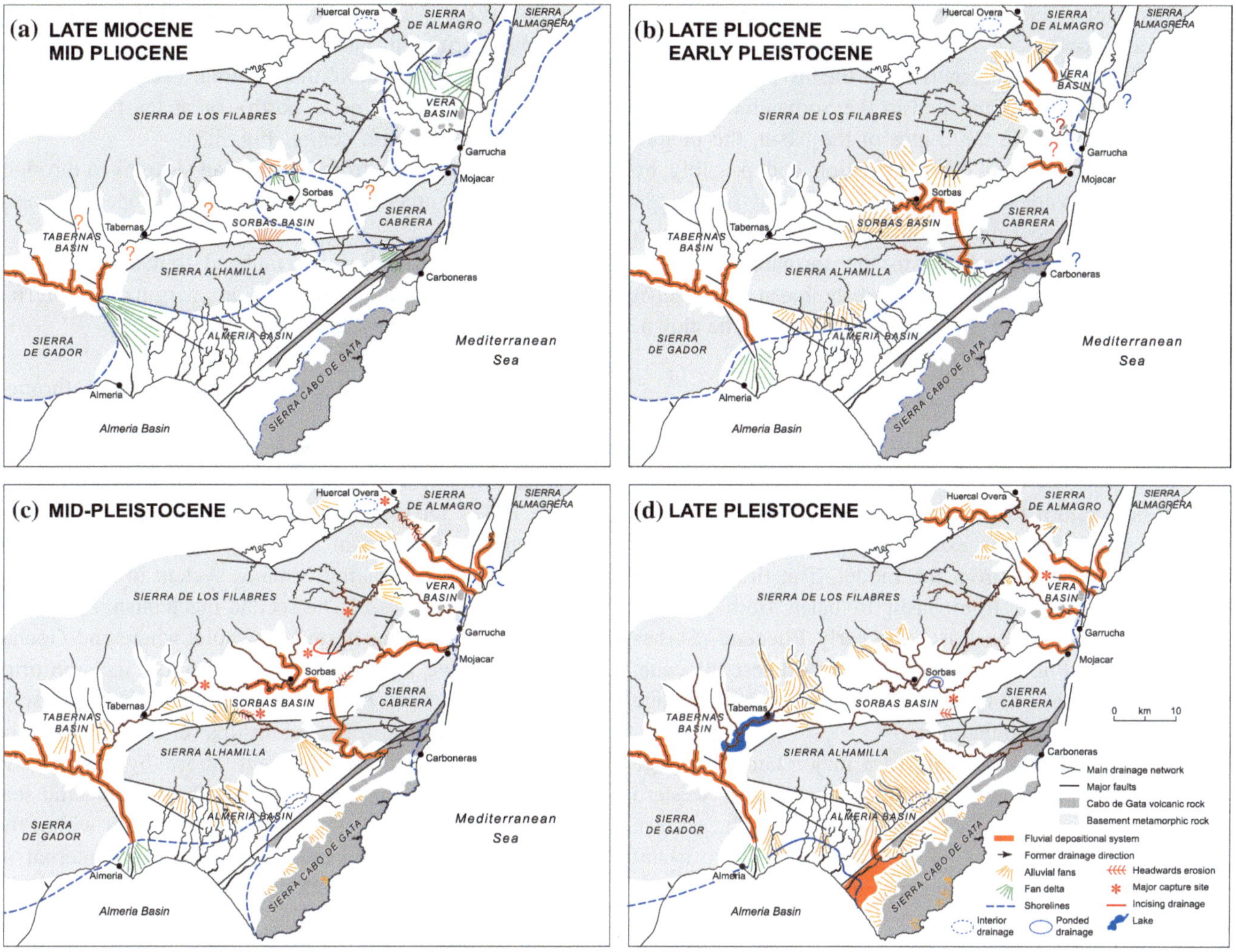

Fig. 3.8 Summary of the palaeogeographic evolution by four time slices

late Miocene and mid-Pliocene, the sea withdrew successively from the Tabernas and Sorbas basins, forming a coastline to the south of the Sierras Gador, Alhamilla, and Cabrera (Fig. 3.2b), with the exception of the earliest Pliocene encroachment into the Sorbas basin. Further north, the Vera basin formed a marine gulf. The Cabo de Gata range was an offshore island. There were major fan deltas coincident with large fluvial sediment inputs from the Tabernas basin into the western Almería basin, within the Vera basin, and smaller fan deltas in the northeast of the Almeria basin. Fluvial sedimentation is identifiable along the Andarax valley to the west of the Tabernas basin, and as small fringing fans in the Sorbas basin. Elsewhere inland, there is little or no preserved evidence and we assume that some degree of fluvial incision was the rule.

By the end-Pliocene/early Pleistocene, the sea had retreated, first from the Sorbas basin, then the Vera basin, leaving only the Almería basin under marine conditions. There, fan deltas were present in the west (the prograded Andarax fan delta) and at the outlets of the Lucainena and Aguas/Feos drainages in the north of the basin. In the Vera basin, alluvial fans had replaced the earlier fan deltas, feeding a basin-centre zone of interior drainage. In the Sorbas basin, large fans fed an axial fluvial system (the Aguas/Feos) that exited the basin to the south to form a fan delta on the margins of the still marine Almería basin (see above).

By the mid-Pleistocene, the coastline had retreated to more or less its present position, with the exception of a marine gulf at the western end of the Almería basin. The main river systems had now become established and incision into the underlying Neogene sediments had begun. The main river systems were the Andarax from the Tabernas basin to the western end of the Almeria basin, the Aguas/Feos from the Sorbas basin to the northern end of the Almería basin and the main rivers of the Vera basin, the lower Aguas, the Antas, and the Almanzora systems. There was probably interior drainage in the centre of the Almería basin. Also during this period, there were a number of headwater and basin-marginal river captures diverting drainage between basins (see below).

By the late Pleistocene, incision of the fluvial system was the dominant trend, accentuated in the Sorbas basin by the Aguas/Feos capture. In the centre of the basin, the picture was further complicated by deformation and possibly by karst-induced subsidence of the gypsum, causing ponding in the Urra area. In the Tabernas basin, uplift within the basin ponded the drainage and led to the formation of large coalescent alluvial fans in the east of the basin. Coalescent alluvial fans dominated the geomorphology of the down-sagged centre of the Almeria basin.

3.4.2 The Model

We can now return to the general model presented earlier (Fig. 3.3) and assess the extent to which the four basins accord with or depart from the model. The timing of the marine retreat varied from basin to basin from the late Messinian (Tabernas Basin) to the early Pliocene (Sorbas Basin), the late Pliocene (Vera Basin), the latest Pliocene, and the Pleistocene (Almería basin). The final marine stage involved one of two situations:

1. a large fluvial input leading to a major fan-delta stage [e.g. during the early (?) Pliocene, in the western Almería basin; during the late Pliocene, in the eastern Almería basin/Argamasson area, and in the Vera basin; and during the Plio-Pleistocene, in the eastern Almería basin in the Polopos area];
2. a lower fluvial sediment input leading to marginal beaches and small fringing alluvial fans (e.g. during the early Pliocene in the Sorbas basin, represented by the Cariatiz formation).

The timing of marine withdrawal from the Tabernas basin is uncertain. During deposition of the Pliocene Gador formation, the western part of the basin hosted a large fluvial depositional system which fed the fan delta in the Rioja corridor between the Tabernas and Almería basins (Figs. 3.5, 3.8). Within the Tabernas basin, there is no evidence for a fan-delta phase during marine withdrawal. The same is true for the Sorbas basin (Figs. 3.4, 3.8). In both cases, fan deltas were formed only at the basin outlets where the drainage entered the still marine Almería basin (Figs. 3.7, 3.8). Fan deltas do, however, also occur within the Vera basin (Figs. 3.6, 3.8), both basins essentially acting as receiving basins for larger fluvial systems.

A phase of alluvial fan sedimentation then followed marine withdrawal, either by large alluvial fans succeeding the fan deltas, as in the Vera and Almería (north) basins (Figs. 3.6, 3.7, 3.8), or by large fans fed by the increased sediment input induced by further uplift of the mountain source areas, as in the Sorbas basin (Gochar formation: Figs. 3.4, 3.8). Along minor mountain fronts, elsewhere small basin-fringing fans were deposited. The drainage systems at the time were either external (Tabernas, Sorbas/Gochar Formation, Almería north), or at least for a time, internal (Vera, Almería centre: Fig. 3.8).

The transformation from alluvial fan systems to through-fluvial drainage may have been initiated in one of several ways:

1. by fan dissection through fanhead trenching and progradation, or through an incising marginal or interfan channel (see Harvey 1996, 2011);
2. by basin-wide incision following aggradation;
3. by distally induced dissection brought about through base-level change.

The first appears to have been the case on the Río Alias, in the transition from Terrace A to Terrace B (Whitfield and Harvey 2012). Dissection by a marginal channel appears to have been the mechanism initiating end-Gochar dissection at the head of the Cariatiz/Castanos system in the northern part of the Sorbas basin. The second mechanism appears to characterise the Sorbas basin as a whole, where end-Gochar aggradation gave way to basin-wide shallow incision prior to the aggradation of Terrace A, presumably in response to changes in critical power relationships (cf. Bull 1979). The third mechanism, though brought about by tectonically induced base-level change rather than through eustatic sea-level change, appears to have sustained incision within the Tabernas basin and effected the switch from internal to external drainage in the Vera basin (Stokes 2008).

The switch from an aggrading to an incising river system in some cases occurred simultaneously with the establishment of through-drainage (especially in the case of base-level-induced incision; cf. the Tabernas and the Vera basins). In other cases, it was delayed (e.g. the Sorbas basin), responding to later changes in critical power relationships. Tectonically induced incision in the Tabernas basin had developed by the early Pleistocene, but elsewhere developed in mid-Pleistocene or later and has hardly been effective at all in the central section of the Almería basin.

During drainage incision, many modifications to the drainage network have taken place. The consequent drainage (Fig. 3.9a) was initiated from mountain to basin or basin to coast as marine withdrawal took place. In some cases, this drainage became locally transverse to structure by superimposition onto underlying structures or locally antecedent as neotectonic activity continued (Harvey and Wells 1987).

Strike-parallel subsequent streams (Fig. 3.9a) developed by the preferential incision into the outcrops of weaker rocks (e.g. the lower Aguas incising into the weak Messinian Abad marl). Basin-marginal strike-streams became common (e.g. Ramblas de Lucainena and Mizala into Tortonian marls and sandstones along the southern margin

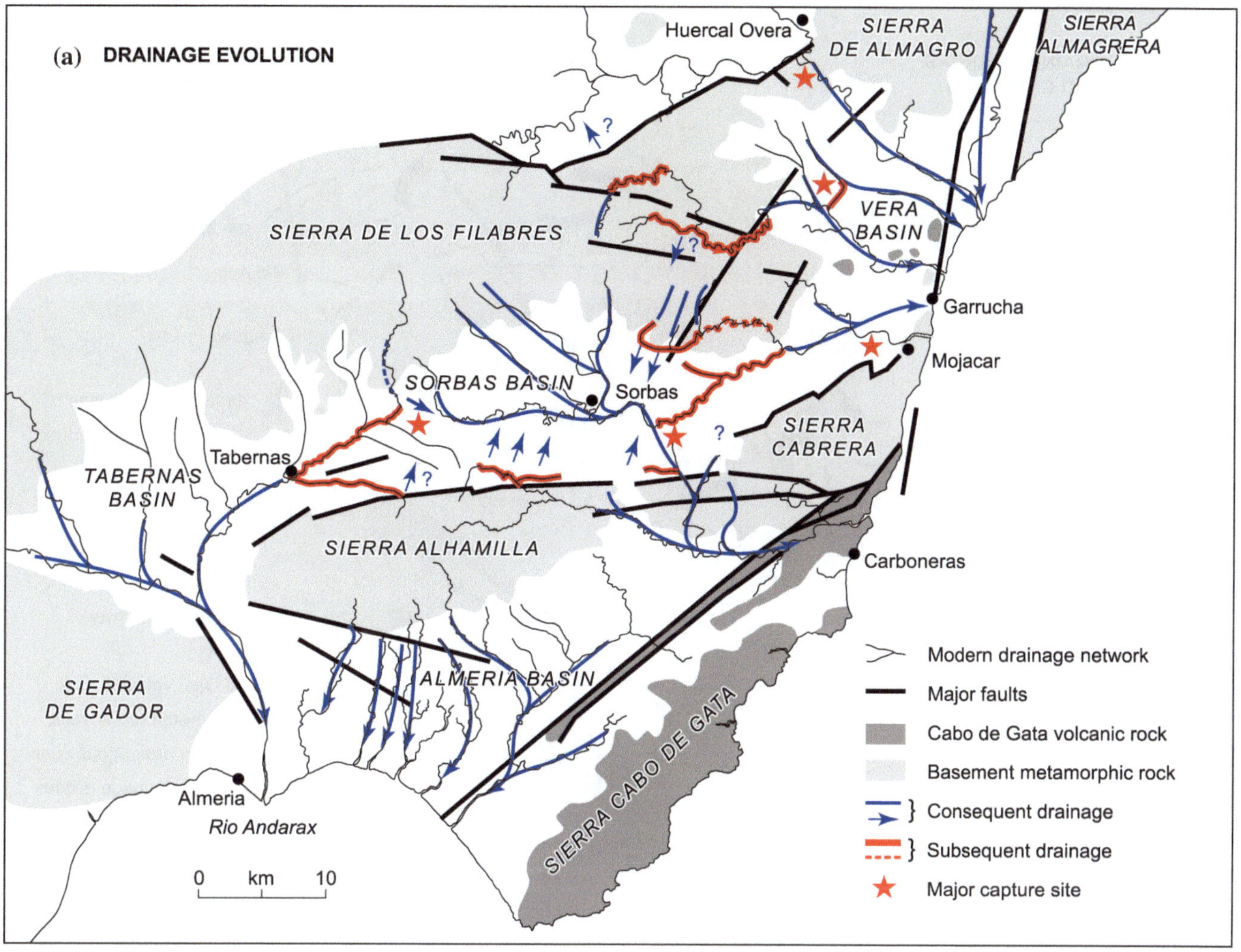

Fig. 3.9 The influence of river capture**a** on drainage net development; **b** on drainage areas (*Note* This figure does not include the effects of the possible Pliocene loss of the headwaters of the Gafares tributary of the Alias)

of the Sorbas basin; the Cariatiz Rambla along the northern margin of the Sorbas basin (Fig. 3.9a; for locations see Fig. 3.2a). As the subsequent streams developed, they brought about captures, modifying the drainage network (Fig. 3.9a). Some captures were relatively minor, modifying the drainage pattern only within a basin (e.g. Mather 2000b; Whitfield and Harvey 2012). Other captures were more important, effecting basin-to-basin transfers (Fig. 3.9b). In this way, the Tabernas basin gained drainage from the west of the Sorbas basin. The Sorbas basin lost this drainage to the Tabernas basin and also via the Lucainena to the Almería basin (Mather 2000a). It lost drainage to the Vera basin via the Cariatiz system and of course by the Aguas/Feos capture. The Vera basin gained drainage via the head of the Jauro/Antas system from the proto-Sorbas basin (Stokes and Mather 2000; Mather et al. 2000), and later from the Sorbas basin via the Cariatiz/Castanos and the Aguas/Feos captures. It also gained the drainage of the Huercal-Overa basin by the headwards erosion of the Almanzora (Stokes and Mather 2003). The Alias system in the northern part of the Almería basin gained drainage from the Sorbas basin through headwards erosion by the Lucainena and later by the Mizala, but lost a substantial part of its headwaters through the Aguas/Feos capture.

To return now to the model; how far do the four sedimentary basins accord with or depart from the model proposed above? The Tabernas basin is aberrant in that the fan-delta stage relates only to the basin outlet. Then, tectonically induced dissection was dominant, until the late Pleistocene, when again tectonic deformation intervened to produce ponding in the basin centre and alluvial fan formation in at the basin head. In this basin, the high uplift rate, together with tectonic deformation, override any other general trends. The Sorbas basin, not surprisingly, more or less accords with the model, but again with the fan-delta stage occurring later in the sequence than proposed and

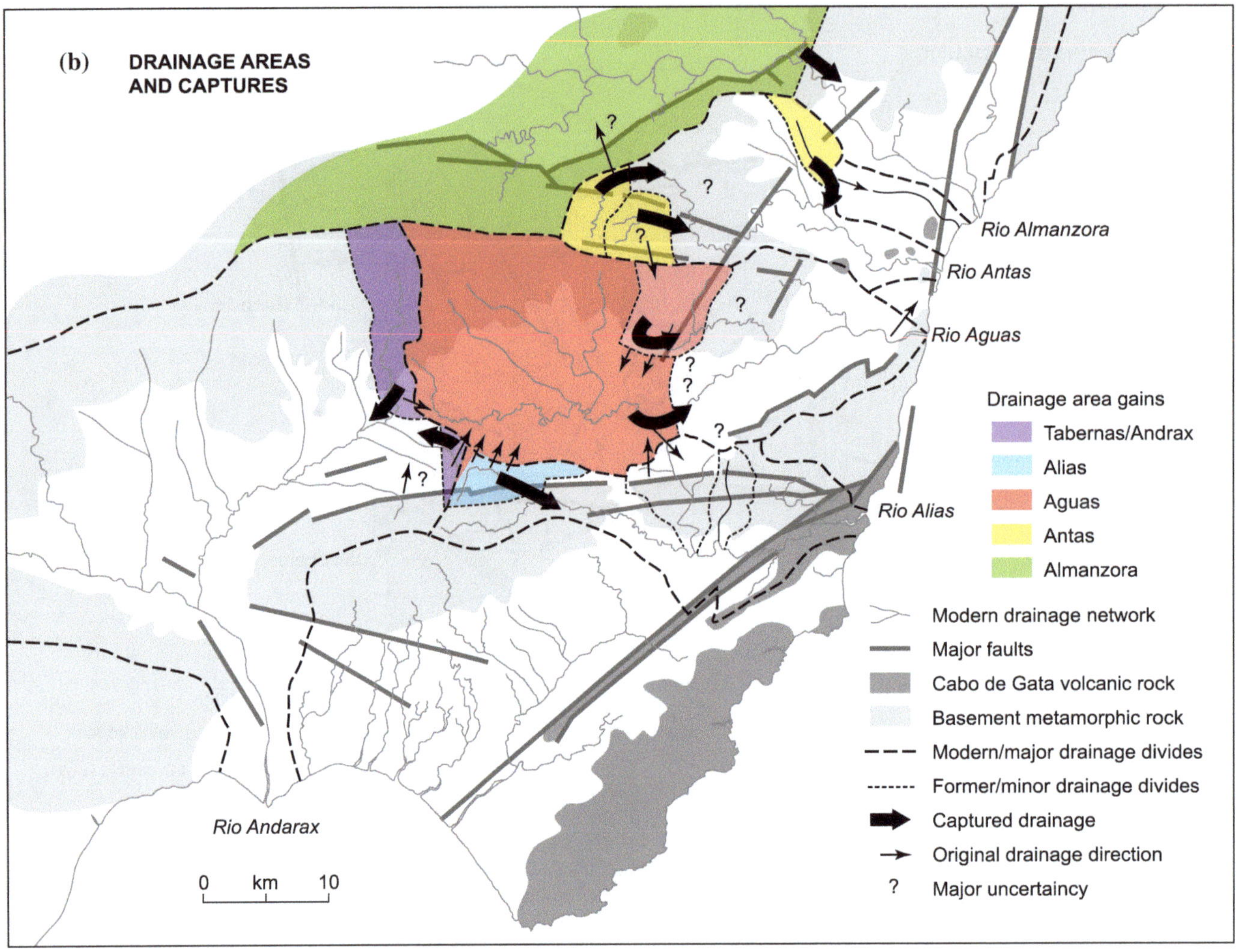

Fig. 3.9 (continued)

relating to the basin outlet rather than to fluvial input. The basin switches from short-lived internal drainage (Zorreras member) to through-drainage (Gochar formation) to dissection (Pleistocene terrace sequence). The major complications are brought about by river capture, particularly by the late Pleistocene Aguas/Feos capture. The Vera basin broadly accords with the model, but is modified by substantial increases in drainage area. The Almería basin is complex, with the four segments operating independently and all showing some elements which accord with the model. The northern (Alias) segment accords more-or-less with the model, albeit with complications related to ongoing tectonics, river capture, and base-level change. The central section has barely developed beyond the coalescent alluvial fan stage. The west central segment is immature and has been dominated by uplift, marine offlap and minor incision throughout the Pleistocene, and the western sector by the progradation of the Andarax fan delta during the overall fall of Pleistocene relative sea level.

The final consideration relates to the timing of the successive stages in the proposed model sequence. The timing clearly reflects the degree and rates of post-Pliocene epeirogenic uplift (Fig. 3.2b), with the Tabernas basin (maximum uplift rates) effectively by-passing the fan-delta and early alluvial fan stages, and being dominated by incision until the mid–late Pleistocene deformation that complicated the later development. In the Sorbas basin (moderate-high uplift rates) marine regression was early, there was no major fan-delta phase within the basin and only a short-lived phase of internal drainage (Zorreras member: end-Messinian/early Pliocene). Otherwise the early establishment of through drainage (Gochar formation; late Pliocene to early Pleistocene) and the mid-Pleistocene initiation of basin-wide incision reflect the moderate to high uplift rates. The incisional sequence was modified especially in the late Pleistocene by river capture. In the Vera basin (moderate to low uplift rates), a sequence can be identified that accords well with the model, with a late Pliocene fan-delta phase, an early Pleistocene alluvial fan and internal drainage phase, and mid-Pleistocene establishment of through drainage. The Almería basin (lowest uplift rates) is immature and shows differential development between the four basin segments. Only the northern segment shows a moderately complete sequence, similar to that in the Vera basin, but lacking a phase of internal drainage. The central sector has hardly progressed beyond the coalescent alluvial fan phase with, until very recently, internal drainage. The other segments exhibit only limited stages of the model. The dependence of the model sequence on uplift rates can be corroborated by the work of Silva et al. (2008) in the Lorca area, to the north of our study region, where uplift rates along the Guadalentin trough are very low. There, late Pleistocene trough-marginal fans became trenched and replaced by through-drainage only during the Holocene.

3.4.3 Interacting Controls

We had previously considered the interplay between the controls on drainage evolution and how they vary through time in relation to one study area, the Rio Alias in the northern Almería basin (Whitfield and Harvey 2012). By considering all four basins, we are now able to assess the wider applicability of those trends.

The overriding long-term controls of both the spatial and temporal aspects of drainage evolution are the pattern and rate of post-Pliocene epeirogenic uplift, and how they vary among the four basins. The pattern affected the positions of the major consequent streams (Fig. 3.9a). The rate affected the timing of marine retreat and drainage initiation. It also affected the timing of the various stages identified in the model. The effects of differential uplift are effective over the whole of the area and over the whole timescale since marine withdrawal from individual basins.

A second regional signal is that provided by Quaternary climatic fluctuations, particularly affecting sediment supply, which as elsewhere in Mediterranean Europe involve terrace and fan aggradation during glacials and incision during interglacials (Fuller et al. 1998; Macklin and Woodward 2009; Bridgeland and Westaway 2008; Mather 2009; Whitfield et al. 2013). This signal becomes effective once dissection has begun, from mid-Pleistocene onwards. It is effective regionally in the river terrace and alluvial fan sequences.

The other controls operate over more limited temporal and spatial scales. Localised tectonic disturbance may influence both basin-filling and dissection stages. During basin-filling, it may modify sediment accumulation patterns. For example: early Pleistocene faulting in the Vera basin modified fluvial sediment thicknesses (Stokes 2008); continued uplift of the Serrata del Marchante growth fold in the Tabernas basin, affected the location of alluvial fans (Harvey 2006); and differential deformation within the Sorbas basin caused thickness variations in the Zorreras member (Mather 1991).

During the dissection phases, local tectonic deformation particularly affects stream gradients and local base levels and may therefore modify the aggradation or incisional regime. In the Tabernas basin, mid-late Pleistocene uplift of the growth fold in the west of the basin ponded the drainage, allowed fan deposition in the upper basin and then on dissection triggered an incision wave to progress into the upper basin (Harvey et al. 2003); in the Sorbas basin, a growth fold at Moras produced contrasts in incised meander forms upstream and downstream of the fold axis (Harvey 2007); and tectonic deformation, accentuated perhaps by the presence of gypsum, caused the late Pleistocene ponding at Urra (Mather et al. 1991; Harvey 2001, 2007). In the Vera basin, early–mid-Pleistocene tectonic activity triggered the switch from internal to external drainage (Stokes and Mather 2000). In the Almería basin near Argamasson, movement on the Carboneras fault triggered local incision of the Alias, complicating the terrace sequence (Maher and Harvey 2008), and at Lucainena fault movement triggered a local capture.

The seaward ends of the main rivers have been affected by Quaternary eustatic sea-level changes, lowstands during global glacials and highstands during interglacials. Interestingly, incision could be triggered during glacial lowstands at the same time as the direct climate-driven sediment-excess regime of aggradation. On the Alias (Whitfield and Harvey 2012), during glacials this interplay produced early incision followed by later burial. On the Andarax, there is little evidence of incision, merely of progradation of the fan delta seawards. The high sediment yields and perhaps low offshore gradients appear to have prevented incision. This is similar to the situation reported by Harvey et al. (1999) for the small alluvial fan drainages on the Cabo de Gata coast, where fan progradation took place during glacials, but incision triggered by coastal erosion was characteristic of interglacials.

The other important local mechanism that modifies the fluvial sequence is river capture. Not only does it alter the routing of water and sediment within and between drainage basins, but also it radically alters stream power and local base levels (Bishop 1995; Calvache and Viseras 1997). The most important captures affecting our study area have been itemised above. The effects have impact downstream on both beheaded and receiving systems and on the source area drainage, but the effects may diminish with distance away from the capture site. Although a capture is a "once-off" event, its effects are permanent.

An important aspect of how the fluvial system responds to change in the controlling factors is how the effects of change propagate through the system.

Responses to increases in the critical power needed (cf. Bull 1979) (i.e. to relative increases in sediment supply as opposed to flood power), whether generated by source-area uplift, climatically, or in relation to land-cover changes tend to propagate downstream by burial of the pre-existing topography. The response of the fluvial system to the reverse case, a decrease in relative sediment supply or an increase in flood power, in response perhaps to long-term erosion-related decrease in sediment availability (ageing), or climatic or land-cover changes triggering incision, is less certain. Does incision take place at a series of critical locations throughout the network and propagate locally both upstream by headcut recession and downstream through flow convergence?

The spatial effects of base-level fall whether regional in response to sea-level change or more local in response to tectonics or river capture tend to be limited to the vicinity of the structure itself, the capture site, or to the coastal area, but may involve some propagation upstream of an incisional nickpoint. The nickpoint generated on the Alias at Argamasson by movement on the Carboneras fault propagated ca. 4 km upstream twice, following terraces C and D, but both times was held up there by the outcrop of a resistant sandstone (Fig. 3.1g) (Maher and Harvey 2008). In contrast, in the Tabernas basin, the incision wave generated by the breach through the growth fold has propagated upstream by ca. 10 km in the last 20 ka. (Harvey 2002b), but there the incision has been generally into weak Tortonian marls. The nickpoints generated on the upper Aguas by the Aguas/Feos capture have propagated some 13 km upstream (Harvey 2002b) in about 60 ka, though there incision was complicated by the late Pleistocene deformation at Urra. Nickpoints generated by the last glacial sea-level lowstand have receded by about 5 km on the Alias and about 10 km on the much larger lower Aguas, but in both cases these distances are from the present coast, rather than from a location now somewhere offshore where the nickpoints would have been generated. Three factors would influence the rate of nickpoint recession: the magnitude of the base-level fall, assuming that the gradients generated are sufficient to trigger incision, the size and power of the river involved, and the weakness of the underlying strata. The effects of a base-level rise do not propagate significantly upstream; they would merely be the deposition of a local wedge of sediment.

3.5 Conclusions

In summary, the effects of the various controls operate over different temporal and spatial scales, and as the scales change so does the nature of the interaction.

The overriding long-term control is epeirogenic uplift, its spatial pattern and rates. This control operates over the whole region and over the timescale of landform evolution from the late Neogene onwards. It controls the timing of marine withdrawal and the steepness of tectonically generated regional gradients, which within the context of the erosional resistance of the underlying strata, govern the switch from basin-filling to dissection and the overall incision rates.

Within that context, Quaternary climatic cycles control sediment supply and therefore, during an overall incisional sequence, govern river behaviour and terrace development. The mid–late Quaternary alluvial fan and terrace sequences give regional signals, recognisable throughout the area in all basins with the possible exception of the Tabernas basin. There, not only is the dating of the fluvial sequence uncertain, but also the climatic signal may have been overridden by the dominance of the tectonic signal.

Other, more local controls operate within the regional contexts of the uplift patterns and the Quaternary climatic sequence, usually modifying but not overriding regional signals. Of these controls, sea-level change only affects the distal reaches; the response to ongoing tectonics tends to be local, but the effects of river capture are widespread within the systems involved and permanent, especially when major captures are involved.

One characteristic we have observed on the Alias (Whitfield and Harvey 2012), but that probably also applies elsewhere, is that as incision progresses the continuity through the system decreases and the sedimentology and fluvial morphology become increasingly locally determined. We wonder if this is a common characteristic of incising drainages? Or is it particular to tectonically active semi-arid landscapes?

Acknowledgments We thank the editors of this book, Francisco Gutierrez and Mateo Gutiérrez for their encouragement and attention to detail. We acknowledge the support received from the Universities of Liverpool and Plymouth and from Liverpool, John Moores University. We also thank the many people with whom we have worked in Almería, from whom we have learned much. They include Roy Alexander, Astrid Blum, Juan-Carlos Braga, Pat Brenchley, Adolfo Calvo, Trevor Elliott, Hazel Faulkner, Gez Foster, Derek France, Jim Griffiths, Jack Hannam, Andy Hart, Dave Hodgson, Janet Hooke, Suzanne Hunter (nee Miller), Andreas Lang, Luna Leopold, José Martín, Barbara Mauz (who is also thanked for preliminary results of OSL dating of Terrace D sediments on the Rio Alias), Chris Meikle, Pablo Silva, José Goy, Steve Wells, Keith Westhead, and Cari Zazo. We are grateful to Lindy, Joe, and Jill Walsh from the Cortijo Urra field centre in Sorbas, for hospitality during most of our field campaigns. Finally, we thank Sandra Mather, formerly of the cartographics unit of the School of Environmental Sciences at the University of Liverpool for her cartographic expertise in producing the illustrations.

References

Alexander RW, Harvey AM, Calvo A, James PA, Cerda A (1994) Natural stabilisation mechanisms on badland slopes, Tabernas, Almeria, Spain. In: Millington AC, Pye K (eds) Environmental change in drylands: biogeographical and geomorphological perspectives. Wiley, Chichester, pp 85–111

Alexander RW, Calvo-Cases A, Arnau-Rosalen E, Mather AE, Lazaro-Suau R (2008) Erosion and stabilisation sequences in relation to base level changes in the El Cautivo badlands, SE Spain. Geomorphology 100:83–90

Bishop P (1995) Drainage re-arrangement by river capture, beheading and diversion. Prog Phys Geogr 19:449–473

Blum A (2007) Controls on long-term drainage development of the Carboneras basin, SE Spain. PhD thesis. University of Plymouth

Bousquet JC (1979) Quaternary strike-slip faults in southeastern Spain. Tectonophysics 52:277–286

Braga JC, Martín JM, Quesada C (2003) Patterns and average rates of late Neogene–Recent uplift of the Betic Cordillera, SE Spain. Geomorphology 50:3–26

Bridgeland DR, Westaway R (2008) Preservation patterns of Late Cenozoic fluvial deposits and their implications: results from IGCP 449. Quatern Int 189:5–38

Bull WB (1979) The threshold of critical power in streams. Geol Soc Am Bull 90:453–464

Calaforra JM, Pulido-Bosch A (2003) Evolution of the gypsum karst of Sorbas (SE Spain). Geomorphology 50:173–180

Calvache ML, Viseras C (1997) Long-term mechanisms of stream piracy in southeast Spain. Earth Surf Proc Land 22:93–105

Candy I, Black S, Sellwood BW, Roan JS (2003) Calcrete profile development in Quaternary alluvial sequences, southeast Spain: implications for using calcretes as a basis for landform chronologies. Earth Surf Proc Land 28:169–185

Candy I, Black S, Sellwood BW (2004a) Quantifying time series of pedogenic calcrete formation using U-series disequilibria. Sed Geol 170:177–187

Candy I, Black S, Sellwood BW (2004b) Interpreting the response of a dryland river system to Late Quaternary climate change. Quatern Sci Rev 23:2513–2523

Candy I, Black S, Sellwood BW (2005) U-series isochron dating of immature and mature calcretes as a basis for constructing Quaternary landform chronologies. Quatern Res 64:100–111

Delgado-Castilla L (1993) Estudio sedimentológico de los cuerpos sedimentarios Pleistocenos en la Rambla Honda, al N. de Tabernas, provincia de Almería (SE de Espana). Cuaternario y Geomorfología 7:91–100

Delgado-Castilla L, Pascual-Molina A (1993) Geology and micromammals of the Serra-1 site (Tabernas basin, Betic Cordillera). Estud Geol 49:361–366

Frostick LE, Reid I (1989) Climatic versus tectonic controls of fan sequences: lessons from the Dead Sea, Israel. J Geol Soc Lond 146:527–538

Fuller IC, Macklin MG, Lewin J, Passmore DG, Wintle AG (1998) River response to high frequency climate oscillations in southern Europe over the past 200 ky. Geology 26:275–278

García AF, Zhu Z, Ku TL, Sanz de Galdeano C, Chadwick OA, Chacón Montero J (2003) Tectonically driven landscape development within the eastern Alpujarran Corridor, Betic Cordillera, SE Spain. Geomorphology 50:83–110

García AF, Zhu Z, Ku TL, Chadwick OA, Chacón Montero J (2004) An incision wave in the geologic record, Alpujarran Corridor, southern Spain (Almería). Geomorphology 60:37–72

Giaconia F, Booth-Rea G, Martínez-Martínez JM, Azañón JM, Pérez-Peña JV, Pérez-Romero J, Villegas I (2012) Geomorphic evidence of active tectonics in the Sierra Alhamilla (eastern Betics, SE Spain). Geomorphology 145–146:90–106

Goy JL, Zazo C (1986) Synthesis of the Quaternary in the Almería littoral, neotectonic activity and its morphologic features, Western Betics Spain. Tectonophysics 130:259–270

Hart AB, Griffiths JS, Mather AE (2000) The role of landsliding in landscape development in the Rio Aguas catchment, SE Spain. In: Bromhead E, Dixon N, Ibsen ML (eds) Landslides in research theory and practice. Thomas Telford, London, pp 701–706

Harvey AM (1984) Aggradation and dissection sequences on Spanish alluvial fans: influence on morphological development. Catena 11:289–304

Harvey AM (1987) Patterns of Quaternary aggradational and dissectional landform development in the Almería region, southeast Spain: a dry-region tectonically-active landscape. Erde 118:193–215

Harvey AM (1996) The role of alluvial fans in the mountain fluvial systems of southeast Spain: implications of climatic change. Earth Surf Proc Land 21:543–553

Harvey AM (2001) Uplift, dissection and landform evolution: the Quaternary. In: Mather AE, Martin JM, Harvey AM, Braga JC (eds) A field guide to the Neogene sedimentary basins of the Almería Province, South–East Spain. Blackwell Science, Oxford, pp 225–322

Harvey AM (2002a) Factors influencing the geomorphology of alluvial fans: a review. In: Pérez-González A, Vegas J, Machado MJ (eds) Aportaciones a la Geomorfología de España en el Inicio del Tercer Milenio. Instituto Geológico y Minero de España, Madrid, pp 59–75

Harvey AM (2002b) Effective timescales of coupling within fluvial systems. Geomorphology 44:175–201

Harvey AM (2006) Interactions between tectonics, climate and base level: Quaternary fluvial systems of Almería, southeast Spain. In: Alberti AP, Bedoya JP (eds) Geomorfología y Territorio. Actas de la IX Reunión Nacional de Geomorfologia, Universidade de Santiago de Compostella, Santiago de Compostella, Spain, pp 25–48

Harvey AM (2007) High sinuosity bedrock channels: response to rapid incision—examples in SE Spain. Revista de Cuartenario y Geomorfologia 21(3–4):21–47

Harvey AM (2011) Dryland alluvial fans. In: Thomas DSG (ed) Arid zone geomorphology, 3rd edn. Wiley-Blackwell, Chichester, pp 333–371

Harvey AM, Wells SG (1987) Response of Quaternary fluvial systems to differential epeirogenic uplift: Aguas and Feos river systems, southeast Spain. Geology 15:689–693

Harvey AM, Miller SY, Wells SG (1995) Quaternary soil and river terrace sequences in the Aguas/Feos river systems: Sorbas basin, SE Spain. In: Lewin J, Macklin MG, Woodward JC (eds) Mediterranean Quaternary river environments. Balkema, Rotterdam, pp 263–282

Harvey AM, Silva P, Mather AE, Goy J, Stokes M, Zazo C (1999) The impact of Quaternary sea-level and climatic change on coastal alluvial fans in the Cabo de Gata ranges, southeast Spain. Geomorphology 28:1–22

Harvey AM, Foster GC, Hannam J, Mather AE (2003) The Tabernas alluvial fan and lake system, southeast Spain: applications of mineral magnetic and pedogenic iron oxide analyses towards clarifying the Quaternary sediment sequences. Geomorphology 50:151–171

Harvey AM, Whitfield (nee Maher) E, Mather AE, Stokes M (2012) The transformation of alluvial fans into continuous fluvial systems: the Quaternary evolution of the Neogene sedimentary basins of the eastern Betic Cordillera, Almeria. In: González-Diaz A. (coord)

et al (eds) Avances de la Geomorfología en Espana 2010–2012. PubliCan, Santander, pp 13–17
Haughton P (2001) Tectonics and sedimentation: the evolving turbidite systems of the Tabernas basin. In: Mather AE, Martin JM, Harvey AM, Braga JC (eds) A field guide to the Neogene sedimentary basins of the Almería Province, South–East Spain. Blackwell Science, Oxford, pp 89–115
Hsu KJ, Montadert L, Bernoulli D, Cita MB, Erickson A, Garrison KE, Kidd KB, Melieres F, Muller C, Wright R (1977) History of the Mediterranean salinity crisis. Nature 267:399–403
Keller JVA, Hal SH, Dart CJ, McClay KR (1995) The geometry and evolution of a transpressional strike-slip system: the Carboneras fault, SE Spain. J Geol Soc London 152:339–351
Kelly M, Black S, Rowan JS (2000) A calcrete-based U/Th chronology for landform evolution in the Sorbas basin, southeast Spain. Quatern Sci Rev 19:995–1010
De Larouziere F, Bolze J, Bordet P, Hernández J, Montenat Ch, Ott d'Estevou Ph (1988) The Betic segment of the lithospheric Trans-Alboran shear zone during the Late Miocene. Tectonophysics 152:41–52
Macklin MG, Woodward J (2009) River systems and environmental change. In: Woodward J (ed) Physical geography of the mediterranean region. Oxford University Press, Oxford, pp 319–352
Maher E (2005) The Quaternary evolution of the Rio Alias, southeast Spain, with emphasis on sediment provenance. PhD thesis, University of Liverpool
Maher E, Harvey AM (2008) Fluvial system response to tectonically induced base-level change during the late-Quaternary: the Río Alias southeast Spain. Geomorphology 100:180–192
Maher E, Harvey AM, France D (2007) The impact of a major Quaternary river capture on the alluvial sediments of a beheaded river system, the Río Alias SE Spain. Geomorphology 84:344–356
Martín-Suárez E, Freudenthal M, Krijgsman W, Rutger Fortuin A (2000) On the age of the continental deposits of the Zorreras Member (Sorbas Basin, SE Spain). Geobios 33:505–512
Mather AE (1991) Cenozoic drainage evolution of the Sorbas Basin SE Spain. PhD thesis, University of Liverpool
Mather AE (1993a) Evolution of a Pliocene fan delta: links between the Sorbas and Carboneras basins, SE Spain. In: Frostick LES (ed) Tectonic controls and signatures in sedimentary successions. Blackwell Scientific Publications, Oxford, pp 277–290
Mather AE (1993b) Basin inversion: some consequences for drainage evolution and alluvial architecture. Sedimentology 40:1069–1089
Mather AE (2000a) Impact of headwater river capture on alluvial system development: an example from SE Spain. J Geol Soc Lond 157:957–966
Mather AE (2000b) Adjustment of a drainage network to capture induced base-level change: an example from the Sorbas basin, SE Spain. Geomorphology 34:271–289
Mather AE (2009) Tectonic setting and landscape development. In: Woodward J (ed) Physical geography of the mediterranean region. Oxford University Press, Oxford, pp 5–32
Mather AE, Harvey AM (1995) Controls on drainage evolution in the Sorbas basin, southeast Spain. In: Woodward J, Lewin J, Macklin MG (eds) Mediterranean Quaternary river environments. Balkema, Rotterdam, pp 65–76
Mather AE, Stokes M (2001) Marine to continental transition. In: Mather AE, Martin JM, Harvey AM, Braga JC (eds) A field guide to the Neogene sedimentary basins of the Almería Province, South–East Spain. Blackwell Science, Oxford, pp 186–224
Mather AE, Westhead RK (1993) Plio/Quaternary strain of the Sorbas Basin, SE Spain: evidence from sediment deformation structures. Quat Proc 3:57–65
Mather AE, Harvey AM, Brenchley PJ (1991) Halokinetic deformation of Quaternary river terraces in the Sorbas Basin, South–East Spain. Zeitschrift fur Geomorphologie (Suppl 82): 87–97
Mather AE, Harvey AM, Stokes M (2000) Quantifying long term catchment changes of alluvial fan systems. Geol Soc Am Bull 112:1825–1833
Mather AE, Martin JM, Harvey AM, Braga JC (2001) A field guide to the Neogene sedimentary basins of the Almería Province, South–East Spain. Blackwell Science, Oxford 350
Mather AE, Stokes M, Griffiths JS (2002) Quaternary landscape evolution: a framework for understanding contemporary erosion, SE Spain. Land Degrad Manage 13:1–21
Mather AE, Griffiths JS, Stokes M (2003) Anatomy of a 'fossil' landslide from the Pleistocene of SE Spain. Geomorphology 50:135–149
Meikle CD (2009) The Pleistocene drainage evolution of the Río Almanzora, Vera Basin, SE Spain. PhD thesis, University of Newcastle, UK
Meikle CD, Stokes M, Maddy D (2010) Field mapping and GIS visualisation of Quaternary river terrace landforms: an example from the Río Almanzora, SE Spain. Journal of Maps 2110:531–542. doi:10.4113/jom.2009.1100
Nash DJ, Smith RF (1998) Multiple calcrete profiles in the Tabernas basin, southeast Spain: their origins and geomorphic implications. Earth Surf Proc Land 23:1009–1029
Ovejero G, Zazo C (1971) Niveles marinos pleistocenos en Almería (SE España). Quaternaria 15:145–159
Pope RJJ, Wilkinson KN (2005) Reconciling the roles of climate and tectonics in Late Quaternary fan development on the Sparta piedmont, Greece. In: Harvey AM, Mather AE, Stokes M (eds) Alluvial fans: geomorphology, sedimentology, dynamics. Geological Society, London, pp 133–152 (Special Publication 251)
Postma G (1984) Mass-flow conglomerates in a submarine canyon: Abrioja fan-delta, Pliocene, Southeast Spain. In: Koster EH, Steel R (eds) Sedimentology of gravels and conglomerates, vol 10. CSPG Special Publications, Calgary, pp 237–258
Postma G, Roep ThB (1985) Resedimented conglomerates in the bottomsets of Gilbert-type gravel deltas. J Sediment Petrol 55:874–885
Ritter JB, Miller JR, Enzel Y, Wells SG (1995) Reconciling the roles of tectonism and climate in Quaternary alluvial fan evolution. Geology 23:245–248
Santisteban JI, Schulte L (2007) Fluvial networks of the Iberian Península: a chronological framework. Quatern Sci Rev 26:2378–2757
Sanz de Galdeano C (1990) Geologic evolution of the Betic Cordilleras in the Western Mediterranean, Miocene to the present. Tectonophysics 172:107–119
Sanz de Galdeano C, Shanov S, Galindo-Zaldívar J, Radulov J, Nikolov G (2010) A new tectonic discontinuity in the Betic Cordillera deduced from active tectonics and seismicity in the Tabernas Basin. J Geodyn 50:57–66
Schulte L, Julià R, Burjachs F, Hilgers A (2008) Middle Pleistocene to Holocene geochronology of the Río Aguas terrace sequence (Iberian Península): fluvial response to Mediterranean environmental change. Geomorphology 98:13–33
Silva PG, Bardaji T, Calmel-Avila M, Goy J-L, Zazo C (2008) Transition from alluvial to fluvial systems in the Guadalentín

depression (SE Spain): Lorca fan versus Guadalentin River. Geomorphology 100:140–153
Stokes M (1997) Plio-Pleistocene drainage evolution of the Vera Basin, SE Spain. PhD thesis, University of Plymouth
Stokes M (2008) Plio-Pleistocene drainage development in an inverted sedimentary basin: Vera basin, Betic Cordillera, SE Spain. Geomorphology 100:193–211
Stokes M, Mather AE (2000) Response of Plio-Pleistocene alluvial systems to tectonically induced base-level changes, Vera Basin, SE Spain. J Geol Soc Lond 157:303–316
Stokes M, Mather AE (2003) Tectonic origin and evolution of a transverse drainage: the Río Almanzora, Betic Cordillera, Southeast Spain. Geomorphology 50:59–81
Stokes M, Mather AE, Harvey AM (2002) Quantification of river-capture-induced base-level changes and landscape development: Sorbas basin, SE Spain. In: Jones SJ, Frostick LE (eds) Sediment flux to basins: causes, controls and consequences. Geological Society, London, pp 23–35 (Special Publication 191)
Stokes M, Cunha PP, Martins AA (2012a) Techniques for analysing Late Cenozoic river terrace sequences. Geomorphology 165–166:1–6
Stokes M, Griffiths JS, Mather AE (2012b) Palaeoflood estimates of Pleistocene coarse grained river terrace landforms (Río Almanzora, SE Spain). Geomorphology 149–150:11–26
Thurber DL, Stearns CE (1965) Th230-U234 dates of late Pleistocene fossils of the Mediterranean and Moroccan littorals. Quaternaria 7:29–42
Völk HR (1967) Relation between Neogene sedimentation and late orogenic movements in the Eastern Betic Cordillera (SE Spain). Geol Mijnbouw 46:471–474
Weijermars R (1991) Geology and tectonics of the Betic zone, SE Spain. Earth-Sci Rev 31:153–236
Whitfield (nee Maher) E, Harvey AM (2012) Interaction between the controls on fluvial system development: tectonics, climate, base-level and river capture—Rio Alias, southeast Spain. Earth Surf Processes Land 37:1387–1397
Whitfield RG, Macklin MG, Brewer PA, Lang A, Mauz B, Whitfield (nee Maher) E (2013) The nature, timing and controls of the Quaternary development of the Rio Bergantes, Ebro basin, northeast Spain. Geomorphology 196:106–121
Zazo C, Goy JL, Hoyos M, Dumas B, Porta J, Martinell J, Baena J, Aguirre E (1981) Ensayo de síntesis sobre el Tirreniense Peninsular Español. Estudios Geológicos 37:257–262
Zazo C, Goy JL, Dabrio CJ, Bardají T, Hillaire-Marcel C, Ghaleb B, González-Delgado JA, Soler V (2003) Pleistocene raised marine terraces of the Spanish Mediterranean and Atlantic coasts: records of coastal uplift, sea-level highstands and climate changes. Mar Geol 194:103–133

Granite Landforms in Galicia

4

Juan Ramón Vidal-Romaní, Marcos Vaqueiro, and Jorge Sanjurjo

Abstract

Galicia, in the NW of the Iberian Peninsula, is dominated by igneous rocks, mostly granitoids intruded during the Variscan orogeny. These granitoids can be grouped into four types: post- and syn-tectonic tonalite granites, and post- and syn-tectonic leucogranites. Granite landforms in Galicia have been largely controlled by endogenous features defined during their intrusion. Subsequently, tectonics associated with the Alpine orogeny between the Eocene and the beginning of the Late Miocene resulted in a dense network of faults and fractures. These structures delimit a heterogeneous mosaic of blocks in many cases formed by granite rocks, which were affected by differential tectonic movements during the Palaeogene, controlling the development of mountain ranges and depressions. However, the final subaerial exposure of the granite bedrock is mainly related to a wide range of erosion processes since Palaeogene times. In spite of the limited extent of the granitic outcrops in Galicia, they display a broad variety of landforms.

Keywords

Inselbergs • Pseudobedding • Polygonal cracking • Sheet structure • Tafone • Granite caves • Speleothems

4.1 Introduction

The development of the landscape in Galicia started around 200 million years ago (Mesozoic), significantly later than the formation of the rocks (granite, slates and schist predominantly) on which it is developed, dating back to the Palaeozoic (542–299 Myr) (Johnston and Gutiérrez-Alonso 2010) (Fig. 4.1). During the Mesozoic, the Pangea mega-continent split into several tectonic plates, one of them including the Galicia region. Consequently, most of the morphological features observable at the present time formed during the Cenozoic (De Vicente and Vegas 2009; Martín-González 2009).

Since the time the Galician territories emerged, they have been affected by erosion mainly related to fluvial, marine and glacial processes. During the Mesozoic, the rivers in Galicia began to flow towards the Atlantic coast carving the valleys that millions of years later became one of the most characteristic geomorphic features of its Atlantic coast; the "rías" (lower reaches of fluvial valleys submerged by the sea) (Viveen et al. 2013). Nevertheless, the most important landmark in the landscape evolution of Galicia dates back to the Palaeogene (65–34 Myr), when the Iberian Peninsula was defined as a small plate between two converging tectonic plates: the Eurasian Plate and the African Plate. The continental collision led to a dense fracturing and the

J. R. Vidal-Romaní (✉) · M. Vaqueiro · J. Sanjurjo
Instituto Universitario de Geología, University of Coruña, Coruña, Spain
e-mail: juan.vidal.romani@udc.es

J. R. Vidal-Romaní · M. Vaqueiro · J. Sanjurjo
Club Espeleolóxico Trapa, Vigo, Spain

F. Gutiérrez and M. Gutiérrez (eds.), *Landscapes and Landforms of Spain*, World Geomorphological Landscapes,
DOI: 10.1007/978-94-017-8628-7_4, © Springer Science+Business Media Dordrecht 2014

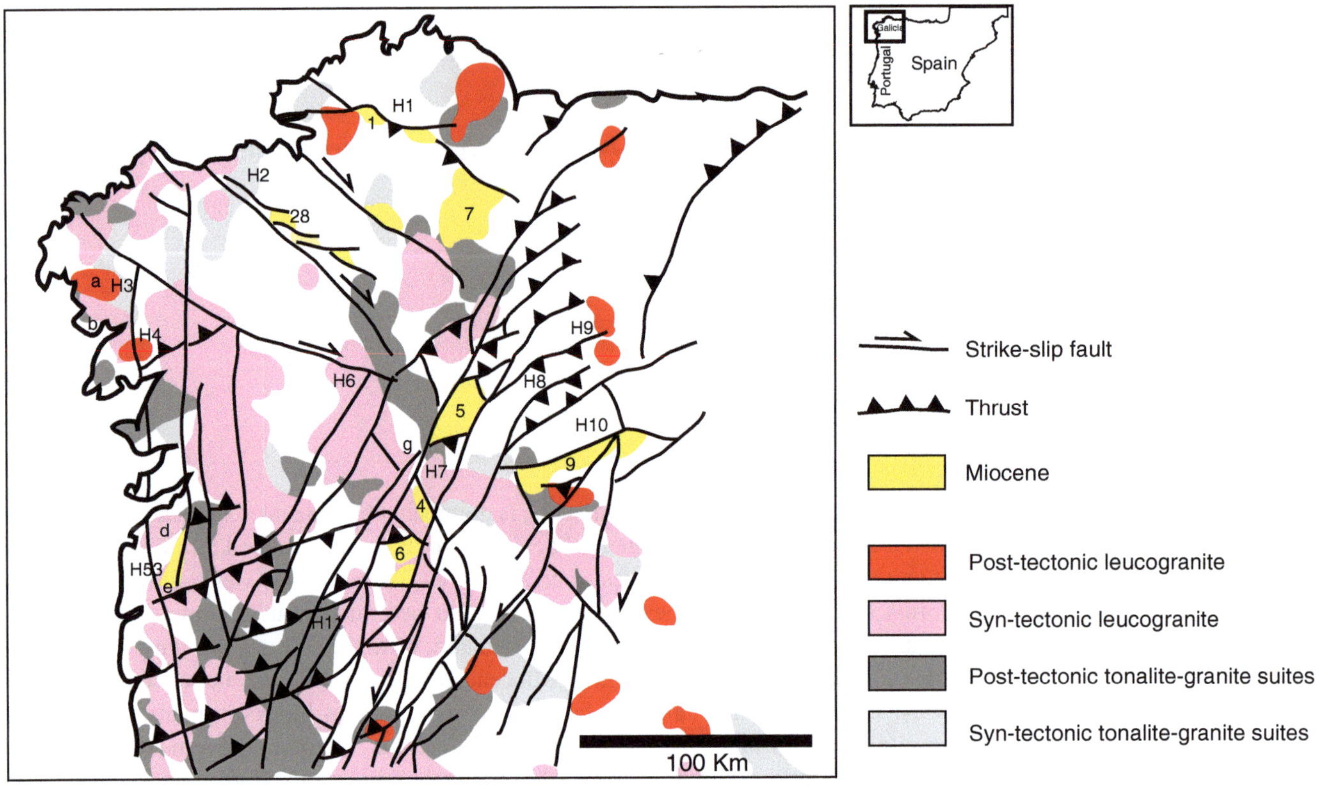

Fig. 4.1 Distribution of the main granitoid groups and Tertiary faults in NW Spain. Cenozoic Basins: (*1*) As Pontes, (*2*) Meirama, (*3*) Budiño, (*4*) Maceda, (*5*) Monforte, (*6*) Xinzo, (*7*) Villalba-Abadín, (*8*) Boimorto and (*9*) El Bierzo. (*H*) Horsts: (*1*) Xistral, (*2*) Xalo, (*3*) Pindo, (*4*) Barbanza, (*5*) Galiñeiro, (*6*) Faro, (*7*) Manzaneda, (*8*) Courel, (*9*) Ancares, (*10*) Trevinca and (*11*) Peneda Gêrez. Block caves: (*a*) Pindo, (*b*) Louro, (*c*) Barbanza, (*d*) Folón, (*e*) Trapa, (*f*) Albarellos, (*g*) Adeghas [modified from De Vicente and Vegas (2009), Johnston and Gutiérrez-Alonso (2010)]

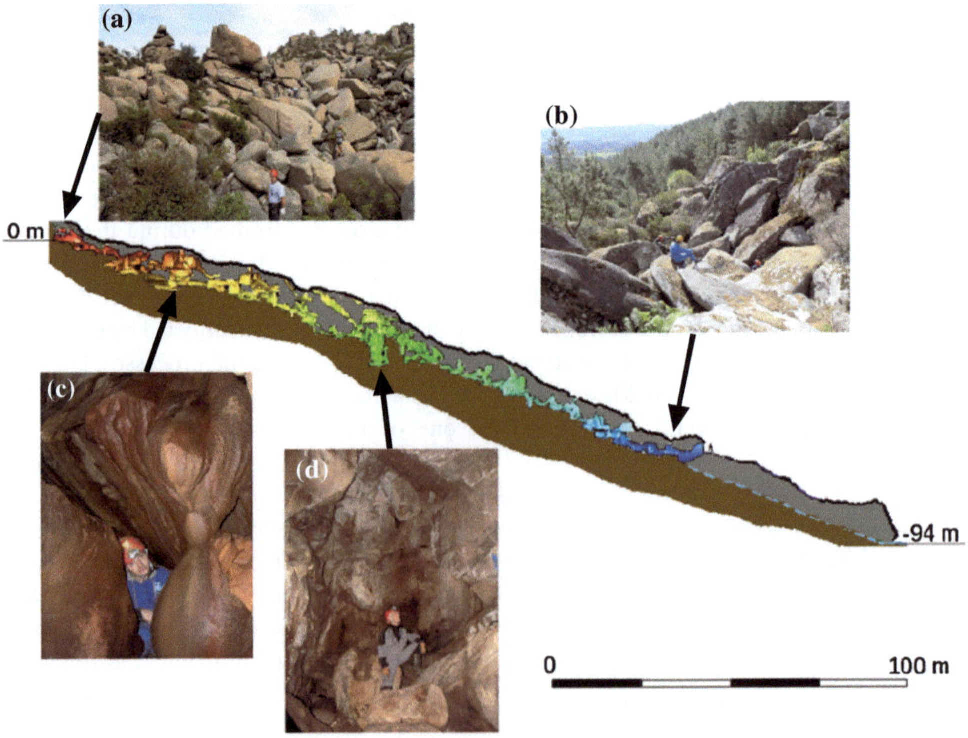

Fig. 4.2 Longitudinal profile of the granitic cave of A Trapa (Tui, Galicia, Spain). Colours from *red* to *blue* indicate the zonation of the cave from drier to wetter parts. **a** and **b** Big blocks on the surface, **c** Pigotite speleothem 3 kyr old in A Trapa Cave, **d** View of one of the chambers in a granitic cave

Fig. 4.3 Top of granite dome with sheet structure (Oleiros, Salvaterra de Miño). **a** Detail of boudinage. **b** and **c** Shearing with dense foliation

definition of blocks affected by differential tectonic movements throughout the Tertiary (Fig. 4.1). Tectonic activity produced the mountain ranges of Galicia, which correspond to the uplifted blocks and are largely underlain by granitoids (O Pindo, Barbanza, O Galiñeiro, Manzaneda-Invernadoiro, Os Ancares, Gêrez-Xurés, Peneda, etc.). A secondary effect of this uplifting process was the generation of important accumulations of blocks in the bottom of the depressions by slope movement processes, which currently form spectacular block streams (Fig. 4.2a, b). These are the most recent granite features, of particularly high interest when associated with pseudokarstic systems including speleothems. These caves host troglodyte sites from the Epipalaeolithic to the Middle Ages (Vidal-Romaní and Vaqueiro 2007) (Fig. 4.2c).

Finally, the mountain ranges uplifted during the Tertiary favoured the development of small glaciers above 1,000 m a.s.l., during the Quaternary (2.58 Ma–15 ka). In the numerous glaciated areas of Galicia underlain by granitoids (Os Ancares, Manzaneda, Peneda Gêrez-Xurés, etc.) (Vidal-Romaní et al. 1990), Pleistocene erosional glacial features are superimposed on the previous granite landforms.

4.2 Granite Landforms

When dealing with granite terrains, geomorphologists tend to describe landforms without paying much attention to their genesis. It is the case of the so-called pseudobedding (Twidale 1982) that has been frequently considered as an exogenic feature. However, their analysis strongly suggests that they are related to tectonic shear bands (Fig. 4.3a–c). In granite rocks, the discontinuities of intrusive endogenous origin are generally the main factor that controls the surface morphology. The dilemma of considering a granitic landform endogenous or exogenous is especially outstanding in Galicia, given its complex geological history. Granite landforms are classified into three major groups: megaforms, mesoforms and microforms (Twidale and Vidal-Romaní 2005). Megaforms are those whose minimum dimensions are at least 100 m; mesoforms range between 1 and 100 m; and microforms are typically smaller than 1 m.

4.3 Megaforms. Bornhardts or Rock Domes, Castle Rocks, Nubbins and Tors

Generally, megaforms correspond to the different types of residual hills (Twidale 1982). In Galicia, the most frequent types are bornhardts or rock domes of large dimensions. The geometry of these landforms has been defined by the primary intrusive structure of the granitic rock, without significant modifications after their exposure by erosion of the country rocks in which they intruded. In Galicia, these landforms lack flared or overhanging walls or the characteristic counter-slopes in the pediments surrounding them. Most cases cited coincide with coastal areas characterized by an intense dissection during the Cenozoic. Thus, they can be considered as recent landforms. In Galicia, they have been named as "moa" (molar) because of their rounded and sticking out morphology. Probably, the best-developed example is O Pindo, a mountain group described by Nonn (1966) as a complex inselberg (Fig. 4.4). The O Pindo was formed by two successive intrusions, the first one of leucogranites and the next one of post-tectonic tonalites.

The other types of residual hills like castle rocks, nubbins and tors, of smaller dimensions than the previous type, are also frequent in Galicia. Castle rocks are convex reliefs controlled by systems of sub-vertical discontinuities which are called "castelos" in Galicia (Fig. 4.5a). Nubbins, considered by some authors as the initial stages of exposure of the rock domes (Twidale 1982), are common landforms. They are found in all types of granites, mainly inland, but they also occur as granite rocks emerging from the sea. Finally, tors are associated with all types of granites and are found in very different geomorphic contexts, including coastal areas and inland sectors with periglacial conditions.

4.4 Microforms

Microforms are features with maximum dimensions lower than 1 m. There are discrepancies when assigning an exogenous or endogenous origin to these forms.

Fig. 4.4 Panoramic view of the granite inselberg in the O Pindo Massif

4.4.1 Linear Concave Forms (Rills, Grooves, Flutings, Clefts, Runnels and Gutters)

The prototype corresponds to channel-shaped features that concentrate run-off over rock surfaces. Rills, runnels and gutters develop on horizontal or gently sloping surfaces, whereas grooves and flutings correspond to similar features associated with more inclined surfaces (Fig. 4.5b). In Galicia, these types of microforms are particularly frequent in coastal areas and sometimes connected with gnammas (weathering pits) (Fig. 4.5d).

4.4.2 Rounded Convex Forms (Gnammas, Pits, Pans, Vasques)

The gnammas, designated with the local name pía, are almost ubiquitous in the Galician granite landscape. Two genetic types of gnammas have been proposed. One of subaerial or epigenic origin, in which the accumulation of surface water on a concavity leads to the formation of a hollow by physical and/or chemical weathering (Fig. 4.5d). Other gnammas have a more complex genesis related to the concentration of stresses in particular points of the granite massif during the emplacement of the intrusion and, therefore, can be considered as endogenous features (Vidal-Romaní 1985).

4.4.3 Tafoni (Cavernous Weathering)

Tafone (plural tafoni) is a Corsican word meaning perforation or window. The geomorphological term refers to a shallow cavern or hollow. It was first described by Casiano de Prado in 1864 from the Sierra de Guadarrama, central Spain, later by Reusch in 1883 from Corsica, and by Hult in 1888 in Galicia, NW Spain (see Twidale and Vidal-Romaní 2005). However, the most well-known citation is the one by Penck in 1894 from Corsica (see Klaer 1956). Tafoni are especially well developed in granite outcrops where the rock is compartmentalized into blocks by sheet-type fracture systems. The clusters of closely spaced alveoles developed on the inner walls of tafoni are designated by many authors as honeycomb weathering (Fig. 4.5c). At a local scale, the distribution of tafoni is puzzling, with some blocks hollowed, and others, immediately adjacent and apparently identical, intact (Twidale and Vidal-Romaní 2005). In Galicia, tafoni are a very well-developed morphological feature. They show an unusually high spatial frequency and are typically associated with the zones of most active morphological evolution. Some authors interpret tafoni as features of subaerial origin (Klaer 1956; Migon 2006), whereas others propose that their formation is initiated in the subsurface (Vidal-Romaní 1985; Twidale and Vidal-Romaní 2005; Roqué et al. 2012).

Fig. 4.5 Different types of granite landforms from Galicia. **a** Castle rock at Quilmas, Coruña. **b** Incised rills at Roncudo, Corme, Coruña. **c** Tafone with honeycomb development at Ézaro, Coruña. **d** Gnamma showing lichen colonization at Corme, Coruña

Fig. 4.6 Concordant valley controlled by sheet fractures with a synclinal structure in the northern sector of the granitic massif of O Pindo. Note the sheet fractures dipping towards the valley axis

4.4.4 Sheet Structure, Pseudobedding, Boudinage and Polygonal Cracking in Granites: A Continuum Sequence of a Strain Process

In granites with a well-developed deformation by shearing, weathering is preferentially controlled by foliation planes, resulting in a pseudostratified granite. The discontinuous layers may be up to one metre thick (sheet structure) and are typically a few centimetres thick (pseudobedding) (Figs. 4.3a–c, 4.6). In some cases, closely spaced foliation produces thin partings some millimetres thick (Twidale and Vidal-Romaní 1994, 2005; Migon 2006). An intermediate type is the granite boudinage (Vidal-Romaní 2008), in which thin foliation bands alternate with thick ones, forming granite boudins (Fig. 4.3a–c). The dip of these foliation surfaces can vary considerably from sub-vertical to sub-horizontal. These microforms are very abundant in Galicia.

4.4.5 Polygonal Cracking

It corresponds to a shallow network of cracks perpendicular to the sheet structure that produces a mosaic of fragments and is genetically related to boudinage (Ramsay and Huber 1987; Twidale and Vidal-Romaní 2005) in both sedimentary rocks and granites. Some authors attribute them to cracking due to thermal expansion related to insolation or to the accumulation of Fe–Mn-oxides and carbonates (Klaer 1956; Twidale 1982; Migon 2006). However, in the cases studied in Galicia, they are due to shearing caused by regional stress fields (Ramsay and Huber 1987; Plotnikov 1994; Twidale and Vidal-Romaní 2005) as shown by their alternation with the shearing planes of the granite and the frequent infill of the polygonal cracking by quartz or pegmatite injections of clear endogenous origin (Vidal-Romaní 2008).

4.4.6 Caves Associated with Rock Fall Accumulations and Associated Speleothems

In areas with intense fracturing and accumulations of large boulders related to slope instability, kilometre-size caves may be found (Vidal-Romaní and Vaqueiro 2007) (Fig. 4.2a–d). In these settings, generally with high gradient, underground waters may have a high erosive capability, resulting in the formation of potholes (Fig. 4.2d). When run-off infiltration is slow, the sedimentation of hydrosuspensions (slurry) of grains of quartz, feldspar and mica, or the growth of authigenic minerals in the cave at room temperature prevail. They are called speleothems due to their morphology and sedimentary environment (Twidale and Vidal-Romaní 2005). So far, opal-A, evansite, struvite, pigotite, allophane, haematite, goethite, and nanocrystals of halite, gypsum, plumboaragonite and calcite have been reported (Vidal-Romaní et al. 2010) (Fig. 4.2c).

4.5 Conclusions

The first problem to understand the genesis of rocky landforms developed on granitic rocks is to determine to which extent they are controlled by exogenous, endogenous or both features. In Galicia, the granite landscape has a strong endogenous inheritance due to the complex tectonic history of the region, including intrusion phases during the Late Palaeozoic and brittle fracturing in the Palaeogene. None of these stages must be excluded to fully understand the true origin of the granitic geomorphology of Galicia.

Acknowledgments We thank Ana Martelli for the English translation of the chapter.

References

De Vicente G, Vegas R (2009) Large-scale distributed deformation controlled topography along the western Africa-Eurasia limit: Tectonic constrains. Tectonophysics 474:124–143

Johnston ST, Gutiérrez-Alonso G (2010) The North American Cordillera and West European Variscides: contrasting interpretations of similar mountain systems. Gondwana Res 17:516–525

Klaer W (1956) Verwiterungsformen im granit auf Korsika. V. E. B. Hermann Haack Geographysch-kartographische Anstalt, Gotha

Martín-González F (2009) Cenozoic tectonic activity in a Variscan basement: evidence from geomorphological markers and structural mapping (NW Iberian Massif). Geomorphology 107:210–225

Migon P (2006) Granite landscapes of the World. Oxford University Press, Oxford

Nonn H (1966) Les régions cotiéres de la Galice (Espagne). Etude Géomorphologique. Pub. Fac. Lelt. Univ. Strasbourg. Ph.D. Thesis

Plotnikov LM (1994) Shear structures in layered geological bodies. Russian translated series 104. A.A. Balkema, Brookfield, USA

Ramsay JG, Huber M (1987) The techniques of modern structural geology: folds and fractures, vol 2. Academic Press, London

Roqué C, Zarroca M, Linares R (2012) Subsurface initiation of tafoni in granite terrains. Geophysical evidence from NE Spain: geomorphological implications. Geomorphology 196:94–105

Twidale CR (1982) Granite landforms. Elsevier, Amsterdam

Twidale CR, Vidal-Romaní JR (1994) On the multistage development of etch forms. Geomorphology 11:157–186

Twidale CR, Vidal-Romaní JR (2005) Landforms and geology of granite terrains. Balkema, Amsterdam

Vidal-Romaní JR (1985) El Cuaternario de la provincia de La Coruña. Modelos elásticos de formación de cavidades. Servicio de Publicaciones. Universidad Complutense de Madrid, Serie Tesis Doctorales, Madrid. 283 pp

Vidal-Romaní JR (2008) Forms and structural fabric in granite rocks. Cad Lab Xeol de Laxe 33:175–198

Vidal-Romaní JR, Vaqueiro M (2007) Types of granite cavities and associated speleothems: genesis and evolution. Nature Conserv 63:41–46

Vidal-Romaní JR, Brum A, Zezere J, Rodrigues L, Monge C (1990) Evolución cuaternaria del relieve granítico en la Serra de Gerêz-Xurés, (Minho, Portugal Ourense, Galicia). Cuaternario y Geomorfología 4:3–12

Vidal-Romaní JR, Sanjurjo J, Vaqueiro M, Fernández D (2010) Speleothem development and biological activity in granite cavities. Geomorphologie: Rel Proc Environ 4:337–346

Viveen W, Schoorl JM, Veldkamp JMA, Van Balen RT, Desprat S, Vidal-Romaní JR (2013) Reconstructing the interacting effects of base level, climate and tectonic uplift in the Lower Miño River terrace record: a gradient modelling evaluation. Geomorphology 186:96–118

Geomorphology of La Pedriza Granitic Massif, Guadarrama Range

5

Javier de Pedraza, Rosa M. Carrasco, and David Domínguez-Villar

Abstract

The Pedriza de Manzanares, located in the Guadarrama Range, Iberian Central System, is a Variscan massif formed by I-type peraluminous leucogranites intruded in the late Paleozoic (~307 ma). The morphostructure of the massif is largely the result of the reactivation of faults during the Alpine Orogeny (Paleogene–Pliocene) and the associated etching/exhumation processes. The former produced a stair-stepped topography (block faulting), and the latter gave rise to granitic domes and crests related to differential erosion. The domes, and to a lesser extent the crests with tors and widespread chaotic blocks, are the essential features of the landscape in the Pedriza de Manzanares. The low chemical weathering susceptibility of the granites in this area, together with the fracture system, favour the development of structural landforms. Moreover, on bare rock exposures, minor landforms formed by diverse weathering processes are frequent.

Keywords

Granite landforms • Etchplanation • Stair-stepped morphology • Iberian Central System

5.1 Introduction

Granitic rocks are often used as ornamental and construction materials in the Iberian Peninsula. Many villages, walled precincts, cathedrals, aqueducts, and monasteries have been built with this rock type since the Iron Age (1st millennium BC), and some are catalogued as historical monuments or designated UNESCO World Heritage Sites. In many parts of Spain, these rocks are known as *piedras berroqueñas* (crag rocks), and toponyms such as *berrueco* or *bolo* (granite boulder), *berrocal* (granite tor), or *yelmo* (helmet or granite dome) are frequently used in granitic areas. These include the Pedriza (stony site) de Manzanares, traditionally considered by scientists, mountaineers, and hikers as one of the most remarkable granite landscapes in the Iberian Peninsula. Indeed, the granite landforms considerably increase the natural and scientific interest of the area, which was designated Natural Protected Area long time ago (Hernández-Pacheco 1931). Since 1978, the Pedriza de Manzanares has been part of the protected area included in the Parque Regional de la Cuenca Alta de El Manzanares (Pedraza et al. 1989).

J. de Pedraza (✉)
Department of Geodynamics, Complutense University of Madrid, Madrid, Spain
e-mail: javierp@geo.ucm.es

R. M. Carrasco
Department of Geological and Mining Engineering, University of Castilla-La Mancha, Castilla-La Mancha, Spain
e-mail: Rosa.Carrasco@uclm.es

D. Domínguez-Villar
National Research Centre on Human Evolution (CENIEH), Burgos, Spain
e-mail: david.dominguez@cenieh.es

F. Gutiérrez and M. Gutiérrez (eds.), *Landscapes and Landforms of Spain*, World Geomorphological Landscapes,
DOI: 10.1007/978-94-017-8628-7_5,

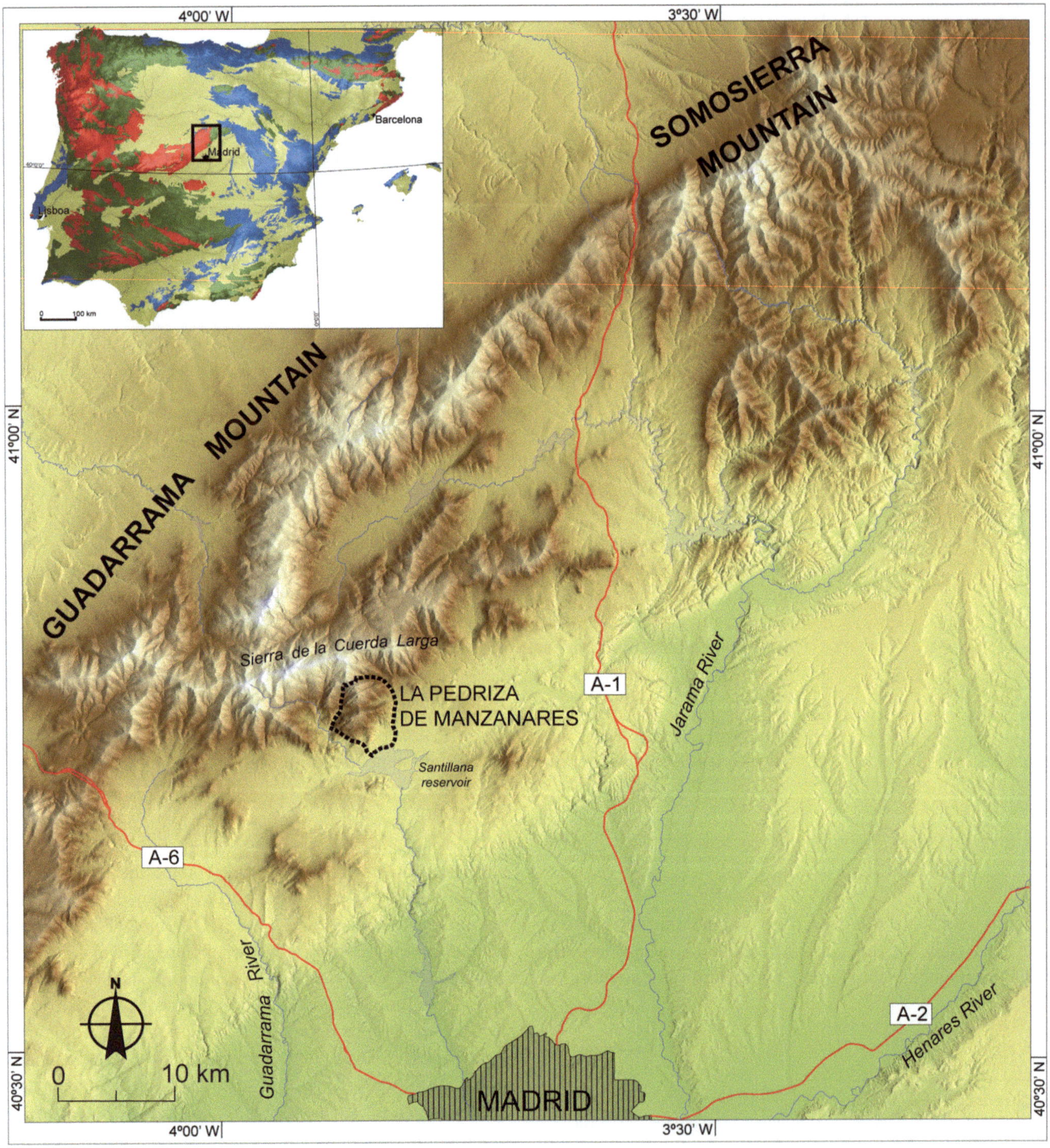

Fig. 5.1 Location of La Pedriza de Manzanares in the Guadarrama Range (Iberian Central System). Inset map depicts the main geological units of the Iberian Peninsula. *Green* and *red* Variscan or Iberian Massif (Pre-Paleozoic and Paleozoic); *red* represents the igneous rocks. *Blue* Alpine areas (Mesozoic and Lower Cenozoic). *Yellow* Cenozoic sedimentary basins

5.2 Geological and Geographical Setting

The Pedriza de Manzanares (PdM) is located in the Guadarrama Range (Iberian Central System), 50 km NE of Madrid (Fig. 5.1). From the topographic perspective, it is a N–S-oriented spur of the ENE-WSW trending Sierra de la Cuerda Larga, whose summit reaches 2,380 m a.s.l. at Hierro. This spur runs from the crest of the main ridge at 2,126 m a.s.l. to the piedmont of the Manzanares el Real-Guadalix de la Sierra intramontane depression, lying at

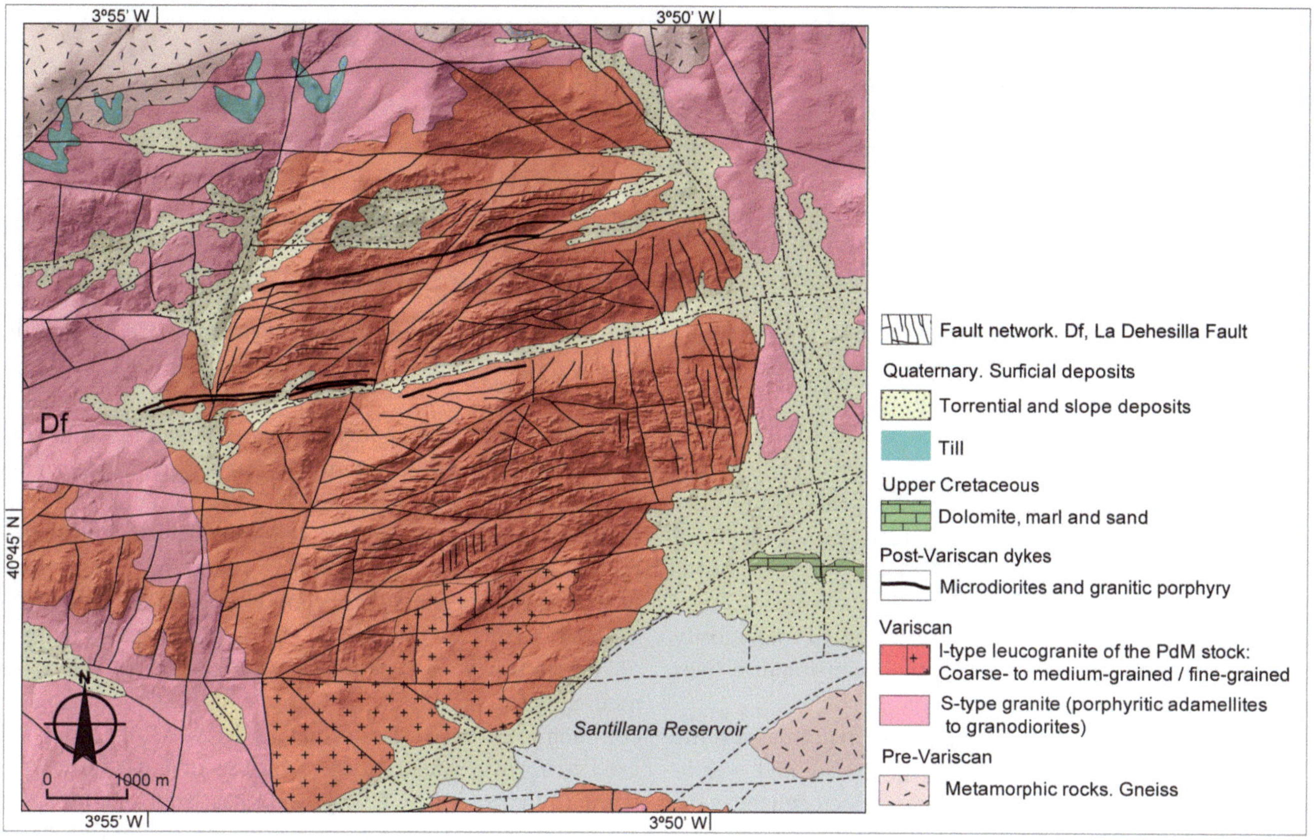

Fig. 5.2 Geological map of La Pedriza de Manzanares (*PdM*)

900 m a.s.l. with a NE-SW orientation. The average local relief of its slopes is around 800 m, and the stepped ridgeline descends from 2,126 m to 1150 m a.s.l.

The PdM forms a part of an intraplate system of uplifted tectonic blocks involving the Variscan basement (Iberian Central System). The faults that bound the blocks originated in the late-Variscan orogeny (Carboniferous) and were reactivated during the Alpine orogeny, essentially during the Miocene, determining the current morphostructural configuration. The Dehesilla Fault (Df) divides the PdM into two main orographic units with some distinctive geological, geomorphological, and physiographic features. Traditionally, these units have been called La Pedriza Anterior, to the S, and La Pedriza Posterior, to the N (De Prado 1864).

Geologically, the PdM is a granite stock dating from ~307 Ma that intruded pre-Variscan rocks as well as older post-tectonic (late-Variscan) peraluminous granite rocks (S-type granite). The lithology of the PdM stock corresponds to an I-type peraluminous leucogranite of medium to coarse grain size. Aplitic rocks also appear locally. Its mineralogy comprises K-feldspar, quartz, and plagioclase as essential minerals, and biotite, zircon, apatite, and monazite as accessory minerals (Pérez-Soba and Villaseca 2010). Local intrusions cross-cutting the granitoids include veins of porphyritic granite and microdiorites. Quaternary sediments are not very extensive and correspond to glacial, periglacial, slope, and fluvio-torrential deposits (Pérez-González and Ruíz-García 1990; Pérez-González et al. 1990) (Fig. 5.2).

Overall, the climate of the PdM is of mountain temperate type. However, due to its location in the central sector of the Iberian Peninsula, far from the maritime influence (continental factor), but within the Mediterranean biogeographical region (Mediterranean factor), the climate shows marked seasonal differences, with dry summers.

5.3 Landforms

The granitic landforms of the Iberian Central System have been used as indicators to infer the genesis and recent evolution of the landscape (Gutiérrez-Elorza and Rodríguez-Vidal 1978; Pedraza 1978, 1989). The "major landforms" are good indicators of past morphogenetic processes and environmental conditions (relict landforms), while "minor landforms" provide evidence of recent and active processes. Consequently, the granitic morphology of the PdM is analysed differentiating these two landform groups.

Fig. 5.3 The general dome-shaped structure of La Pedriza de Manzanares stock and the individualized dome morphologies. Df: La Dehesilla fault (see Fig. 5.2). (*Photo* courtesy of Dr. Luis Carcavilla)

5.3.1 Major Landforms

In the Iberian Central System, leucocratic granites typically form residual reliefs due to their high resistance to chemical weathering. The PdM clearly illustrates this pattern, forming a fresh rock granitic massif. Here, the rocks transformed into grus by chemical weathering are limited and mostly occur along bands associated with the main fracture lines and surficial deposits (Fig. 5.2).

The absence of widespread thick regolith/saprolite mantles indicates restricted etching processes and weathering concentrated along fracture zones, which enhances the development of major structurally controlled landforms. Granitic domes are one of the most characteristic landforms in the PdM (Pedraza et al. 1989). In fact, the overall topography of the area is that of a dome-shaped megastructure, which allows a fairly good reconstruction of the geometry of the intrusive stock. This megastructure is divided into two main sectors, each of them including dome-shaped features. Finally, each of these sectors can be subdivided into smaller zones in which individual dome, half-dome, or crest-shaped morphologies can be observed (Fig. 5.3). This sequence of landforms may be considered as a fractal structure of complex genesis, which supports the hypothesis of tectonic involvement in the primary origin of the sheet joints (Twidale et al. 1996; Vidal-Romaní and Twidale 1999).

The dome-shaped landforms of the PdM can be designated as bornhardts, although they display complex and asymmetrical geometries. This is because they are controlled by various superimposed sheet joint systems and are often bevelled by secondary fracture systems. Thus, different morphologies can be observed in the same dome depending on its geometry in plan view (circular, elliptical, or irregular) and the relationship among its axes (Fig. 5.4).

Together with the archetypal domes controlled by sheet jointing, or bornhardts, other dome-like landforms can also be found in an incipient, advanced, or very advanced stage of disintegration. The incipient stage is characterized by subvertical radial fractures and exfoliation joints widened by weathering, although with scarce relative block displacement or individualization. At this stage, the original dome geometry remains, but the sheet structure is compartmentalized forming slabs (compartmented-dome physiognomy). As disintegration progresses, the slabs release blocks that form massive accumulations at the foot of the slopes resulting in a landscape of chaotic blocks (Fig. 5.4a, c). Nubbins and tors rarely develop as part of this intermediate process. Instead, in this evolutionary stage, it is common to observe series of blocks derived from a slab that used to form a sheet structure on top of a new emergent bornhardt-type dome. This process has been interpreted as an evolutionary sequence, including the successive "rejuvenation" of the dome-shaped landforms (Pedraza et al. 1996). In the PdM, tors with castle koppie-type morphology are also found. These landforms develop on medium- to fine-grained granites with predominantly orthogonal joints. Rocky crests also develop on granites with similar structure (Fig. 5.5).

In some areas with closely spaced domes, rock corridors forming a labyrinthic landscape may be found. Initially, the corridors are excavated by preferential joint-guided weathering. The current processes are the rock fall (blocks and large blocks) and rock block (slabs) slides causing chaos of blocks and labyrinthic landscapes (Fig. 5.4c).

5.3.2 Minor Landforms

The most common minor landforms in the PdM are structural features (clefts, split and craked blocks, and fractured slabs), weathering pits, and microchannels. Cavernous weathering and flared forms are also present but are less common. Finally, other minor structural landforms such as nervations, pseudo-bedding, or polygonal cracking, among others, are irregularly distributed.

Weathering pits or gnammas. Most of these rock holes correspond to the pan-morphological type. These are flat-bottomed depressions 10–50 cm deep, 0.25–2 m across, and with gentle to overhanging sides. They are mostly oval or

Fig. 5.4 Dome-shaped landforms. **a** Half-dome type (El Yelmo, E slope). **b** Cone or asymmetrical bell type (El Yelmo, SW slope). **c** System of truncated domes in El Pájaro-Las Buitreras massif (S slopes). **d** Pyramid-like relief (SE slope of La Pared de Santillana)

Fig. 5.5 **a** Las Torres rock crests. **b** Cancho de los Muertos castle koppies

circular in plan view, although irregular or angular outlines are not rare. The gnammas are frequently aligned along joints and interconnected through a shallow microchannel. However, most of the gnamma clusters are not controlled by the joint network and display a random distribution on subhorizontal surfaces or gentle slopes.

Pit and armchair type gnammas are less common and associated with specific local contexts. The spillway of gnammas can be the origin of gutters, runnels, and channels connected with the general drainage network. In the case of gnammas developed over long periods (e.g. $>10^4$–10^5 years) on large boulders or tor summits, the coalescence of the depressions together with the deepening of the drainages results in chaotic surfaces (pseudo-karren). Additionally, although rare, there are cases of differential weathering around the walls and bottoms of relict gnammas (i.e. those that do not retain rainwater anymore due to spillway over-deepening or wall breaching), producing plate-shape forms and rock doughnuts (Fig. 5.6).

The surface weathering and erosion processes involved in the development of these morphologies are in agreement with the most generally accepted ideas: (1) increase in porosity by chemical weathering, (2) granular disaggregation in the walls and in some cases in the bottom, and (3) abrasion of the bottom by remobilization of accumulated debris due to the wind action and the rainwater (Twidale 1982; Twidale and Vidal-Romaní 2005; Domínguez-Villar et al. 2008, 2009).

Fig. 5.6 Weathering pits or gnammas. **a** Isolated circular gnammas and partially coalesced irregular, joint-related gnammas. The vegetation-covered band underlain by grus in the upper part of the image corresponds to La Gran Cañada fault zone. **b** System of coalescent gnammas and runnels on a deeply weathered block. Notice the armchair- and cylindrical-type gnammas in the proximal edge of the block (El Mirador de El Tranco area). **c** Plate-shaped features and rock doughnuts (El Tranco area)

Fig. 5.7 **a** Grove-type microchannels on the walls of block within a dome in advanced stage of disintegration. The image shows various types of fractured blocks. (El Mirador de El Tranco area). **b** Flute-type microchannels on Cancholosillo dome. **c** Rock pavement and gutter-type microchannels (El Tranco area)

Bedrock microchannels (flutes, gutters, pseudo-karren forms): Fluting (or pseudo-*rillenkarren*) generally forms on the walls of landforms with steep slopes. Grooves (or pseudo-*rundkarren*) are characteristic of landforms with relatively extensive flat summits and the presence of gnammas. Flat summits and gnammas are capable of retaining rainwater or snow, which enhance the widening of microchannels by weathering producing grooves. Gutters and runnels (or shallow pseudo-*rinnenkarren* and eroded pseudo-*rinnenkarren*) are far less common than those described above, especially runnels. These microchannels are often linked to gnammas, acting as their drains. Where the topography is favourable, gutters, and to lesser extent runnels, connect directly with the grooves developed in the walls.

The origin of these landforms in the PdM is explained by granular disintegration due to physical weathering (temperature changes, frost-shattering, and crystal growth), preceded by chemical weathering processes that increase the surficial porosity of the rock. These are essentially

Fig. 5.8 Cavernous weathering and flared forms. **a** Block with advanced basal corrosion evolved into a tafonization process and highly degraded gnammas on the top (El Tranco area). **b** Poorly preserved flared slope with superimposed tafoni controlled by the joint network (El Mirador de El Tranco area). **c** Honeycomb weathering on a wall of the El Pájaro-Las Buitreras truncated dome (see Fig. 5.4c). **d** Tafoni weathering at the base of "La Falsa Bola de los Navajuelos" balancing block

hydration/dehydration and oxidation of the ferromagnesian minerals that involve the release of Fe ions and the disturbance of the structure of the crystals. The flutes and grooves, due to their morphological characteristics and distribution within the PdM, are the granitic landforms which best illustrate the convergence between these landscapes and karst (Fig. 5.7). In fact, these features are often called pseudokarren, graniterrillen, or silikatrillen (Twidale 1982).

Cavernous weathering and flared forms: Tafoni, referring to shallow caves or partly closed holes, are almost nonexistent in the PdM. Well-developed flared slopes are uncommon, although traces of them are ubiquitous around boulders, pedestals, and mushroom rocks. Differential corrosion at the base of walls and blocks may lead to the formation of open tafoni or alveoli and locally flared-like laterally continuous concavities. Some large walls and overhanging slopes show honeycomb weathering. According to some studies (e.g. Mellor et al. 1997), these features represent the initial stage of tafoni and cavity development. This may indicate that there is active cavernous weathering in the PdM, although gravitational processes may limit or interrupt its development (Fig. 5.8).

Structural features: The rocks in the PdM are intensely fractured, facilitating physical weathering and mass wasting processes (Centeno-Carrillo and García-Rodríguez 2005). Vertical walls, overhanging slopes, natural bridges, visors and shelters, as well as split and cracked blocks are frequent landforms in the area (Fig. 5.7a).

Natural shelters are common in the PdM, and some of them constitute important archaeological sites (e.g. Los Aljibes shelter with Neolithic cave paintings from ~4,000 year BP). Conversely, caves are very scarce. The

Fig. 5.9 **a** Pseudostratification developed in leucogranite (Collado de las Dehesillas). **b** Meso-pseudostratification developed in the contact zone between leucogranite and the porphyritic adamellites/granodiorites (Torreta de los Porrones area)

Fig. 5.10 Peculiar landforms. In the PdM, the local names given to the rocks are inspired by their shapes, e.g. the bird, the hammer, the face, the snail, the camel, etc. Here, we show two examples: a perched block called the chalice and a tor named the little elephant

toponymy of the area includes two features called "Cueva del Ave María" and "Cueva de la Mora". The former is a pseudo-cave 3 m long associated with a rock fall deposit. The later, 23 m long, is an underground gallery formed along a shear band in which the granitic rock is intensely fractured. The origin of this structural cave may be attributed to granular and block disintegration, as well as to the removal of the resulting detritus by run-off derived from water seepage. However, preparatory subsurface processes similar to those described in other granite massifs cannot be ruled out (Roqué et al. 2012).

Most of the nervations and pesudobedding are related to differential erosion controlled by textural differences (microgranite venules) and shear planes, respectively (Fig. 5.9). Orthogonal cracks occur on indurated surfaces associated with exposed shear planes. However, polygonal and orthogonal cracks developed in the surface of the walls, boulders or large granite slabs, usually are controlled by textural variations in the rocks (specially the grain-size) and, consequently, are the result of differential erosion processes. The differential erosion caused by interconnected microchannels (fluting and grooves) on these walls locally generates residual microreliefs convergent with the mushroom-shaped pillars. These microreliefs are called in this area *setas de pared* (wall mushrooms) or *orejones* (big-ears) because of their position and morphology.

Other minor landforms: A reach of the Manzanares River channel carved on granite bedrock with microdiorite veins displays numerous potholes ranging from centimetres to metres in diameter. Locally, the coalescence of these depressions has resulted in the development of large basins several metres in diameter.

Pedestals and plinth rocks, which are normally associated with one of the groups of minor landforms mentioned above, have a fairly irregular morphology and complex origin involving a structural control (Fig. 5.10).

Fig. 5.11 General view of the western flank of the PdM spur showing its stepped topography. S1 summit surface (Sierra de la Cuerda Larga); S2 upper intermediate surface (Pedriza Posterior); S3 lower intermediate surface (Pedriza Anterior); and S4 piedmont surface

5.4 Evolution of the Relief

The relief of the PdM is mainly related to the Alpine reactivation, essentially during the Miocene, of older late-Variscan faults, resulting in the stepped block mountain topography. The morphogenesis can be summarized in three basic stages.

(a) Formation of planation surfaces and the associated weathering mantles (post-Variscan and pre-Alpine morphogenetic cycle). Various pre-Cenozoic planation surfaces have been identified in the Iberian Meseta (Schwenzner 1936). These are explained as polygenic and heterochronic peneplain-type surfaces cut across Variscan structures (Birot and Solé-Sabarís 1954; Pedraza 1978). Planation processes were accompanied by intense weathering of the basement rocks, resulting in thick kaolinic and lateritic mantles (Centeno and Brell 1987; Molina-Ballesteros et al. 1997).

(b) Principal morphogenetic cycle—Alpine tectonic reactivation (Paleogene-Pliocene): The current morphostructure of the PdM and the Iberian Central System can be envisaged as a stair-stepped topography or piedmont-treppen system comprising four main surfaces. The two oldest ones (S1 and S2) correspond to the ancient peneplain-type morphology, deformed by block tectonics and preserved on the summits. The other two are complex surfaces developed coevally to the Alpine tectonic cycle (Pedraza 1978, 1989). The oldest records a stage (Paleogene-Miocene) of etchplanation processes (S3). Later within this stage, the rocks underlain by thick weathering mantles and the etchplains were exposed, resulting in the development of the major landforms of the PdM. The later stage (Pliocene–Pleistocene) started with neotectonic deformations that compartmentalized the relief and defined the current piedmont (S4). This stage also included reshaping of the previous landforms, and in the piedmont, the etchplain acquired the characteristic pediment morphology (Fig. 5.11).

(c) Current morphogenetic cycle (Quaternary): This corresponds to the recent and current morphogenesis, dominated by rilling, torrential erosion, mass movements, and periglacial processes. Evidence of local glacial processes is also found in the highest sectors of the PdM, but with limited imprint on the granitic morphology. The minor landforms observable nowadays were formed during this stage, and the progressive erosion of many of the major landforms continues.

5.5 Conclusions

The current relief of the PdM is the result of a long and complex geomorphological history, controlled by three key factors: the type of intrusive body, the composition of granite, and intense fracturing. Since the granites correspond to a small intrusive body (stock) of relatively late emplacement, it may have been subjected to significant confining stress, which could support a primary origin for the sheet joints and the proliferation of bornhardt-type dome landforms. The rocks of the PdM correspond to peraluminous leucogranites, poor in ferromagnesian minerals. This factor has limited the effectiveness of chemical weathering and explains why the PdM can be classified as a fresh rock massif with very limited chemical corrosion. Finally, the dense network of fractures and joints has facilitated the development of physical weathering and gravitational processes. These processes have produced chaotic landscapes dominated by blocks derived from the disintegration of the domes.

References

Birot P, Solé-Sabarís Ll (1954) Investigaciones sobre la morfología de la Cordillera Central Española. Instituto "Juan Sebastián Elcano" (CSIC), Madrid

Centeno JD, Brell JM (1987) Características de las alteraciones de la Sierras de Guadarrama y Malagón (Sistema Central Español). Features of alterations of igneous rocks from Sierras of

Guadarrama and Malagón (Spanish Central Range). Cuaderno Laboratorio Xeolóxico de Laxe, 12: 79–87. http://www.iux.es/index.php?page=5&serie=1
Centeno-Carrillo JD, García-Rodríguez M (2005) El papel de los procesos gravitacionales en los relieves graníticos: el derrumbe de Peña Sirio (Pedriza de Manzanares, Madrid). Tecnologí@yDesarrollo, 3:1–18. http://www.uax.es/publicaciones/archivos/TECMAD05_001.pdf
De Prado C (1864) Descripción Fisica y Geológica de la província de Madrid. Junta General Estadística, Madrid
Domínguez-Villar D, Arteaga C, García-Giménez R, Smith EA, Pedraza J (2008) Diurnal and seasonal water variations of temperature, pH, redox potential and conductivity in gnammas (weathering pits): implications for chemical weathering. Catena 72:37–48
Domínguez-Villar D, Razola L, Carrasco RM, Jennings CE, Pedraza J (2009) Weathering phases recorded by gnammas developed since last glaciation at Serra da Estrela, Portugal. Quatern Res 72:218–228
Gutiérrez-Elorza M, Rodríguez-Vidal J (1978) Consideraciones sobre la morfogénesis del Sistema Central. Boletín Geológico y Minero 89:109–113
Hernández-Pacheco E (ed) (1931) Guía de los Sitios Naturales de Interés Nacional. Número 1. Sierra de Guadarrama. Junta de Parques Nacionales y Patronato Nacional de Turismo, Madrid
Mellor A, Short J, Kirkby SJ (1997) Tafoni in the El Chorro area, Andalucia, Southern Spain. Earth Surf Proc Land 22:817–833
Molina-Ballesteros E, García-Talegón J, Vicente-Hernández MA (1997) Palaeoweathering profiles developed upon the Iberian Hercynian Basement: their relationship to the oldest Tertiary surface in Central and Western Spain. In: Widdowson M (ed), Tertiary and pre-tertiary Palaeosurfaces: recognition, reconstruction and environmental implications. Geological Society of London, Geological Society of London, pp 175–185 (Special Publication 120)
Pedraza J (1978) Estudio geomorfológico de la zona de transición entre las sierras de Gredos y Guadarrama (Sistema Central Español). Tesis Doctoral (PhD document). Universidad Complutense, Madrid
Pedraza J (1989) La morfogénesis del Sistema Central y su relación con la morfología granítica. Morphogenesis of the central range (Spain) and its relation with granite morphologies. Cuaderno Laboratorio Xeolóxico de Laxe 13:31–46
Pedraza J, Sanz MA, Martín A (1989) Formas graníticas de la Pedriza. Agencia de Medio Ambiente, Comunidad de Madrid, Madrid
Pedraza J, Carrasco RM, Díez-Herrero A, Martín-Duque JF, Martín-Ridaura A, Sanz-Santos MA (1996) Geomorfología. Principios Métodos y Aplicaciones. Rueda, Madrid
Pérez-González A, Ruíz-García P (eds) (1990) Mapa Geológico de Cercedilla. Mapa Geológico Nacional 1:50000, n° 508. IGME. Madrid
Pérez-González A, Rodríguez-Fernández LR, Ruíz-García P (eds) (1990) Mapa Geológico de Torrelaguna. Mapa Geológico Nacional 1:50000, n° 508. IGME. Madrid. http://www.igme.es/internet/cartografia/cartografia/magna50.asp
Pérez-Soba C, Villaseca C (2010) Petrogenesis of highly fractionated I-type peraluminous granites: La Pedriza pluton (Spanish Central System). Geologica Acta 8(2):131–149
Roqué C, Zarroca M, Linares R (2012) Subsurface initiation of tafoni in granite terrains—geophysical evidence from NE Spain: geomorphological implications. Geomorphology. doi:10.1016/j.geomorph.2012.06.015
Schwenzner JE (1936) Zur Morphologie das Zentralspanischen Hochlandes. Geogr. Abhandl., 3ª ser., T. X, 3 Helft, Stuttgart
Twidale CR (1982) Granite landforms. Elsevier, Amsterdam
Twidale CR, Vidal-Romaní JR, Campbell EM, Centeno JD (1996) Sheet fractures: response to erosional offloading or tectonic stress? Zeitschrift für Geomorphologie Supplementband 106:1–24
Twidale CR, Vidal-Romaní JR (2005) Landforms and geology of granite terrains. Balkema, Leiden
Vidal-Romaní JR, Twidale CR (1999) Sheet fractures, other stress forms and some engineering implications. Geomorphology 31:13–27

6 Conglomerate Monoliths and Karst in the Ebro Cenozoic Basin, NE Spain

J. Guerrero, F. Gutiérrez, and M. Gutiérrez

Abstract

The sedimentary fill of the Ebro Cenozoic Basin, NE Spain, includes thick conglomerate successions in the marginal sectors associated with the surrounding Alpine orogens. These commonly cemented alluvial fan and fan delta conglomerates grade rapidly into less resistant fine-grained facies. Differential excavation of the basin fill, together with erosion processes controlled by vertical fractures in the massive and indurated conglomerates, has resulted in the development of monoliths, locally known as mallos, with vertical walls that may reach more than 300 m in height. The cemented and fractured conglomerates in some sectors of the Catalan margin of the basin, mostly composed of calcium carbonate, display features characteristic of well-developed karst terrains, including sinkholes, karst springs, and multilevel cave systems several kilometers long with spelothems.

Keywords

Structural geomorphology • Conglomerate pinnacles • Conglomerate karst • Mallo

6.1 Introduction

The sedimentary fill of foreland basins typically includes significant continental successions (molasse) related to the syntectonic unroofing of the surrounding mountain belts (Allen and Allen 1990). Thick alluvial fan sediments consisting of massive conglomerates and breccias commonly occur along the margins of these basins. Fluvial dissection of the basin fill, together with structurally controlled differential weathering and erosion of these coarse-grained detrital rocks, may result in the development of towers, pinnacles, and fins hundreds of meters high. These structural landforms are widely developed in subhorizontally lying calcareous conglomerates at the margins of the Ebro Cenozoic Basin (Gutiérrez and Peña 1994), where they are usually designated as mallos (meaning thick stick or cudgel; Biarge 2004). The most outstanding examples are located in the following: (1) Tobía-Matute, Anguiano, and Viguera-Islallana in La Rioja, at the boundary with the Iberian Chain; (2) Agüero, Murillo-Riglos, and Salto del Roldán, along the Pyrenean margin of the Ebro Basin in Aragón; and (3) Sant Llorenç de Munt, Montserrat, and L'Espluga de Francolí, in the basin border associated with the Catalan Coastal Chain (Fig. 6.1).

The conglomeratic monoliths may occur as isolated towers with rounded summits and pinnacles (Fig. 6.2), or may form a maze of monoliths and narrow corridors controlled by the fracture pattern (Fig. 6.3). The development of these landscapes is related to the concurrence of a set of lithological, structural, and mechanical factors, within a geomorphic context of fluvial entrenchment and differential erosion (e.g., Benito 1986, 1993; García-Ruiz 2007).

J. Guerrero (✉) · F. Gutiérrez (✉) · M. Gutiérrez
Department of Earth Sciences, University of Zaragoza, Zaragoza, Spain
e-mail: jgiturbe@unizar.es

F. Gutiérrez
e-mail: fgutier@unizar.es

F. Gutiérrez and M. Gutiérrez (eds.), *Landscapes and Landforms of Spain*, World Geomorphological Landscapes,
DOI: 10.1007/978-94-017-8628-7_6,

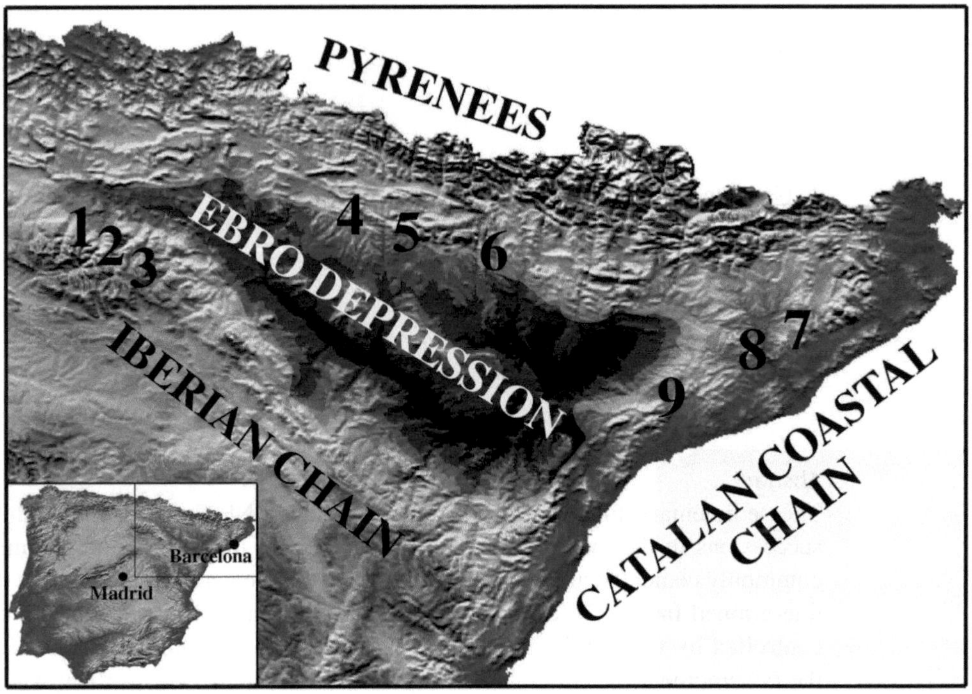

Fig. 6.1 Shaded relief model of the NE sector of the Iberian Peninsula indicating the location of the main sites with conglomerate monoliths along the margin of the Ebro Depression. Tobía-Matute (*1*), Anguiano (*2*), Viguera-Islallana (*3*), Agüero (*4*), Murillo-Riglos (*5*), Salto del Roldán (*6*), Sant Llorenç de Munt (*7*), Monserrat (*8*), L'Espluga de Francolí (*9*)

In some sectors (e.g., Riglos, Salto del Roldán), the conglomerate formations are crossed by major transverse drainages coming from the adjacent ranges and flowing along deeply entrenched canyons (Fig. 6.4). These scenic landscapes have attracted human beings from prehistoric times to satisfy their material, spiritual, and recreation needs. In the karstified conglomerates of Cataluña, caves, and rock shelters were used as settlement sites in Upper Paleolithic and Neolithic times (Bergadá et al. 1997). The conglomeratic massif of Montserrat, declared Natural Park, includes the Montserrat Monastery, one of the most popular pilgrimage sites of Cataluña. The archetypal Mallos de Riglos in Aragón is one of the most frequented sites in Spain by rock climbers. It includes several monoliths with vertical walls more than 300 m high and the mythic El Puro (the cigar), a spire first climbed in 1953 after several deadly attempts (Fig. 6.2). An additional feature of the calcareous conglomerates in the SE margin of the basin (Freixes 1987; Bergadá et al. 1997) is the development of cave systems and internal karst hydrology comparable with those characteristic of carbonate and evaporite terrains (e.g., Goeppert et al. 2011). Conglomerate monoliths are also found in other regions of Spain, such as Vadiello, Pyrenees (Rodríguez-Vidal 1986) or Peracense Castle, Iberian Chain (Gutiérrez et al. 2005; Lozano et al. 2007). Outstanding examples of conglomeratic monoliths in other countries include the World Heritage sites of Meteora (Greece), Danxiashan (China), Kata Tjuta (the Olgas, Australia; Twidale 2010), and the Putangirua Pinnacles Scenic Reserve (New Zealand). This chapter reviews the processes and controlling factors involved in the genesis of the conglomeratic monoliths of the Ebro Basin and describes their most important karst features documented in the literature.

6.2 Geological and Geomorphological Setting

The Ebro Cenozoic Basin in NE Spain constitutes the southern foreland basin of the Pyrenees. The development of these two major structural units is related to the convergence and collision of the Iberian and European plates from the end of the Cretaceous to the Miocene (Anadón and Roca 1996). The Ebro Depression, which essentially coincides with the Ebro foreland basin, is a structural and erosional depression drained by the Ebro River and surrounded by Alpine mountain belts, the Pyrenees to the N, the Iberian Chain to the SW and the Catalan Coastal Chain to the SE (Fig. 6.1). During the initial evolutionary stages (Paleocene and Eocene), the Ebro Basin was largely dominated by marine environments, with deposition of thick fan delta and alluvial fan sequences in the eastern sector (Muñoz et al. 2002; Pardo et al. 2004). In late Eocene times, with the Priabonian regression, the basin became a land-locked

Fig. 6.2 The Pisón Mallo (Riglos, Pyrenean margin), a conglomerate monolith with vertical walls more than 300 m high. The pinnacle attached to the cliff is called El Puro (the cigar), a classical challenge for rock climbers (Photograph taken by F. Gutiérrez)

depression with continental sedimentation (Riba et al. 1983). The paleogeography during this endorheic stage essentially comprised alluvial fans in the marginal areas of the basin grading into evaporite and carbonate lakes in the most subsiding sectors. The distribution of the main evaporite and carbonate lacustrine formations records a progressive southward migration of the depositional axis of the basin related to the propagation of the Pyrenean orogenic wedge and the forebulge (Riba et al. 1983; Ortí 1997). The marginal conglomerates, mainly deposited by sheet floods and debris flows (Blair and McPherson 1994), typically display a sharp lateral facies change into less resistant sandstones and argillaceous rocks. In Middle–Late Miocene times, the basin was captured by a proto-Ebro River and opened toward the Mediterranean Sea (Vázquez-Urbez et al. 2002; García-Castellanos et al. 2003; Pérez-Rivarés et al. 2004; Arche et al. 2010). During this stage, a new drainage network developed and dissected the basin fill by headward expansion, generating stepped sequences of mantled pediments and terraces. The topography within the Ebro Depression is mainly controlled by differential erosion related to the distribution of lithofacies. The central lacustrine successions, commonly capped by limestone units, and the marginal conglomerates form prominent reliefs, while erosional depressions with extensive mantled pediment sequences occur in the areas dominated by argillaceous sediments.

6.3 Controlling Factors and Origin

The development of conglomerate monoliths requires the concurrence of a number of litho-structural and geomorphic factors (e.g., Benito 1986, 1993; García-Ruiz 2007; Fig. 6.5).

- The existence of a thick succession of massive conglomerates with a sharp lateral change into more erodible fine-grained detrital facies, as is commonly the case in alluvial fan environments. This lithological change favors the formation of conglomerate escarpments by differential erosion (Fig. 6.2). Massive or poorly bedded conglomerates are more suitable for the development of monoliths. The towers and pinnacles developed on relatively well-bedded conglomerates, including fine-grained beds, tend to reach lower heights and display more irregular scarps with ledges and notches related to stratigraphically controlled differential recession (e.g., Mallos de Agüero; Fig. 6.6).
- High degree of induration of the conglomerates related to carbonate cementation. This feature grants cohesion and compressive strength values high enough to withstand towers several hundred meters high (Young et al. 2009) and favors surface and subsurface dissolution processes, especially when the clasts are mainly composed of limestone. García-Ruiz (2007) compares the landforms developed on indurated calcareous conglomerates versus low cohesion siliceous conglomerates along the Iberian Chain margin in La Rioja. The distribution of these lithofacies is controlled by the rocks exposed in the drainage basins (Muñoz-Jiménez 1992) that used to feed the alluvial fans in which they were deposited. The cemented conglomerates form excellent examples of monoliths with vertical walls (Tobía-Matute, Anguiano, and Viguera-Islallana), whereas gully systems and deep-seated landslides are the main erosional landforms developed in the loose siliceous conglomerates.
- A critical predisposing factor is related to the attitude, spacing, and orientation of the facture system. Subvertical fractures are required for the development of high- and steep-walled monoliths (Fig. 6.7). Fractures whose dip show a slight deviation for a vertical attitude limit the

Fig. 6.3 Gently dipping conglomerates in the vicinity of Matute village (Iberian Chain margin), deeply dissected by a fluvial canyon (Manzanar Creek) and fracture-controlled corridors (Photograph taken by F. Gutiérrez)

Fig. 6.4 Conglomerates at the northern margin of the Ebro Basin dissected by the transverse Flumen River canyon (Salto del Roldán, Guara Natural Park). Image of the Amán Mallo taken from San Miguel Mallo. To the left, Mesozoic and Paleogene formations of the Pyrenees overthrusting the subhorizontally lying Tertiary conglomerates (Photograph taken by F. Gutiérrez)

Fig. 6.5 Sequence of block diagrams illustrating the development of monoliths on thick and massive conglomerate successions affected by widely spaced orthogonal fractures (Illustration produced by Santiago Alberto-Moralejo)

Fig. 6.6 The Mallos de Agüero (Huesca Province) carved on relatively well-bedded cemented conglomerates affected by a widely spaced network of vertical fractures. These Tertiary sediments where deposited on the proximal sector of alluvial fans developed at the Pyrenean margin of the Ebro Basin. Agüero village at the foot of the cliffs (Photograph taken by F. Gutiérrez)

height of the monoliths and favor the development of upward-tapering pinnacles and spires. The density and spacing of the fractures should reach values within a range adequate to compartmentalize upright parallelepipedic rock masses that may evolve into stable monoliths. A high density of fractures leads to a rapid erosion of the conglomerate massif (Fig. 6.8), whereas massive reliefs and mesas tend to develop on conglomerates with a low density of joints. Benito (1986, 1993), comparing the spatial distribution of conglomerate landforms in Aguero-Riglos area and the fracture density, infers that in that sector of the Pyrenean margin, the mallo landform is best developed with fractures densities of around 75 fractures/km^2. Fracture orientation also plays a significant role in the development and geometry of the monoliths. Orthogonal fracture systems control the formation of towers and pinnacles, whereas conglomerates affected by a more penetrative set of parallel fractures promote the formation of elongated monoliths (fins), like in Montserrat (Figs. 6.8, 6.9) and Sant Llorenç de Munt.

- Relatively rapid entrenchment of the fluvial network favors (1) the differential erosion of the more distal fine-grained alluvial fan facies and the development of conglomerate escarpments; (2) the headward expansion and incision of gully systems, which contributes to the preferential erosion of conglomerates along fracture-controlled drainages and the evacuation of sediments accumulated at the foot of the scarps. In fact, some of the best examples of monoliths occur associated with major fluvial systems, whose incision has led to a continuous rejuvenation of the topographic gradient (e.g., Riglos, Salto del Roldán, Viguera-Islallana; Figs. 6.4, 6.7). In the Ebro Basin, this long process of base-level lowering and differential erosion started in Miocene times, when the basin was captured by the external drainage network and changed to exorheic conditions.

Figure 6.5 illustrates the development of conglomerate monoliths. In an initial phase, the entrenchment of the drainage network in the basin and differential erosion of fine-grained alluvial fan facies result in the development of

Fig. 6.7 The Mallo Firé in Riglos, an association of elongated monoliths controlled by *vertical fractures*. The Gállego River valley with well-developed terraces in the *bottom*, and the conglomerates of Murillo de Gállego on the *upper-right corner*. The saddle situated to the *right* of the Mallo Firé corresponds to the thrust that defines the contact between the Ebro Basin (conglomerates) and the Pyrenees (folded carbonate rocks) (Photograph taken by F. Gutiérrez)

Fig. 6.8 Oblique aerial view of the conglomerates in Montserrat affected by a dominant NNE–SSW fracture set with variable spacing, controlling the development of fin-like elongated monoliths. The monoliths with higher relief are associated with the areas where the fracture spacing is wider

a conglomeratic massif bounded by an escarpment. A conjugate system of widely spaced vertical fractures controls focused weathering and erosion processes resulting in steep-sided gullies, locally hanging, and corridors that individualize rock prisms. Rock falls are the dominant processes acting on the competent calcareous conglomerate walls, promoted by mechanical (e.g., frost shattering), chemical (e.g., dissolution), and biological (e.g., root wedging) weathering. Fluvial erosion, mostly related to severe storm events in the gully bottoms, contributes to the evacuation of the deposits supplied from the hillslopes. Long-sustained base-level lowering related to fluvial entrenchment leads to the development of progressively higher monoliths with vertical walls and rounded summits.

6.4 Karst in Conglomerate Monoliths

Fractured conglomerates consisting of limestone clasts with a cemented calcareous matrix are suitable formations for the development of karst features. Subsurface water circulation through fractures and interstitial pores causes the

Fig. 6.9 Les Agulles, Montserrat, elongated monoliths controlled by a penetrative system of parallel and vertical fractures (Photograph taken by C. Miñarro)

Fig. 6.10 Images of Cuberes cave. **a** Column. **b** Flowstone. **c** Canyon-like subhorizontal passage (courtesy of L. Almela)

progressive enlargement of discontinuity planes and the dissolution of the cement, matrix, and limestone clasts, resulting in a permeability increase. This is a self-accelerating process that fosters higher flow and karstification. Over time, a mature karst may develop in conglomerates, displaying many of the surface and underground landforms and hydrogeological characteristics of carbonate and evaporite karst systems (e.g., Goeppert et al. 2011).

The scientific literature dealing with karst features developed in conglomerates is rather limited. Conglomerate karst has been reported in Kuruköprü Basin, Turkey (Degirmenci and Günay 1993), southern France (Bès 1994), the northern foreland basin of the Alps, Germany (Scholz and Strohmenger 1999), Russia (Filippov 2004), northern Italy (Ferrarese and Sauro 2005), Slovenia (Gabrovsek 2005; Lipar and Ferk 2011 and references therein), Jura Mountains, NW Switzerland (Lapaire et al. 2006), the northern Alps, Austria-Germany (Goeppert et al. 2011). Ferrarese and Sauro (2005) mapped more than 2,000 sinkholes and blind valleys in the Miocene Montello Conglomerate in Italy. Bol'shaya Cave, Russia, with more than 47 km of surveyed passages, is the longest conglomerate cave in the world (Filippov 2004).

In Spain, karst features developed on carbonate-rich conglomerates have been documented in a number of areas (e.g., Martín-Algarra et al. 1989; Durán and López-Martínez 1999; Antón 1992). However, the most outstanding examples are found in conglomerates associated with the Catalan margin of the Ebro Basin. Karst landforms have been described in calcareous Paleogene conglomerates at Sant Llorenç del Munt, Sant Miguel Montclar, L'Espluga de Francolí, Montserrat Massif, Montsen, Serradell, and Sant Llorenç de Morunys (Freixes 1989). Surface karst landforms are typically restricted to solutionally enlarged fractures, swallow holes, and microkarren developed on large limestone boulders. The main dissolution-related features correspond to well-developed caves and the characteristic subsurface karst hydrology (Bergadá et al. 1997). The Terrasa Excursionist Center has inventoried around 600 caves in Sant Llorenç del Munt, some of them with archaeological and paleontological sites (Jorquera 1970; Nebot and Hernández 2007). The 12.8-km-long Cuberes Cave in L'Espluga de Francolí is one of the longest conglomerate caves in the world (Freixes 1989; Fig. 6.10).

According to Bergadá et al. (1997), the cave systems developed in the calcareous conglomerates of the Catalan margin of the Ebro Basin consist of short and vertical conduits that connect with subhorizontal passages that may reach several hundred meters long. Vertical conduits result from dissolution along vertical joints, while the horizontal galleries are often associated with more fractured conglomerates or impervious beds. These passages, commonly developed at different levels, display circular or ellipsoidal sections modified by basal incisions (keyhole section) and lateral notches, recording an initial phreatic phase and subsequent vadose entrenchment related to the episodic entrenchment of the drainage network. Freixes (1987, 1989) studied a number of temporal and permanent springs in the Sant Llorenç del Munt conglomerate massif. He found retarded downward vadose circulation before reaching a primary flow pathway. He classified these peculiar karst aquifers as diffuse flow hydrogeological systems with relative long water residence time. Precipitation from oversaturated water results in the formation of spelothems in the caves (Fig. 6.10) and tufa deposits in some springs.

6.5 Conclusions

The development of monoliths (mallos) in the calcareous conglomerates along the margins of the Ebro Basin results from the concurrence of several factors, related to the sedimentary, diagenetic, and erosional history of the sediments. These are: (1) Deposition of thick calcareous conglomerates in alluvial fans and fan deltas developed at the tectonically active margins of the basin; (2) Sharp lateral change between the conglomerates and more distal fine-grained facies; (3) Significant induration of the deposits by cementation processes; (4) Development of a network of vertical fractures with adequate spacing; (5) Progressive entrenchment of the drainage network after the capture of the basin and differential erosion of the fine-grained facies; (6) Preferential weathering and erosion processes acting on the fractures, leading to the compartmentalization and individualization of monoliths and in some cases the formation of mazes of corridors. These calcareous conglomerates cemented by calcium carbonate and affected by fractures, constitute a suitable terrain for the development of geomorphic and hydrological karst systems, like those reported in the Catalan margin of the basin. These conglomerate massifs display grikes, swallow holes, and well-integrated cave systems consisting of vertical fracture-controlled conduits and different levels of subhorizontal passages formed under phreatic conditions.

Acknowledgments This work has been supported by the Spanish national project CGL2010-16775 (Ministerio de Ciencia e Innovación and FEDER) and the Regional project 2012/GA-LC-021 (DGA-La Caixa).

References

Allen PA, Allen JR (1990) Basin analysis. Principles and applications. Blackwell, Oxford, p 451

Anadón P, Roca E (1996) Geological setting of the Tertiary basins of the Northeast Spain. In: Friend PF, Dabrio CI (eds) Tertiary basins of Spain, the stratigraphic record of crustal kinematics. World and Regional Geology, vol 6. Cambridge University Press, Cambridge, pp 43–48

Antón FJ (1992) Karst conglomerático en la vertiente sur de la sierra de Urbión (Duruelo, Soria). Cuadernos de Sección, Historia 20:3–16

Arche A, Evans G, Clavell E (2010) Some considerations on the initiation of the present SE Ebro river drainage system: post- or pre-Messinian? J Iberian Geol 36:73–85

Benito G (1986) Génesis del modelado tipo mallo. Cuadernos de Investigación Geográfica 12:25–37

Benito G (1993) Genesis and evolution of the mallo pinnacle landform. In: Gutiérrez M, Sancho C, Benito G (eds) Second intensive course on applied geomorphology: Arid regions. Copistería Kronos, Zaragoza, pp 179–183

Bergadá MM, Cervello JM, Serrat D (1997) Karst in conglomerates in Catalonia (Spain): morphological forms and sedimentary sequence types recorded on archaeological sites. Quaternaire 8:267–277

Bès C (1994) Fracturation et karstification dans les Hautes-Corbières. Spélé Aude 3:28–52

Biarge F (2004) Mallos, un relieve emblemático. Ediciones del Mallo. Huesca, p 238

Blair TC, McPherson JG (1994) Alluvial fans and their natural distinction from rivers based on morphology, hydraulic processes, sedimentary processes, and facies assemblages. J Sediment Res A64:450–489

Degirmenci M, Günay G (1993) Origin and catchment area of the Olukköprü karst springs. Hydrogeological processes in Karst Terranes. In: Proceedings of the Antalya symposium and field seminar. IAHS Publ. 207, pp 97–105

Durán JJ, López-Martínez J (1999) El karst en Andalucía. Instituto Tecnológico Geominero de España, Madrid, p 192

Ferrarese F, Sauro H (2005) The Montello hill: the "classical karst" of the conglomerate rocks. Acta Carsologica 34:439–448

Filippov A (2004) Siberia, Russia. In: Gunn J (ed) Encyclopedia of caves and karst science. Fitzroy Dearborn, New York, pp 645–647

Freixes A (1987) Características del funcionamiento y la estructura de los sistemas hidrológicos karstificados de los conglomerados de la Serra de L'Obac (Depresión Terciaria del Ebro). Geogaceta 2:49–51

Freixes A (1989) Principales ejemplos de karst en conglomerados y areniscas. In: Durán JJ and López-Martínez J (eds) El karst en España, Monografía, vol 4. Sociedad Española de Geomorfología, Madrid, pp 295–298

Gabrovsek F (2005) Caves in conglomerate: case of Udin Borst, Slovenia. Acta Carsologica 34:507–519

García-Castellanos D, Vergés J, Gaspar-Escribano J, Cloetingh S (2003) Interplay between tectonics, climate and fluvial transport during the Cenozoic evolution of the Ebro Basin (NE Iberia). J Geophys Res 108:1–18

García-Ruiz JM (2007) El relieve de los conglomerados de borde de cuenca de la Rioja. In: Arnaez J, García-Ruiz JM (eds) Espacios naturals y paisajes en la Rioja. Instituto de Estudios Riojanos, Logroño, pp 73–85

Goeppert N, Goldscheider N, Scholz H (2011) Karst geomorphology of carbonate conglomerates in the Folded Molasse zone of the Northern Alps (Austria/Germany). Geomorphology 130:289–298

Gutiérrez F, Gutiérrez M, Gracia FJ (2005) Karst, neotectonics and periglacial features in the Iberian Ranges. Field Trip C-5. In: Sixth international conference on geomorphology. Zaragoza, p 58

Gutiérrez M, Peña JL (1994) Depresión del Ebro. In: Gutiérrez M (ed) Geomorfología de España. Rueda, Madrid, pp 305–349

Jorquera M (1970) Yacimiento paleontológico en la cova Simanya (Sant Llorentç del Munt). Mediterránea 6:7–8

Lapaire F, Becker D, Chrite R, Luetscher M (2006) Karst phenomena with gas emanations in Early Oligocene conglomerates: risks within a highway context (Jura Switzerland). Bull Eng Geol Environ 66:237–250

Lipar M, Ferk M (2011) Eogenetic caves in conglomerate: an example from Udin Borst, Slovenia. Int J Speleol 44:53–64

Lozano MV, Fabregat C, López S, González JM (2007) Albarracín rodeno landscape (Spain). In: Härtel H, Cílek V, Jackson A, Williams R (eds) Sandstone landscape. Academia, Praha, pp 368–371

Martín-Algarra A, Soria Mingorance J, Vera JA (1989) Paleokarst mesozoicos y terciarios en la Cordillera Bética. In: Durán JJ, López-Martínez J (eds) El karst en España, Monografía, vol 4. Sociedad Española de Geomorfología, Madrid, pp 299–308

Muñoz-Jiménez A (1992) Análisis tectosedimentario del Terciario del sector occidental de la Cuenca del Ebro (Comunidad de La Rioja). Instituto de Estudios Riojanos. Logroño, p 347

Muñoz A, Arenas C, González A, Pardo G, Pérez A, Villena J (2002) Ebro Basin (Northeastern Spain). In: Gibbons W, Moreno T (eds) The geology of Spain. Geological Society of London, London, pp 301–309

Nebot M, Hernández T (2007) Mamífers trobats a les cavitats de Sant Llorent del Munt i L'Obac. VI Trobada d'Estudiosos de Sant Llorentç del Munt i L'Obac, Diputació de Barcelona, Barcelona, pp 121–124

Ortí F (1997) Evaporitic sedimentation in the South Pyrenean Foredeeps and the Ebro Basin during the Tertiary: a general view. In: Busson G, Schreiber BCh (eds) Sedimentary deposition in rift and foreland basins in France and Spain. Columbia University Press, New York, pp 319–334

Pardo G, Arenas C, González A, Luzón A, Muñoz A, Pérez A, Pérez-Riverés FJ, Vázquez-Urbez M, Villena J (2004) La Cuenca del Ebro. In: Vera JA (ed) Geología de España. Sociedad Geológica de España-IGME, Madrid, pp 533–543

Pérez-Rivarés F, Garcés M, Arenas C, Pardo G (2004) Magnetostratigraphy of the Miocene continental deposits of the Montes de Castejón (central Ebro Basin, Spain): geochronological and paleoenvironmental implications. Geologica Acta 2:221–234

Riba O, Reguant S, Villena J (1983) Ensayo de sintesís estratigráfica y evolutiva de la Cuenca Terciaria del Ebro. Libro jubilar JM Ríos. IGME, Madrid, pp 131–159

Rodríguez-Vidal J (1986) Geomorfología de las Sierras Exteriores oscenses y su piedemonte. Instituto de Estudios Altoaragoneses. Huesca, p 172

Scholz H, Strohmenger M (1999) Dolinenartige Sackungsstrukturen in den Molassebergen des südwestbayerischen Alpenvorlandes. Jahresberichte und Mitteilungen des Oberrheinischen Geologischen Vereins 81:275–283

Twidale CR (2010) Uluru (Ayers Rock) and Kata Tjuta (The Olgas). In: Migon P (ed) Geomorphological landscapes of the World. Springer, Dordrecht, pp 321–332

Vázquez-Urbez M, Arenas C, Pardo G (2002) Facies fluvio-lacustres de la unidad superior de la Muela de Borja (Cuenca del Ebro): modelo sedimentario. Revista de la Sociedad Geológica de España 15:41–54

Young RW, Wray RAL, Young ARM (2009) Sandstone landforms. Cambridge University Press, Cambridge, p 304

7 The Karst of the Tramuntana Range, Mallorca Island

Ángel Ginés and Joaquín Ginés

Abstract
Exokarst landforms, as well as caves and shafts, are common in the Tramuntana Range (Mallorca Island, western Mediterranean) owing to the presence of extensive limestone outcrops and suitable bioclimatic environmental conditions. The mountain range has excellent examples of polje-like depressions, dolines, karrenfields and karst gorges, especially in the northern sector. Karrenfields are the most remarkable landforms because of their morphological variety and widespread occurrence. They illustrate the impact of climatic gradients and soil erosion on their development and distribution.

Keywords
Karst landscape • Pinnacle topography • Karrenfields • Mediterranean karst

7.1 Introduction

The geology of Mallorca Island, largely composed of limestone rocks, together with its typical Mediterranean bioclimatic conditions, favours the development of distinctive karst landscapes. The most outstanding examples are located in the Tramuntana Range, which is the main mountainous area in Mallorca (Fig. 7.1a). It exhibits a remarkable variety of solutional landforms, including widespread karrenfields. This diversity is largely related to a broad variety of environmental conditions, mainly controlled by climatic gradients linked to the altitude, ranging from sea level to above 1,400 m. The impact of human activity on the environment over the last five millennia, together with natural deforestation (Ginés 1995), have controlled the complex evolution that records the karren features within a mid-mountain bioclimatic and geomorphological framework (Ginés and Ginés 1995, 2009).

The NE–SW-trending Tramuntana Range has a mediterranean climate characterized by a significant summer drought from June to September. Average annual rainfall reaches 1,400 mm in the central and highest sector of the range, decreasing rapidly towards the lower and peripheral zones of the mountain chain (<500 mm/year in the SW and NE ends). Thus, the rainfall pattern follows that of the elevation (Fig. 7.1a, c). Snowfalls are scarce and restricted to the highest sectors during a few winter days. High-intensity rainfall storm events (over 250 mm in 24 h) are not exceptional, particularly in autumn, due to sudden irruptions of cold air in the middle and upper parts of the troposphere over a still hot water mass in the Mediterranean Sea. Mean annual temperature ranges from 12 °C in the central and highest part of the mountains, to 17 °C in the NE and SW terminations (Formentor and Andratx) (Fig. 7.1c). Seasonal variability is noticeable, with winter and summer mean temperatures below 10 °C and close to 25 °C, respectively. In general terms, the climate in Mallorca Island is that of mid-latitudes but modified by azonal

Á. Ginés (✉) · J. Ginés
Departament de Ciències de la Terra, Universitat de les Illes Balears, Palma, Spain
e-mail: agines.gracia@yahoo.es

F. Gutiérrez and M. Gutiérrez (eds.), *Landscapes and Landforms of Spain*, World Geomorphological Landscapes,
DOI: 10.1007/978-94-017-8628-7_7,

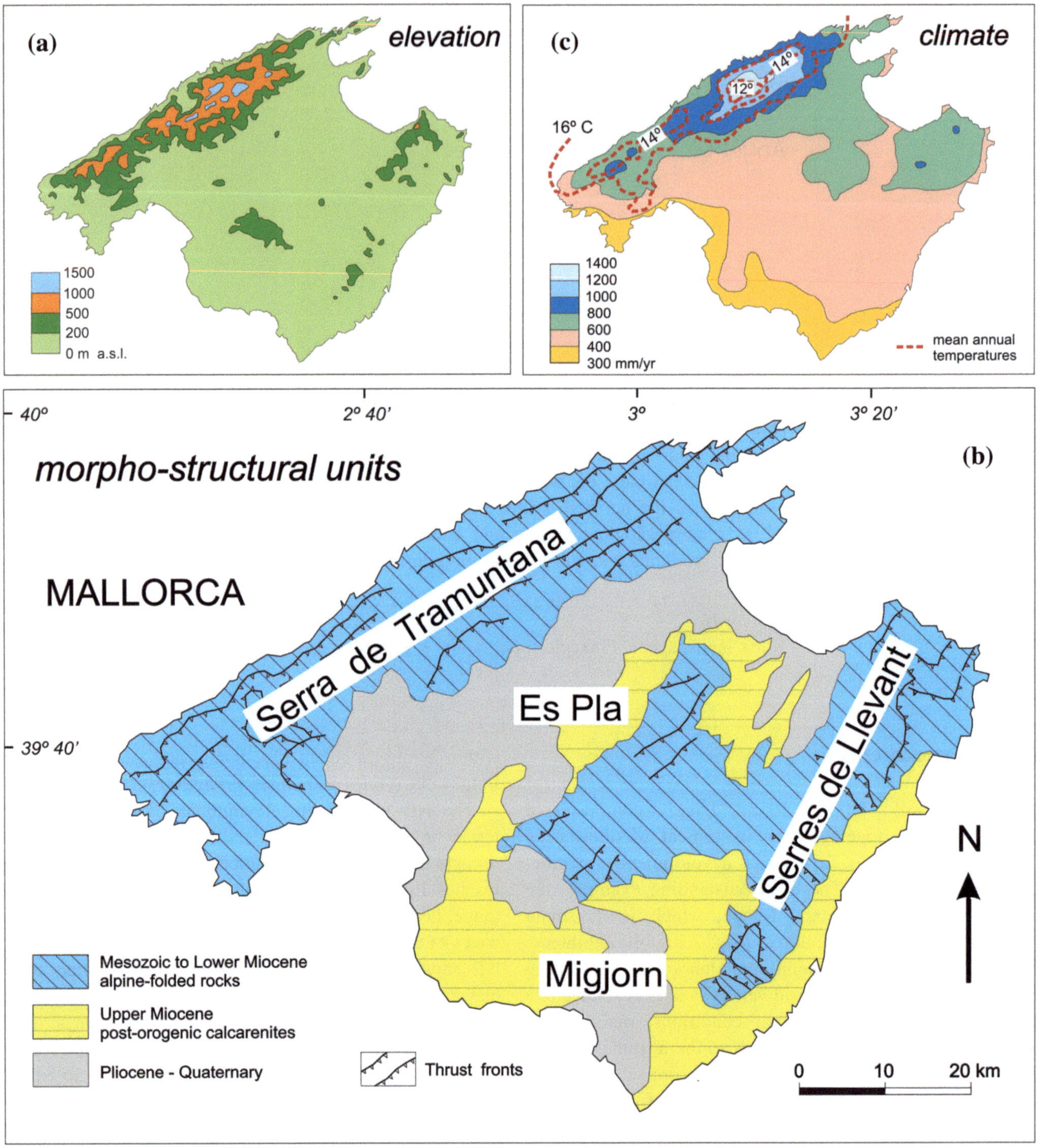

Fig. 7.1 Geographic features of Mallorca Island. **a** Simplified elevation map; **b** Main morphostructural units controlled by the geological structure; **c** Average annual rainfall and temperature

features related to the Mediterranean Sea. Within this general context, diverse microclimatic conditions related to the elevation, and ranging from humid to semiarid, control the distribution of the different karren assemblages along the Tramuntana Range.

7.2 Geological History

Mallorca is the largest and central island of the Balearic Archipelago (Fig. 7.2). The Balearic Islands are located in the western sector of the Mediterranean basin. Mallorca,

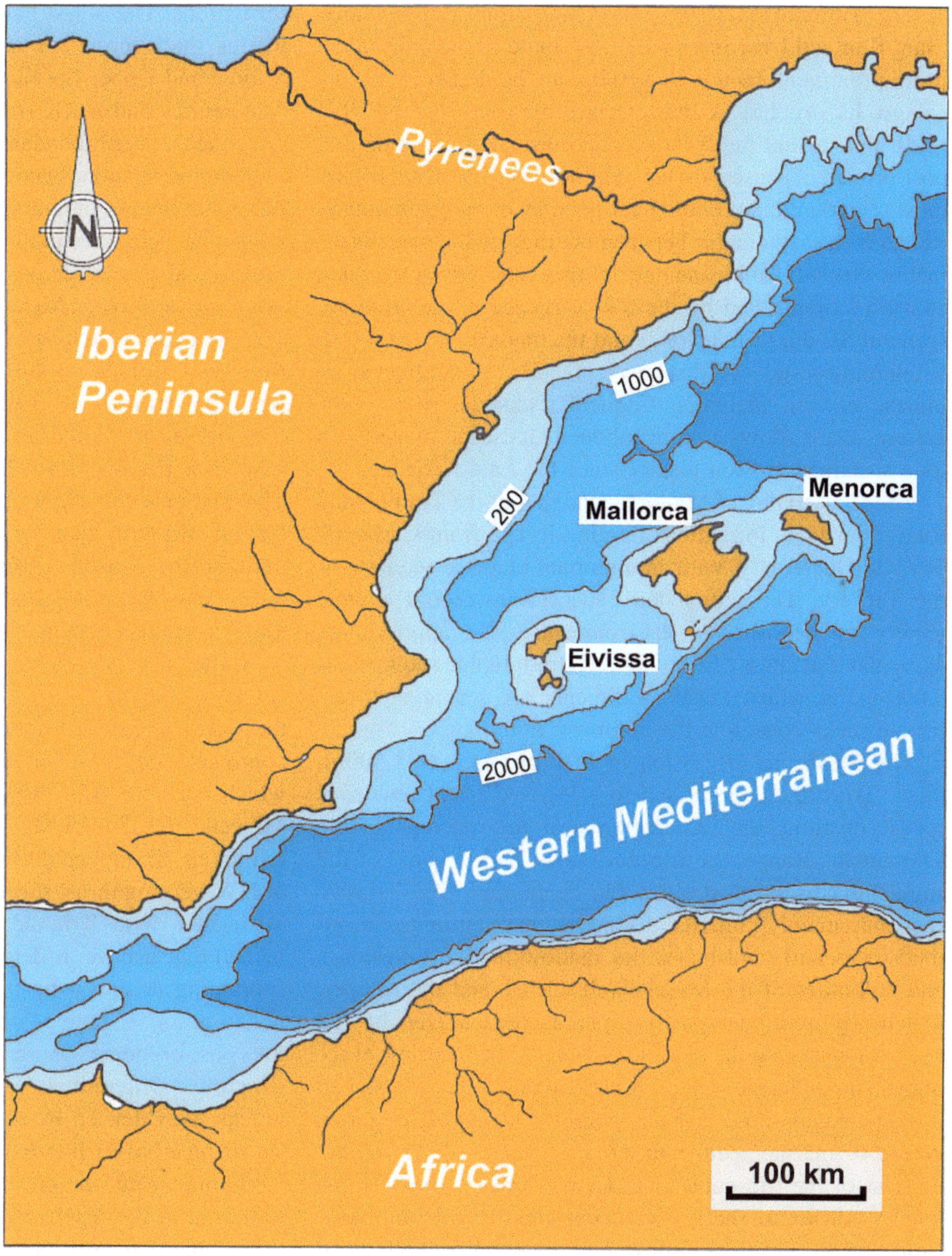

Fig. 7.2 Location of Mallorca Island in the Western Mediterranean basin as a part of the Balearic Promontory, a partially submerged prolongation of the Alpine Betic orogen

with a perimeter of approximately 560 km and a surface area of about 3,650 km^2, is the seventh largest island in the Mediterranean, and together with Menorca, is the most remote with respect to any continental landmass. The island of Mallorca exhibits a rhomboidal shape (96 × 78 km), with its vertices oriented to the four cardinal points. The northernmost point is Cap de Formentor, to the east is Punta de Capdepera, to the south Cap de ses Salines and to the west Sant Elm.

Regarding the geomorphology and tectonics, three main morphostructural units can be differentiated: the Tramuntana Range, Es Pla depression and the Llevant Ranges (Fig. 7.1b). They are related to a complex NE–SW horst and graben structure superimposed on previous compressional structures. The mountain ranges correspond to horsts: Tramuntana Range, Llevant ranges, as well as some ridges in the central sector of the island. The central Es Pla depression, dominated by a flat topography, corresponds to an overall graben structure, including several sub-basins at Palma, Sa Pobla and Campos. The NE–SW-trending Tramuntana Range, 90 km long and 15 km in average width, constitutes the north-western mountainous portion of the island. Here, several peaks reach more than 1,000 m a.s.l. (e.g. Puig Major 1,445 m).

The Tramuntana Range, stretching from Andratx area (Sant Elm and Dragonera islet) to Cap de Formentor, is the most prominent mountain system in the Balearic Archipelago. Its structure is characterized by a series of NE-SW folds, thrusts and faults developed during the orogenic and post-orogenic phases of the Alpine orogeny. One of the most remarkable morphological features of the Tramuntana Range is the asymmetry between the rugged and steep NW-facing coastal flank, and the more gentle SE side. Such contrast is controlled by the NW vergence of the structure (general SE dip of the strata) and the maturity of the relief. While the coastal side displays a juvenile relief, affected by intense erosion including frequent rockslides, the south-eastern sector shows a more subdued landscape related to a long-sustained erosion history since the Late Neogene.

The stratigraphy of Mallorca, and that of the Tramuntana Range, includes formations ranging in age from Carboniferous to Quaternary, with an important hiatus at the base of the Tertiary. The sedimentary successions record a wide variety of depositional environments. The approximate aggregate thickness of the exposed stratigraphic sequence is 3,000 m, in which clastic sediments are scarce and carbonate rocks constitute the dominant lithology.

The Mesozoic succession in Mallorca is over 1,500 m thick. Triassic, Jurassic, and to a lesser extent Cretaceous rocks constitute the vast majority of the outcrops in the Tramuntana Range, the Llevant ranges and in some of the small hills in the central Es Pla depression. The Triassic sequence mainly consists of continental Buntsandstein red mudstones and sandstones, the shallow marine limestones and dolomites of the Muschelkalk facies, and the Keuper, which represents a regressive phase characterized by red and yellowish argillaceous sediments, halite-bearing evaporites and volcanic rocks. Rhaetian dolomites, deposited during the end of the Triassic, point the beginning of a long period of marine sedimentation that continued throughout the rest of the Mesozoic. The Lower Jurassic rocks (Lias) mainly consist of shallow platform massive micritic limestones ca. 400 m thick. These limestones form the bulk of the intensely karstified summits at the Tramuntana Range. The variegated Middle Jurassic (Dogger) and Upper Jurassic (Malm) rocks record a progressive transition towards hemipelagic and pelagic environments. The pelagic sedimentation commenced during the Upper Jurassic, and water depth increased during the Lower Cretaceous with deposition of marls and white marly limestones. The Cretaceous rocks are scarce in Mallorca, although they can reach 150 m in thickness.

The Cenozoic successions exceed 1,500 m in thickness. The Palaeocene and Lower Eocene in the Balearics correspond to a long erosional hiatus related to the emersion of the area, which accounts for the absence of rocks of these ages, as well as for the erosion of Upper Cretaceous formations. The oldest Palaeogene sediments, a few tens of metres thick, crop out in the Llevant ranges and are of Middle and Upper Eocene age. They are formed chiefly by calcarenites and marls rich in nummulites. The Oligocene rocks are more abundant, consisting of polygenic conglomerates, siltstones and limestones with algal concretions. Later, sediments including limestone conglomerate Burdigalian in age, accumulated in Lower-Middle Miocene times, and are affected by folding and thrusting. The synorogenic Lower-Middle Miocene sequences found in Mallorca comprise lacustrine and pyroclastic rocks (earliest Miocene), turbidites (during Burdigalian to Langhian tectonic pulsations), as well as lacustrine and alluvial fan deposits accumulated in terrestrial fault-bounded basins related to the final Late Miocene compressional episodes. The post-orogenic Upper Miocene rocks form a tabular area (called Migjorn) that surrounds the mountain ranges and underlie the coastal platform at the south and east of the island (see Fig. 7.1b); but without significant outcrops in the Tramuntana Range.

Mallorca is the most extensive emerged portion of the so-called Balearic Promontory (Fig. 7.2). This is part of the Betic orogen, a fold and thrust belt resulting from the continental collision between the African and Iberian plates, which occurred from the Upper Cretaceous to the Middle Miocene (ca. 84–15 Ma). The main compressional structures were developed during the Alpine orogeny and consist of NW-verging thrust sheets. The thrust planes are generally associated with Triassic marls and evaporites (Keuper facies) that behave as detachment horizons. The amount of shortening achieved by the thrust structure has been estimated at 44 %. Late Miocene extensional tectonics during the post-orogenic phase generated a horst and graben structure. Differential vertical displacement of tectonic blocks is evidenced by uplifted Pleistocene shorelines and subsidence basins like Palma, Inca-Sa Pobla and Campos.

Summing up, the present-day geological architecture of Mallorca is the result of a complex evolution comprising three main phases: (1) Mesozoic sedimentary phase; (2) Alpine compressional tectonics; and (3) Post-orogenic extensional neotectonics (Late Neogene to Quaternary). The latter phase produced the main morphostructural units, namely the Tramuntana Range, Es Pla depression and Llevant ranges (Rose et al. 1978; Ginés et al. 2012).

7.3 General Landforms

Mallorca exhibits an almost unparalleled variety of landforms within a relatively small territory: steep mountains, deep ravines, precipitous cliffs, broad fertile plains, sandy bays and charming coves (Rose et al. 1978). The topography of Mallorca is mainly determined by late Cenozoic

Fig. 7.3 The impressive gorges of Sa Fosca and Torrent de Pareis are surrounded by vast and almost impassable karrenfields, characterized by an extremely jagged terrain, comprising sharp ridges and pinnacles separated by deep clefts (*Photo* J. Ginés)

extensional tectonics, responsible for the gentle hills of the central plains, as well as the rugged mountains of the Tramuntana Range and Llevant ranges. The rest of the island is characterized by subdued rolling terrains surrounded by a platform of horizontally lying Upper Miocene carbonates, covered locally by Pliocene to Pleistocene sediments. No perennial watercourses are found in Mallorca. Run-off is drained by flash ephemeral streams locally designated as "torrents".

In the Tramuntana Range, the fold and thrust structure is transversally dissected by discordant streams flowing through spectacular narrow gorges that cut across the massive limestone successions. The most distinctive geomorphic features of the Tramuntana Range are closely related to lithology. The alternation of soft rocks (marls, clays, volcanics) and resistant limestones determine many of the landforms observed in the landscape. Since about 65 % of the Tramuntana Range corresponds to limestone outcrops, karst landforms are the most outstanding ones (Ginés 1998). Polje-like depressions, dolines, extensive karrenfields and karstic gorges are widely distributed over the entire mountain range (Bär 1989; Ginés et al. 1989; Ginés and Ginés 1995, 2009). The impressive gorge of Torrent de Pareis, with walls 300 m high, is a remarkable example of such a rugged terrain (Fig. 7.3). Generally, the current landscape of the Tramuntana Range is a particular mixture of karstic wilderness and anthropogenic features such as terraces, cultivated areas and farmhouses, whose economical upkeep is nowadays uncertain (Ginés 1999).

The vast majority of the karst landforms in the Tramuntana Range occur on Lower Jurassic micritic limestones, which are by far the most common outcropping lithology in the mountain range. Large karrenfields, dolines, pinnacles, caves and shafts are ubiquitous features on these rocks, characterized by their remarkable purity and high strength. Minor outcrops of Rhaetian dolomites and Lower Miocene (Burdigalian) limestone conglomerates also show good examples of karst landforms in the Tramuntana Range: deep shafts and karren pinnacles, respectively.

7.4 Karst Geomorphology

Karst landforms are common in the Tramuntana mountains. Large karst depressions with flat floors up to 2 km across, resembling poljes but generally smaller or drained by surface drainage can be found, especially in the northern part of the range (Ginés and Ginés 1995). Some of them are not closed depressions because they have been captured by streams or karst gorges. Their elongated shape is in most cases controlled by the geological structure (Fig. 7.4), especially thrust faults that place in contact impervious marly rocks against karstifiable limestones.

Dolines only show a relatively high density in some intensively karstified sectors in the central part of the mountain range (Escorca municipality), although they have a lower imprint on the landscape than karren landforms. Dolines typically show a subcircular or ellipsoidal perimeter

Fig. 7.4 Water collected into Coma de Son Torrella feeds the Font des Verger cave-spring, located in the nearby transversal valley of Biniaraix. This elongated polje-like depression (1.5 km long and 250 m wide) shows a structurally controlled NE-SW trend (*Photo* J. Ginés)

Fig. 7.5 Typical doline of the central sector of the Tramuntana Range belonging to a cluster of four dolines called Es Clots Carbons. The flat floor of the doline shows a particular fire-controlled plant community dominated by heather; *Erica arborea* (*Photo* A. Ginés)

(Fig. 7.5), and their area ranges from 200 to 15,000 m^2 (major axial lengths between 20–150 m) (Ginés et al. 1989). They tend to form clusters, and their flat bottom is typically underlain by silty soils. Nonetheless, the total area covered by dolines is negligible compared with that of the surrounding karrenfields.

Karren are widespread on large limestone outcrops devoid of soil cover, forming extensive karrenfields. These highly corroded bedrock exposures, up to several km^2 in area, behave as groundwater input areas with no surface drainage. Well-developed karrenfields constitute an almost impassable terrain, with sharp ridges, clefts and spectacular

Fig. 7.6 Typical karst landscape of the Tramuntana Range (Lluc, Escorca municipality). This characteristic karren assemblage is a combination of rillenkarren, trittkarren and rinnenkarren features, sculpturing the sides of pinnacles which emerge over a cleared holm-oak (*Quercus ilex*) forest. Elevation of this site is 550 m a.s.l., and average annual rainfall values exceed 1,000 mm (*Photo* A. Ginés)

pinnacles resembling some tropical karren landscapes. The most impressive karrenfields are located on the NW slopes of the northern sector of the range, between Sóller and Pollença (Fig. 7.6), at rather moderate elevations. Here, the occurrence of karren precludes any kind of cultivation.

Observation of karren features throughout the Tramuntana Range allows recognition of several distinctive karren assemblages, whose distribution shows a clear correlation with the altitudinal gradient. For example, solutional forms present in the highest summits of the range are very different from those located at the lowest elevations, with semi-arid climate (Ginés and Ginés 1995; Ginés 1998). Likewise, the best-developed karrenfields in Mallorca are found in quite specific environmental conditions, characterized by precipitation >800 mm/year, mean annual temperature >15 °C, and elevation between 200 and 700 m a.s.l.

Karrenfields are vast bare rock areas dominated by small-scale karren sculpturing (Ginés 2004, 2009). Straight solutional shapes are typically the most conspicuous and abundant features in most karrenfields, namely rillenkarren (Fig. 7.7) and rinnenkarren (Fig. 7.8). Both karren types, which differ significantly in size and genesis, have given a characteristic furrowed appearance to rock outcrops. Rinnenkarren are remarkably wider runnels and result from solution on rock slopes by channel-collected waters, whereas straight rillenkarren flutes rarely exceed 2 cm in width and are generated by direct-rainfall impact processes. Some morphological and evolutionary features of rillenkarren have been intensively investigated at the Lluc site (Lozano 1884; Bär et al. 1986; Ginés 1998), in the central sector of the Tramuntana Range (520 m a.s.l.; mean annual precipitation >1,000 mm). Recent investigations reveal that the morphometric characteristics of rillenkarren are related to the environmental conditions, reflecting altitude-controlled climatic gradients and increasing rainfall and decreasing temperature with elevation (Ginés and Ginés 2009; Lundberg and Ginés 2009). An additional interesting aspect of rillenkarren is the contribution of biokarstic processes to the development of these elementary forms. Fiol et al. (1996) demonstrate that the mechanical removal of small limestone particles, detached by the impact of raindrops, is an efficient process involved in rillenkarren growth. This detachment of particles is greatly favoured by the presence of endolithic algae that have previously corroded the rock surface, with the subsequent weakening of its crystalline structure.

The spectacular karren assemblages dominated by pinnacles show clear evidence of subsoil dissolution during the first stages of evolution of these karrenfields (Ginés 1995). Such subcutaneous inheritance is evident from the rounded geometry of the pinnacle ridges and the rather smooth appearance of most of the rundkarren runnels. The inherited subsoil features are in many cases obliterated by typical bare karren types (rainpits, kamenitzas, rillenkarren, sharp wandkarren), as described by Smart and Whitaker (1996) and Ginés et al. (2010). It is worth to mention the high density of pinnacles (more than 10 pinnacles per hectare), the substantial proportion of surface area occupied by small pinnacles, as well as by larger residual rocky hills with sharp pointed forms. More than 10 % of the pinnacles are more than 10 m high, and their slopes predominantly show

Fig. 7.7 Straight rillenkarren (solution flutes), a typical feature in the Tramuntana Range. The average width of the flutes is around 1.6 cm (*Photo* A. Ginés)

Fig. 7.8 Vertical rinnenkarren (solution runnels) wider than 20 cm, commonly furrowing the flanks of pinnacles that were formed by subsoil dissolution (*Photo* J. Ginés)

gradients exceeding 45°, which contributes to make these karst landscapes extremely rough and wild.

Caves and shafts are quite abundant in the rugged and deeply entrenched karst areas of Mallorca (Ginés and Ginés 2011), with over one thousand of subterranean cavities inventoried up to date. Most of the caves are less than 1 km in long; only two caverns are more than 300 m deep. The dominant features are vadose shafts, which often include vertical pits more than 100 m deep (Fig. 7.9). Some small shafts located in the high country seem to be genetically related to corrosion favoured by snow accumulation. A small number of sub-horizontal hydrologically active

Fig. 7.9 The impressive Avenc de Femenia shaft, a single pit 120 m deep (*Photo* J. Ginés)

passages (water-table caves) are occasionally found. On the other hand, most of the sub-horizontal caves are composed of presumably old collapse chambers. As a general trend, the main speleogenetic phases must be pre-Quaternary, as the Pliocene endemic vertebrate remains found in some cave sediments indicate (Ginés et al. 2012). It seems that apart from intense speleothem growth, the majority of the caves have undergone limited morphological modifications during the Middle and Late Pleistocene.

7.5 Conclusions

The Tramuntana Range in Mallorca Island is a representative area for the study of karst landforms and processes in a mountainous Mediterranean environment. Jurassic and Lower Miocene limestones, affected by folding and thrusting during the Alpine orogeny, show evidence of intense karstification, including large karst depressions, deep gorges, dolines and extensive karrenfields with a remarkable diversity of solutional micro- and meso-forms. Abundant shafts and caves also characterize the karst landforms of these mountains.

The distribution and morphometry of the different karren assemblages are controlled by the climatic gradients, mainly determined by the elevation, ranging from sea level to 1,445 m. Biokarst processes have a significant contribution in karren development, especially in some specific ecological situations, like the semi-arid peripheral sectors of the range and the north-facing slopes of the highest summits. Some spectacular karren-pinnacle landscapes, characteristic of the central part of the range, have evolved from subsoil karren exhumed by various soil-loss processes.

The Tramuntana Range is an excellent natural laboratory for karst studies and particularly for karren investigation in mid-latitudes, owing to the great variety of environmental conditions, as well as to the significant human impact on the area over the last five millennia.

References

Bär WF (1989) Atlas Internacional del Karst. Hoja 5: Lluc/Sierra Norte (Mallorca). Endins 14–15:27–42

Bär WF, Fuchs F, Nagel G (1986) Lluc/Sierra Norte (Mallorca)—Karst einer mediterranen Insel mit alpidischer Struktur (UIS International Atlas of Karst Phenomena, sheet 5). Zeitschrift für Geomorphologie N.F., Suppl. Bd. 59:27–48 + 1 map

Fiol LA, Fornós JJ, Ginés A (1996) Effects of biokarstic processes on the development of solutional rillenkarren in limestone rocks. Earth Surf Proc Land 21:447–452

Ginés A (1995) Deforestation and karren development in Mallorca, Spain. In: Bárány-Kevei I (ed) Environmental effects on karst terrains. Special issue of Acta Geographica Szegediensis. Homage to László Jakucs, Szeged, pp 25–32

Ginés A (1998) L'exocarst de la serra de Tramuntana de Mallorca. In: Fornós JJ (ed) Aspectes geològics de les Balears. Universitat de les Illes Balears, Palma de Mallorca, pp 361–389

Ginés A (1999) Agriculture, grazing and land use changes at the Serra de Tramuntana karstic mountains. Int J Speleol 28 B (1/4):5–14

Ginés A (2004) Karren. In: Gunn J (ed) Encyclopaedia of caves and karst science. Fitzroy Dearborn, New York, pp 470–473

Ginés A (2009) Karrenfield landscapes and karren landforms. In: Ginés A, Knez M, Slabe T, Dreybrodt W (eds) Karst rock features, Karren sculpturing. Zalozba ZRC, Ljubljana, pp 13–24

Ginés A, Ginés J (1995) Les formes exocàrstiques de l'illa de Mallorca/The exokarstic landforms of Mallorca island. In: Ginés A, Ginés J (eds) El carst i les coves de Mallorca/Karst and caves in Mallorca. Endins, 20/Mon. Soc. Hist. Nat. Balears 3, Palma de Mallorca, pp 59–70

Ginés A, Fiol LA, Pol A, Rosselló JA (1989) Morfologia i vegetació d'un grup de dolines de la Serra de Tramuntana (Mallorca). Endins 14–15:43–52

Ginés A, Ginés J, Miralles PM (2010) Anàlisi morfomètrica del carst de pinacles mediterrani de sa Mitjania (Escorca, Mallorca). Endins 34:109–124

Ginés A, Ginés J, Fornós JJ, Bover P, Gómez-Pujol L, Gràcia F, Merino A, Vicens D (2012) An introduction to the Quaternary of Mallorca. In: Ginés A, Ginés J, Gómez-Pujol L, Onac BP, Fornós JJ (eds) Mallorca: a Mediterranean benchmark for Quaternary studies. Mon. Soc. Hist. Nat. Balears 18, Palma de Mallorca, pp 13–53

Ginés J, Ginés A (2009) Mid-mountain karrenfields at Serra de Tramuntana in Mallorca Island. In: Ginés A, Knez M, Slabe T, Dreybrodt W (eds) Karst rock features, Karren sculpturing. Zalozba ZRC, Ljubljana, pp 375–390

Ginés J, Ginés A (2011) Classificació morfogenètica de les cavitats càrstiques de les Illes Balears. In: Gràcia F, Ginés J, Pons GX, Ginard A, Vicens D (eds) El carst: patrimoni natural de les Illes Balears. Endins, 35/Mon. Soc. Hist. Nat. Balears 17, Palma de Mallorca, pp 85–102

Lozano R (1884) Anotaciones físicas y geológicas de la isla de Mallorca. Excma. Diputación Provincial de Baleares. Imprenta de la Casa de Misericordia. Palma de Mallorca

Lundberg J, Ginés A (2009) Rillenkarren. In: Ginés A, Knez M, Slabe T, Dreybrodt W (eds) Karst rock features, Karren sculpturing. Zalozba ZRC, Ljubljana, pp. 185-210

Rose J, Crabtree K, Cuerda J, Osmaston AH (1978) The Quaternary of Mallorca. Quaternary Research Association, Bedfordshire

Smart PL, Whitaker FF (1996) Development of karren landform assemblages—a case study from Son Marc, Mallorca. In: Fornós JJ, Ginés A (eds) Karren landforms. Universitat de les Illes Balears, Palma de Mallorca, pp 111–122

Atapuerca Karst and its Palaeoanthropological Sites

8

Ana Isabel Ortega, Alfonso Benito-Calvo, Alfredo Pérez-González, Eudald Carbonell, José María Bermúdez de Castro, and Juan Luis Arsuaga

Abstract

The Sierra de Atapuerca caves are located in the southern flank of an anticline formed by Upper Cretaceous limestones and dolomites. These caves are mainly sub-horizontal passages or water table caves recording palaeodrainage from south to north, roughly parallel to the anticline axis. In the south, groundwater recharge is mainly associated with fractures at the contact between Mesozoic carbonates and the overlying Miocene marls, while the discharge area is located to the north, in the headwaters of the Pico River. The passages are arranged in three main levels interconnected by shafts and chambers. These cave levels are perched around +90, +70 and +60 m above the Arlanzón River, coinciding with the relative heights of fluvial terraces. Episodic fluvial downcutting led to the formation of successively lower karst levels and the entrenchment of the upper conduits under vadose conditions. Accessible dry caves were used by fauna and hominids, preserving an exceptional archaeo-palaeontological record spanning from ~1.2 Myr until the end of the Middle Pleistocene. The sites of Elefante, Gran Dolina, Galería and Sima de los Huesos have provided exceptional findings for understanding the first steps of human evolution in Europe. These sites relate to the occupation of the ancient cave entrances and areas inside the cave.

A. I. Ortega (✉) · A. Benito-Calvo · A. Pérez-González · J. M. Bermúdez de Castro
Centro Nacional de Investigación sobre la Evolución Humana (CENIEH), Burgos, Spain
e-mail: anaisabel.ortega60@gmail.com

A. Benito-Calvo
e-mail: alfonso.benito@cenieh.es

A. Pérez-González
e-mail: alfredo.perez@cenieh.es

J. M. Bermúdez de Castro
e-mail: josemaria.bermudezdecastro@cenieh.es

A. I. Ortega
Grupo Espeleológico Edelweiss (GEE), Burgos, Spain

E. Carbonell
Institut Català de Palaeoecologia Humana i Evolució Social, Tarragona, Spain
e-mail: eudald.carbonell@urv.cat

J. L. Arsuaga
Centro UCM-Carlos III de Evolución y Comportamiento Humanos, Madrid, Spain
e-mail: jlarsuaga@isciii.es

F. Gutiérrez and M. Gutiérrez (eds.), *Landscapes and Landforms of Spain*, World Geomorphological Landscapes, DOI: 10.1007/978-94-017-8628-7_8,

Keywords
Karst • Multilevel caves • Fluvial incision • Archaeological site • Pleistocene

8.1 Introduction

The Sierra de Atapuerca, which constitutes the northernmost range of the Iberian Chain, is an inlier of Mesozoic formations located in the north-eastern sector of the Duero Basin, 15-km east of the city of Burgos (North Spain). This range covers 12 km^2 and reaches a maximum elevation of 1,085 m a.s.l. (San Vicente Hill). It contains a multilevel cave system, whose evolution has made possible the preservation of many prehistoric and palaeontological remains. The Pleistocene sites have been excavated systematically since 1978, when Emiliano Aguirre set in motion the Atapuerca Project (Aguirre 2007). Elefante, Gran Dolina, Galería and Sima de los Huesos sites provide crucial clues on the human history of Eurasia from ~1.2 Myr until the end of the Middle Pleistocene. The site was listed as Spanish Cultural Heritage in 1991 and an UNESCO World Heritage Site in 2000.

8.2 Geological and Geomorphological Framework

The Sierra de Atapuerca is located in the north-east sector of the Cenozoic Duero Basin, which is bounded to the north and south-east by the western Pyrenees (Cantabrian Mountains) and the Iberian Chain, respectively, while to the north-east it connects with the Ebro Cenozoic Basin through the Bureba Corridor (Fig. 8.1a). In this area, the Sierra de Atapuerca constitutes positive relief underlain by a NW–SE anticline composed of Mesozoic rocks, mainly Upper Cretaceous limestones and dolostones.

The long-term landscape evolution in this region is dominated by the development of planation surfaces related to erosion–sedimentation/uplift cycles during the Neogene (Benito-Calvo and Pérez-González 2007) and the predominance of the fluvial incision of the current valleys during the Quaternary (Benito-Calvo et al. 2008). In the Sierra de Atapuerca, four planation surfaces were recognised forming a high plateau in the summit of the Sierra (1,080 m). The youngest planation surface develops widely in Upper Miocene sediments of the Duero Basin near the Sierra de Atapuerca, forming the Lower Páramo Plateau. This surface is incised by fluvial valleys (Fig. 8.1b), whose main course, the Arlanzón valley, contains the most complete fluvial record, consisting of 14 fluvial terraces developed from +92 to 97 m above the active channel.

8.3 The Atapuerca Cave System

The caves in the Sierra de Atapuerca have mainly developed in a 40–70-m-thick sequence of Upper Cretaceous limestones and dolomites, next to the contact with Neogene sediments of the Duero Basin. The known caves are located in San Vicente Hill (1,085 m a.s.l.), forming a sequence of sub-horizontal inactive passages perched above the modern course of the Arlanzón River (Ortega et al. 2013). The cave system has 4.7 km of explored passages, including the Cueva Mayor-Cueva del Silo system and the Elefante, Dolina and Galería cave entrances, completely filled by sediments (Figs. 8.2, 8.3). These entrances and their detrital fill were discovered at the beginning of the twentieth century, when a deep trench was excavated in the limestones on the SW flank of the Sierra de Atapuerca for the construction of a railway.

8.3.1 Upper Level

The 1,000-m-long upper level is the largest one (Figs. 8.2, 8.3a) and has an average cross section 10 m wide and 15 m high. This passage in Cueva Mayor exhibits subcircular-shaped roofs at 1,015–1,020 m a.s.l., while some dissolution chimneys can reach 1,025–1,030 m a.s.l. Vadose entrenchment has produced keyhole cross sections, whose base reaches the intermediate level at the end of the conduit.

The Salón del Coro displays evidence of major breakdown processes in the walls and the roofs. This chamber is the largest volume in the system (about 53,000 m^3) and is related to the collapse of the current entrance of Cueva Mayor. The enlargement of this chamber also affects the intermediate and lower levels. Speleothems reach large sizes and a strongly eroded flowstone formed at 1,014–1,010 m a.s.l. marks the position of the former floor in this passage before vadose entrenchment (Fig. 8.4a).

8.3.2 Intermediate Level

The intermediate level consists of a sub-horizontal passage situated at 1,000–1,005 m a.s.l. In Cueva Mayor, this level is 700 m long, 6–10 m wide and 2–4 m high, and is shifted towards the west (Figs. 8.2, 8.3). This passage begins in the south associated with a fracture-controlled conduit at the

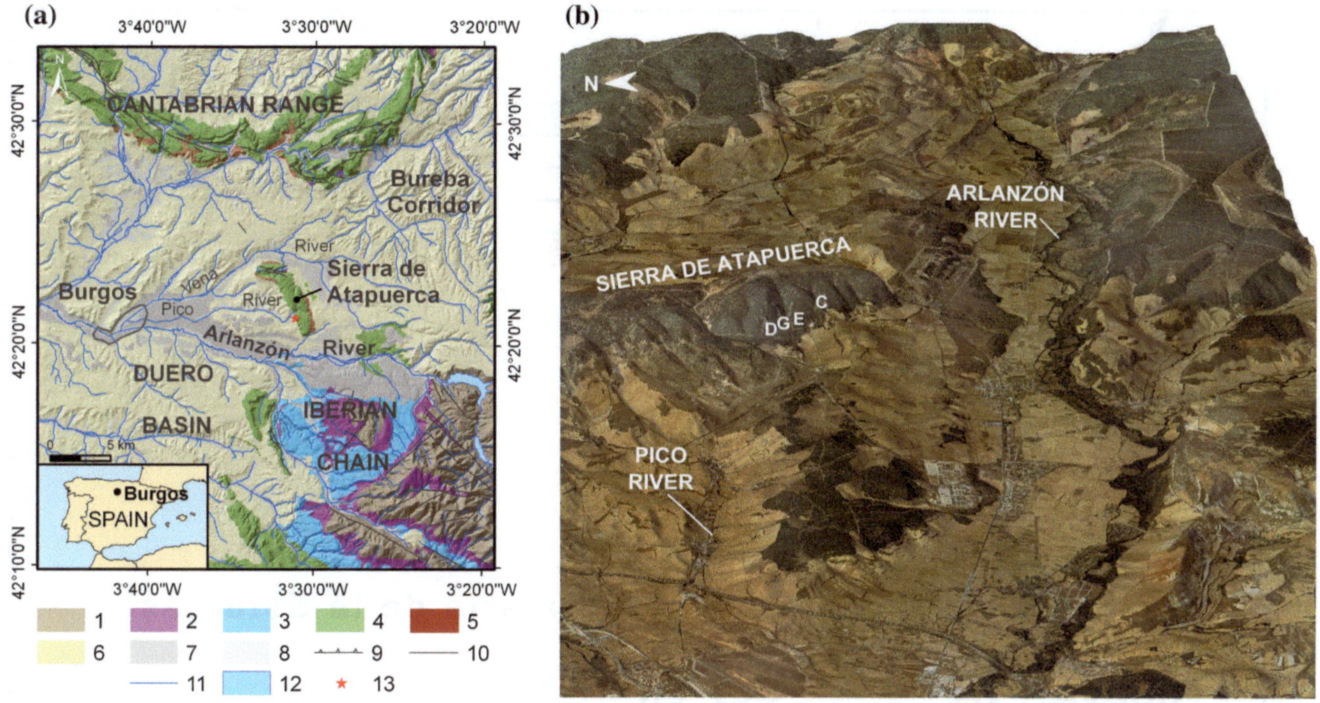

Fig. 8.1 **a** Geographical and geological framework. *Legend 1* Palaeozoic; *2* Triassic; *3* Jurassic; *4* Cretaceous; *5* Palaeogene; *6* Neogene; *7* Pleistocene; *8* Holocene; *9* Thrust; *10* Fault; *11* Drainage network; *12* Reservoir; *13* Sierra de Atapuerca cave system. **b** 3D perspective of the Sierra de Atapuerca and de Arlanzón Valley, showing the location of the archaeo-palaeoanthropological sites. *Legend D* Dolina; *G* Galería; *E* Elefante; *C* Cueva Mayor

contact between Cretaceous carbonates and Miocene marls. The passage continues to the confluence of the Propiedad and Valhondo valleys, where it intersects the ground surface (Elefante Cave). As in the upper level, vadose entrenchment developed at the downstream end of the passage (Fig. 8.4b).

The Trinchera caves belonging to this level are Dolina-Penal, Galería Complex and Elefante (Figs. 8.3b, 8.5a–d), which contain important Early and Middle Pleistocene deposits. Dolina-Penal and Galería Complex are located on the northern slope of Propiedad valley and constitute the entrance sectors of the caves, which were completely filled by sediments. These clastic deposits currently block the entry to the passages, which continue eastward. Dolina-Penal is the entrance sector of a WNW-ESE-oriented passage, with a keyhole section and a subcircular roof at 1,000–1,001 m. The Galería complex corresponds to a large elongated chamber with a phreatic roof above 995 m a.s.l. Roof collapses facilitated the accumulation of allochthonous sediments and the access of humans and animals.

8.3.3 Lower Level

This is the smallest level, and its location also displays a westward shift (Figs. 8.2, 8.3a), near the contact with the Miocene sediments. At the southern and northern ends of this level, there are two mazes: Cueva del Silo and Cueva del Compresor, respectively. Between them, the passage in Cueva Peluda formed at the same level as the principal conduit in Cueva del Silo. The Sima de los Huesos in Cueva Mayor, a 15-m-deep shaft with scallops indicating upward flow, completes this level (Fig. 8.5e). The lower level developed at 985–990 m a.s.l. and is perched at +58–51 m above the Arlanzón River. The lower level includes numerous dissolution chimneys with evidence of rising flow that reached 1,000–1,005 m a.s.l. and 990–995 m a.s.l. in the Silo and Compresor caves (Fig. 8.4c), respectively. Thick fluvial fill consisting of gravels of metamorphic rocks as well as sand and silt facies (Figs. 8.3a, 8.4e), situated at +40–53 m above the Arlanzón River, is preserved inside Cueva del Silo, Cueva Peluda and Sima de los Huesos.

8.4 Archaeo-Palaeoanthropological Sites

The Pleistocene sites in Sierra de Atapuerca record the occupation of ancient cave entrances associated with the intermediate karst level (Elefante, Dolina and Galería Complex) and areas inside the cave, such as the Sima de los

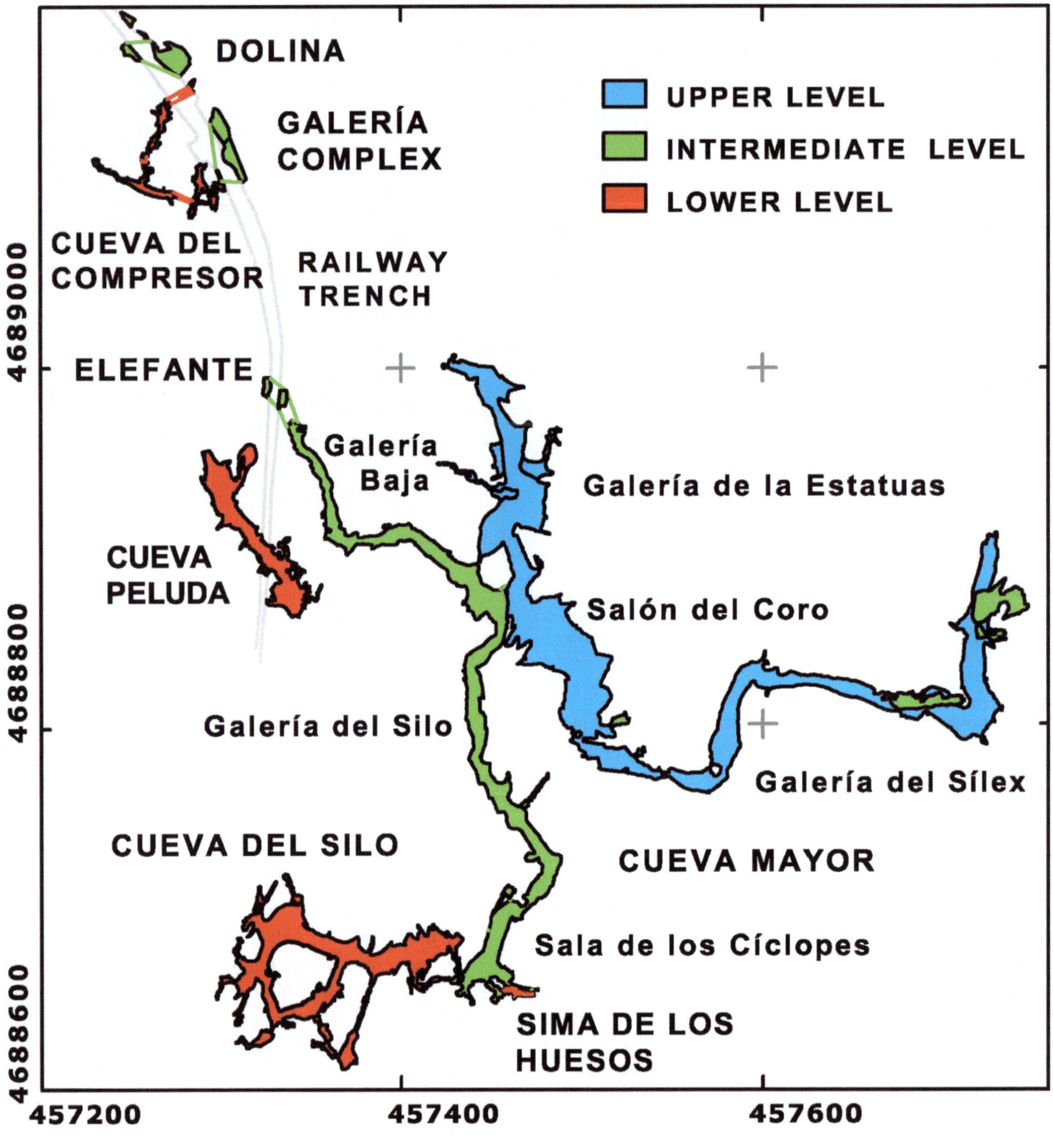

Fig. 8.2 Map of the Atapuerca multilevel cave system (Modified from Ortega et al. 2013)

Huesos. These sites host a very continuous Early and Middle Pleistocene archaeo-palaeoanthropological record, offering the possibility of analysing environmental, faunal, human and technological evolution throughout the last one million years.

8.4.1 Elefante Site

In Elefante site, the cave has a typical keyhole section and displays a stratigraphic sequence 18 m thick comprising 16 lithostratigraphic units (Fig. 8.6a). Facies consist mostly of

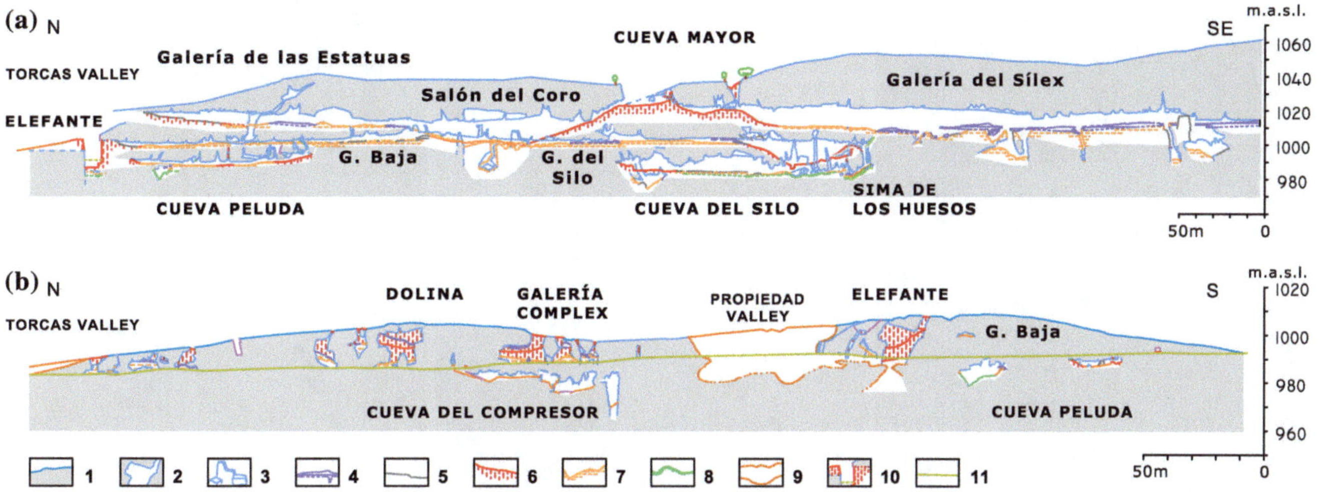

Fig. 8.3 Longitudinal profiles of the caves of the Sierra de Atapuerca. **a** Cueva Mayor—Cueva del Silo System and Cueva Peluda. **b** Caves and passages exposed in the railway trench. *Legend 1* Land surface; *2* Caves; *3* Blocks; *4* Eroded speleothem; *5* Speleothem; *6* Allochthonous facies; *7* Autochthonous facies; *8* Fluvial gravels; *9* Valley; *10* Railway trench; *11* Base of railway trench

debris flow deposits and to a lesser degree of sands, gravels and laminated clays. Lateral facies changes are common, and the lower units are affected by normal faults due to the collapse of the sediments (Rosas et al. 2006). The sequence can be divided into three sedimentary packages. The lower package (TE7–TE16), Early Pleistocene in age according to the cosmogenic burial dating of TE7 (1.13 ± 0.18 Ma) and TE9 (1.22 ± 0.16 Ma), has yielded the oldest European hominid fossils and evidence of the earliest human culture, together with a rich faunal assemblage earlier than the Jaramillo Subchron (Carbonell et al. 2008; Bermúdez de Castro et al. 2011). The intermediate package (TE17–TE19), dated to the Middle Pleistocene, has yielded an abundant assemblage of large mammals and stone tools included in the calcareous breccias of TE18 and TE19 (Cuenca-Bescós et al. 2010; Rodríguez et al. 2011). Finally, the youngest package (TE20 and TE21) records the final fill of the cavity and soil formation processes, possibly in the Upper Pleistocene.

8.4.2 Gran Dolina Site

Dolina Cave formed in the northern side of Propiedad valley and displays a keyhole section filled by a sedimentary sequence 16 m thick, in which interior and exterior facies can be divided into 11 units (Fig. 8.6b). TD1 and TD2 consist of reverse polarity interior facies (silty clays, breakdown facies and eroded speleothems) that display features indicative of a closed cave (Pérez-González et al. 2001). The overlying allochthonous sequence represents an open cave with the input of external sediments, including Pleistocene animal and human remains. This sequence is formed by Units TD3–4 to TD11 and comprises mainly debris gravity flows (mud and clasts), tractive facies, laminated silts and some layers of speleothems. This sequence is important for the hominid fossils found in Unit TD6, which have been proposed as a new species, *Homo antecessor* (ca. 800–900 ka), associated with primitive Mode 1 stone tools, and rich fauna characteristic of the end of the Early Pleistocene (Bermúdez de Castro et al. 1997; Carbonell et al. 1999a; Berger et al. 2008). This record may be attributed to an intensive occupation of the cave with complex behaviours, including the oldest evidence of cannibalism. Palaeomagnetic data indicate that the Matuyama-Brunhes boundary is situated at the top of TD7 (Parés and Pérez-González 1995). Unit TD8 represents the last Early Pleistocene fauna and TD8–9 and TD10 record the occurrence of Middle Pleistocene fauna (Cuenca-Bescós and García 2007). TD10 provides evidence of intense human activity in a new cultural phase of Mode 2 to Mode 3 technology (Ollé et al. 2013), dated from about 430 to 250 ka (Falguères et al. 2001; Berger et al. 2008).

8.4.3 Galería Site

The Galería complex is about 50 m south of Gran Dolina. Its sedimentary fill is formed by five lithostratigraphic units (GI to GV) and a soil developed on Unit GV (Fig. 8.6c). The oldest unit GI consists mainly of interior facies (laminated silts, limestone breccias, as well as reworked and in situ speleothems) with bioturbation and some external influence at the top (Pérez-González et al. 2001), where the Matuyama-Brunhes boundary was detected. The GII to GIV units are allochthonous and are composed of calcareous

Fig. 8.4 Examples of morphological features and sediments of the Atapuerca caves. **a** Incision in the Galería de Sílex (*upper level*) has eroded the deposits accumulated in its floor, leaving a perched flowstone (*left*). **b** The Sala de los Cíclopes (*intermediate level*) shows a typical keyhole section with significant erosion. **c** The solution chimneys of Cueva del Compresor, with scallops indicating upward flow (*lower level*). **d** Partially eroded sediments in the wall of the Sala de los Cíclopes. **e** *Lower* gallery of the Cueva Peluda fluvial deposits (*Photo* M. A. Martín, Edelweiss Speleological Group)

fluvial gravels and debris flow facies grading towards the north into laminated sandy clay-loam. Human presence has been confirmed in Units GII and GIII, between about 500 and 250 ka years ago (Berger et al. 2008; Falguères et al. 2013). A rich assemblage of Mode 2 lithic artefacts is associated with two fossil human remains and an abundant Middle Pleistocene palaeontological record (Carbonell et al. 1999b; Ollé et al. 2013).

8.4.4 Sima de los Huesos Site

The Sima de los Huesos site is located in a shaft inside Cueva Mayor. Its stratigraphic sequence has not been exposed to the outside environment and displays three sedimentary episodes, interrupted by at least one erosional phase (Bischoff et al. 1997, 2007). At the bottom, there are reworked marls overlain by sand and clay units with reverse polarity (Parés et al. 2000). At the top, two Middle Pleistocene fossiliferous units develop in clay breccias, capped by flowstone and a guano layer (Fig. 8.6d). The fossiliferous units contain the skeletal remains of hundreds of bears (*Ursus deningeri*), some carnivores, and the most representative sample of fossil hominids for the Middle Pleistocene worldwide, with remains of at least 28 *Homo heidelbergensis* individuals (Arsuaga et al. 1997; García 2003; Martinón-Torres et al. 2012). In addition, a single Mode 2 hand-axe is proposed to be evidence of symbolic rituals (Carbonell and Mosquera 2006), distant from the site of human occupation.

Fig. 8.5 Sediment-filled passages of the intermediate level exposed in the railway trench. **a** Aerial view of the railway trench. **b** Elefante site. **c** Gran Dolina site. **d** Galería site. **e** Shaft-chimney of Sima de los Huesos, with scallops indicating rising flows (*Photo* M. A. Martín, Edelweiss Speleological Group)

8.5 Speleogenesis and Human Occupation

The Atapuerca cave system exhibits three sub-horizontal levels of inactive passages. These conduits record prolonged periods of base-level stability. In addition, the palaeocurrents inferred from the scallops indicate a flow from the south towards the north in the three levels. In the south, fracture-controlled chimneys with scallops are indicative of rising flows, along which maze cave systems formed. These features suggest rising groundwater inputs in the southern sector, near the Arlanzón valley, especially for the lower and intermediate levels. The main portion of the passages was formed by groundwater flowing towards the north, and their sub-horizontal attitude was controlled by the water table. These underground channels intersect the ground surface at the headwaters of the Pico valley, where resurgences or springs were active during the formation of each cave level (Fig. 8.1b).

The upper level is situated at +80–90 m above the present-day base level, coinciding with the Lower Páramo Plateau (1,020–1,025 m a.s.l) and terrace T2 of the Arlanzón River (+80–88 m). This indicates long-sustained base-level stability of the water table over a long time span, favouring the formation and enlargement of sub-horizontal passages. As a consequence of water table lowering, an extensive flowstone formed under vadose conditions on the floor of the passages in the upper level. The base-level drop recorded by the development of terrace T3 of the Arlanzón River (+70–78 m) led to the formation of passages with keyhole cross sections in the distal sectors of the upper-level passages. In the Salón del Coro, this erosion process favoured the collapse of the roof and the walls (Ortega et al. 2013).

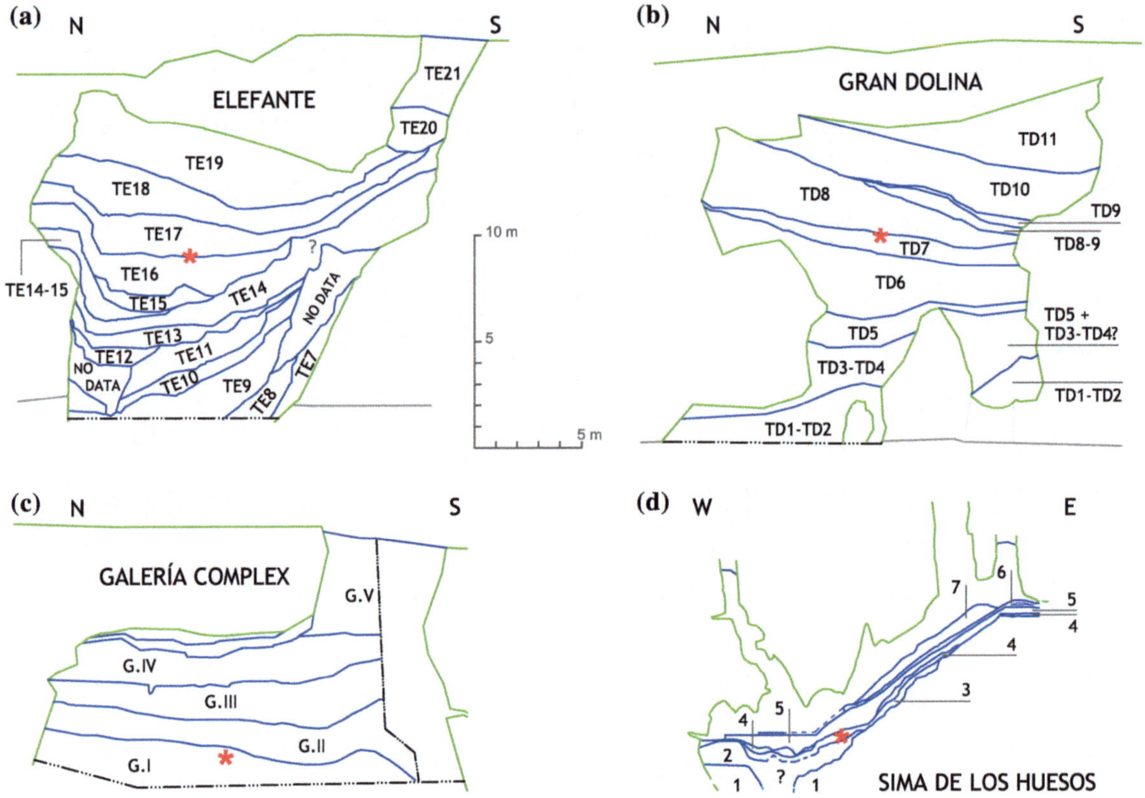

Fig. 8.6 Stratigraphic sequences of the Pleistocene Atapuerca sites. **a** Elefante site. **b** Gran Dolina site (Modified from Rodríguez et al. 2011). **c** Galería site (Modified from Ortega 2009). **e** Sima de los Huesos. *1* Reworked marls; *2* Sands and silts; *3* Red clay; *4* and *5* Fossiliferous clay breccias; *6* Speleothem; *7* Mud brecccia with bat guano (Modified from Bischoff et al. 2007). The *asterisk* marks the position of the Matuyama-Brunhes boundary

The intermediate level has been related to the period of base-level stability corresponding to the Arlanzón River terrace T3 (+70–78 m), which developed at about 1.14 ± 0.13 Ma (ESR date, Moreno et al. 2012). This drop in base level caused a change to vadose conditions in the second level, favouring the entry of surface sediments in their distal sectors (Cíclopes, Elefante and Dolina) (Ortega et al. 2013). Cíclopes Chamber, corresponding to this phase, has a sedimentary fill consisting of breccias and fluvial sands and silts with reverse polarity attributed to the Matuyama Chron (Parés et al. 2010; Fig. 8.4d). In addition, the lowest units in the Elefante fill (TE9, TE7) yielded a ca. 1.2 Ma age (Carbonell et al. 2008) and represents the first entry of hominids, animals and sediments from the surface. New occupations ascribed to *H. antecessor* and the associated sedimentary phases took place at the end of the Early Pleistocene (Carbonell et al. 2001; Ollé et al. 2013). The infill of Dolina and Elefante continued with the massive entry of debris flows and alluvial sediments, recording the Brunhes-Matuyama reversal in Dolina (TD7) and Elefante (TE16–TE17) (Parés and Pérez-González 1995; Carbonell et al. 2008). The period of base-level stability recorded by terrace T4 (+60–67 m) led to the formation of the Galería Complex.

The roofs of the lower level largely coincide with terrace T5, which would have controlled the development of this level (Sima de los Huesos, Cueva del Silo, Cueva Peluda and Cueva del Compresor), in which fluvial sediments incorporated by the Arlanzón fluvial system have been described (Ortega 2009). This input is associated with the enlargement of the lower-level conduits, causing the subsidence and deformation of the Lower Pleistocene Elefante sedimentary units (TE7 to TE17). This suggests a Middle Pleistocene age for the fluvial sediments in Cueva del Silo, which can be altitudinally correlated with terrace T6 (+44–46 m).

The youngest water table-controlled cave level is related to terrace T7 (+38–40 m), which produced a small cave sublevel represented by the lower passages in the Silo, Compresor and Peluda caves (978 m a.s.l.) and caused the incision of the fluvial deposits accumulated inside Cueva del Silo-Peluda. With the base-level drop recorded by T8 terrace (+26–35 m), the multilevel cave system became totally disconnected from and perched above the Arlanzón River network.

A third cultural phase corresponds to the Middle Pleistocene, when *H. heidelbergensis* used the cave entrances of the intermediate level at Elefante (TE19), Dolina (TD10), Galería Complex (TGII–III) and Sima de los Huesos inside Cueva Mayor (Ollé et al. 2013). All this evidence reveals the significance of the human occupation. Stratigraphic and biostratigraphic studies have shown that entrance sectors in the intermediate level were filled to the roof in the late Middle Pleistocene (Carbonell et al. 2001; Cuenca-Bescós et al. 2010; Rodríguez et al. 2011) when the Sima de los Huesos became isolated from the outside environment (Bischoff et al. 1997; García and Arsuaga 2011). In the Upper Pleistocene, the karst of San Vicente Hill became inactive and was completely perched, with only minimal animal and human activity being documented. A fourth phase of human occupation took place in recent prehistoric times, when all the open caves in San Vicente Hill were used for diverse purposes (Ortega 2009).

8.6 Conclusions

The Atapuerca cave system consists of a series of subhorizontal passages (water table caves) with drainage directions from south to north, oriented roughly parallel to the strike of the folded carbonate formations. The passages intersect the surface in the northern sector, where resurgences or springs were active at the head of the Pico River during the Pleistocene. The passages are arranged in three main levels, connected to each other by shafts and chambers, and perched around +90, +70 and +60 m above the modern Arlanzón River. These water table caves can be correlated altitudinally with different terrace levels. The different passages show a progressive westward shift and correlate with the episodic downcutting of the regional base level. Throughout their evolution, the caves have changed progressively into vadose conditions, allowing the entrance of fauna and hominids, whose remains have been preserved in Early and Middle Pleistocene sedimentary sequences filling the cave entrances, such as the Trinchera sites (Dolina, Galería and Elefante), or in the interior of the caves, like the Cueva Mayor sites (Sima de los Huesos).

Acknowledgments This work was supported by the DGICYT projects CGL 2009-12703-C03-01, CGL2012-38434-C03-02 and CGL2010-21499, as well as by the Junta de Castilla y León. The authors wish to thank the multidisciplinary Research Team of the Sierra de Atapuerca (EIA) and the Grupo Espeleológico Edelweiss for constant scientific and logistic support. We thank James Bischoff for comments on this paper. Peter Smith revised the English in the final draft of the manuscript.

References

Aguirre E (2007) Atapuerca (Burgos, Northern Spain)—potential, progress and questions. Cour Forsch-Inst Senckenberg 259:11–120

Arsuaga JL, Bermúdez de Castro JM, Carbonell E (eds) (1997) The Sima de los Huesos Hominid Site. J Hum Evol 33:105–421

Benito-Calvo A, Pérez-González A (2007) Erosion surfaces and Neogene landscape evolution in the NE Duero Basin (north-central Spain). Geomorphology 88:226–241

Benito-Calvo A, Pérez-González A, Parés JM (2008) Quantitative reconstruction of Late Cenozoic landscape: a case study in the Sierra de Atapuerca (Burgos, Spain). Earth Surf Proc Land 33:196–208

Berger GW, Pérez-González A, Carbonell E, Arsuaga JL, Bermúdez de Castro JM, Ku TL (2008) Luminiscence chronology of cave sediments at the Atapuerca paleoanthropological site, Spain. J Hum Evo 55:300–311

Bermúdez de Castro JM, Arsuaga JL, Carbonell E, Rosas A, Martínez I, Mosquera M (1997) A hominid from the Lower Pleistocene of Atapuerca, Spain: possible ancestor to Neandertals and modern human. Science 276:1392–1395

Bermúdez de Castro JM, Martinón-Torres M, Gómez-Robles A, Prado-Simón L, Sarmiento S, Martín-Francés L, Olejniczak A, Carbonell E (2011) Early Pleistocene human mandible from Sima del Elefante (TE) cave site in Sierra de Atapuerca (Spain): a comparative morphological study. J Hum Evol 61:12–25

Bischoff JL, Fiztpatrick JA, León L, Arsuaga JL, Falguères C, Bahain JJ, Bullen T (1997) Geology and preliminary dating of the hominid-bearing sedimentary fill of the Sima de los Huesos Chamber, Cueva Mayor of the Sierra de Atapuerca, Burgos, Spain. Spain. J Hum Evol 33:129–154

Bischoff JL, Wiliams RW, Rosenbauer RJ, Aramburu A, Arsuaga JL, García N, Cuenca-Bescós G (2007) High-resolution U-series dates from the Sima de los Huesos hominids yields 600+ infinito/−66 kyrs: implications for the evolution of the early Neanderthal lineage. J Archaeol Sci 34:763–770

Carbonell E, Mosquera M (2006) The emergence of symbolic behavior: the sepulchral pit of Sima de los Huesos, Sierra de Atapuerca, Burgos, Spain. Comptes Rendus Paleovol 5:155–160

Carbonell E, Bermúdez de Castro JM, Arsuaga JL (eds) (1999a) Gran Dolina Site: TD6 Aurora Stratum (Burgos, Spain). J Hum Evol 37:309–700

Carbonell E, Rosas A, Díez JC (eds) (1999b) Atapuerca: Ocupaciones humanas y paleoecología del yacimiento de Galería. Arqueología en Castilla y León 7:1–390

Carbonell E, Arsuaga JL, Bermúdez de Castro JM (eds) (2001) Atapuerca. L'anthropologie 105:1–312

Carbonell E, Bermúndez de Castro JM, Parés JM, Pérez-González A, Cuenca-Bescós G, Ollé A, Mosquera M, Huguet R, van der Made J, Rosas A, Sala R, Vallverdú J, García N, Granger DE, Martinón-Torres M, Rodríguez XP, Stock GM, Vergès JM, Allue E, Burjachs F, Cáceres I, Canals A, Benito A, Díez C, Lozano M, Mateos A, Navazo M, Rodríguez J, Rosell J, Asuaga JL (2008) The first hominin of Europe. Nature 425:465–470

Cuenca-Bescós G, García N (2007) Bioestratigraphic succession on the Early and Middle Pleistocene mammals faunas of the Atapuerca cave site (Burgos, Spain). Courier Forschung-Institut Senckenberg 259:99–110

Cuenca-Bescós G, Rofes J, López-García JM, Blain H-A, de Marfá RJ, Galindo-Pellicena MA, Bennásar-Serra LM, Melero-Rubio JM, Arsuaga JL, Carbonell E, Bermúdez de Castro JM (2010)

Biochronology of Spanish Quaternary small vertebrate faunas. Quat Int 212:109–119
Falguères C, Bahain JJ, Yokohama Y, Bischoff JL, Arsuaga JL, Bermúdez de Castro JM, Carbonell E, Dolo JM (2001) Datation par RPE et U-Th des sites pléistocènes d'Atapuerca: Sima de los Huesos, Trinchera Dolina et Trinchera Galería. Bilan géochronologique. L'anthropologie 105:71–81
Falguères C, Bahain JJ, Bischoff JL, Pérez-González A, Ortega AI, Ollé A, Quiles A, Ghaleb B, Moreno D, Dolo JM, Shao Q, Vallverdú J, Carbonell E, Bermúdez de Castro JM, Arsuaga JL (2013) Combined ESR/U-Series chronology of Acheulian Hominid-bearing layers at Trinchera Galería site, Atapuerca, Spain. J Hum Evol 65(2):168–184
García N (2003) Osos y otros carnívoros de la Sierra de Atapuerca. Fundación Osos de Asturias, Spain
García N, Arsuaga JL (2011) The Sima de los Huesos (Burgos, northern Spain): palaeoenviroment and hábitats of homo heidelbergendis during the Middle Pleistocene. Quat Sci Rev 30:1413–1419
Martinón-Torres M, Bermúdez de Castro JM, Gómez-Robles A, Pardo-Simón L, Arsuaga JL (2012) Morphological description and comparison of the dental remains from Atapuerca-Sima de los Huesos site (Spain). J Hum Evol 62:7–58
Moreno D, Falguères C, Pérez-González A, Duval M, Voinchet P, Benito-Calvo A, Ortega AI, Bahain JJ, Sala R, Carbonell E, Bermúdez de Castro JM, Arsuaga JL (2012) ESR chronology of alluvial deposits in the Arlanzón valley (Atapuerca, Spain): contemporaneity with Atapuerca Gran Dolina site. Quat Geochr 10:418–423
Ollé A, Mosquera M, Rodríguez XP, Lombera-Hermida A, García-Antón MD, García-Medrano P, Peña L, Menéndez L. Navazo M, Terradillos M, Bargalló A, Márquez B, Sala R, Carbonell E (2013) The Early and Middle Pleistocene technological record from Sierra de Atapuerca (Burgos, Spain). Quat Inter 295:138–167
Ortega AI (2009) Evolución geomorfológica del Karst de la Sierra de Atapuerca (Burgos) y su relación con los yacimientos pleistocenos que contiene. Ph.D. Thesis, University of Burgos, Burgos, Spain, 627 pp
Ortega AI, Benito-Calvo A, Pérez-González A, Martín-Merino MA, Pérez-Martínez R, Parés JM, Aramburu A, Arsuaga JL, Bermúdez de Castro JM, Carbonell E (2013) Evolution of multilevel caves in the Sierra de Atapuerca (Burgos, Spain) and its relation to human occupation. Geomorphology 196:122–137
Parés JM, Pérez-González A (1995) Paleomagnetic ages for hominids at Atapuerca Archaeological site, Spain. Science 269:830–832
Parés JM, Pérez-González A, Weil AB, Arsuaga JL (2000) On the age of the Hominid Fossil at the Sima de los Huesos, Sierra de Atapuerca, Spain: Paleomagnetic Evidence. Am J Phys Anthropol 111:451–461
Parés JM, Pérez-González A, Arsuaga JL, Bermúdez de Castro JM, Carbonell E, Ortega AI (2010) Characterizing sedimentary history of cave deposits, using archaeomagnetism and rockmagnetism, Atapuerca (N Spain). Archeometry 52:882–898
Pérez-González A, Parés JM, Carbonell E, Aleixandre T, Ortega AI, Benito A, Martín MA (2001) Géologie de la Sierra de Atapuerca et stratigraphie des remplissages karstiques de Galería et Dolina (Burgos Espagne). L'Anthopologie 105:27–44
Rodríguez J, Burjachs F, Cuenca-Bescós G, García N, van der Made J, Rosas A, Pérez-González A, Blain HA, Expósito JM, López-García JM, García-Antón M, Allué E, Cáceres I, Huguet R, Mosquera M, Ollé A, Rosell J, Parés JM, Rodríguez XP, Díez C, Rufes J, Sala R, Saladié P, Vallverdú J, Bennasar ML, Blasco R, Bermúndez de Castro JM, Carbonell E (2011) One million years of cultural evolution in a stable environment at Atapuerca (Burgos, Spain). Quat Sci Rev 30:1396–1412
Rosas A, Huguet R, Pérez-González A, Carbonell E, Bermúdez de Castro JM, Valverde J, van der Made J, Allué E, García N, Martínez-Pérez R, Rodríguez J, Sala R, Saladie P, Benito A, Martínez-Maza C, Bastir M, Sánchez A, Parés JM (2006) The "Sima del Elefante" cave site at Atapuerca (Spain). Estudios Geológicos 62:327–348

Evaporite Karst in Calatayud, Iberian Chain

9

Francisco Gutiérrez

Abstract

The Cenozoic sedimentary fill of Calatayud Graben, an intramontane basin within the Iberian Chain, includes an evaporitic sequence around 500 m thick with significant halite and glauberite units in the subsurface. Interstratal dissolution of the salt-bearing evaporites has generated megacollapse structures covering up to 12 km^2 in which Neogene sediments have subsided as much as 200 m. The Quaternary alluvium related to the present-day fluvial systems shows sharp changes in thickness locally reaching more than 100 m. The thickened terrace deposits fill basins several kilometres long generated by dissolution-induced synsedimentary subsidence. The area offers the opportunity to examine excellent exposures of subsidence structures and paleosinkholes that illustrate the mechanisms involved in sinkhole development (sagging, collapse, and suffosion). Dissolution and hydrocompaction subsidence has caused extensive structural damage in Calatayud city, including outstanding historical buildings. On November 2003, a collapse sinkhole undermined the foundation of a five-storey building, leading to its demolition, involving around 5 million euro of direct losses.

Keywords

Subsidence • Sinkhole hazard • Salt karst • Non-tectonic deformation • Heritage loss

9.1 Introduction to the Evaporite Karst in Spain

Evaporite outcrops in Spain mostly occur in Alpine orogens and Cenozoic basins, covering more than 35,000 km^2, approximately 7 % of the total country area (Fig. 9.1). These figures explain the great diversity of evaporite karst features and the significant impact of the environmental problems related to evaporite dissolution in "Alpine Spain" (Gutiérrez et al. 2001, 2008a). Marine evaporite sedimentation covers a wide time span, from the Triassic to the present day, whereas the majority of the continental evaporites were deposited in lake environments during Paleogene and Neogene times. Most of the evaporitic formations are made up of Ca sulphate (gypsum and anhydrite) or Ca sulphate and halite. Some marine and continental formations also include K–Mg chlorides and Na sulphates (glauberite and thenardite), respectively. Geochemical and isotopic studies demonstrate that the Tertiary lacustrine evaporites were derived from the recycling (dissolution and reprecipitation) of Mesozoic marine formations (Utrilla et al. 1992). Gypsum, with an equilibrium solubility of 2.4 g/l, is commonly the only evaporitic mineral exposed at the surface. The highly soluble chloride salts and Na sulphates rarely crop out; the equilibrium solubilities of halite and glauberite ($Na_2Ca[SO_4]_2$) in distilled water are 360 and 118 g/l, respectively. However,

F. Gutiérrez (✉)
Department of Earth Sciences, University of Zaragoza, Zaragoza, Spain
e-mail: fgutier@unizar.es

F. Gutiérrez and M. Gutiérrez (eds.), *Landscapes and Landforms of Spain*, World Geomorphological Landscapes,
DOI: 10.1007/978-94-017-8628-7_9,

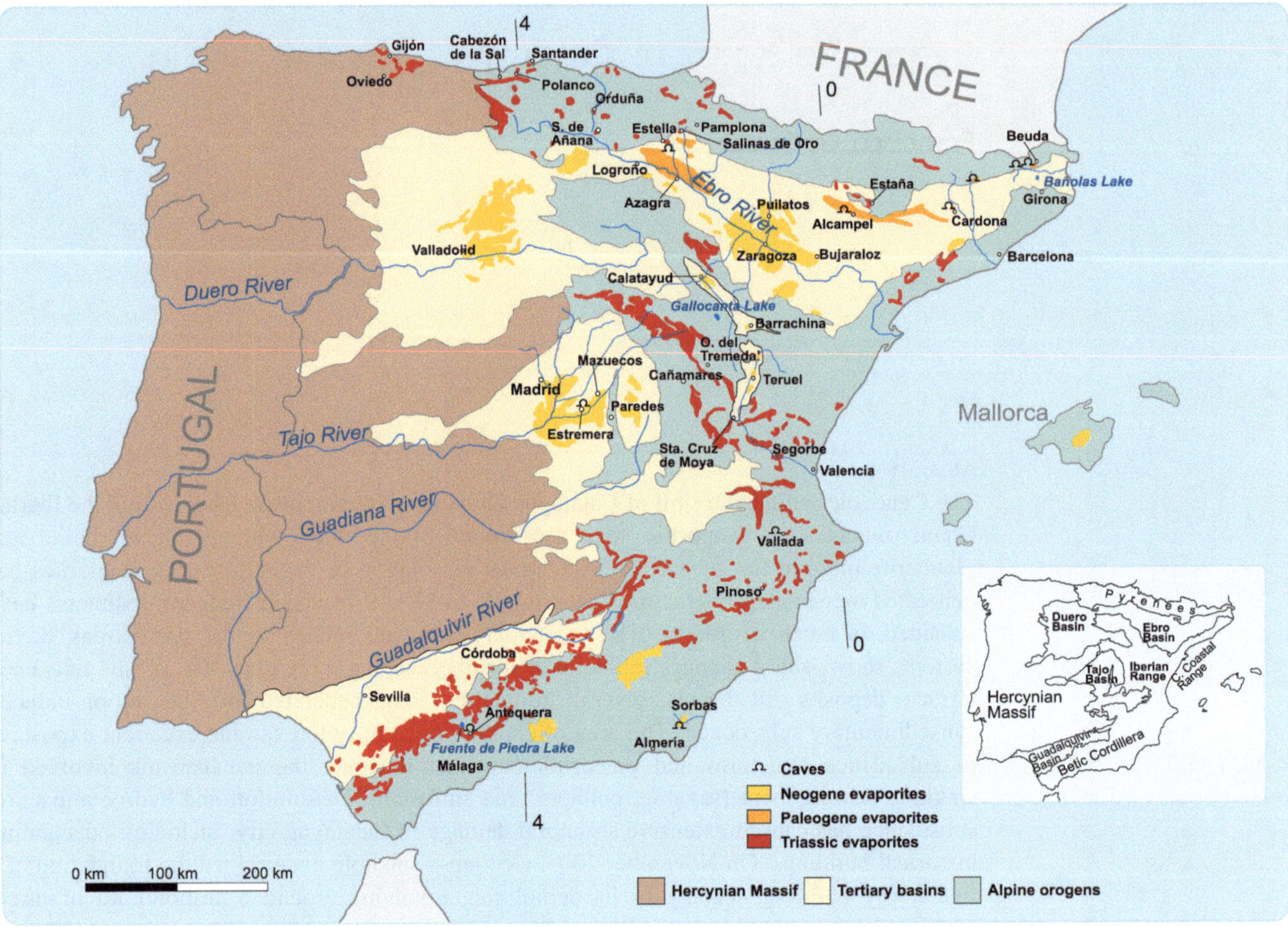

Fig. 9.1 Distribution of the main evaporite outcrops in Spain and local names of the most relevant evaporite karst areas (adapted from Gutiérrez et al. 2008a)

their dissolution by groundwater at depth, frequently unnoticed, has played an instrumental role in the development of karstic phenomena in numerous areas.

The most widespread episode of evaporitic sedimentation is recorded by Triassic shallow-marine successions up to several 100 m thick, consisting of Ca sulphate and halite units embedded in variegated marls and mudstones, mainly the Keuper facies. This formation occurs in numerous outcrops across the Alpine orogens, locally forming diapiric structures (Fig. 9.1). Evaporite karst landforms in these Triassic terrains are not frequent, most probably because the remaining salts are situated at significant depth, associated with impervious units and capped by thick karstic residues developed over long periods of time. The presence of various types of subsidence sinkholes is rather common in halokinetic structures, where groundwater flows interact with uprising salts at shallow depth: Antequera region, western Betics (Calaforra and Pulido-Bosch 1999); Pinoso Diapir, eastern Betics; Salinas de Oro, Estella, Orduña, and Polanco diapirs in the western Pyrenees. The 210-m-deep El Sumidor Cave, developed within a halokinetic structure in Vallada, eastern Betics, is one of the deepest gypsum caves in the world (Calaforra and Pulido-Bosch 1996). A relatively frequent feature in the areas underlain by the Triassic Keuper facies is the occurrence of permanent and ephemeral lakes in enclosed basins associated with either collapse sinkholes or broad subsidence depressions, including some of the largest lakes in Spain: Fuente de Piedra Lake (13.6 km^2) and the numerous lakes of Antequera area, western Betics; Gallocanta Lake (14.5 km^2), Iberian Chain (Gracia et al. 2002; Chap. 11); Estaña Lakes (López-Vicente et al. 2009), or Montcortés Lake, Pyrenees (Gutiérrez et al. 2012c). Moreover, interstratal karstification of Triassic evaporites has resulted in or contributed to the development of striking gravitational morpho-structures, including prominent fault scarps (Zenzano fault, Iberian Chain; Carbonel et al. 2013), monoclinal flexures with crestal grabens (Río Seco monocline, Iberian Chain; Gutiérrez et al. 2012a), and graben systems (grabens of Peracals, Pyrenees; Gutiérrez et al. 2012c).

The most important Tertiary marine evaporite formations from the karst perspective are (1) the halite-bearing Middle Eocene Beuda Gypsum, tectonically incorporated in the eastern sector Pyrenean orogenic wedge; (2) the Cardona Saline Formation, dominated by halite with a substantial amount of K–Mg chlorides, mainly sylvite and carnallite. This formation is exposed in the outstanding Cardona Diapir, north-eastern sector of the Ebro Basin, constituting the only significant salt outcrop (0.9 km^2) in Western Europe; and (3) the Late Messinian gypsum of the intramontane Sorbas Basin, in the eastern Betics. The Banyoles Lake is related to a cluster of coalescent collapse sinkholes generated by hypogenic dissolution of the Beuda Gypsum in a groundwater discharge zone (Canals et al. 1990). Structurally controlled sinkholes are relatively common in the alluvial surfaces developed over the Beuda Gypsum (e.g. Fluvià Valley, Sant Miquel de Camp Major), and subsidence has caused significant damage in houses of Besalú and Beuda villages. The Cardona diapir has a well-developed cave system, some of them formed in historical times in relation to anthropogenic changes induced by mining operations. The 680-m-long Forat Mico Cave is one of the longest salt caves in the world. The 335-m-long Riera Salada Cave was carved in the halite debris of a slag heap accumulated between 1925 and 1972. In March 1998, the interception of a phreatic conduit by a shallow mine led to an inrush of fresh water from the Cardener River into the underground excavations, resulting in the generation of the 4,300-m-long Salt Meanders Cave, and the development of a large number of damaging sinkholes, with the consequent abandonment of the mine (Cardona and Viver 2002; Lucha et al. 2008). The exposed salt rock, as well as the halite-rich slag heaps, displays magnificent and rapidly evolving karren and pedestals (Mottershead et al. 2008). The remarkable karst landforms developed in the Sorbas Basin are illustrated in Chap. 10.

The most significant karst features developed in Tertiary continental evaporites occur in the central sector of the Ebro Basin (Zaragoza area), Madrid Basin, and Calatayud Graben within the Iberian Chain (Gutiérrez et al. 2008a). A common characteristic of these evaporites is the presence of thick halite and glauberite units in the subsurface. The topic that has received wider attention in the Tertiary basins is the development of dissolution-induced subsidence phenomena, including sinkholes and the associated hazards. The sectors where the evaporitic bedrock is overlain by Quaternary alluvium are the most prone to subsidence. This alluvial karst occurs in reaches of the main Spanish fluvial systems where they traverse evaporitic outcrops (Fig. 9.1) and are particularly intense where the bedrock includes highly soluble salts (e.g. halite, glauberite). Commonly, in these areas, the alluvial deposits are locally thickened as much as 100 m. The thickened alluvium fills complex solution basins up to several tens of kilometres long generated by synsedimentary subsidence. Guerrero et al. (2008, 2013) present updated reviews of the alluvial systems affected by evaporite karst subsidence. Another common feature is the presence of abundant gravitational deformations that may affect the alluvial mantle (rockhead karstification), or both the cover and the bedrock (interstratal karstification). The latter is particularly common in areas where the evaporitic sequence bears halite and glauberite units in the subsurface (Guerrero et al. 2008, 2013). Interstratal dissolution of salts may also generate perched kilometre-sized enclosed depression at the valley margins that behave as starved basins (Guerrero et al. 2013). These deformation structures, together with dissolutional features found in paleokarst exposures, are the best source of information to understand the subsidence processes involved in the generation of sinkholes: sagging, suffosion, and collapse (Gutiérrez et al. 2008b). The current activity of dissolution and subsidence processes, frequently induced or accelerated by human activities, results in the formation and reactivation of sinkholes, which may constitute a geohazard of great socio-economic impact. Generally, sinkholes show a higher probability of occurrence in the lower alluvial levels, coinciding with areas where development and human activity tend to concentrate, resulting in high risk situations. The most severely affected areas include Zaragoza city and its surroundings (e.g. Galve et al. 2009; Gutiérrez et al. 2009a), Calatayud city (Gutiérrez and Cooper 2002; Gutiérrez et al. 2004), the south-eastern sector of Madrid metropolitan area, and Oviedo city. As an example, in Oviedo, dewatering for the construction of a car park induced the reactivation of an old sinkhole in 1998, resulting in the demolition of 362 apartments and direct losses estimated at 18 million euro (Pando et al. 2013).

This chapter illustrates the geological, geomorphological, and environmental implications of evaporite karst in Calatayud Graben. This is the Tertiary basin in Spain in which dissolution-induced subsidence phenomena display a wider diversity and has produced some of the most dramatic features. Here, long-sustained evaporite dissolution has generated large collapse structures in Neogene sediments, thickenings and gravitational deformation in Quaternary alluvium, and costly subsidence damage in the historical city of Calatayud.

9.2 Geological Setting

Calatayud Graben is a Cenozoic basin located within the Iberian Chain, an intraplate Alpine orogen generated by the tectonic inversion of Mesozoic extensional basins from Paleogene to Early-Middle Miocene times (orogenic stage). During the subsequent postorogenic stage, extensional tectonics has produced grabens superimposed on the previous contractional structures. A first rifting episode, which started

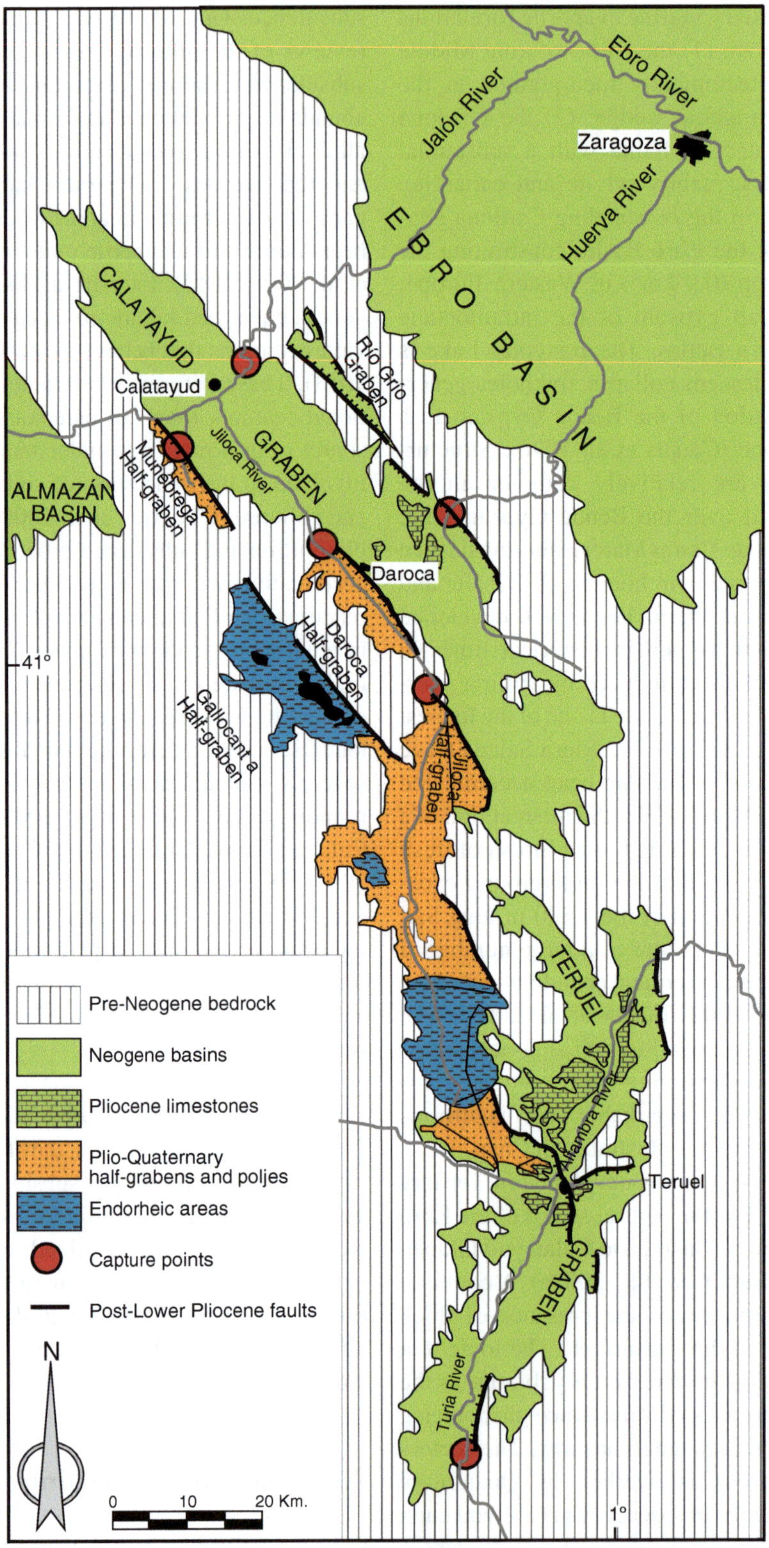

Fig. 9.2 Sketch showing the distribution of Neogene and Plio-Quaternary grabens in the central sector of the Iberian Chain

in the Lower-Middle Miocene, generated the two largest intramontane basins of the central sector of the Iberian Chain, Calatayud Graben, and Teruel Graben. Since the Late Pliocene, the second extensional episode produced new grabens locally superimposed or inset with respect to the previous basins (Gutiérrez et al. 2008c, 2012b) (Fig. 9.2).

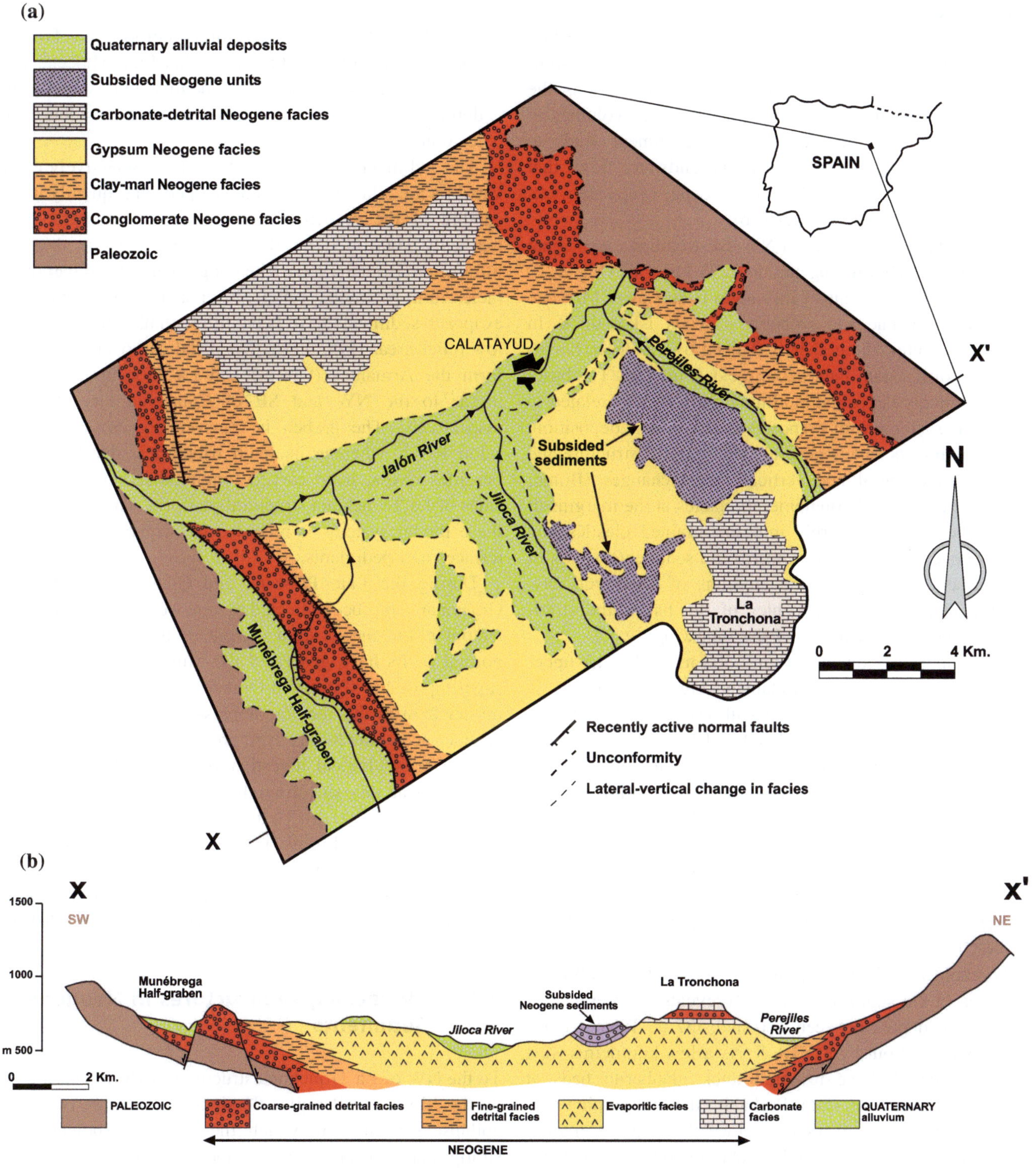

Fig. 9.3 a Geological sketch of a portion of Calatayud Graben in Calatayud city area, showing the facies distribution of the Cenozoic infill and the Plio-Quaternary neotectonic extensional structures developed in the south-western margin of the basin. **b** Geological cross-section across Calatayud Graben (see trace in **a**)

The change in the Neogene and Plio-Quaternary structural depressions from endorheic-aggradational to exorheic-incisional conditions has taken place through the progressive capture of the basins by the external drainage network (Fig. 9.2). Once each basin was captured, the new intrabasinal drainage incised the endorheic infill and expanded by headward erosion, developing stepped sequences of alluvial levels (pediments and terraces). This change has played an instrumental role in the development of evaporite dissolution subsidence phenomena. Under exorheic conditions, the new fluvial systems are able to evacuate large volumes of evaporites in solution, and their entrenchment leads to higher hydraulic gradients for the groundwater flows that may circulate through progressively deeper evaporite rocks (e.g. Gutiérrez et al. 2008c, 2012b).

The Calatayud Graben is a NW–SE-trending extensional basin around 110 km long and up to 25 km wide (Fig. 9.2). The age of the sedimentary fill ranges from Lower Miocene to Lower Pliocene and probably exceeds 1 km thick. In Calatayud sector, the graben is flanked by mountain ranges made up of resistant Palaeozoic rocks (Fig. 9.3). The basin fill was mainly deposited in alluvial fans distally related to lacustrine systems with evaporite and carbonate deposition. These sediments have a general subhorizontal structure and show abrupt lateral and vertical facies changes (Bomer 1960). The proximal conglomeratic facies at the margins of the graben change sharply into fine-grained clastics and evaporite–carbonate facies towards the sedimentary axis of the graben (Fig. 9.3). In the environs of Calatayud city, located close to the depocenter of the basin, the stratigraphic succession consists of the following:

- An evaporitic sequence around 500 m thick. The upper 200 m that crop out in Calatayud area is composed of laminar and nodular gypsum with thin interbedded marl partings. The gypsum ($CaSO_4{\cdot}2H_2O$) is a secondary diagenetic facies resulting from the hydration of anhydrite ($CaSO_4$) and the incongruent dissolution of glauberite (Ortí and Rosell 1998, 2000). According to Collantes and Griffo (1982), in the surroundings of Calatayud, gypsum constitutes around 85 % of the exposed evaporitic sequence. In addition to gypsum, borehole data indicate a significant proportion of halite (NaCl) and glauberite ($Na_2Ca[SO_4]_2$) at depth. The borehole described by Marín (1932) in Paracuellos de Jiloca (5 km south of Calatayud) reveals the presence of halite at depths between 170 and 537 m. Recent boreholes carried out by MYTA S. A., a few kilometres west of Calatayud, have revealed thick glauberite beds at depths higher than 15 m. Probably, the outcropping secondary gypsum is restricted to a weathered zone, grading into anhydrite, halite, and glauberite towards the unaltered inner zone of the massif, as it has been documented in the very similar Zaragoza Gypsum Formation (Salvany 2009).
- To the south of Calatayud city, the evaporite formation is overlain by a carbonate–detrital sequence around 100 m thick (Hernández et al. 1983; Sanz-Rubio et al. 2003). This succession comprises two fluvio-lacustrine limestone units with a reddish detrital unit in between (Fig. 9.3b). The upper tufaceous limestone forms the caprock of a large NW–SE-trending structural platform (La Tronchona) slightly tilted to the NW. This Late Miocene unit constitutes a useful marker to identify and assess recent deformation related to dissolution-induced subsidence and tectonics.

The sediments of the basin have been selectively excavated by the alluvial systems developed subsequently to the capture of the graben. Consequently, the topography is markedly controlled by the distribution of the different lithofacies (Bomer 1960). The conglomeratic and carbonate facies form prominent reliefs, whereas the argillaceous and evaporite sediments have been differentially eroded to form low relief areas (Fig. 9.3b). The distal carbonate sediments form the Armantes (968 m) and La Tronchona (870 m) mesas to the NW and SE of Calatayud city (533 m), respectively. The graben is transversally crossed by the Jalón River, whereas its main tributaries, the Jiloca and Perejiles rivers, have carved longitudinal valleys at both sides of La Tronchona structural platform. The evolution of these fluvial systems is recorded by stepped sequences of terraces and pediments.

During the Late Pliocene and Quaternary, Calatayud Graben area has been affected by renewed extensional block tectonics. On the western margin of the graben, NW–SE normal faults have generated the Munébrega Half-graben, superimposed to Calatayud Graben (Figs. 9.2 and 9.3). The master fault of this tectonic depression has offset a Pleistocene terrace of the Jalón River, producing an antislope scarp. A paleoseismological investigation on this fault has been conducted at this site through the excavation of a trench (Gutiérrez et al. 2009b). To the NE of Calatayud Graben, the Río Grío Graben (Fig. 9.3), traditionally interpreted as an erosional fluvial valley, corresponds to a recently discovered Plio-Quaternary graben, whose alluvial fill more than 90 m thick has been deeply incised (Gutiérrez et al. 2009c, 2012b).

9.3 Megacollapse Structures in Neogene Sediments

To the NW of La Tronchona structural platform and framed by the Jiloca and Perejiles floodplains, there are two large collapse structures in which the detrital–carbonate units stratigraphically above the evaporite sequence have subsided due to the interstratal karstification of the underlying soluble sediments (Gutiérrez 1996, 1998) (Fig. 9.3). These two zones, designated as the Maluenda and the Perejiles subsidence areas, cover 4.4 and 12 km^2, respectively (Gutiérrez 1996). Here, the supra-evaporitic subsided units are strongly deformed and locally foundered within and juxtaposed to the evaporites (Fig. 9.4). They show abundant gravitational deformation, including ductile structures

Fig. 9.4 Oblique aerial view showing strongly deformed carbonate–detrital Neogene sediments (*dark* vegetated area) foundered within the subhorizontal gypsum strata (vegetation-free light terrain) in Maluenda subsidence area

(synforms, antiforms, and flexures) with a preferential NW–SE trend and brittle structures (faults and breccias). The strike and dip of the deformed strata are very chaotic, showing sharp changes between nearby locations. In contrast, the adjacent older gypsum strata are solely affected by a joint network and maintain a subhorizontal structure. The cartographical boundary between the deformed units and the undeformed gypsum has a very irregular and interdigitated pattern (Figs. 9.3 and 9.4).

In both areas, the supra-evaporitic sediments locally show synforms and basin structures in which certain units thicken towards the core. These thickness changes have been attributed to synsedimentary dissolution-induced basins (Sanz-Rubio et al. 1997; Gutiérrez 1998). However, most of the subsidence and deformation is postsedimentary. Probably, its initiation took place once the basin became exorheic and the new drainage network started to incise the sedimentary infill. In Maluenda area, the Neogene subsided sediments are locally overlain by the deposits of a Jiloca River terrace located at 100–105 m above the channel. This alluvium, dated magnetostratigraphically as Early Pleistocene, is thickened by synsedimentary karstic subsidence, locally reaching more than 100 m in thickness. This implies that subsidence in the Neogene sediments was active before and during the sedimentation of the thickened terrace deposit (Lower Pleistocene). The fact that the floodplain deposits of the Jiloca and Perejiles rivers overlap the subsided sediments in Maluenda and Perejiles areas, respectively (Fig. 9.3), and that the deposits of a Lower Pleistocene terrace fossilize collapsed units in Maluenda area indicates that subsidence has progressed vertically downwards below the base level. Considering that the floodplains are at an altitude of 560–570 m, and that the base of the supra-evaporitic sequence is above 770 m in elevation, the maximum karst subsidence in both areas has reached at least 200 m (Gutiérrez 1996, 1998).

The intense karstification of the evaporitic formations located to the NW of the Tronchona structural platform is related to the convergence of several factors:

- The Maluenda and Perejiles subsidence areas are located in the sector where the most soluble evaporitic facies were deposited (depocenter), including a large proportion of halite and glauberite in the subsurface. Very likely, interstratal karstification has operated at several stratigraphic levels, affecting preferentially the most soluble beds.
- The NW–SE-oriented Tronchona structural platform, about 20 km long and 3 km wide, is tilted to the NW, towards the collapse areas and the main base level (Jalón River). This extensive and karstified structural surface constitutes a recharge zone, which has its main discharge area in its north-western edge, where the Maluenda and Perejiles areas are situated. This idea is supported by the presence of tufa deposits located to the NE of Paracuellos de Jiloca, which have been related to a paleospring (Gutiérrez 1996, 1998).
- The entrenchment of the fluvial systems in the sedimentary fill of the basin from the capture of the graben has favoured the circulation of underground waters through progressively deeper stratigraphic levels.
- Subsidence structures in the Neogene sediments show a prevalent NW–SE orientation. The NW–SE-trending Munébrega and Río Grío Plio-Quaternary grabens reveal

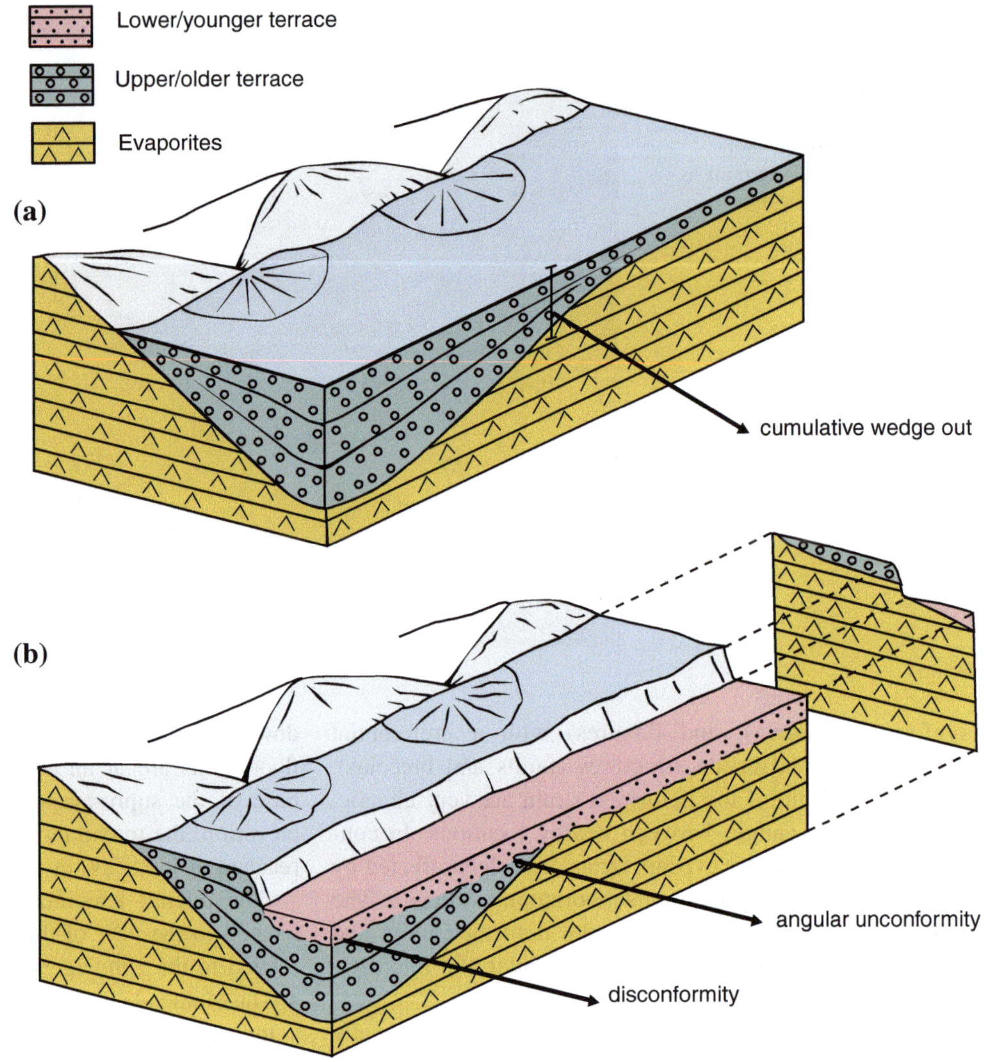

Fig. 9.5 Morpho-stratigraphic arrangement of terraces affected by synsedimentary dissolution subsidence. **a** The thickened terrace deposits fill a dissolution-induced basin showing centripetal dips and cumulative wedge-outs at the margins. **b** The deposits of the subsequent terrace are inset in the bedrock upstream of the area affected by subsidence and superposed by angular unconformity or paraconformity onto the thickened sediments of the previous alluvial level. The morphogenetic surfaces of both terraces are stepped along the whole valley reach (adapted from Gutiérrez 1998)

that the area is affected by active extensional tectonics. Probably, this extension has generated or dilated previously existing NW–SE fractures in the Neogene sediments favouring karstification along these structures. The formation and opening of fractures may be also favoured by the debutressing effect produced by the entrenchment of the Jiloca and Perejiles valleys (Jennings 1985).

- On the NE margin of the Jiloca River, in Paracuellos de Jiloca village and next to the Maluenda area, there is a spa with a perched spring that permanently issues water of the sodium chloride hydrochemical facies. This fact supports the hypothesis that karstification may be partially related to upward hypogenic groundwater flows coming from detrital aquifer units situated beneath the evaporites. Groundwater that recharges at the margins of the basin circulates through coarse-grained detrital units confined by argillaceous facies towards the axis of the basin, ultimately discharging through upward transformational flows.

9.4 Synsedimentary Subsidence and Paleosinkholes in Quaternary Alluvium

The Jalón–Jiloca–Perejiles alluvial system has been affected by synsedimentary and postsedimentary subsidence phenomena related to the karstification of the evaporitic bedrock throughout its Quaternary evolution (Gutiérrez 1996, 1998). Ten stepped alluvial levels have been differentiated within the Calatayud Graben. The height of the terrace levels above the river channels are as follows: T10: 115 m; T9: 105–100 m; T8: 90–85 m; T7: 75–70 m; T6: 65–60 m; T5: 55–50 m; T4: 45 m; T3: 35–30 m; T2: 25–20 m; T1 (floodplain): 5–3 m. Terrace levels T9 and T7 have been dated as Lower Pleistocene through magnetostratigraphic analyses. All these levels are represented by aggradation terraces. Degradation terraces developed in the deposits of older terraces (fill-cut terraces) have been recognized for some levels. The identified pediment levels are P10, P9, P8,

Fig. 9.6 Cutting of the Madrid–Zaragoza highway (A–2) next to Calatayud Hospital, showing a non-tectonic angular unconformity between dipping alluvium of terrace T8 and subhorizontal deposits of terrace T4

Fig. 9.7 Synsedimentary synform with cumulative wedge-outs affecting fine-grained terrace deposits accumulated in a palustrine area associated with a paleosinkhole (30T 612951 4577253; coordinate system ETRS89)

P7, and P4 (Px correlative to Tx). These may be aggradation surfaces (mantled pediments) or degradation surfaces developed either in older alluvium or in Miocene gypsiferous bedrock. The degradation surfaces of each level occur upstream and in the margins of the sectors where the correlative Quaternary alluvium is thickened due to synsedimentary subsidence. These spatial relationships suggest that degradation processes have been related to gradient changes in the alluvial system caused by karstic subsidence (Gutiérrez 1996, 1998).

Quaternary alluvial deposits overlying the salt-bearing evaporitic bedrock show conspicuous thickenings, recording dissolution-induced synsedimentary subsidence. Abrupt thickness variations from less than 10 m to more than 100 m may be observed in a single terrace. The thickened alluvium locally fills dissolution-induced basins with centripetal dips and cumulative wedge-outs at the margins (Fig. 9.5). These thickened terrace deposits show a high proportion of flood-plain facies, whereas in areas where the alluvium has not been affected by synsedimentary subsidence, channel gravel facies dominate. As a consequence of the local thickening of the alluvium, the deposits of a younger terrace may be inset into the bedrock or superposed onto older thickened and deformed deposits. Nonetheless, the surfaces of both terraces may show a stepped arrangement along the whole valley reach (Fig. 9.5). These morpho-stratigraphic features, very common in fluvial valleys carved in Tertiary evaporites in Spain, can be observed on the eastern margin of the Jiloca River valley, next to the confluence with the Jalón River. Here, the deposits of several terraces fill a dissolution-induced

Fig. 9.8 Dissolution and subsidence structures exposed in cuttings of the Madrid–Zaragoza highway (A–2) along the southern margin of Jalón Valley. **a** Sagging in terrace deposits related to the dissolutional lowering of the evaporite rockhead. Note the marly karstic residue between the bedrock and the cover. **b** Dissolutional conduits filled with alluvium derived from the overlying terrace deposits. **c** Sagging and collapse on bedrock related to deep-seated interstratal karstification. The collapse structure is controlled by outward-dipping antithetic normal faults and brecciation. **d** Collapse structure with dome-shaped failure surfaces related to the upward propagation (stoping) of a deep-seated cavity

trough more than 4 km long and 0.7 km wide where the alluvium thickness exceeds 100 m. The southern cutting of the A-2 Madrid–Zaragoza motorway next to N-234 road shows the superposition of two sedimentary units bounded by angular unconformity (Fig. 9.6). The thickened and deformed deposits of the lower unit have been ascribed to the Early Pleistocene terrace level T8 (90–85 m above the channel). The deposits of this terrace reach more than 90 m in vertical thickness (the stratigraphic thickness is higher), and its 25° dip progressively attenuates towards the top of the sequence. The upper unit corresponds to terrace T4 (45 m above the channel). The deposits of this terrace, capped by a petrocalcic horizon, show a progressive thickness increase towards the NE, and its aggradation surface is tilted in the same direction, indicating that the terrace has undergone long-sustained synsedimentary and postsedimentary dissolution-induced subsidence.

The alluvial cover overlying the evaporite bedrock shows numerous ductile and brittle gravitational deformation structures (Hoyos et al. 1977; Gutiérrez 1996, 1998). Some of these dissolution-induced deformations also affect the evaporitic bedrock (interstratal karstification). Excellent examples are found in the cuttings of the A-2 Zaragoza–Calatayud motorway, along the southern margin of the Jalón Valley. Many of these subsidence structures correspond to paleodolines that were filled with marl and carbonate palustrine–lacustrine facies, recording the development of swampy environments and ponds in closed depressions in the floodplain. Remains of aquatic gastropods, amphibian, and fishes have been found in some of these paleosinkhole fills (Fig. 9.7). These structures and the associated sediments provide valuable information on the subsurface processes involved in the currently active subsidence. Four main mechanisms of sinkhole generation have been identified through the study of the subsidence structures identified in the exposed paleokarst (Gutiérrez et al. 2008b; Gutiérrez and Cooper 2013) (Fig. 9.8): (a) progressive lowering of the evaporitic rockhead by dissolution and gradual sagging of the alluvial cover (cover sagging sinkholes) (Fig. 9.8a); (b) development of dissolutional conduits and fissures (grikes) at the top the bedrock and downward migration (ravelling) of detrital particles from the alluvial cover (Fig. 9.8b); this process may cause the progressive settlement of the alluvial mantle (cover suffosion sinkholes), or the upward propagation of cavities through cohesive alluvium that may eventually produce catastrophic collapses (cover collapse sinkholes); (c) sheet-like dissolution within the evaporitic bedrock, typically focused on salt beds, and progressive sagging of the overlying bedrock and alluvial cover (bedrock and cover sagging sinkholes) (Fig. 9.8c); (d) generation of dissolutional cavities within the bedrock and upward propagation of the void by progressive roof collapse (stoping), culminating in the formation of bedrock collapse sinkholes and bedrock and cover collapse sinkholes (Fig. 9.8c and d).

One of the best examples of a paleosinkhole may be examined in a gravel pit excavated in the T4 terrace NW of Calatayud Hospital (30T 612951 4577253; coordinate system ETRS89). The paleosinkhole displays several superimposed ductile and brittle subsidence structures, as well as soft-sediment deformation, whose relative chronology may be inferred on the basis of cross-cutting relationships (Fig. 9.7): (a) Synform in section and basin structure in 3D made up of marls and fines deposited in a ponded and/or swampy doline developed by bending in the floodplain. The thickening of the strata towards the core and the upward dip attenuation (cumulative wedge-out) indicate that subsidence was coeval to deposition. (b) The sagging structure is cross-cut by subvertical normal faults (ring faults) recording the development of a collapse sinkhole. Ductile structures and soft-sediment deformation associated with these failure planes suggest that the collapse occurred during or soon after deposition, when the alluvium was still in a soft and water-saturated state. This temporal evolution pattern has relevant applied implications, revealing that slow subsidence in a sinkhole may eventually transform into more dangerous catastrophic collapse. Soft-sediment deformation includes irregular masses of gravel (gravel pockets) embedded in fine-grained facies, spatially associated with the collapse faults. Similar structures have been interpreted by several authors as liquefaction–fluidization features (Postma 1983; Johnson 1986; Nocita 1988). These gravel pockets may be explained as liquefaction structures caused by local sediment shaking and fluid overpressure induced by the generation of a catastrophic sinkhole (Gutiérrez 1998).

9.5 Subsidence Hazard in Calatayud Historical City

Calatayud city was founded by the Muslims in 716 AD at an important junction of communication routes. In 1120 AD, it was conquered by King Alfonso I "the Battler". This attractive city from the historical and artistic point of view

Fig. 9.9 Tilted tower (11th–12th C) of San Pedro de los Francos Church (14th C) in La Rúa Street. The *upper* 5-m-high portion of the tower was removed in 1840 to avoid its collapse over the Wersage Palace located in the opposite side of the street, where the Royal family used to be hosted

was declared Historical Monument in 1967 and currently has a population of around 22,000 inhabitants. Although the urban area is located in an advantageous location from the economic perspective, its geomorphological setting has posed severe problems since old times. The city is located on the NW margin of the Jalón River valley (Fig. 9.3), at the foot of a nearly vertical gypsum escarpment approximately 100 m high. It is positioned partly on an alluvial fan fed by La Rúa and Las Pozas streams and partly on the Jalón River floodplain. The city also extends onto the gypsum cliff, with cave dwellings excavated in the slopes. Since its foundation, urban development has been largely constrained by geohazards, including (a) floods caused by the Jalón River, and La Rúa and Las Pozas streams, which have caused severe damage in the floodplain and the alluvial fan, respectively. Run-off of the flashy La Rúa and Las Pozas streams has been progressively diverted through the construction of dams, artificial channels, and tunnels since Muslim times

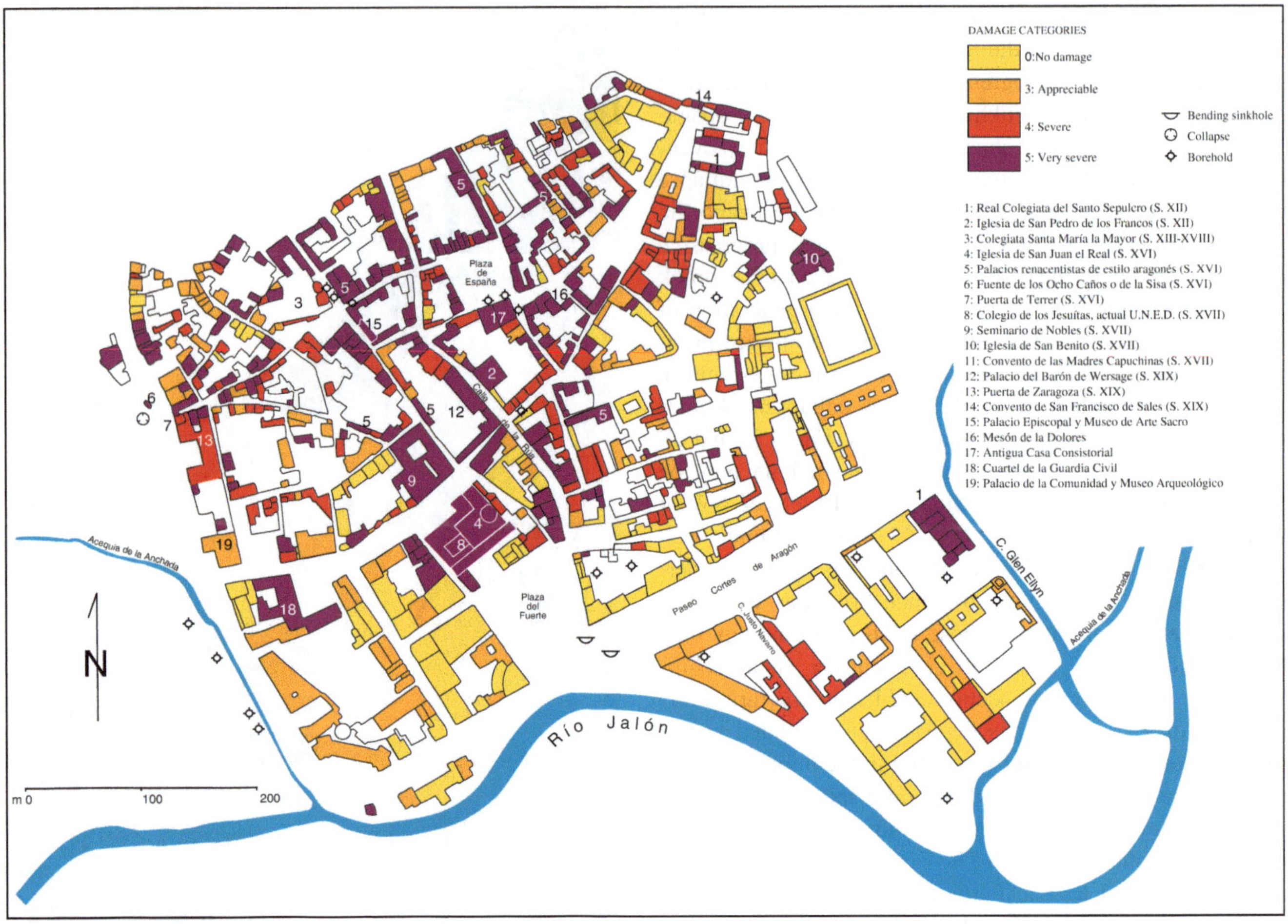

Fig. 9.10 Map showing the distribution of subsidence damage in a sector of Calatayud city (modified from Gutiérrez 1998). The *green asterisk* indicates the location of the Blue House, undermined by a collapse sinkhole in 10 November 2003

(see Gutiérrez and Cooper 2002); (b) rockfalls and topples have constrained development on the gypsum scarp areas. In 1988, a person was killed by a rockfall. These rapid slope movements have caused the destruction of several buildings and frequent cuts in the roads running along the foot of the cliff; (c) subsidence causes severe detrimental effects on both the floodplain and fan areas.

Most of the old buildings in the city have been damaged by subsidence, and some important monuments have been demolished. Calatayud is considered from a geotechnical perspective the most problematic city of Aragón Region, both for remediation of historical buildings and for modern development. Subsidence damage in the buildings includes tilting and cracking (Fig. 9.9), sloping floors, sheared door and window openings, and collapsed roofs. Peripheral damage includes broken pipes, with the consequent extra water supply to the subsurface, and pavement sagging and collapse. The full cost of the damage is difficult to estimate, but the losses to the irreplaceable artistic and historical heritage are large. The spatial distribution of the subsidence and its causes have been analysed on the basis of the geotechnical characterization of the materials beneath Calatayud and a systematic building damage assessment (Gutiérrez 1998; Gutiérrez et al. 2000; Gutiérrez and Cooper 2002).

The stratigraphic and geotechnical characteristics of the rocks and soils beneath Calatayud city were studied from 43 boreholes, 18 of which reached the evaporitic bedrock. From the surface downwards, these materials include the following:

- A top layer of unsorted and unconsolidated anthropogenic ground whose thickness decreases towards the Jalón River. Compaction of the made-ground rubble may locally affect buildings resting directly on the fill.
- Alluvial fan deposits mainly consisting of gypsiferous silts with scattered gypsum and limestone clasts. This unit is more than 12 m thick in the proximal fan area and wedges out towards the floodplain. These soils are characterized by a very loose packing with the silt particles bound by gypsum "bridges". This deposit may undergo a rapid reduction in volume with the addition of water due to dissolution of the interparticle gypsum bonds and the consequent collapse of the granular framework

(hydrocollapse or hydrocompaction). This type of subsidence may affect structures in the fan area.

- Fluvial deposits underlie the gypsiferous silts in the fan area and the made-ground in the floodplain. In the floodplain area, the thickness of these deposits ranges from 7 to 24 m. The thickness variations are largely related to synsedimentary subsidence phenomena. They are made up of channel gravels and floodplain fines, including palustrine facies deposited in swampy subsidence depressions that locally reach more than 6 m in thickness. The consolidation of these deposits may cause subtle settlement of the ground surface.
- A karstic residue up to 9 m thick, composed of soft, dark grey marls with scattered gypsum particles between the alluvium and the evaporitic bedrock.
- The karstic residue grades downwards into the evaporitic bedrock with a well-developed tertiary porosity (solutionally enlarged discontinuity planes).

The spatial distribution and intensity of the subsidence were indirectly analysed through the construction of a building damage map of a broad sector of the city covering around 35 ha (Fig. 9.10). The selected area includes buildings varying in age from twelfth century to modern and includes a wide range of morpho-stratigraphic settings, from the proximal fan area to the riverbank. The damage was assessed in 1996–1997 by examination of the building façades. A damage category was assigned to each building on a scale of 0–5, based on the *Subsidence Engineers' Handbook* ranking system established by the British National Coal Board (NCB) for the evaluation of mining subsidence damage (NCB 1975). Peripheral effects and internal damage of the buildings were not surveyed in the study, so only four levels of damage 0, 3, 4, and 5 were used. Thus, level 0 on the map also includes damage attributable to levels 1 and 2 in the NCB ranking. Although this methodology has clear limitations, mainly because of the heterogeneity of the "structural markers", the damage map provides a rough idea about the distribution of the subsidence effects and allows inferring the main causative processes.

The map of the subsidence categories shows that buildings affected by severe damage (level 5) are found in all the sectors of the selected area (Fig. 9.10). However, the proximal fan area has a higher concentration of buildings with the highest damage category. In this area, the blocks flanking La Rúa Street are the most severely affected. The presence of buildings affected by severe damage throughout the whole study area, including sites without anthropogenic rubble or gypsiferous silts, indicates that karstification of the bedrock is one of the main causes of the subsidence. The relatively greater impact of subsidence in the proximal fan area suggests that hydrocollapse and possibly dissolution of the gypsiferous silts contribute significantly to subsidence in this area. The higher subsidence activity in La Rúa Street may be

Fig. 9.11 Collapse sinkhole occurred catastrophically in 10 November 2003 next to and beneath the *Blue* House. *Photograph* taken in the morning of November 10

explained by the high thickness of the anthropogenic rubble and the existence of preferential underground flows along the buried La Rúa channel (Fig. 9.9). Other processes, such as compaction of man-made ground, consolidation of fluvial deposits, or the collapse of old cellars, may also play a relevant role at specific sites, but do not seem to contribute significantly to the subsidence that affects a great part of the urban area. Additional information on subsidence damage and remediation history of some remarkable historical buildings (e.g. Santa María la Mayor, San Pedro de los Francos Church, Spain Square, Ocho Caños Fountain) may be found in Gutiérrez and Cooper (2002).

To our knowledge, the most infamous subsidence event occurred in Calatayud corresponds to the 2003 collapse sinkhole that led to the demolition of the so-called Blue House (Fig. 9.11). This was a five-storey building with basement built in the 1970s on the Jalón River floodplain (location indicated with an asterisk in Fig. 9.10). On the night of 9 November 2003, some of the tenants of the 52 flats perceived strange noises and the occurrence of cracks. At around 3:30 a.m., subsequent to the evacuation of the building, a sinkhole 6 m long formed suddenly and noisily in the pavement next to the Blue House (Fig. 9.11). The water table, located at a depth of 3 m, made it impossible observing the bottom of the depression. On the morning of November 10, the sinkhole started to be filled with gravel to prevent the widening of the hollow by mass wasting processes. After dumping 350 m^3 of aggregate, a volume much greater than the expected, the technical party of the City Council suspected that the cavity could extend underneath the basement of the building. This hypothesis was corroborated by means of perforations conducted in the basement; they detected a funnel-shaped cavity with a 10-m-deep apex located beneath the Blue House. An additional 250 m^3 of concrete was needed to fill the void. A cavity, rooted in the cavernous bedrock, propagated upwards through the alluvial cover by progressive roof collapse. Once it reached the rigid base of the building and the pavement, it enlarged laterally generating a funnel-shaped void. Undermining of the foundations of the building caused the brittle deformation of the structure. Subsequently, a portion of the pavement bridging the cavity failed, resulting in the observed collapse sinkhole (Gutiérrez et al. 2004). In subsequent days, intense geotechnical investigations and stabilization works were carried out. A few months later, the Blue House was demolished. Considering estimated costs of 3 million euro for the flats, 1 million for the stabilization works and geotechnical investigations, and 0.8 million for the demolition, the direct economic losses caused by this unforeseeable single subsidence event exceed 4.8 million euros. Additional costs have been derived from housing rentals, the closure of businesses, moves, and land depreciation (Gutiérrez et al. 2004). The owners demanded a compensation to the *Consorcio de Compensación de Seguros* (a consortium integrated and funded by all insurance companies), which covers losses related to extraordinary events like some geohazards or terrorist attacks. Unfortunately, this institution does not cover sinkhole-related damage, heading the owners to a state of neglect; mortgages of non-existing flats are being paid.

Acknowledgments This work has been supported by the Spanish national project CGL2010-16775 (Ministerio de Ciencia e Innovación and FEDER) and the regional project 2012/GA-LC-021 (DGA-La Caixa).

References

Bomer B (1960) Aspectos morfológicos de la Cuenca de Calatayud-Daroca y de sus zonas marginales. Estudios Geográficos 80:393–402

Calaforra JM, Pulido-Bosch A (1996) Some examples of gypsum karst and the most important gypsum caves in Spain. Int J Speleol 25:225–237

Calaforra JM, Pulido-Bosch A (1999) Gypsum karst features as evidence of diapiric processes in the Betic Cordillera, Southern Spain. Geomorphology 29:251–264

Canals M, Got H, Juliá R, Serra J (1990) Solution-collapse depressions and suspensates in the limnocrenic lake of Banyoles (NE Spain). Earth Surf Proc Land 15:243–254

Carbonel D, Gutiérrez F, Linares R, Roqué C, Zarroca M, McCalpin J, Guerrero J, Rodríguez V (2013) Differentiating between gravitational and tectonic faults by means of geomorphological mapping, trenching and geophysical surveys. The case of the Zenzano fault (Iberian Chain, N Spain). Geomorphology 189:93–108

Cardona F, Viver J (2002) Sota la Sal de Cardona. Espeleo Club de Gràcia, Barcelona 129 p

Collantes LP, Griffo JL (1982) Panorámica general del yeso en España. Los yesos de la Cubeta de Calatayud en la Provincia de Zaragoza. Tecniterrae 8(46):53–63

Galve JP, Gutiérrez F, Lucha P, Bonachea J, Remondo J, Cendrero A, Gutiérrez M, Gimeno MJ, Pardo G, Sánchez JA (2009) Sinkholes in the salt-bearing evaporite karst of the Ebro River valley upstream of Zaragoza city (NE Spain). Geomorphological mapping and analysis as a basis for risk management. Geomorphology 108:145–158

Gracia FJ, Gutiérrez F, Gutiérrez M (2002) Origin and evolution of the Gallocanta polje (Iberian Range, NE Spain). Zeitschrift für Geomorphology 46:245–262

Guerrero J, Gutiérrez F, Lucha P (2008) The impact of halite dissolution subsidence on fluvial terrace development. The case study of the Huerva River in the Ebro Basin (NE Spain). Geomorphology 100:164–179

Guerrero J, Gutiérrez F, Galve JP (2013) Large depressions, thickened terraces, and gravitational deformation in the Ebro River valley (Zaragoza area, NE Spain): evidence of glauberite and halite interstratal karstification. Geomorphology 196:162–176

Gutiérrez F (1996) Gypsum karstification induced subsidence: effects on alluvial systems and derived geohazards (Calatayud Graben, Iberian Range, Spain). Geomorphology 16:277–293

Gutiérrez F (1998) Fenómenos de subsidencia por disolución de formaciones evaporíticas en las fosas neógenas de Teruel y Calatayud. Ph. D. thesis, University of Zaragoza, 569 p

Gutiérrez F, Cooper AH (2002) Evaporite dissolution subsidence in the historical city of Calatayud, Spain: damage appraisal and prevention. Nat Hazards 25:259–288

Gutiérrez F, Cooper AH (2013) Surface morphology of gypsum karst. In: Frumkin A (ed) Treatise on geomorphology. Karst Geomorphology, vol 6. Elsevier, Amsterdam, pp 425–437

Gutiérrez F, Cooper AH, García-Hermoso F (2000) Spatial assessment, mitigation and prevention of evaporite dissolution subsidence damage in the historical city of Calatayud, Spain. In: Carbognin L, Gambolati G, Johnson AI (eds) Land subsidence. Proceedings of the 6th international symposium on land subsidence, vol 1, pp 237–248

Gutiérrez F, Ortí F, Gutiérrez-Elorza M, Pérez-González A, Benito G, Gracia-Prieto J, Durán JJ (2001) The stratigraphical record and activity of evaporite dissolution subsidence in Spain. Carbonates Evaporites 16:46–70

Gutiérrez F, Lucha P, Guerrero J (2004) La dolina de colapso de la casa azul de Calatayud (noviembre de 2003). Origen, efectos y pronóstico. In: Benito G, Díez-Herrero A (eds) Riesgos naturales y antrópicos en Geomorfología. Actas de la VIII Reunión Nacional de Geomorfología, Toledo, pp 477–488

Gutiérrez F, Calaforra JM, Cardona F, Ortí F, Durán JJ, Garay P (2008a) Geological and environmental implications of evaporite karst in Spain. Environ Geol 53:951–965

Gutiérrez F, Guerrero J, Lucha P (2008b) A genetic classification of sinkholes based on the analysis of evaporite paleokarst exposures in Spain. Environ Geol 53:993–1006

Gutiérrez F, Gutiérrez M, Gracia FJ, McCalpin JP, Lucha P, Guerrero J (2008c) Plio-Quaternary extensional seismotectonics and drainage network development in the central sector of the Iberian range (NE Spain). Geomorphology 102:21–42

Gutiérrez F, Galve JP, Lucha P, Bonachea J, Jordá L, Jordá R (2009a) Investigation of a large collapse sinkhole affecting a multi-storey building by means of geophysics and the trenching technique (Zaragoza city, NE Spain). Environ Geol 58:1107–1122

Gutiérrez F, Masana E, González A, Guerrero J, Lucha P, McCalpin JP (2009b) Late Quaternary paleoseismic evidence on the Munébrega half-graben fault (Iberian range, Spain). Int J Earth Sci 98:1691–1703

Gutiérrez F, Lucha P, Jordá L (2009c) The Río Grío depression (Iberian Range, NE Spain). Neotectonic graben vs. fluvial valley. In: Pérez-López R, Grützner C, Lario J, Reicherter K Silva P (eds) Archeoseismology and palaeoseismology in the Alpine-Himalayan collisional zone. 1st INQUA-IGCP-567 international workshop on earthquake archaeology and palaeoseismology, Baelo Claudia, Spain, pp 43–46

Gutiérrez F, Carbonel D, Guerrero J, McCalpin JP, Linares R, Roque C, Zarroca M (2012a) Late Holocene episodic displacement on fault scarps related to interstratal dissolution of evaporites (Teruel Neogene Graben, NE Spain). J Struct Geol 34:2–19

Gutiérrez F, Gracia FJ, Gutiérrez M, Lucha P, Guerrero J, Carbonel D, Galve JP (2012b) A review on Quaternary tectonic and nontectonic faults in the central sector of the Iberian Chain, NE Spain. J Iberian Geol 38:145–160

Gutiérrez F, Linares R, Roqué C, Zarroca M, Rosell J, Galve JP, Carbonell D (2012c) Investigating gravitational grabens related to lateral spreading and evaporite dissolution subsidence by means of detailed mapping, trenching, and electrical resistivity tomography (Spanish Pyrenees). Lithosphere 4:331–353

Hernández A, Del Olmo P, Aragonés E (1983) Memoria y Mapa Geológico de España, E. 1:50.000. Ateca (437). IGME, Madrid

Hoyos M, Zazo C, Goy JL, Aguirre E (1977) Estudio geomorfológico en los alrededores de Calatayud. Actas de la III Reunión Nacional de Cuaternario Ibérico, pp 149–160

Jennings JN (1985) Karst geomorphology. Blackwell, Oxford 239 p

Johnson SY (1986) Water-escape structures in coarse-grained, volcaniclastic, fluvial deposits of the Ellensburg formation, South-Central Washington. J Sediment Petrol 56(6):905–910

López-Vicente M, Navas A, Machín J (2009) Geomorphic mapping in endorheic catchments in the Spanish Pyrenees: an integrated GIS analysis of karstic features. Geomorphology 111:38–47

Lucha P, Cardona F, Gutiérrez F, Guerrero J (2008) Natural and human-induced dissolution and subsidence processes in the salt outcrop of the Cardona Diapir (NE Spain). Environ Geol 53:1023–1035

Marín A (1932) Boletín de Sondeos. IGME, Madrid 3(1):29–99

Mottershead DN, Duane WJ, Inkpen RJ, Wright JS (2008) An investigation of the geometric controls on the morphological evolution of small-scale salt terrains, Cardona, Spain. Environ Geol 53:1091–1098

NCB (1975) Subsidence engineers' handbook. National Coal Board Mining Department, United Kingdom 111 p

Nocita BW (1988) Soft-sediment deformation (fluid escape) features in a coarse-grained pyroclastic-surge deposit, north-central New Mexico. Sedimentology 35:275–285

Ortí F, Rosell L (1998) Unidades evaporíticas de la Cuenca de Calatayud (Mioceno inferior-medio, Zaragoza). Geogaceta 23:111–114

Ortí F, Rosell L (2000) Evaporite systems and diagenetic patterns in the Calatayud basin (Miocene, Central Spain). Sedimentology 47:665–685

Pando L, Pulgar JA, Gutiérrez-Claverol M (2013) A case of man-induced ground subsidence and building settlement related to karstified gypsum (Oviedo, NW Spain). Environ Earth Sci 68:507–519

Postma G (1983) Water escape structures in the context of a depositional model of a mass flow dominated conglomeratic fan-delta (Abrioja Formation, Pliocene, Almería Basin, Spain). Sedimentology 30:91–103

Salvany JM (2009) Geología del yacimiento glauberítico de Montes de Torrero. Universidad de Zaragoza, Zaragoza 80 p

Sanz-Rubio E, Hoyos M, Calvo JP, Sánchez-Moral S, Cañaveras JC, Sesé C (1997) Paleokarstificaciones y evolución de los sistemas deposicionales miocenos de la Cuenca de Calatayud. In: Alcalá L, Alonso-Zarza AM (eds) Itinerarios Geológicos en el Terciario del Centro y Este de la Península.). Excursiones del III Congreso del Grupo Español del Terciario, Cuenca, pp 109–134

Sanz-Rubio E, Sánchez-Moral S, Cañaveras JC, Abdul-Aziz H, Calvo JP, Cuezva S, Mazo AV, Rouchy JM, Sesé C, van Dam J (2003) Síntesis de la cronoestratigrafía y evolución sedimentaria de los sistemas lacustres evaporíticos y carbonatados neógenos de la cuenca de Calatayud-Montalbán. Estud Geol 59:83–105

Utrilla R, Pierre C, Ortí F, Pueyo JJ (1992) Oxygen and sulphur isotope compositions as indicators of the origin of Mesozoic and Cenozoic evaporites from Spain. Chem Geol (Isotope Geoscience Section) 102:229–244

The Gypsum Karst of Sorbas, Betic Chain

10

Fernando Gázquez and José María Calaforra

Abstract

The gypsum karst of Sorbas is one of the most well-known gypsiferous areas in the world from the speleological point of view. The Yesares formation comprises an alternating sequence of gypsum and marly strata deposited during the Messinian salinity crisis. The development of the cave systems is closely linked to this alternation of sediments. The caves have been mainly formed by erosion of the marl units, whilst gypsum dissolution was especially active during the initial stages. On the surface, striking karst landforms are found, such as gypsum tumuli, a great variety of dolines formed by different mechanisms, several types of karren, as well as a 30-m-high gypsum scarp. Proto-conduits on the ceilings of the galleries, as well as some sedimentary features, provide evidence of the initial phreatic speleogenetic phases. In addition, the gypsum caves of Sorbas have a great deal of speleothems, some of them unique worldwide (e.g. gypsum balls, hollow stalagmites, gypsum trees and trays, and deflated stalagmites). The protection of the gypsum karst of Sorbas, due to the wide variety of surface and subsurface geomorphological features, should be considered of top priority for the administration.

Keywords

Kart • Gypsum • Caves • Speleothems • Speleogenesis

10.1 Introduction

Although carbonate terrains have focused the attention of most karst investigations; the interest in gypsum karst areas has gradually increased over the past 40 years (Klimchouk et al. 1996; Gutiérrez and Cooper 2013). Probably, the increasing attention that the study of gypsum karst areas has received was fostered by the discovery in the 1970s of extensive maze caves in Western Ukraine, reaching more than 200 km of linear development (Klimchouk and Aksem 2002), and the frequent environmental problems associated with these areas (e.g. Cooper and Gutiérrez 2013).

Other relevant gypsum karst areas are located in the fore-Ural region, the Eastern European Platform (Andrejchuk and Klimchouk 1996), the southern Siberian platform (Trzcinski 1996), China (Yaoru and Cooper 1996), and the USA (Johnson 1996). In the context of Western Europe, gypsum karst areas can be found in Italy (Forti and Sauro 1996), France (Chardon and Nicod 1996), Germany (Kempe 1996), Great Britain (Cooper 1996), and Spain (Calaforra and Pulido-Bosch 1996; 1997; Gutiérrez et al. 2008). The gypsum karst of Sorbas (Almería, SE Spain) is one of the most well-known in Spain, mainly due to its cave

F. Gázquez · J. M. Calaforra (✉)
Water Resources and Environmental Geology Research Group, University of Almería, Almería, Spain
e-mail: jmcalaforra@ual.es

F. Gázquez
e-mail: f.gazquez@ual.es

F. Gázquez
Unidad Asociada UVA-CSIC al Centro de Astrobiología, University of Valladolid, Valladolid, Spain

F. Gutiérrez and M. Gutiérrez (eds.), *Landscapes and Landforms of Spain*, World Geomorphological Landscapes,
DOI: 10.1007/978-94-017-8628-7_10,

systems (Calaforra and Pulido-Bosch 2003). Nearly one thousand caves are concentrated in only 12 km^2, forming one of the most important gypsum karst areas in the world from the cave science perspective.

The subterranean network, extending for tens of kilometres beneath the arid gypsiferous landscape, contains the largest cave complex in Andalusia, the Cueva del Agua Cave, which comprises 8 km of galleries and is the longest gypsum cave in Spain.

10.2 Geological and Geographical Setting

The gypsum karst of Sorbas is located in the Tabernas–Sorbas Basin, SE Spain. This is an intramontane Neogene basin within the Betic Cordillera (see Chap. 3 in this volume), including significant gypsiferous Messinian evaporites in its sedimentary fill (Dronkert 1977). It lies within a topographic depression bounded to the north by the Filabres Range and to the south by the Alhamilla and Cabrera ranges (Fig. 10.1). The region has a semi-arid climate, with a mean annual precipitation lower than 210 mm.

The karstified Messinian gypsum (Yesares Member) occurs within a 120-m-thick cyclic sequence consisting of alternating gypsum and marly units (Dronkert 1977). The selenitic gypsum units reach 30 m in thickness. The thickest and most complete sequence of the Yesares Member crops out in the Río Aguas canyon, where 12 cycles of alternating gypsiferous and carbonate–pelitic-laminated sediments were initially described by Dronkert (1976). Subsequently, Krijgsman et al. (2001) recognized 14 sedimentary cycles. Such cyclic stratigraphy has been attributed to precessional cycles of the Earth with impact on the global climate (Krijgsman et al. 2001). Details of typical sedimentary structures related to gypsum growth can be observed in each of the massive gypsum bed, particularly the "supercones" (Dronkert 1976). These are large tree-like gypsum sedimentary structures which can be observed in the caves, where the base of a gypsum layer is fully visible on a cave ceiling.

The Messinian sequence is scarcely deformed in the centre of the basin, barely affected by minor tilting and faulting. Fracturing and stratification frequently play a decisive role in the linear development and configuration of the cave levels (Calaforra and Pulido-Bosch 2003).

The Yesares Member is overlain by (1) marine sandstones with some intercalated siltstones of the Sorbas Member (defined by Roep et al. 1979); (2) coastal plain silts and sands of the Zorreras Member; and (3) continental conglomerates of the Góchar Formation (Mather and Harvey 1995). The semi-pervious nature of the overlying sequence was of great importance for the development of the gypsum karst, as it creates semi-confined conditions on the gypsum strata at the western edge of the basin and controls the hydrogeological regime of the main spring of Los Molinos (Calaforra 1998). The impervious base of the gypsum aquifer comprises silts and clays of the Abad Member.

10.3 Surface Geomorphological Features

The gypsum karst of Sorbas displays a wide variety of surface features, including tumuli (Calaforra and Pulido-Bosch 1999) (Fig. 10.2) and numerous dolines. More than 1,000 cave entrances, many of them corresponding to collapse sinkholes, have been discovered in an area of about 12 km^2 (Calaforra and Pulido-Bosch 2003).

10.3.1 Dolines and Blind Valleys

The dolines are one of the main surface expressions of the endokarst. They collect the run-off acting as inlet points for the underlying karst aquifer. Water recharge also occurs through slow seepage among the gypsum crystals, following their cleavage planes and intercrystalline spaces (Sanna et al. 2012).

More than a thousand of small dolines have been inventoried in the Sorbas karst. In the area of the Cueva del Agua, more than one hundred dolines have been mapped in an area covering 1 km^2, which represents the highest density of dolines reported in Spain (Fig. 10.2b).

Dolines are usually classified according to their genesis. Collapse dolines are the most common type, related to the collapse of the cave roofs. Solution dolines result from differential dissolution on outcropping gypsum. Sinkholes are also related to fractures or joints in the rock, especially linked to the evolution of the gypsum scarp (Calaforra 2003).

Blind valleys are also very common in the gypsum karst of Sorbas. The Cueva del Yeso, located in the Barranco del Infierno, is an excellent example where the cave entrance has captured de main stream remaining an old beheaded and perched gypsum gully over the galleries of the cave.

10.3.2 Tumuli

The tumuli are one of the most unusual and least-commonly described surface features, which are frequently associated with bare gypsum karst landscapes (e.g. Gutiérrez and Cooper 2013; Fig. 10.2a). They are caused by upward-doming of the uppermost gypsum layer forming hemispherical mounds that may reach several metres in diameter. Their formation, among other factors, is especially

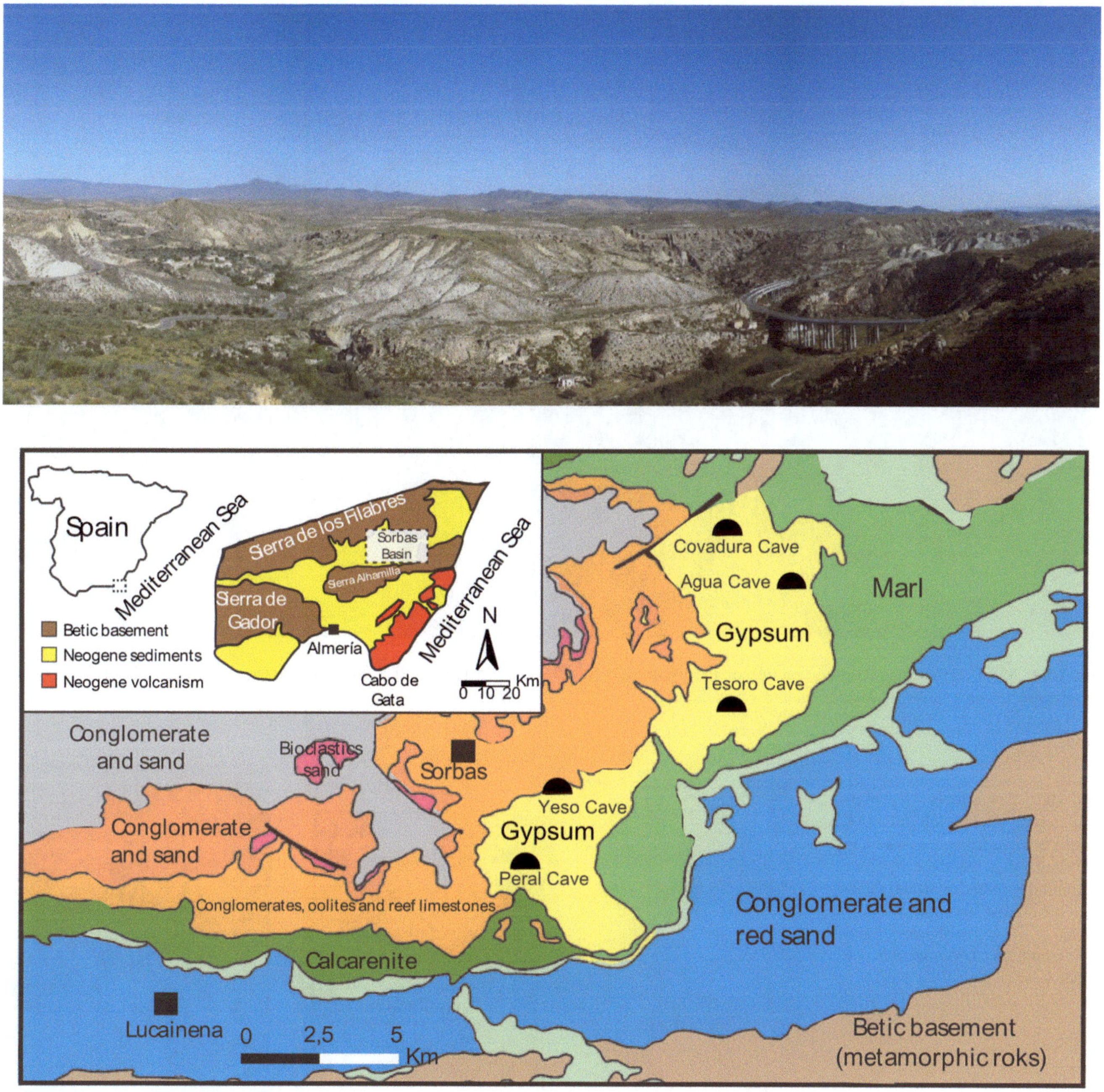

Fig. 10.1 Panoramic view of the gypsum karst of Sorbas (*above*) and geological sketch of the Sorbas Basin (*below*)

favoured by semi-arid conditions, so that Sorbas is an adequate area for their development (Calaforra and Pulido-Bosch 1999).

The genesis of tumuli is linked to currently active dissolution and reprecipitation processes (e.g. Artieda 2013). The porosity of the macrocrystalline gypsiferous rock enables rain and condensation water to flow though interstitial spaces and discontinuities between and within the large gypsum crystals. Initially, this water, with low mineral content, dissolves the gypsum. Subsequently, the high evaporation rate leads to saturation and recrystallization of the gypsum in the inter- and intracrystalline spaces. The precipitation of secondary gypsum is accompanied by crystal growth pressure, which induces an increase in volume and the gentle upward-doming of the gypsum layer situated at the ground surface (Fig. 10.3).

Fig. 10.2 Main surface geomorphic features of the gypsum karst of Sorbas: **a** tumuli; **b** dolines; **c** karren; **d** doline wallkarren; **e** gypsum escarpment (*Photographs* by Jabier Les and Laura Sanna)

Fig. 10.3 Evolution of the gypsum tumuli. The development of tumuli is a cyclic process which is favoured by a sequence of short wet and dry intervals which facilitate the almost simultaneous processes of dissolution and reprecipitation (*Photograph* by Jabier Les)

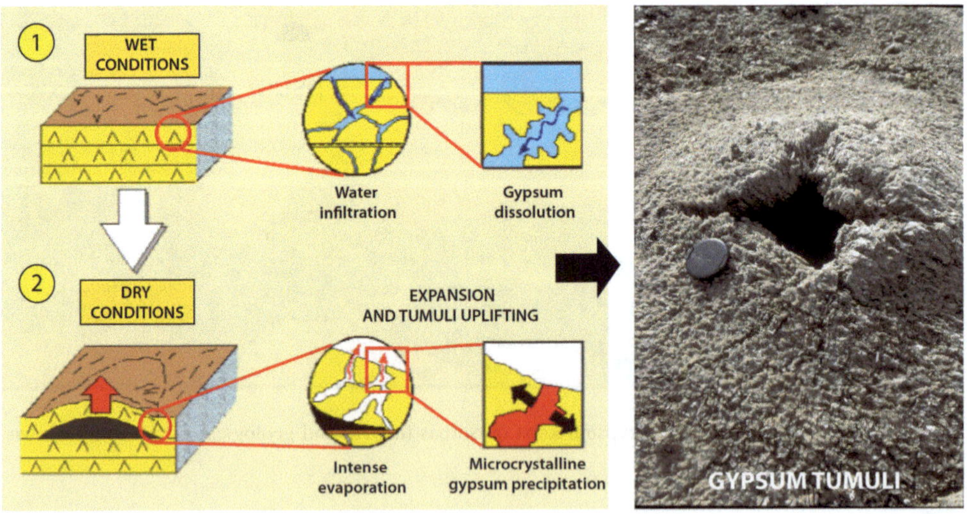

10.3.3 Karren

A wide variety of karren features may be found in gypsum karst areas, like in limestone terrains, but with some morphological differences partially related to factors such as the size and orientation of the crystals and the development of crusts (Gutiérrez and Cooper 2013).

Solution flutes consist of small grooves separated by sharp crests. Parallel ridges commonly occur on microcrystalline gypsum (Fig. 10.2c). Karren with large ridges and grooves are more common on outcrops of macrocrystalline gypsum crystals. Like in carbonate rocks, where dissolution has occurred below a soil cover, the karren tend to be rounded (roundkarren) and where the ground is very

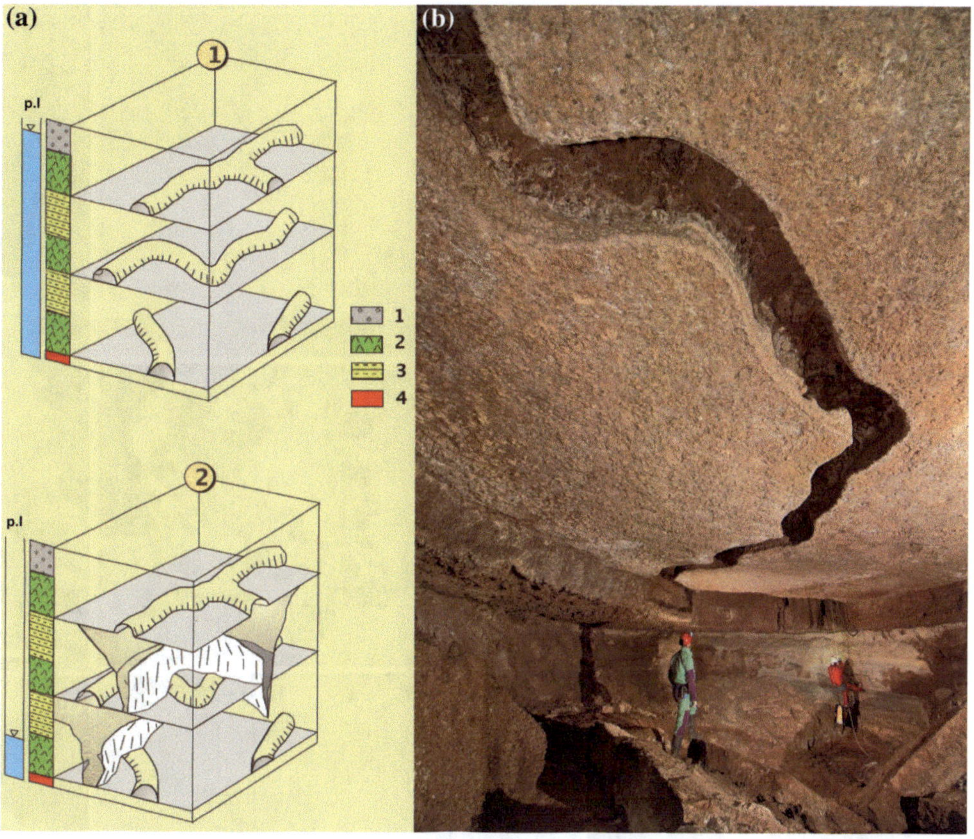

Fig. 10.4 a Evolutionary diagram of the gypsum karst of Sorbas: *1* gypsum karst proto-channels in multilayer aquifer under confined conditions; 2 development of large galleries by vadose erosion of marl strata. *Legend 1* units overlying the gypsiferous sequence, *2* gypsum, *3* interbedded marls, *4* basement, pl: piezometric level. **b** Proto-channels on the ceiling of gallery in Covadura Cave (*Photograph* by Victor Ferrer)

flat, solution features frequently display a sinuous form (meanderkarren). Nevertheless, the role of CO_2 is not important in gypsum dissolution and solubility and slope are the key factors for the formation of gypsum karren. Locally, water run-off on the walls of the dolines produces large vertical grooves, which often lead to the interior of caves (Fig. 10.2d).

10.3.4 Gypsum Escarpment

The gypsum landscape of Sorbas shows strong contrasts related to the distribution of the gypsum-bearing formation and the surrounding rocks. On a large scale, the gypsum karst landscape forms a scarp-edged platform (up to 30 m height) related to differential erosion of the softer surrounding marls and clays (Fig. 10.2e). This inverted relief is caused by the geological development of the karst itself. Whilst gypsum tends to dissolve and support caves at depth, surface erosion in the marl and clay outcrops occurs at a much faster rate. The scarp has a length over 4 km and is developed along the edge of the gypsum outcrop. The processes involved in the formation of the gypsum escarpment can be observed along the Río Aguas, where large rockfalls of gypsum blocks cause the gradual recession of the scarp.

10.4 Speleogenesis and Cave Evolution

Up to six passage levels have been recognized in the gypsum caves controlled by the stratification planes between the impervious marls and the soluble gypsum units (Fig. 10.4a). This arrangement is related to the hydrogeological history of the area. Initially, the gypsum karst evolved as a semi-confined multilayer aquifer under phreatic conditions, enabling the formation of small proto-conduits in individual gypsum beds, whilst the intervening marls and clays acted as impervious barriers (Fig. 10.4b).

During a subsequent stage, the decline of the piezometric level in relation to fluvial incision led to vadose conditions. Mechanical erosion in the underlying marls and clays became the dominant processes (Fig. 10.4a). This genetic duality indicates that the Sorbas gypsum karst can be considered an example of interstratal karstification, in which underground erosion processes and the resulting features should be considered products of the hydrogeological development.

Fig. 10.5 Examples of gypsum speleothems in the caves of Sorbas: **a** gypsum stalactites; **b** hollow stalagmites; **c** gypsum Christmas tree; **d** gypsum rims; **e** gypsum spikes in clayey sediments; **f** gypsum balls; **g** inactive ("fossil") gypsum trees; **h** airflow-controlled gypsum rims (*Photographs* by Jabier Les, Laura Sanna, Manuel Gutiérrez and Fernando Gázquez)

10.5 Peculiar Speleothems

Many of the galleries of the gypsum karst of Sorbas show a predominantly horizontal attitude, have relatively low vertical development (less of 10 m in many cases), as well as multiple entrances (Calaforra and Pulido-Bosch 2003). These features give rise to relatively intense airflow within the galleries that favours evaporation and gypsum precipitation in the form of speleothems (Calaforra et al. 2008; Gázquez et al. 2011; Gázquez 2012).

10.5.1 Gypsum Balls

One of the speleothems formed by capillary processes and water exudation from the gypsiferous bedrock are the gypsum balls (Fig. 10.5f). The water, with high calcium sulphate concentration, emerges into the cave through small cracks and pores and leads to evaporation-induced gypsum precipitation (Calaforra 2003). The resulting precipitates have a subspherical geometry with concentric structure and are frequently filled with clayey material derived from the gypsum bedrock. Occasionally, it is possible to observe their internal structure with a hollow, a clay-filled centre and an overlying gypsum crust.

10.5.2 Hollow Stalagmites

Hollow stalagmites are one of the most exceptional speleothems documented in the Sorbas gypsum karst (Figs. 10.5b and 10.6). Their formation involves simultaneous precipitation and dissolution of gypsum, but with the peculiarity that their origin is linked to calcite stalactites. A central orifice runs from apex to base of these stalagmites. The water dripping from the stalactites remains undersaturated with respect

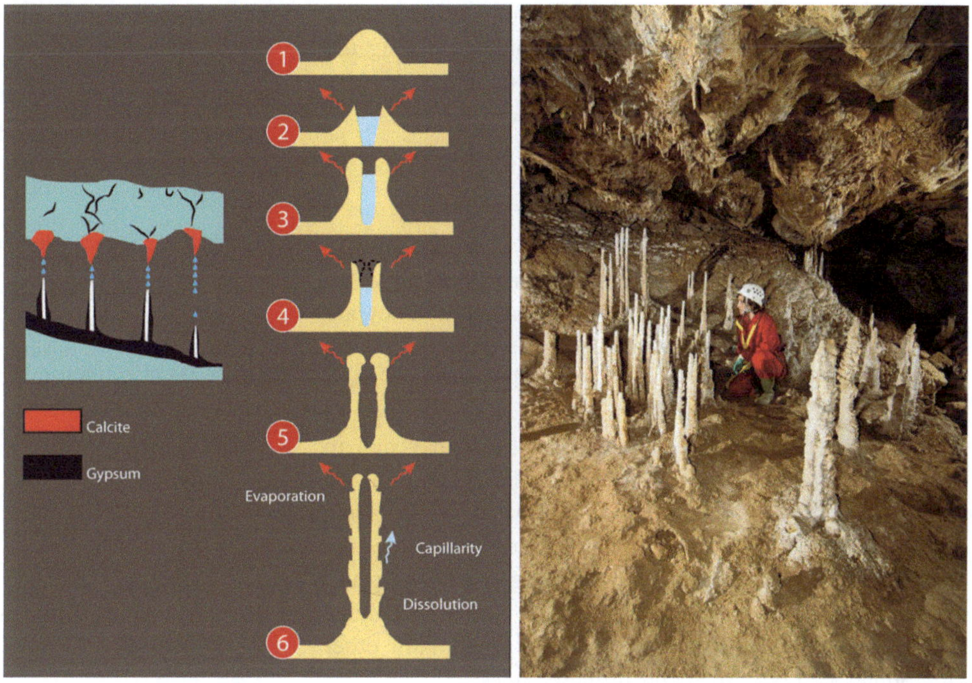

Fig. 10.6 Genesis of the gypsum hollow stalagmites. They form by dripping water coming from carbonate stalactites. The drops remain undersaturated due to the common ion effect (calcite–gypsum). Stalagmites develop a hole, whilst at the same time they are growing by capillarity and evaporation processes (1–6) (*Photograph* by Victor Ferrer)

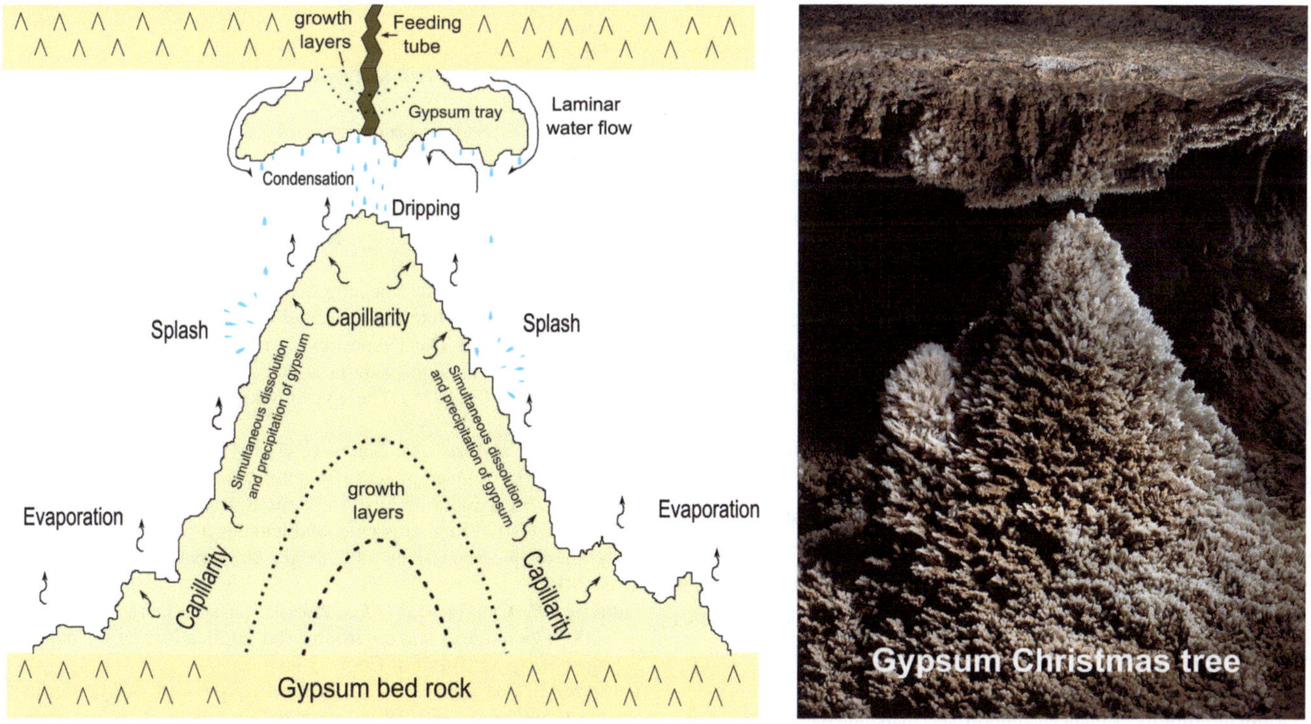

Fig. 10.7 Sketch of the gypsum trees and trays. This speleothem could be ascribed to the generic group called "coraloid speleothems". The complexity of processes involved in their formation can be summarized as (1) intense drip flow; (2) the impact of drops causes water splashing on the surface of the speleothem; (3) water rises by capillarity between speleothem crystals; (4) slight air currents cause the evaporation of water favouring capillarity and precipitation; and (5) condensation of evaporated water on the roof (the tray speleothem) with consequent corrosion (Photograph by Manuel Gutiérrez)

to gypsum due to calcite precipitation (common ion effect). Consequently, the central orifice is maintained during the growth of the stalagmite. Evaporation and capillarity give rise to these mineralogical natural jewels (Calaforra 1998).

10.5.3 Gypsum Trees and Trays

Gypsum trees and trays (Calaforra and Forti 1994) are other peculiar speleothems of the Sorbas gypsum karst. They are composed of aggregated gypsum acicules and centimetric crystals, resembling a Christmas tree (Fig. 10.5c). As a general rule, these speleothems are crowned by gypsum trays that function as a feeder that supplies water rich in calcium sulphate to the gypsum tree, where precipitation occurs during evaporation periods (Fig. 10.7). In some cases, these gypsum trees evolve into stalagmite-like structures when dripping from the ceiling ceases. Gypsum rims (Fig. 10.5d), gypsum spikes within clayey sediments (Fig. 10.5e), gypsum crusts (Fig. 10.5h), and many other precipitates can be also found in these caves.

10.5.4 Airflow-Controlled Speleothems

In addition to these unique speleothems and thanks to the multiple forms that gypsum can adopt as it precipitates, there are many other sorts of gypsiferous speleothems. Stalactites are the most common gypsum speleothems in the gypsum karst. In places, deflected stalactites can be found (Fig. 10.5a), which usually grow controlled by the main airflow direction in the caves as a result of higher evaporation rate in the windward side (Gázquez et al. 2011). Although the presence of gypsum speleothems is very common in the majority of the caves of the Sorbas karst, Covadura Cave, located on the north-western area of the gypsum massif (Fig. 10.1), hosts the wider variety of gypsum features mentioned above.

In order to preserve its caves, geomorphological and biological features, a total of 2,375 hectares were declared the Protected Natural Space of the Gypsum Karst of Sorbas in 1989 by the Andalucía Government (Calaforra 2003). In addition, the area received the status of Special Protection Area for Birds in 2002 and currently is proposed as a Site of Community Importance (Sanna et al. 2012).

Acknowledgments Financial support was made available through the Spanish Science Grant AP-2007-02799, funds from the Water Resources and Environmental Geology Research Group (University of Almería) and the "GLOCHARID" Project of the Junta de Andalucía Regional Government. The authors want to thank Dr Piotr Migon by his useful comment during the review of the manuscript, special thanks to Víctor Ferrer, Jabier Les, Manuel Gutiérrez and Laura Sanna for kind permission to use their photographs.

References

Andrejchuk V, Klimchouk A (1996) Gypsum karst of the Eastern-European plain. Int J Speleol 25:251–261

Artieda O (2013) Morphology and micro-fabrics of weathering features on gyprock exposures in a semiarid environment (Ebro Tertiary Basin, NE Spain). Geomorphology (in press)

Calaforra JM (1998) Karstología de yesos. PhD thesis, University of Granada (Spain), p 388

Calaforra JM (2003) El Karst en Yeso de Sorbas. Un recorrido subterráneo por el interior del yeso. Ed. Publicaciones Calle Mayor S.L., p 83

Calaforra JM, Forti P (1994) Two new types of gypsum speleothems from New México: gypsum trays and gypsum dust. Natl Speleol Soc Bull 56:32–37

Calaforra JM, Pulido-Bosch A (1996) Some examples of gypsum karst and the most important gypsum caves in Spain. Int J Speleol 25:225–237

Calaforra JM, Pulido-Bosch A (1997) Peculiar landforms in the gypsum karst of Sorbas (Southeastern Spain). Carbonates Evaporites 12:110–116

Calaforra JM, Pulido-Bosch A (1999) Genesis and evolution of gypsum tumuli. Earth Surf Proc Land 24:919–930

Calaforra JM, Pulido-Bosch A (2003) Evolution of the gypsum karst of Sorbas (SE Spain). Geomorphology 50:173–180

Calaforra JM, Forti P, Fernández-Cortés A (2008) Speleothems in gypsum caves and their palaeoclimatological significance. Environ Geol 53:1099–1105

Chardon M, Nicod J (1996) Gypsum karst of France. Int J Speleol 25:203–208

Cooper AH (1996) Gypsum karst of Great Britain. Int J Speleol 25:195–202

Cooper AH, Gutiérrez F (2013) Dealing with gypsum karst problems: hazards, environmental issues and planning. In: Frumkin A (ed) Treatise on geomorphology. Karst geomorphology, vol 6. Elsevier, San Diego, pp 451–461

Dronkert H (1976) Late Miocene evaporites in the Sorbas basin and adjoining areas. Memoria Sociedad Geologica Italiana 16:341–362

Dronkert H (1977) The evaporites of the Sorbas Basin. Revista de Investigación Geológica de la Diputación de Barcelona 33:55–76

Forti P, Sauro U (1996) The gypsum karst of Italy. Int J Speleol 25:239–250

Gázquez F, Calaforra JM, Sanna L, Forti P (2011) Espeleotemas de yeso: ¿Un nuevo proxy paleoclimático? Boletín de la Real Sociedad Española de Historia Natural 105 (1–4):15–24

Gázquez F (2012) Registros paleoambientales a partir de espeleotemas yesíferos y carbonáticos. PhD thesis, University of Almería, Spain, p 381

Gutiérrez F, Calaforra JM, Cardona F, Ortí F, Durán JJ, Garay P (2008) Geological and environmental implications of the evaporite karst in Spain. Environ Geol 53:951–965

Gutiérrez F, Cooper AH (2013) Surface morphology of gypsum karst. In: Frumkin A (ed) Treatise on geomorphology. Karst geomorphology, vol 6. Elsevier, San Diego, pp 425–437

Johnson KS (1996) Gypsum karst in the United States. Int J Speleol 25:183–193

Kempe S (1996) Gypsum karst of Germany. Int J Speleol 25:209–224
Klimchouk A, Forti P, Cooper A (1996) Gypsum karst of the world: a brief overview. Int J Speleol 25:159–181
Klimchouk AB, Aksem SD (2002) Gypsum karst in the Western Ukraine: hydrochemistry solution rates. Carbonates Evaporites 17(2):142–153
Krijgsman W, Fortuin AR, Hilgen FJ, Sierro FJ (2001) Astrochronology for the Messinian Sorbas basin (SE Spain) and orbital (precessional) forcing for evaporite cyclicity. Sed Geol 140:43–60
Mather AE, Harvey AM (1995) Controls on drainage evolution in the Sorbas Basin, SE Spain. In: Lewin J, Macklin MG, Woodward JC (eds) Mediterranean Quaternary river environments. Balkema, Rotterdam, pp 65–76
Roep TB, Beets DJ, Dronkert H, Pagnier H (1979) A prograding coastal sequence of wave-built structures of Messinian age, Sorbas, Almeria, Spain. Sed Geol 22:135–163
Sanna L, Gázquez F, Calaforra JM (2012) A geomorphological approach in the study of hydrogeology of gypsum karst of Sorbas (SE Spain). Geografia Fisica e Dinamica Quaternaria 35:153–166
Trzcinski Y (1996) Gypsum karst in the South of the Siberian platform, Russia. Int J Speleol 25:293–295
Yaoru L, Cooper AH (1996) Gypsum karst in China. Int J Speleol 25:297–307

Gallocanta Saline Lake, Iberian Chain

11

F. Javier Gracia

Abstract

Gallocanta Lake, covering 14.5 km^2, is the greatest ephemeral saline lake in Europe. It is located in the Iberian Chain, NE Spain, in the bottom of a karst polje. The Gallocanta saline lake formed once Jurassic limestones were almost completely corroded, and the floor of the depression was underlain by Triassic clays and evaporites. The late Quaternary evolution of the lake can be reconstructed from the deposits underlying the lake bottom and from different levels of lacustrine terraces located in its downwind side. Different phases of flooding and desiccation can be deduced from both sources of data. The current dynamics of the lake is controlled by water-level fluctuations and wind action. Wind-driven waves and longshore currents transport sediments to the downwind zone and generate barrier islands, spits and submerged bars, with a dynamic behaviour very similar to that of marine coastal environments. Lake segmentation due to cuspate foreland growth has divided the original lake into minor ones. Segmentation is still active at present and tends to isolate a minor lacustrine body. Progressively decreasing rainfall, together with sediment supply to the lake, enhanced by extensive agricultural practices in the basin, have frequently led to lake desiccation over the last decades. Extensive polygonal soils and salt crusts cover the bottom during drying-up periods.

Keywords

Saline lake • Holocene evolution • Coastal dynamics • Iberian Chain

11.1 Introduction

Gallocanta Lake, covering 14.5 km^2, is the greatest ephemeral saline lake in Europe, and probably one of the best preserved (Fig. 11.1). It is 7.7 km long, 2.8 km wide, and has a maximum depth of 2.5 m (reached in 1917; Rodó et al. 2002), although during dry periods, it becomes completely desiccated. Water salinity is between 100 and 1,000 times higher than the salinity of the fresh meteoric water entering the lake (Comín et al. 1990). Salts in the lake waters are entirely supplied by underground flow from underlying evaporites. The sediments in the centre of the lake consist of carbonate and sulphate muds, while during dry periods, the bottom is covered by a thin and discontinuous salt crust.

In 1972, Gallocanta Lake was declared Zone of Controlled Hunting, in 1984 National Hunting Refuge, and in 1995 Wildlife Refuge of International Interest. Since 1988, it is included in the list of wetlands with international interest (Ramsar Convention). It is also Zone of Special Protection by the European Birds Directive. In 2006, the lake and a peripheral zone were declared Nature Reserve by the regional government. The lake constitutes an extraordinary

F. J. Gracia (✉)
Department of Earth Sciences, University of Cádiz, Cádiz, Spain
e-mail: Javier.gracia@uca.es

F. Gutiérrez and M. Gutiérrez (eds.), *Landscapes and Landforms of Spain*, World Geomorphological Landscapes,
DOI: 10.1007/978-94-017-8628-7_11,

Fig. 11.1 Views of Gallocanta Lake. *Left* satellite image of Gallocanta Lake taken in July 2010 (Google Earth). *Right* oblique aerial view taken in winter time (*Photo* B. Leránoz)

important station for European migratory birds. Among the numerous species identified every year, the most significant and representative one is the crane (*Grus grus*), which can be present every winter in an average number of individuals higher than 25,000 (Sampietro 2002). Two visitor centres and a Museum of Wild Birds can be visited in the surroundings of the lake.

11.2 Geographical and Geological Setting

The Gallocanta Depression is located in the central sector of the Iberian Chain (NE Spain, Fig. 11.2). This closed topographic basin has a catchment area of 550 km^2, with an elongated shape parallel to the main structural grain of this Alpine orogen (NW–SE). It is flanked by mountain ranges with summits reaching more than 1,400 m a.s.l. The bottom of the depression is located at an average elevation of 1,000 m a.s.l. The endorheic basin hosts more than 20 lakes of variable size, Gallocanta Lake being the greatest. Their nature is variable (e.g. permanent, ephemeral, filled with sediments) and most of them are freshwater lakes. However, Gallocanta Lake and some other minor ponds located in the centre of the basin have saline waters.

The zone has a typically semiarid climate with a mean annual precipitation of around 450 mm and an average annual temperature of 10–11 °C. The depression is dominated by NW winds channelled along the topographic trough, reaching up to 100 km/h.

Geologically, the ranges bounding the depression are composed of Palaeozoic and Lower Triassic siliceous rocks. The NE range close to the lake (Fig. 11.2) is formed by an isoclinal series of Ordovician quartzites. This mountain range flanks an extensive fault-bounded outcrop of deformed Mesozoic sediments. Upper Triassic clays and evaporites (Keuper facies) form the impervious substratum of the Gallocanta Lake. The rest of the Mesozoic units correspond to Jurassic and Upper Cretaceous carbonate rocks. The Alpine deformation structures show a prevalent NW–SE trend. Locally, the carbonate Mesozoic sediments are unconformably overlain by Tertiary detrital sediments and some of them, of probable Late Pliocene age, form isolated mesas.

Quaternary landforms and deposits are mainly represented by pediments, infilled valleys and lacustrine sediments (Fig. 11.2). Pediments develop at the foot of the important mountain fronts and are mantled by a thin alluvial cover. The erosion of the NE quartzitic range produced a sequence of three pediment levels, inset and stepped towards the depression bottom. Wide flat-bottomed valleys develop inside the mountain ranges and are filled by alluvial sediments supplied by the surrounding slopes, many of them characterized by active frost-shattering processes during winter.

The lacustrine sediments around Gallocanta Lake form a stepped sequence of lacustrine terraces. Under the lake, just a ca. 1 m thick sequence of lacustrine sediments has been identified by drilling (Schütt 1998; Pérez et al. 2002, among others). Several stepped Neogene planation surfaces of regional extent can be recognized in the surroundings of Gallocanta Depression (Gutiérrez and Gracia 1997). The lacustrine depression is inset with respect to these regional planation levels, indicating a post-Pliocene age for its generation.

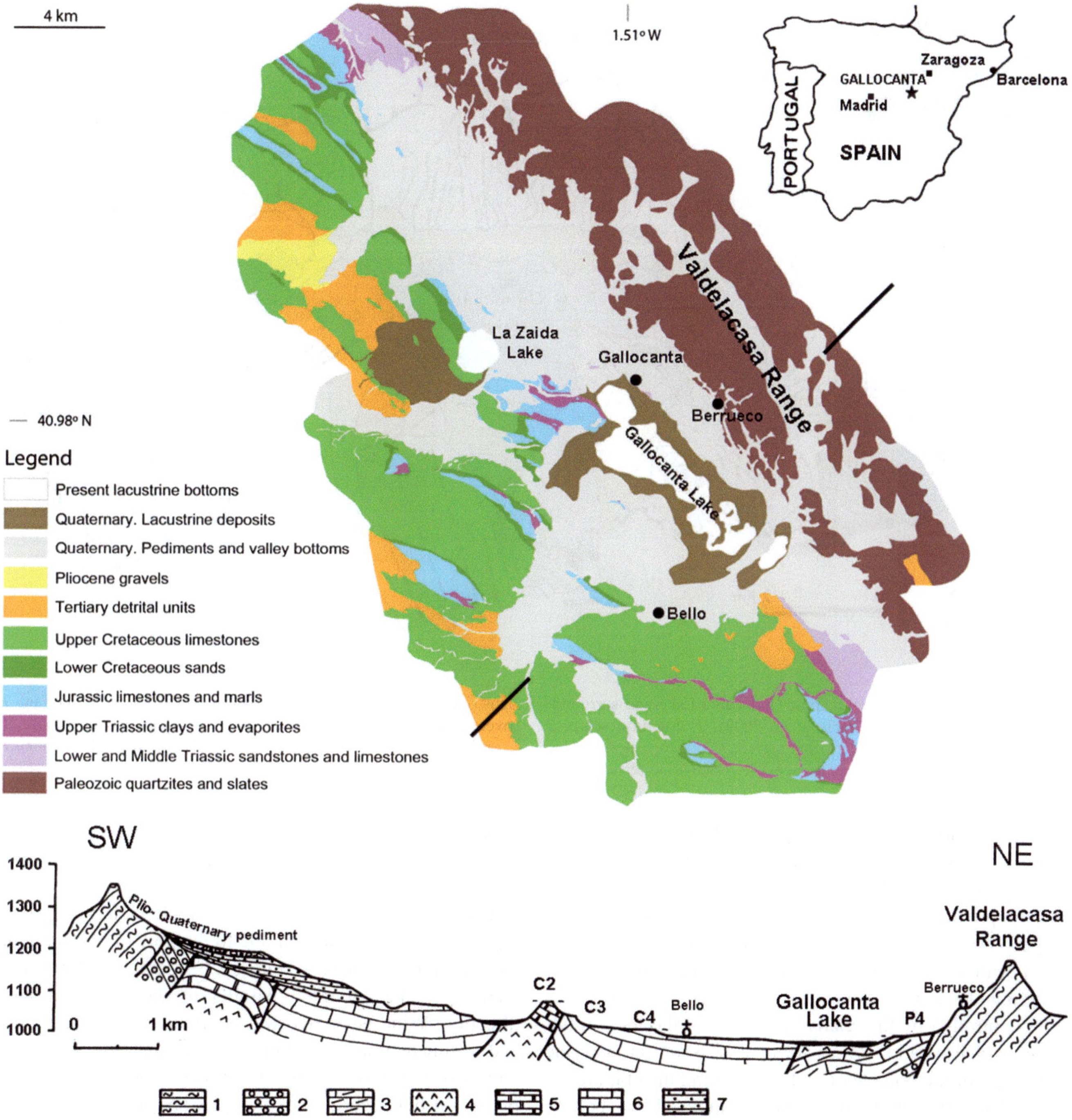

Fig. 11.2 *Top* Location and geological map of Gallocanta lacustrine depression (modified from CHE 2003 by C. Castañeda). *Bottom* Geological cross section of Gallocanta polje (modified from Gracia et al. 2002). Legend: *1* Palaeozoic quartzites and slates, *2* Lower Triassic sandstones, *3* Middle Triassic carbonates, *4* Triassic clays and evaporites (Keuper), *5* Jurassic limestones, *6* Cretaceous limestones, *7* Neogene sands and conglomerates, *C* Karstic corrosion surfaces, *P* pediment

11.3 The Origin and Evolution of the Lacustrine Depression

The origin of the depression was interpreted by Gracia et al. (2002) as a polje developed during the Quaternary. Four corrosion surfaces can be recognized in the basin, the two lowest ones surrounding the main lake (Fig. 11.3), with dolines and abundant karren locally covered by residual clays. The corrosion surfaces show a concentric distribution and are stepped towards the Gallocanta Lake. They develop upon Jurassic limestones, being constrained by the outcrops of Pliocene detrital deposits.

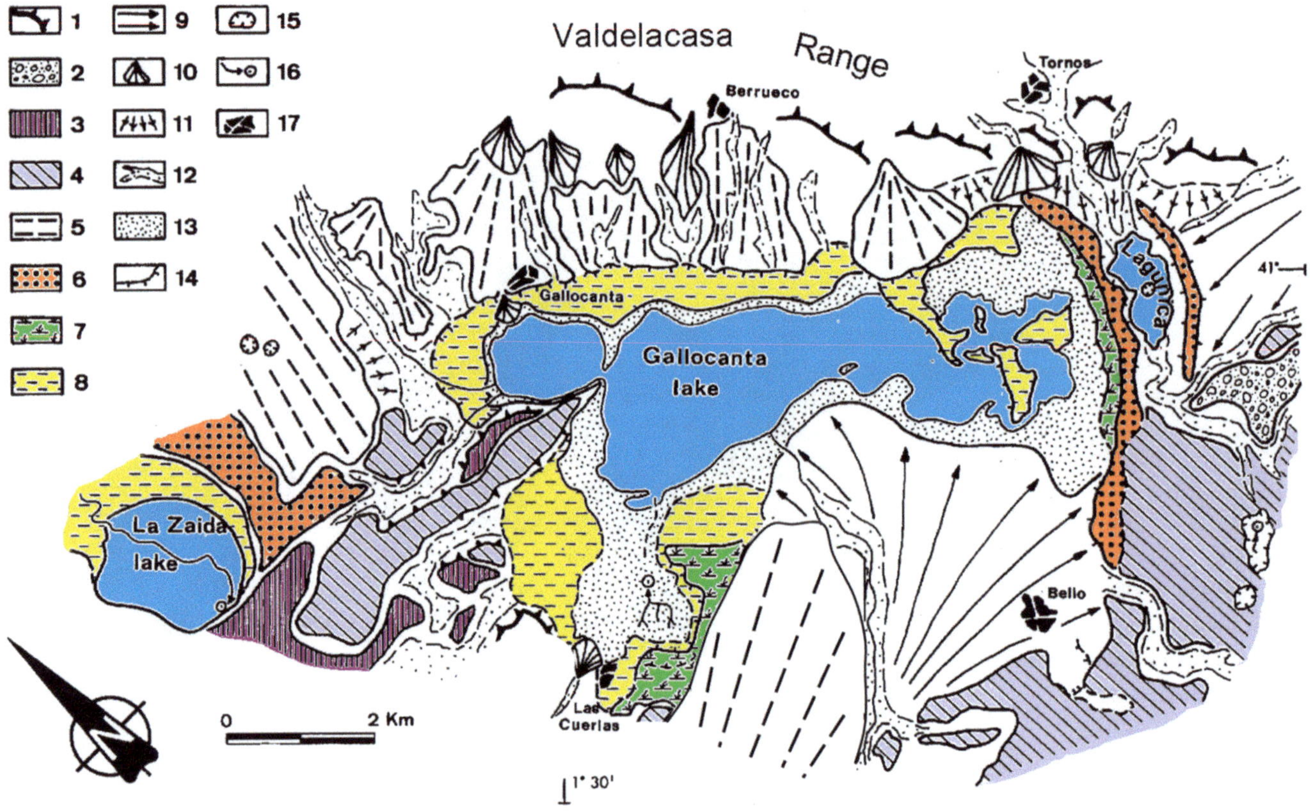

Fig. 11.3 Geomorphological map of Gallocanta Lake and surrounding areas (modified from Gracia et al. 2002). *1* structural scarp, *2* Neogene clastic deposits, *3* corrosion surface C3. *4* corrosion surface C4. *5* pediment P4. *6* lacustrine terrace T4, *7* lacustrine terrace T5. *8* lacustrine terrace T6. *9* mantled pediment P7. *10* alluvial fan, *11* covered slope, *12* flat-bottomed valley, *13* lacustrine floodplain, *14* scarp in Quaternary deposits, *15* Doline, *16* Swallow hole (ponor), *17* village

The formation of the depression and the corrosion surfaces probably occurred after a regional extensional tectonic phase in the Late Pliocene (Gutiérrez et al. 2008). This episode produced differential vertical movements along the range. Very likely, tectonic subsidence generated a shallow basin with internal drainage. The deepening of the polje bottom and the development of stepped corrosion surfaces were controlled by the relative lowering of the local water table, which would favour vertical dissolution and the deepening of the polje bottom until reaching the epiphreatic zone. The alternation of periods dominated by bottom deepening and periods of planation of the polje bottom, both controlled by the position of the water table, resulted in the four stepped corrosion surfaces found in the Gallocanta Polje. The deepening stages of the polje bottom also involved the generation of three stepped levels of mantled pediments around the lake. The deepening of the polje bottom is restricted by the thickness of the soluble materials, about 150 m in this case. Once the polje floor approached the impervious Triassic substratum, a stable lacustrine system developed in its bottom. Consequently, the generation of Gallocanta Lake is linked to the interruption of the polje deepening in the Late Pleistocene (about 12,200 year BP), according to the numerical dates presented by Burjachs et al. (1996).

During its evolution, the Gallocanta Lake has undergone a progressive segmentation due to the growth of paired and cuspate littoral spit bars (Fig. 11.3), a common process in elongated lakes oriented parallel to the dominant wind direction. Wind-generated waves produce shoreline erosion and sediment transport along the lake margins. The dissipation of the wave energy takes place roughly at similar places on both margins of the lake and leads to sediment deposition and the growth of paired spits or cuspate forelands (Lees 1989). The sedimentological and palynological analysis of lake sediments from boreholes shows different stages of lake evolution (Schütt 1998; Luzón et al. 2007a). After an initial development of the lake in the Last Glacial Maximum, during which marginal alluvial fans developed, a second stage began around 10,100 year BP with the establishment of a perennial brackish lake. The maximum lake level was probably reached around 8,010 year BP, with about 10 m water depth. The top surface of the oldest lacustrine terrace (T4), 8 m above the present-day high

Fig. 11.4 Oblique aerial view of the channel between paired spit bars connecting the two main lake portions (*Photo* F. Gutiérrez)

Fig. 11.5 Oblique aerial view of the coastal barrier islands and spits along the southwestern shore of the lake (*Photo* F. Gutiérrez)

water level, connects to the lowest corrosion surface C4 and pediment level P4, and hence, all these surfaces can be considered as roughly coeval. This terrace level constitutes the depositional closure of the original downwind lake embayment and produced the first segmentation episode with the generation of the Lagunica Lake (Fig. 11.3). The sedimentological characteristics of the lacustrine deposit of this terrace level suggest a succession of changing conditions related to lake-level fluctuations (Gracia 1995).

An intermediate episode of water level fall (between 3,405 and 1,510 year BP) could be related to a regional reduction in the hydrological balance (Luzón et al. 2007a). Increasing aridity in historical times (post-Middle Age) produced an ephemeral carbonate-saline lake with frequent oscillations. During this period, new lake segmentation episodes occurred, tending to subdivide the main lacustrine body into minor ones. Although this process is still incomplete, a palustrine zone has formed in the SE border

Fig. 11.6 Lake *bottom* during desiccation episodes. **a** Non-orthogonal desiccation cracks on muds (August 2012), **b** Small pedestals and patterned ground revealing an underlying salt layer, **c** Desiccation of algal mats in a palustrine shore

of the lake and new paired cuspate forelands have grown, which divide the lake into two main portions, connected by a channel with intermittent activity (Figs. 11.3 and 11.4). About 750 years ago, a change is recorded in the lake sediments, from a shallow carbonate lacustrine system with oscillating water level to a saline lake rich in carbonates and organic matter associated with a significant drop in the water level. This change could be related to karst processes developed in the underlying Triassic evaporites (Gracia 1990) or to changes in the underground watershed (Luzón et al. 2007b).

11.4 The Lake

At the present time, the Gallocanta Lake undergoes frequent water-level fluctuations, ranging between 2.5 m of maximum water depth and complete desiccation. Long-term patterns of lake-level variations have been inferred from mineralogical analyses, dating of lacustrine sediments, aerial photogrammetry, historical satellite imagery and in situ continuous measurement of water depth on the deepest sector of the lake. All these records reveal a positive response to the El Niño Southern Oscillation (ENSO), while the lake seems insensitive to the North Atlantic Oscillation (Rodó et al. 1997). Apart from their influence on geochemical and biological processes (Comín et al. 1992), lake-level fluctuations also have important geomorphological effects.

During flooding periods, wind-generated waves and longshore currents travel towards the downwind side of the lake (SE). Clastic sediment supplied by rivers, mainly along the SW lake shore, is transported and deposited throughout the southern coastline. As a consequence, a set of barrier islands and spits develops which enclose small lagoons, in a similar way as those characteristic of marine coasts (Fig. 11.5). When the water level is at its highest stage, some of these depositional landforms are almost completely flooded and become islands (Fig. 11.1).

The progressive sedimentary infill of the SW coastal lagoons is accompanied by an increase in water depth in the northern shore, where waves become higher and more erosive. Different indicators of coastal erosion and shoreline retreat can be observed along the NE lake shore. A retreat of

Fig. 11.7 Salts in the lake bottom during drying-up periods. **a** Salt accumulation on desiccation cracks, **b** Polygons on a gypsum and organic matter-rich ground, **c** Aeolian deposition of salts in the downwind side of the lake (*Photo* P. Vicente), **d** Salt trapping by halophyte plants

about 400 m can be deduced for the eastern border of the central lacustrine body (Gracia 1995), where a 2-m-high microcliff has developed, probably over the last millennium.

During drought periods, the lake becomes completely desiccated and its bottom shows a patterned ground related to the development of different types of desiccation cracks, mainly non-orthogonal systems (Fig. 11.6a). In some cases, stone micro-pedestals and sorted polygons can be observed near the shores, where small stones fill the cracks (Fig. 11.6b). These may be attributed to salt expansion and contraction processes because of thermal changes (Hunt and Washburn 1966). In palustrine areas, algal mats desiccate and discontinuously cover the muddy shores (Fig. 11.6c).

Gallocanta Lake is a hypersaline lake of the Na–Mg–Cl–(SO_4) type (Comín et al. 1990). Intense evaporation and strong winds induce the precipitation of salts. Desiccation cracks are often covered by a thin layer of salt crystals (Fig. 11.7a). The presence of organic matter and gypsum favours the development of deeper cracks and greater polygons (Fig. 11.7b). Prevailing winds transport salts, which accumulate in the downwind margin of the lake (Fig. 11.7c). Most commonly halophyte plants, such as *Salicornia*, act as aeolian salt traps (Fig. 11.7d). Renewed flooding episodes cause mud accumulation in the bottom and silts and fine sands in the shores.

11.5 Conclusions

The evolution, morphology and present dynamics of Gallocanta Lake are strongly controlled by the nature of its substratum and the frequent water-level fluctuations. Karstic corrosion of Jurassic carbonates and the consequent deepening of the Gallocanta polje ended once the underlying Triassic evaporites became the substratum of the lake, causing its salination. Structural control on the distribution of Mesozoic formations led the lacustrine depression to acquire an elongated morphology parallel to the regional structural grain, which coincides with the direction of the prevailing winds channelized along the depression. Climatic oscillations governed the hydrological evolution of the lake during the Holocene, giving rise to a set of stepped lacustrine terraces. Throughout its evolution, prevailing winds generated waves and currents which transported sediments and formed coastal sandy barriers. Wave action also induced the generation and growth of cuspate forelands which segmented the lake in several phases, a process still active. In recent times, climate aridification and hydrogeological processes have changed the lake water geochemistry, leading to the concentration of salts and transforming the originally carbonate lake into a saline lake. Dry periods often lead to the complete desiccation of the lake, with extensive salt precipitation. Climatic trends suggest that this may become a progressively more frequent situation in the near future.

References

Burjachs F, Rodó X, Comín FA (1996) Gallocanta: ejemplo de secuencia palinológica en una laguna efímera. In: Ruiz Zapata B (ed) Estudios Palinológicos. University of Alcalá, Alcalá de Henares, pp 25–29

CHE (Confederación Hidrográfica del Ebro) (2003) Establecimiento de las normas de explotación de la unidad hidrogeológica "Gallocanta" y la delimitación de los parámetros de protección de la laguna. Confederación Hidrográfica del Ebro. Zaragoza

Comín FA, Julià R, Comín MP, Plana F (1990) Hydrogeochemistry of Lake Gallocanta (Aragón, NE Spain). Hydrobiologia 197:51–66

Comín FA, Rodó X, Comín P (1992) Lake Gallocanta (Aragón, NE Spain), a paradigm of fluctuations at different scales of time. Limnetica 8:79–86

Gracia FJ (1990) Evolución morfológica reciente de la Laguna de Gallocanta (Cordillera Ibérica central). 1ª Reun. Nac. Geomorf, S.E.G., Teruel, pp 277–287

Gracia FJ (1995) Shoreline forms and deposits in Gallocanta Lake (NE Spain). Geomorphology 11:323–335

Gracia FJ, Gutiérrez F, Gutiérrez M (2002) Origin and evolution of the Gallocanta polje (Iberian Range, NE Spain). Z. Geomorph. N.F. 46:245–262

Gutiérrez M, Gracia FJ (1997) Environmental interpretation and evolution of the Tertiary erosion surfaces in the Iberian Range (Spain). In: Widdowson M (ed) Palaeosurfaces: recognition, reconstruction and palaeoenvironmental interpretation. Geol Soc Spec Publ 120, London, pp 147–158

Gutiérrez F, Gutiérrez M, Gracia FJ, McCalpin JP, Lucha P, Guerrero J (2008) Plio-Quaternary extensional seismotectonics and drainage network development in the central sector of the Iberian Chain (NE Spain). Geomorphology 102:21–42

Hunt CB, Washburn AL (1966) Patterned ground. In: Hunt CB et al. (eds) Hydrologic basin death valley, California. US Geol Surv Prof. Paper 494B, pp 104–133

Lees B (1989) Lake segmentation and lunette initiation. Z. Geomorph. N.F. 33:475–484

Luzón A, Pérez A, Mayayo MJ, Soria AR, Sánchez Goñi MF, Roc AC (2007a) Holocene environmental changes in the Gallocanta lacustrine basin, Iberian Range, NE Spain. Holocene 5:649–663

Luzón A, Pérez A, Sánchez JA, Soria AR, Mayayo MJ (2007b) Evolution from a freshwater to saline lake: a climatic or hydrogeological change? The case of Gallocanta Lake (northeast Spain). Hydrol Proc 21:461–469

Pérez A, Luzón A, Roc AC, Soria AR, Mayayo MJ, Sánchez JA (2002) Sedimentary facies distribution and genesis of a recent carbonate-rich saline lake: Gallocanta Lake, Iberian Chain, NE Spain. Sedim Geol 148:185–202

Rodó X, Baert E, Comín FA (1997) Variations in seasonal rainfall in Southern Europe during the present century: relationships with the North Atlantic Oscillations and the El Niño-Southern Oscillation. Clim Dyn 13:275–284

Rodó X, Giralt S, Burjachs F, Comín FA, Tenorio RF, Julià R (2002) High-resolution saline lake sediments as enhanced tools for relating proxy palaeolake records to recent climatic data series. Sedim Geol 148:203–220

Sampietro (2002) Las aves. In: Mañas J (coord.) Guía de la naturaleza de Gallocanta. Prames Ed, Zaragoza, pp 72–97

Schütt B (1998) Reconstruction of Holocene paleoenvironments in the endorheic basin of Laguna de Gallocanta, Central Spain by investigation of mineralogical and geochemical characters from lacustrine sediments. Jour Paleolimnology 20:217–234

12 Playa-Lakes and Yardangs in the Bujaraloz-Sástago Endorheic Area, Central Ebro Basin

F. Gutiérrez and M. Gutiérrez

Abstract

The Bujaraloz-Sástago endorheic area occurs on an exhumed structural platform in the central sector of the Ebro Cenozoic Basin, essentially underlain by subhorizontally lying gypsiferous and mudstone units with some limestones. The dominantly flat topography of this structural surface is interrupted by around 150 closed depressions, some of which host playa-lakes of outstanding ecological and geomorphological value. The origin of the depressions is related to subsurface dissolution of the gypsiferous bedrock and aeolian deflation caused by the strong local wind, called *Cierzo*. The leeward side of the largest playas displays yardangs carved on bedrock and unconsolidated Holocene lake terrace deposits. These are the only yardangs documented in Europe so far. Modern and relict lunette dunes also occur on the downwind margin of some playa-lakes. Lacustrine terraces preserved on the margins of the largest basins record alternating periods of aggradation and excavation, attributable to more humid and drier periods, respectively. The available radiocarbon dates from the most extensive terrace, allow us to infer deepening of the largest playa (La Playa) by wind erosion of 6 m over the last 2 ka, yielding an average lowering rate of ca. 3 mm/year. This figure compares well with those calculated in several arid regions of the world, mainly using yardangs carved in Holocene lake deposits.

Keywords

Deflation basins • Lake terraces • Lunette dunes • Gypsum dissolution • Wind erosion rates

12.1 Introduction

The Bujaraloz-Sástago endorheic area includes around 150 closed depressions, some of which host the northernmost playa-lakes in Europe. These fragile semiarid environments, with unique habitats and numerous endemic species, have an outstanding ecological value (Conesa et al. 2011). In 2011, an area of 8,144 ha, including the most representative and best preserved closed depressions and playa-lakes, was declared a RAMSAR site. Moreover, most of the area has been designated by the Regional Government as ZEPA (SPA; Special Protection Area for wild birds) and LIC (SCI; Site of Community Importance). The playa-lakes of Bujaraloz-Sástago are also of remarkable geological interest, with geomorphic features unparalleled in Spain. The development of these basins results from subsurface dissolution of the gypsiferous bedrock and aeolian deflation, which operates when the floor of the playa-lakes is dry. The leeward side of the largest playas displays numerous yardangs oriented parallel to the dominant wind direction and carved in bedrock and lake terrace deposits (Gutiérrez-Elorza et al. 2002). To our knowledge, these are the only

F. Gutiérrez (✉) · M. Gutiérrez
Department of Earth Sciences, University of Zaragoza, Zaragoza, Spain
e-mail: fgutier@unizar.es

F. Gutiérrez and M. Gutiérrez (eds.), *Landscapes and Landforms of Spain*, World Geomorphological Landscapes,
DOI: 10.1007/978-94-017-8628-7_12, © Springer Science+Business Media Dordrecht 2014

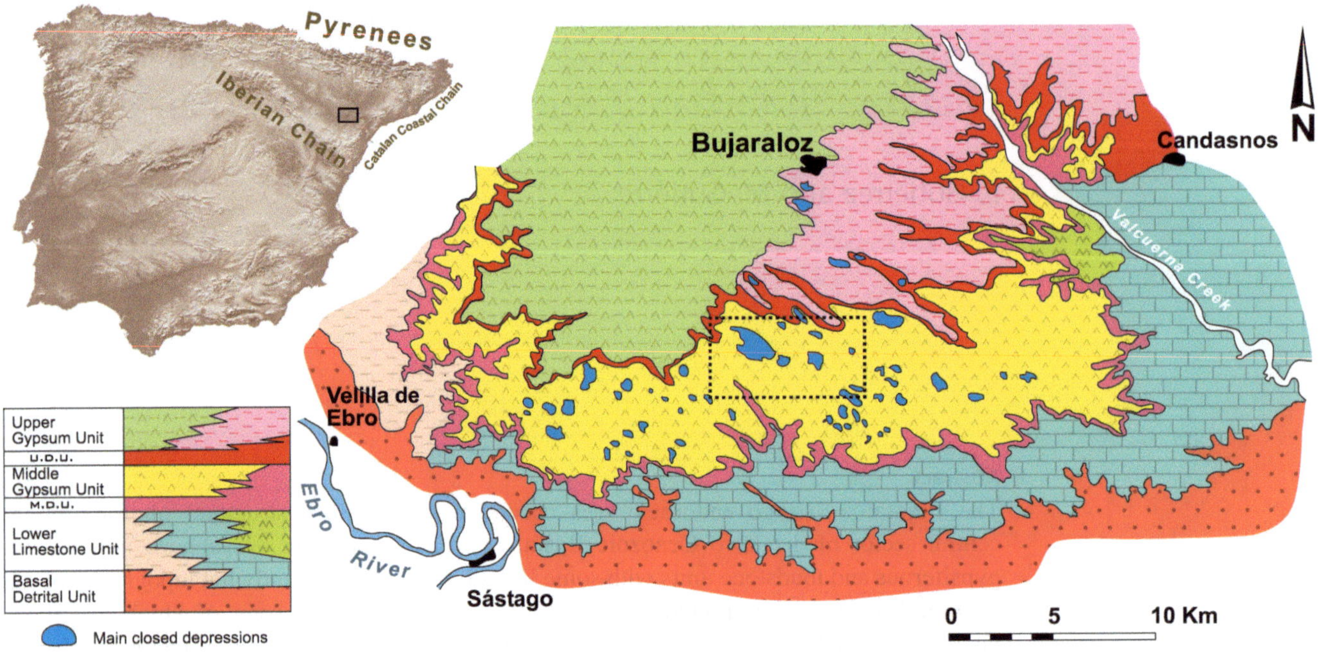

Fig. 12.1 Location and geological map of the Bujaraloz-Sástago endorheic area, developed on a structural surface perched around 200 m above the Ebro River. Note that, most of the closed depressions and playas occur on the unit with a higher proportion of gypsum; Middle Gypsum Unit (based on Salvany et al. 1996). *Dashed line rectangle* indicate area covered by Fig. 12.2

yardangs reported in Europe (Goudie 2007), together with some examples from Hungary (Sebe et al. 2011). The lacustrine terraces mapped in the margins of the largest basins record alternating periods of aggradation and excavation, attributable to more humid and drier periods, respectively (Gutiérrez et al. 2013). Excavation rates estimated on the basis of radiocarbon ages obtained from a Holocene terrace reveal the geomorphic effectiveness of aeolian deflation caused by the strong local wind (*Cierzo*) in this peculiar environment. The playa-lakes in this sector of the Ebro Basin have been some of the most intensely studied lakes in Spain for paleoenvironmental interpretations, based on cores drilled in the bottom of the depressions (González-Sampériz et al. 2008; Gutiérrez et al. 2013 and references therein). However, these investigations have encountered significant limitations in the analysed stratigraphic records, like the low thickness of the deposits, the presence of significant hiatuses related to wind erosion, or the difficulty of finding datable material. Recent studies illustrate that these drawbacks may be partially overcome by incorporating a geomorphological perspective in the investigations and considering the lacustrine terraces preserved in the basin margins. These terraces may be used to (1) establish the morpho-stratigraphic evolution of the basins and identify major paleoenvironmental changes; (2) enlarge the completeness of the stratigraphic record and dating possibilities; and (3) estimate rates of aeolian deflation. The ongoing irrigation plan (Monegros II), affecting ca. 200 km^2 of the endorheic area, will cause significant alterations on the hydrology of these highly sensitive wetlands, with negative consequences on the biocoenosis and some geomorphic processes.

12.2 Geological Setting, Climate and Hydrology

The analysed playa-lake and yardang systems are located in the central sector of the Ebro Cenozoic Basin, NE Spain (Fig. 12.1). This sedimentary basin, deeply dissected by the fluvial network, constitutes a large topographical depression drained by the trunk NW–SE-oriented Ebro River and bounded by the Pyrenees and the Iberian Chain to the north and south, respectively. The Bujaraloz-Sástago internally drained area, characterised by numerous closed depressions with a marked WNW–ESE elongation, has developed on an exhumed structural platform lying at 310–370 m a.s.l., hanging 200 m above the deeply entrenched Ebro River to the south (Fig. 12.1).

The sedimentary fill in this sector of the Ebro Basin is made up of Oligo-Miocene sediments deposited in evaporite and carbonate shallow lakes and in distal alluvial fan environments (Fig. 12.1). The strata, which show a very low (<2°) NW to NE dip, form part of the southern limb of a very open WNW–ESE syncline whose pericline is located SE of Bujaraloz (Quirantes 1978). The structural platform is underlain by the Lower Miocene Bujaraloz-Sariñena Unit, mostly composed of gypsum, mudstone and limestone

(Ramírez 1997; Solá and Costa 1997). Salvany et al. (1994, 1996) differentiate two main gypsum-rich units, each underlain by a detrital unit primarily composed of red mudstones (Fig. 12.1). From base to top, these are Middle Detrital Unit (15–20 m), Middle Gypsum Unit (40 m), Upper Detrital Unit (5–6 m) and Upper Gypsum Unit (100 m). Most of the closed depressions have developed on the Middle Gypsum Unit, exposed in the southern sector of the platform (Fig. 12.1). This unit has a much lower proportion of clay than the thicker Upper Gypsum Unit. Moreover, the size and spatial frequency of the depressions decrease towards the east, consistently with the wedging out of the Middle Gypsum Unit and the lateral change to less soluble facies (Salvany et al. 1994, 1996).

The climate of the central sector of the Ebro Basin is characterised by very hot summers and cold dry winters. The records from Bujaraloz meteorological station indicate that the average annual precipitation and temperature are 360 mm and 14.4 °C, respectively. Precipitation shows a bi-modal seasonal pattern, with the highest rainfall in spring and autumn (Rodó et al. 1997). Annual potential evapotranspiration values reach 788 and 909 mm, according to the Thornthwaite and Blaney–Criddle methods, respectively (García-Vera 1996). The area is characterised by strong winds with prevailing WNW direction, parallel to the dominant trend of the depressions and yardangs, locally designated as *Cierzo*. This cold and dry wind blows mainly in winter and spring, channelled along the Ebro Depression and controlled by the coexistence of anticyclonic conditions in the Cantabrian Sea and low pressure in the Mediterranean Sea (Puicercús et al. 1997). At Zaragoza meteorological station, located 60 km to the NW, the maximum wind speed recorded over the period 1942–2010 reached 135 km/h in February 1954 with a WNW direction. Data recorded in the Bujaraloz anemometric station, restricted to March 1991–January 1993, indicate that the WNW winds reach the highest velocities (<100 km/h) and represent about 75 % of the aeolian energy (Puicercús et al. 1997).

The Upper and Middle Gypsum Units constitute two aquifers separated by the Upper Detrital Unit, which behaves as a leaky aquitard (Fig. 12.1). Permeability in the gypsiferous units is mainly related to solutionally enlarged joints and reaches the highest values in the bottom of the depressions due to enhanced karstification by groundwater flow discharge (García-Vera 1996; Samper-Calvete and García-Vera 1998). The groundwater flow in this internally drained area is controlled by the topography of the platform, an extensive plateau riddled by enclosed depressions. Underground water flows towards and discharges at the bottom of the main basins, forming local and centripetal groundwater flow cells (Sánchez-Navarro et al. 1998). The groundwater salinity increases progressively along the flow path, changing from $Ca–SO_4$ composition in the recharge areas, into an $Na–Mg–Cl–SO_4$ hydrochemical facies in the discharge zones (García-Vera 1996; Salvany et al. 1996). Brines in the playa-lakes lead to the precipitation of salts both at the surface and within the lake sediments (Pueyo 1978/1979; Pueyo and Inglés 1987). Moreover, the extremely flat topography of the playa-lakes is controlled by the water table, which limits erosional lowering in the basins floor by aeolian deflation (Rosen 1994; Yechieli and Wood 2002).

12.3 Geomorphology of the Playa-Lakes and Closed Depressions

12.3.1 General Features and Origin

A total of 149 closed depressions, locally designated as *hoyas*, *clotas* or *saladas*, have been inventoried in the Bujaraloz-Sástago endorheic area (Balsa et al. 1991; Conesa et al. 2011). These topographical basins cover 19.2 km^2, approximately 5 % of the structural platform with internal drainage. The majority of the depressions are markedly elongated and oriented in the WNW–ESE direction, coinciding with the prevailing wind trend (Figs. 12.2 and 12.3). La Playa is the largest lake with 3.5 km in length and covering 1.72 km^2, approximately 9 % of the cumulative area of the bottom of the depressions (Fig. 12.2). This local name might be the source of the term exported by the Spaniards to the SW of the United States, where ephemeral saline lakes are designated as playas (Gutiérrez 2013). The structural platform is also carved by poorly hierarchized flat-bottom infilled valleys with dominant WNW–ESE orientation, most of which flow into closed depressions (Fig. 12.2). A significant proportion of the depressions shows scarped edges, generally controlled by a laterally extensive limestone bed situated in the upper part of the Middle Gypsum Unit (Quirantes 1965). Approximately 20 depressions host ephemeral saline wetlands that get flooded every year (Balsa et al. 1991). These playa-lakes have flat and moist bottoms, indicative of a topography controlled by the water table and capillary fringe (Rosen 1994). Water depth reaches the highest values in winter and rarely exceeds 50 cm (Castañeda 2002; Castañeda et al. 2005). These elongated lakes tend to have asymmetric geometry in plan view, with higher width in the downwind half than in the windward one, which may have a pointed margin (Fig. 12.4). Goudie and Wells (1995) indicate that deflation basins excavated in lake deposits within paleolacustrine basins typically display two types of geometries: (1) ovoid, almost subcircular, with the long axis perpendicular to the formative airflow; and (2) cusp-shaped, with the apex pointing upwind and oriented parallel to the formative wind. A significant proportion of

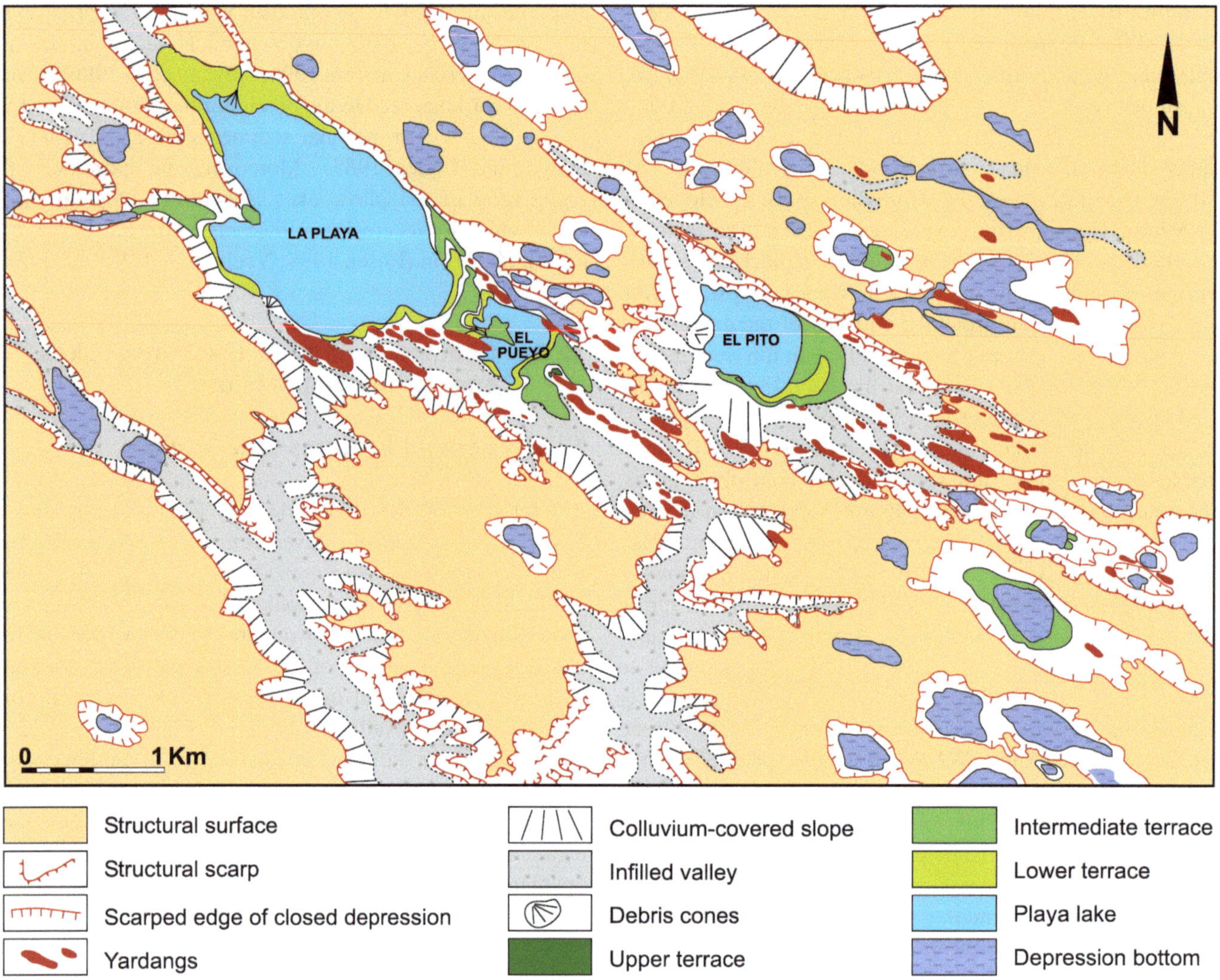

Fig. 12.2 Geomorphological map of the La Playa, El Pueyo and El Pito playa-lakes and surrounding areas, illustrating the distribution of lake terraces, yardangs and other minor closed depressions (redrawn from Gutiérrez-Elorza et al. 2002). The remnants of the Upper terrace are depicted by *small polygons* SE of El Pueyo

the playas in the studied area shows similar characteristics to the latter morphological type.

Evaporation of the brines in the playas leads to precipitation of salts (Pueyo 1978/1979; López et al. 1999). Algal mats 3–4 mm thick develop in the playa-lake floor during flooding periods. Saline efflorescences (bloedite, halite and thenardite) precipitate on the lake floor when the water level recedes. Beneath the surface, there is a black sapropelic mud with a high content in sulphides resulting from reduction of sulphates by bacterial activity. Desiccation cracks develop during dry periods and evolve into saline polygons, eventually framed by tepee-like features as well as bent and overthrusted edges. Saline crusts and algal mats locally show blisters formed by the expelling of gases derived from organic matter decomposition and volume increase related to salt crystallisation (Pueyo 1978/1979).

The margins of the lakes typically display an aureole of halophilous vegetation and wind-transported particles accumulate within and in the leeward side of these plants forming elongated nebkha dunes (Fig. 12.5). These dunes are commonly made up of sand-sized lenticular gypsum crystals. Trains of nebkhas may merge, resulting in linear dunes several tens of metres long. Moreover, some playas display rounded lunette dunes along their leeward side (Hills 1940; Goudie and Wells 1995), currently under investigation (Fig. 12.4). Some lunettes are modern deposits located on the bottom of the depressions, whereas others correspond to relict dunes perched on the downwind margin of the lake basins. Several processes favour wind deflation in the bottom of the playa-lakes during desiccation periods: (1) Drying cycles involve a significant reduction in cohesion of the particles, increasing their susceptibility to wind

Fig. 12.3 Aerial view of El Pito (*foreground*) and La Playa-El Pueyo (*background*) playa-lakes. The WNW–ESE trending elongated hills on the downwind side of the playas correspond to wind-fluted yardangs. *Arrow* indicates prevailing wind direction. Image taken on July 2005

Fig. 12.4 Oblique aerial photograph of Guallar playa, oriented parallel to the prevailing wind. The lake basin has a wider downwind half and a pointed upwind edge. Note the rounded and arcuate modern lunette dune developed along the leeward margin of the playa-lake (*white arrows*). Image taken on July 2005

entrainment; (2) Precipitation of salts at the surface may involve the accumulation of light and loose crystals and the preparation of particles for deflation by salt weathering; (3) Trampling by animals (Thomas 1988; Goudie and Wells 1995); and (4) Accumulation of significant volumes of faecal pellets by worms.

Several hypotheses have been proposed to explain the origin of the closed depressions in the Bujaraloz-Sástago endorheic area: (1) Differential dissolutional lowering of the ground surface controlled by fractures (Mingarro et al. 1981). According to this interpretation, the depressions would correspond to solution sinkholes generated by downward vadose flow in the epikarst zone. However, the playas nowadays behave as groundwater discharge zones; (2) Collapse of large cavities generated by structurally controlled subsurface dissolution in the gypsum bedrock; i.e. bedrock collapse sinkholes (e.g. Quirantes 1965). However, no significant cavities have been found in the boreholes drilled in the area and, according to geophysical surveys conducted in La Playa (Gutiérrez et al. 2013), the lake fill displays a limited thickness and tabular geometry; and (3) Widespread subsurface dissolution and subsidence

Fig. 12.5 Nebkha dunes mostly made up of gypsum crystals trapped by *Salicornia sp*. North-eastern margin of La Playa

in combination with aeolian deflation (Sánchez-Navarro et al. 1998; Gutiérrez et al. 2013). Initially, the depressions may form and evolve as solution sinkholes by percolating water in the vadose zone. Infiltration is favoured by the flat topography of the area. Once the bottom of the dolines reaches the water table zone, local groundwater flow cells that discharge in the depressions are established. That is, a recharge basin is transformed into a hydrologically closed discharge playa (Rosen 1994; Yechieli and Wood 2002). The underground flows that converge in the basins cause widespread dissolution of the bedrock, leading to gradual subsidence. The gypsum removed from the bedrock as solutes precipitates in the lake contributing to vertical accretion. During desiccation periods, the strong WNW winds may cause the erosional lowering of the basins' floor by deflation. These should be also the periods more favourable for the development of yardangs in the leeward sector of the playas.

12.3.2 Yardangs

Yardangs are elongated hills produced by wind erosion in combination with other processes such as weathering and gullying (Laity 1994, 2008; Goudie 1999). The term yardang corresponds to a local word introduced by Hedin (1903) from his study of the Taklimakan Desert, eastern China. The vast majority of the yardangs documented worldwide are located in hyperarid areas (McCauley et al. 1977; Goudie 2007). However, the yardangs associated with the Bujaraloz playas occur in a semiarid environment and are the only yardangs reported in Europe (Goudie 2007).

A total of 50 yardangs with a dominant WNW–ESE (N122E) orientation were mapped in the leeward side of the largest playas, mainly La Playa (1.72 km^2), El Pueyo (0.14 km^2) and El Pito (0.35 km^2) lakes (Gutiérrez-Elorza et al. 2002). This spatial association indicates that the formation of these landforms is related to the increase in the concentration of wind-blown particles in the playas during dry periods, increasing significantly the abrasive capability of the air currents (Figs. 12.2 and 12.3). Forty-four yardangs are developed on Miocene gypsiferous bedrock and six have been recognised in unconsolidated Holocene terrace deposits (Fig. 12.6). There is no consistent relationship between the orientation of the yardangs and the strike of the joints measured in several locations within the study area. The maximum length, width and height of the mapped yardangs are 264, 40 and 17 m, respectively, and the average aspect ratio (length/width) is 4.1. These landforms correspond to meso- and mega-yardangs according to Cooke et al. (1993) and to yardangs and mega-yardangs following the terminology of Livingstone and Warren (1996). The windward slope is always steeper than the leeward side. Following the morphological classification proposed by Halimov and Fezer (1989) from their studies in central Asia, the following yardang types may be differentiated: ridge-yardangs, which constitute elongated appendixes of structural surfaces and are the most frequent ones, mesa-yardangs, cone-yardangs, saw-tooth crest-yardangs and keel yardangs.

12.3.3 Lake Terraces and Morpho-Stratigraphic Evolution

Three lacustrine terraces have been identified in La Playa and El Pueyo by means of detailed geomorphological mapping; upper, intermediate and lower terraces situated at 9, 6 and 0.5 m above the lake bottom, respectively (Gutiérrez-Elorza et al. 2002; Gutiérrez et al. 2013) (Fig. 12.2). The intermediate terrace is the most extensive, and its upper surface merges with the top of the deposits

Fig. 12.6 Yardangs in the downwind side of La Playa (*right*) and within El Pueyo (*left*). The densely vegetated yardangs in El Pueyo have been carved in Holocene lake terrace deposits, whereas the sparsely vegetated elongated hills correspond to yardangs developed on gypsiferous bedrock. The perched flat surface between La Playa and El Pueyo corresponds to the intermediate lake terrace, deposited when both lakes used to form a single and larger lacustrine system

filling flat-bottomed valleys that drain into the playa-lakes. The spatial distribution of this terrace indicates that during its accumulation, La Playa and El Pueyo used to form a single lake around 2.7 km^2, suggestive of more humid conditions. Subsequently, differential aeolian erosion compartmentalised the lake basin into the present-day playas, nowadays separated by a remnant of the intermediate terrace (Figs. 12.2 and 12.6).

The intermediate terrace has been investigated in the NE margin of La Playa by means of two trenches. Here, the 5-m-thick terrace deposits are underlain by a karstic residue 45 cm thick, and mainly consist of tabular and horizontally bedded gypsiferous silts and sands, with a high proportion of lenticular gypsum crystals. Seven radiocarbon dates indicate that the accumulation of the lacustrine deposits took place from ca. 3.9 ka to soon after 2 ka, yielding a maximum average aggradation rate of 2.6 mm/year. Considering the approximate area of La Playa-El Pueyo paleolake during deposition of the intermediate terrace (2.73 km^2) and assuming that 5 m of sediments were accumulated on average during the aggradation phase, a total sedimentary input of 13.65 million m^3 can be estimated. This volume, supplied in a minimum time span of 1.9 ka, yields a maximum sedimentary input rate of 7184 m^3/year and a specific value of 26.31 m^3/ha/year. Subsequently, the lake bottom underwent entrenchment by aeolian deflation, briefly interrupted during the formation of the lower terrace. Differential wind erosion resulted in the segmentation of La Playa and El Pueyo lakes. Considering that the top of the deposits of the intermediate terrace are situated 5.95 m above the lake bottom and that the erosional phase started sometime after 2 ka, a minimum mean erosion rate by wind deflation of 3 mm/year can be estimated. This rate also applies to the yardangs carved in the deposits of the intermediate terrace. Considering the current area of La Playa and El Pueyo (1.72 km^2), aeolian deflation has evacuated a volume of around 11.07 million m^3 in a time period shorter than 2 ka. This means a maximum long-term erosion rate of 5533 m^3/year and a specific rate of 30 m^3/ha/year. The estimated deflation rate for La Playa-El Pueyo lake system compares well with those calculated in several arid regions of the world, mainly using yardangs carved in Holocene lake deposits (Williams 1970; McCauley et al. 1977; Boyé et al. 1978; Cooke et al. 1993; Goudie et al. 1999; Anderson et al. 2002; Washington et al. 2006; Liu et al. 2011).

The morpho-stratigraphic sequence records an overall deepening trend interrupted by net aggradation periods. Aggradation phases are attributed to relatively more humid periods, whilst excavation phases are ascribed to relatively more arid phases, during which the playas remain dry during longer periods, favouring the lowering of the surface by deflation (Gutiérrez et al. 2013 and references therein). These are probably also the evolutionary phases more favourable for the development of yardangs and lunette in the downwind side of the playas. The deepening of the lake bottom is limited by the position of the water table and the capillary fringe, controlling the development of extremely flat surfaces (Stokes 1968; Rosen 1994; Yechieli and Wood 2002). Once the surface meets the water table, the basins cannot get any deeper, but expand laterally through retreat of their margins, especially in easily erodible lake terraces. Further geochronological data may help improving the morpho-stratigraphic model and may contribute to better understanding the temporal relationships between the different landforms (terraces, yardangs, lunettes), as well as the role played by recent environmental changes.

Acknowledgments This work has been supported by the Spanish national project CGL2010-16775 (Ministerio de Ciencia e Innovación and FEDER) and the Regional project 2012/GA-LC-021 (DGA-La Caixa).

References

Anderson RY, Allen BD, Menking KM (2002) Geomorphic expression of abrupt climate change in southwestern North America at the glacial termination. Quatern Res 57:371–381

Balsa J, Pascual ML, Guerrero MC, Montes C (1991) Las saladas de Bujaraloz-Sátago y la Salada de Chiprana: Riqueza natural de Aragón. Empelte 7:3–30

Boyé M, Marmier F, Nelson C, Trécolle G (1978) Le dépots de la Sebka Mellala. Rev Géomorphol Dynam 27:49–62

Castañeda C (2002) El agua de las saladas de Monegros sur estudiada con datos de campo y de satélite. Gobierno de Aragón. Publicación del Consejo de Protección de la Naturaleza 35, Zaragoza, 158 p

Castañeda C, Herrero J, Casterad MA (2005) Facies identification within the playa-lake of the Monegros desert, Spain, from field and satellite data. Catena 63:39–63

Conesa JA, Castañeda C, Pedrol J (2011) Las saladas de Monegros y su entorno. Publicaciones del Consejo de Protección de la Naturaleza de Aragón, Zaragoza, 540 p

Cooke R, Warren A, Goudie A (1993) Desert geomorphology. UCL Press, London, 526 p

García-Vera MA (1996) Hidrogeología de zonas endorreicas en climas semiáridos. Aplicación a Los Monegros (Zaragoza y Huesca). Diputación General de Aragón, Zaragoza, 297 p

González-Sampériz P, Valero-Garcés BL, Moreno A, Morellón M, Navas A, Machín J, Delgado-Huertas A (2008) Vegetation changes and hydrological fluctuations in the Central Ebro Basin (NE Spain) since the Late Glacial period: Saline lake records. Paleogeography, Paleoclimatology, Paleoecology 259:157–181

Goudie AS (1999) Wind erosional landforms: yardangs and pans. In: Goudie AS, Livingstone I, Stokes S (eds) Aeolian environments, sediments and landforms. Wiley, Chichester, pp 167–180

Goudie AS (2007) Mega-yardangs: a global analysis. Geogra Compass 1:65–81

Goudie AS, Wells GL (1995) The nature, distribution and formation of pans in arid zones. Earth Sci Rev 38:1–69

Goudie A, Stokes S, Cook J, Samieh S, El-Rashidi OA (1999) Yardang landforms from Kharga Oasis, south-western Egypt. Zeitschrift für Geomorphologie Supplementband 116:97–112

Gutiérrez M (2013) Geomorphology. CRC Press-Balkema, Leiden, 1017 p

Gutiérrez F, Valero-Garcés B, Desir G, González-Sampériz P, Gutiérrez M, Linares R, Zarroca M, Moreno A, Guerrero J, Roqué C, Arnold LJ, Demuro M (2013) Late Holocene evolution of playa lakes in the central sector of the Ebro Depression based on geophysical surveys and morpho-stratigraphic analysis of lacustrine terraces. Geomorphology 169:177–197

Gutiérrez-Elorza M, Desir G, Gutiérrez-Santolalla F (2002) Yardangs in the semiarid central sector of the Ebro Depression (NE Spain). Geomorphology 44:155–170

Halimov M, Fezer F (1989) Eight yardang types in Central Asia. Zeitschrift für Geomorphologie 33:205–217

Hedin S (1903) Central Asia and Tibet. Scribners, New York

Hills ES (1940) The lunette, a new landform of Aeolian origin. Aust Geogr 3:15–21

Laity JE (1994) Landforms of eolian origin. In: Abrahams AD, Parsom AJ (eds) Geomorphology of desert environments. Chapman Hall, London, pp 506–535

Laity JE (2008) Deserts and desert environments. Willey-Blackwell, Chichester, 342 p

Liu D, Abuduwaili J, Lei J, Wu G, Gui D (2011) Wind erosion of saline playa sediments and its ecological effects in Ebinur Lake, Xinjiang, China. Environ Earth Sci 63:241–250

Livingstone I, Warren A (1996) Aeolian geomorphology: an introduction. Longman, Essex, 211 p

López PL, Auqué L, Mandado J, Vallès V, Gimeno MJ, Gómez J (1999) Determinación de la secuencia de precipitación salina en la Laguna La Playa (Zaragoza, España). I. Condiciones de equilibrio mineral y simulación teórica del proceso. Revista de la Sociedad Geológica de España 55:27–44

McCauley J, Grolier M, Breed C (1977) Yardangs of Peru and other desert regions. USGS Interagency Rep., Astrogeology 81:177 p

Mingarro F, Ordóñez S, López de Azcona MC, García del Cura MA (1981) Sedimentoquímica de las lagunas de los Monegros y su entorno geológico. Boletín Geológico y Minero 92–93:171–195

Pueyo JJ (1978/1979) La precipitación evaporítica actual en las lagunas saladas del área: Bujaraloz, Sástago, Caspe, Alcañiz y Calanda (provincias de Zaragoza y Teruel). Revista del Instituto de Investigaciones Geológicas. Diputación Provincial de Barcelona 33: 5–56

Pueyo JJ, Inglés M (1987) Substrate mineralogy, interstitial brine composition and diagenetic processes in the playa lakes of Los Monegros and Bajo Aragón (Spain). In: Rodríguez-Clemente R, Tardy Y (eds) Geochemistry and mineral formation in the Earth surface, pp 351–372

Puicercús JA, Valero A, Navarro J, Terrén R, Zubiaur R, Martín F, Iniesta G (1997) Atlas Eólico de Aragón. Gobierno de Aragón, Zaragoza, 127 p

Quirantes J (1965) Notas sobre las lagunas de Bujaraloz-Sástago. Geographica 12:30–34

Quirantes J (1978) Estudio sedimentológico y estratigráfico del Terciario continental de los Monegros. Instituto Fernando El Católico, Zaragoza, 200 p

Ramírez JI (1997) Mapa Geológico de España a Escala 1:50000. Gelsa (413). Instituto Tecnológico Geominero de España, Madrid

Rodó X, Baert E, Comín FA (1997) Variations in seasonal rainfall in Southern Europe during the present century: relationships with the North Atlantic Oscillations and the El Niño-Southern Oscillation. Clim Dyn 13:275–284

Rosen MR (1994) The importance of groundwater in playas: a review of playa classifications and the sedimentology and hydrology of playas. In: Rose MR (ed) Paleoclimate and basin evolution of playa systems. Geological Society of America Special Paper, vol 289. pp 1–18

Salvany JM, García Vera MA, Samper J (1994) Influencia del sustrato terciario en el emplazamiento del complejo lagunar de Bujalaroz (Monegros, Cuenca del Ebro). Actas II Congr. Esp. Terciario, Jaca, pp 271–274

Salvany JM, García Vera MA, Samper J (1996) Geología e hidrogeología de la zona endorreica de Bujaraloz-Sástago (Los Monegros, provincia de Zaragoza y Huesca). Acta Geológica Hispánica 30:31–50

Samper-Calvete FJ, García-Vera MA (1998) Inverse modelling of groundwater flow in the semiarid evaporitic closed basin of Los Monegros, Spain. Hydrogeol J 6:33–49

Sánchez-Navarro JA, Pérez A, Coloma P, Martínez-Gil FJ (1998) Combined effects of groundwater and aeolian processes in the formation of the northernmost closed saline depressions of Europe, north-east Spain. Hydrol Process 12:813–820

Sebe K, Csillag G, Ruszkiczay-Rüdiger Z, Fodor L, Thamó-Bozsó E, Müller P, Braucher R (2011) Wind erosion under cold climate: a Pleistocene periglacial mega-yardang system in Central Europe (Western Pannonian Basin, Hungary). Geomorphology 143:470–482

Solá J, Costa JM (1997) Mapa Geológico de España a Escala 1:50,000, Bujaraloz (414). Instituto Tecnológico Geominero de España, Madrid

Stokes WL (1968) Multiple parallel-truncation bedding planes. A feature of wind deposited sandstone formation. J Sediment Petrol 38:510–515

Thomas DSG (1988) The biogeomorphology of arid and semiarid environments. In: Viles A (ed) Biogeomorphology. Blackwell, Oxford, pp 193–221

Washington R, Todd MC, Lizcano G, Tegen I, Flamant C, Koren I, Ginoux P, Engelstaedter S, Bristow CS, Zender CS, Goudie AS, Warren A, Prospero JM (2006) Links between topography, wind, deflation, lakes and dust: the case of the Bodélé Depression, Chad. Geophys Res Lett 33:L09401

Williams GE (1970) Piedmont sedimentation and late quaternary chronology in the Biskra region of the northern Sahara. Zeitschrift für Geomorphologie, Supplementband 10:40–63

Yechieli Y, Wood WW (2002) Hydrogeologic processes in saline systems: playas, sebkhas, and saline lakes. Earth Sci Rev 58:343–365

The Picos de Europa National and Regional Parks

13

Montserrat Jiménez-Sánchez, Daniel Ballesteros, Laura Rodríguez-Rodríguez, and María José Domínguez-Cuesta

Abstract

The E–W trending Cantabrian Mountains, with peaks more than 2,600 m a.s.l., are located along the northern coast of the Iberian Peninsula. After the development of south-verging structures during the Alpine Orogeny, the Cantabrian Mountains were arranged as an asymmetrical relief deeply dissected by the fluvial network, with steep rivers flowing into the Cantabrian Sea in the north and less steep rivers draining towards the Duero Tertiary Basin to the south. The area shows a high geomorphic diversity, including relict Quaternary glacial and periglacial landforms, as well as features related to slope instability, fluvial and karstic processes. This work summarizes the geomorphological features of two different protected areas of the Cantabrian Mountains designated as Picos de Europa: the Picos de Europa National Park and the Picos de Europa Regional Park. Both are representative areas of the high-mountain landscapes of the northern and southern sectors of the Cantabrian Mountains. Moreover, the former hosts good examples of underground alpine karst.

Keywords

Cantabrian Mountains • Picos de Europa • Geomorphology • Glacial • Karst • Caves

13.1 Introduction

The E–W trending Cantabrian Mountains, with peaks higher than 2,600 m a.s.l. (e.g. Torrecerredo Peak, 2,648 m), extend for 480 km next to the northern coast of Spain. It can be considered as the westwards extension of the Pyrenees. The topographical axis of the Cantabrian Mountains roughly corresponds to the divide between the northern catchments, with steep, narrow and short valleys carved by rivers flowing towards the Cantabrian Sea, and the southern catchments, eroded by rivers flowing to the gentle landscapes of the Duero Basin (below 1,000 m a.s.l). This asymmetric topographical configuration is related to the tectonic uplift of the Variscan Massif during the Alpine Orogeny and the development of south-verging thrusts (Alonso et al. 1996). The bedrock comprises a great variety of sedimentary rocks, from Precambrian to Cenozoic in age, whose variable resistance to erosion, among other factors, has controlled the main geomorphological processes and resultant features. The highest areas exhibit well-preserved glacial landforms and active fluvial valleys whose slopes are locally affected by a variety of instability processes, while the limestone areas display striking karst landscapes.

The chronological evolution of the landscape in the Cantabrian Mountains after the Alpine Orogeny is not yet well established. The available data indicate that glaciers

M. Jiménez-Sánchez (✉) · D. Ballesteros ·
L. Rodríguez-Rodríguez · M. J. Domínguez-Cuesta
Department of Geology, University of Oviedo, C/Arias de Velasco, 33005 Oviedo, Spain
e-mail: mjimenez@geol.uniovi.es

D. Ballesteros
e-mail: ballesteros@geol.uniovi.es

L. Rodríguez-Rodríguez
e-mail: laurarr@geol.uniovi.es

M. J. Domínguez-Cuesta
e-mail: mjdominguez@geol.uniovi.es

F. Gutiérrez and M. Gutiérrez (eds.), *Landscapes and Landforms of Spain*, World Geomorphological Landscapes,
DOI: 10.1007/978-94-017-8628-7_13, © Springer Science+Business Media Dordrecht 2014

developed in the highest areas over 700–900 m a.s.l., and that most of the glacial features can be attributed to the Last Glacial Cycle, during which glaciers reached the maximum extent prior to 38 cal ka BP (Jiménez-Sánchez et al. 2013 and references therein). After the retreat of the glaciers, periglacial conditions led to the development of rock glaciers. The debuttressing of the slopes related to the retreat of the glaciers favoured the formation of complex deep-seated landslides. Fluvial erosion, in combination with karst processes, carved spectacular gorges in the more competent rocks, particularly in Carboniferous limestone. The underground landscape constitutes one of the most outstanding geomorphological secrets of the Cantabrian Mountains, with caves, passages and shafts, including some of the deepest, longest and most spectacular examples of the world.

The impressive landscape, together with the ecological value of the Cantabrian Mountains, led to the declaration of several protected areas, five of which have reached the highest international recognition of environmental excellence as Biosphere Reserves. This work summarizes the geomorphological features of two different protected areas: Picos de Europa National Park (PENP) and Picos de Europa Regional Park (PERP) (Fig. 13.1).

13.2 Geographical and geological setting

The PENP covers 647 km^2 north of the main divide of the Cantabrian Mountains, just 20 km south of the Cantabrian Sea. The spectacular relief of the area, with altitudes ranging from 70 to 2,648 m a.s.l., comprises three massifs (Western, Central and Eastern) separated by deep gorges carved by rivers flowing to the north (Sella, Cares, Duje and Deva rivers). The Montaña de Covadonga National Park, in the Western Massif of Picos de Europa, was established in 1918 as the first National Park in Spain. In 1996, its area was enlarged to its present limits, and in 2003, it was declared a Biosphere Reserve.

From a geological point of view (Merino-Tomé et al. 2009; Adrados-González et al. 2012), the area is included within the Picos de Europa Unit, one of the most external structural units of the Cantabrian Mountains. The bedrock mainly consists of a pre-Permian Palaeozoic succession including limestones, shales and sandstones, overlain in several areas by Permian to Triassic sandstones and shales. Variscan thrust tectonics caused the stacking of several thrust sheets, resulting in the piles of Carboniferous limestones more than 3 km thick. This structure was subsequently faulted by reserve faults during the Alpine Orogeny (Merino-Tomé et al. 2009). The presence of a thick and highly fractured pile of limestone is one of the main factors that have conditioned the geomorphology of the area, with karst, glacial, fluvial and slope movement-related landforms. The highest reliefs (over 1,000 m a.s.l.) show evidence of an alpine karst developed before the Middle Pleistocene (Ballesteros et al. 2011).

The PERP, established in 1994 and mainly located south of the main divide of the Cantabrian Mountains, covers a surface area of 1,200 km^2. The northernmost sector of this park overlaps with the PENP in ca. 200 km^2 and includes the highest peaks (e.g. 2,648 at Torrecerredo Peak). Outside of this overlapping region, the altitude ranges between 2,538 and 1,000 m at Peña Prieta Peak and the Duero Tertiary Basin, respectively. The bedrock consists of alternations of detrital (conglomerate, sandstone and shale) and carbonate formations (limestone and dolomite rock), Early Cambrian to Late Carboniferous in age. These formations were affected by thrusts and folding during both the Variscan and Alpine orogenies (Alonso et al. 2009). Lithological variability, together with the wide range of morphogenetic processes (fluvial, glacial, periglacial, nival, gravitational, torrential and karstic), result in a highly diverse landscape. The PERP, where carbonate rocks are less common, contains valuable geomorphic and stratigraphic records for the reconstruction of the glacial evolution of the area.

13.3 Geomorphology

The geomorphology of the Picos de Europa protected areas (both PENP and PERP) will be presented in two parts, including the high-mountain-surface landscape representative of the northern and southern sectors of the Cantabrian Mountains and the caves associated with massifs of Carboniferous limestones.

13.3.1 The High-Mountain Landscape

The present landscape of the Picos de Europa Parks is the result of glacial, periglacial, nival, gravitational, fluvial and karstic processes. Evidence of past glaciations, especially the last one, is clearly recorded in both parks at altitudes above 1,600 m. At lower elevations such evidence may be present, but is commonly less conspicuous.

In the PERP, glacial cirques and peaks with rock slopes affected by frost shattering occur at altitudes above 1,600 m a.s.l. Glacial valleys and moraines are reasonably well preserved as low as 1,140 m a.s.l. Rock glaciers, large rock avalanches and complex landslides provide an evidence of the impact of glacial debuttressing in the slopes once covered by glaciers and the occurrence of periglacial conditions after the Quaternary glaciations. The available data indicate that ice fields developed in some sectors like Fuentes

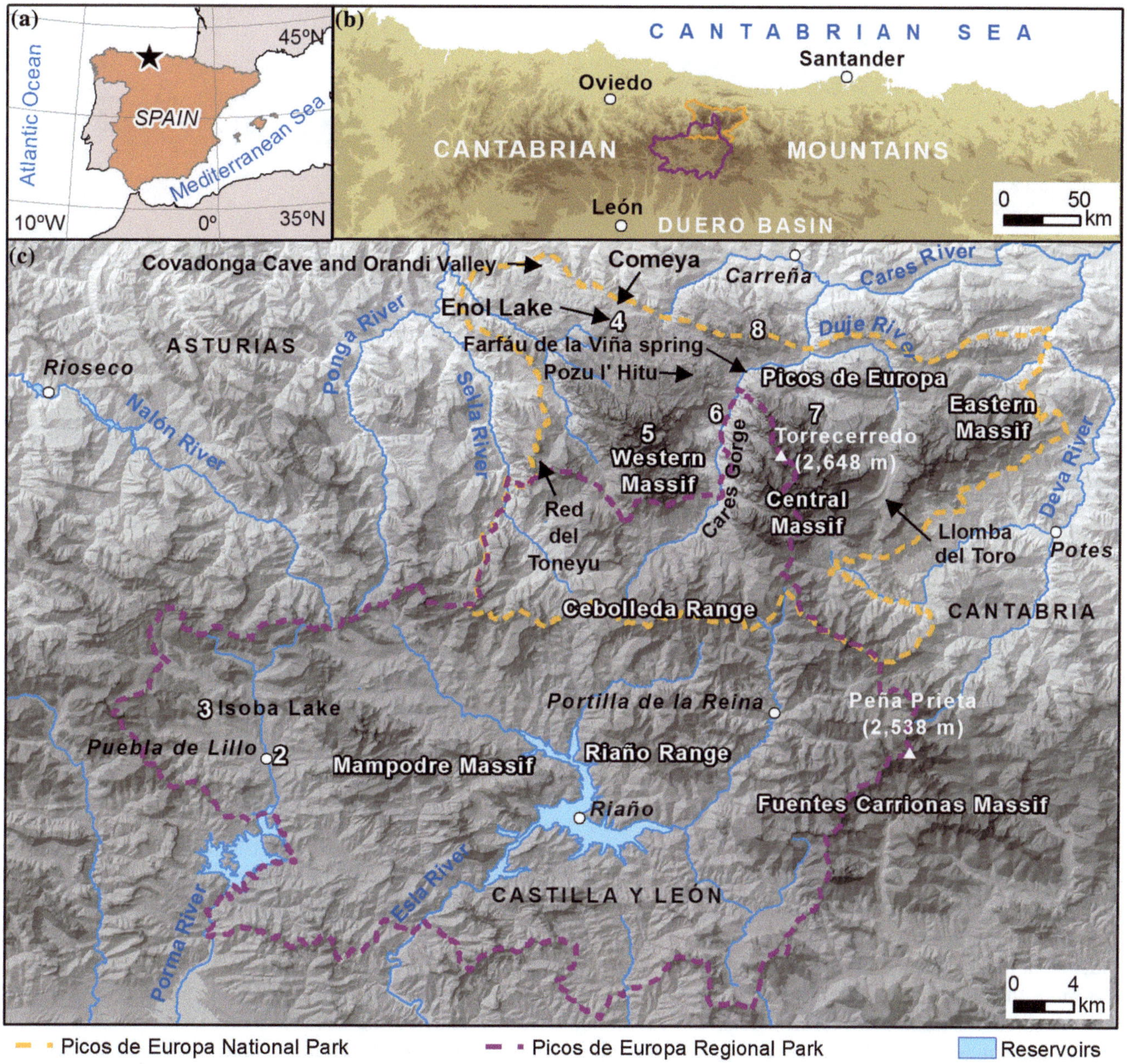

Fig. 13.1 Location of the Picos de Europa National Park and Picos de Europa Regional Park within the Iberian Peninsula (**a**) and the Cantabrian Mountains (**b**). **c** Hillshade model depicting the limits of the parks and the main drainages. *Numbers 2 to 8* indicate the location of Figs. 13.2, 13.3, 13.4, 13.5, 13.6, 13.7, 13.8

Carrionas Massif, Mampodre Massif and Porma-Esla headwaters, and alpine-type glaciers formed in other ranges like Cebolleda or Riaño. Detailed studies carried out in Fuentes Carrionas Massif allowed for the reconstruction of the ice field and the associated glacial tongues during the last glacial maximum (before 36 ka BP); the latter were 3.7–14 km long and flowed down to 1,200–1,450 m a.s.l. (Serrano et al. 2013). Glacial features have been described in other sectors of the PERP, but the extent and morphology of the palaeoglaciers are not well known. Some key localities to understand the spatial and temporal evolution of the glaciers in the PERP area are described below.

The Puebla de Lillo area (Porma River valley) is one of the best sites in the southern sector of the Cantabrian Mountains where moraines and erratic boulders define the position of the glacier fronts during the last local glacial maximum and several subsequent phases. The front of the Porma glacier descended down to 1,130 m a.s.l. during the local glacial maximum, depositing lateral moraines and several erratic boulders of Ordovician quartzite on the water

Fig. 13.2 Panoramic view of Puebla de Lillo valley (*looking west*). The village is located on a glacial outwash plain. On the *left side* of the picture, a large erratic boulder of Ordovician quartzite sandstone is lying on Carboniferous bedrock composed by shale and limestone

Fig. 13.3 Panoramic N–S view of the Isoba Lake developed on a glacially over-deepened depression carved in Carboniferous limestone. The Isoba glacial valley, located at the left side of the picture, shows a well-defined U-shaped cross-profile

divides (Fig. 13.2). The reconstruction of the glacier based on the moraines and erratic boulders indicates that it reached around 19 km long and its thickness was about 60 m at Puebla de Lillo during the local maximum. The evidence in the Porma and Esla valleys suggests that an ice field more than 380 km^2 covered the highest parts of both river catchments during the local maximum, extending beyond the limits of the PERP to the west. The topographical control on the flow pattern of this ice field is evident at the Isoba Lake (1,405 m a.s.l.), an over-deepened depression carved in Carboniferous limestone, where the ice stream split following the Isoba and Silván U-shaped valleys and merged downstream close to the glacial front at Puebla de Lillo (Fig. 13.3).

To the north, past glacial activity in the PENP is recorded by horns, cirques, arêtes, polished surfaces and sheep back rocks mostly carved in Carboniferous bedrock (Alonso 1998). Glaciers covered more than 170 km^2, flowing down to altitudes between 450–800 m during their maximum extent prior to 35.7–40 ka BP (Moreno et al. 2010; Serrano et al. 2012, 2013; Jiménez-Sánchez et al. 2013). Glacial over-deepening controlled by lithological variations and fault-weakened zones resulted in the excavation of hollows, occupied by lakes after glacial retreat. Glacial deposits in Picos de Europa are scarce, but they are essential to reconstruct the position of glacial fronts during the Last Glacial Cycle.

In the Western Massif, the Enol Lake (1,075 m a.s.l, Fig. 13.4) is one of these over-deepened hollows and the biggest lake of the area (0.1 km^2). The U-shaped Enol valley developed by an alpine glacier flowing to the north, towards the Comeya hollow, which is a 1.2 km^2 basin located 700-m downstream of Enol Lake. The Comeya hollow is interpreted as a karst depression, probably enlarged by erosion, filled by an around 60-m-thick sequence of glaciolacustrine, colluvial and torrential deposits (Jiménez-Sánchez and Farias 2002). Till deposits indicate that the Enol glacier front was stabilized at 1,030 m a.s.l., feeding the Comeya hollow (840 m a.s.l.) by melting waters. Samples from cores retrieved from Enol Lake and Comeya hollow were dated by radiocarbon and OSL techniques. The chronological data indicate that at about 43–45 ka, the melting waters of the Enol glacier started to fed the Comeya palaeolake, giving place to the onset of the glaciolacustrine deposition. Subsequently, between 43 and 38 ka, the additional glacial retreat led to the appearance of

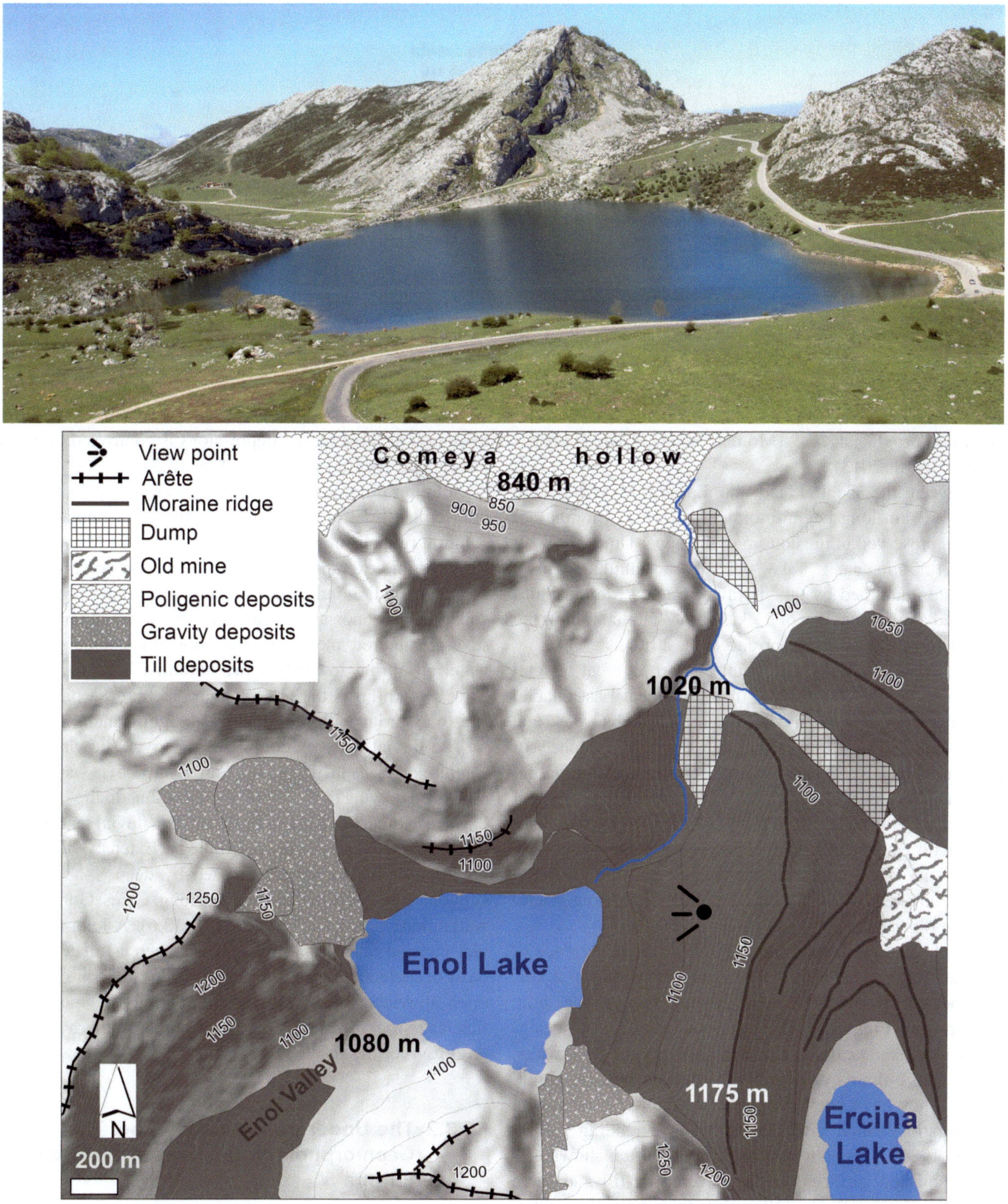

Fig. 13.4 **a** A SE–NW view of the Enol Lake basin, a glacial hollow carved in an alternation of Carboniferous shales and sandstones. At the *left side*, Vega de Enol U-shaped valley allows the reconstruction of a glacier flowing towards the lake during the Last Glacial Cycle. **b** Geomorphological sketch of the area, including the viewpoint from which the picture (**a**) was taken (modified from Moreno et al. 2010; Jiménez-Sánchez et al. 2013)

Fig. 13.5 E–W view of Jou Santu glaciokarstic depression and Peña Santa Peak (2,596 m)

the Enol proglacial lake (Moreno et al. 2010; Jiménez-Sánchez et al. 2013).

Between the Central and the Eastern Massif of the PENP, a 3-km-long lateral moraine, known as Llomba del Toro, is preserved in the bottom of the upper Duje valley, at about 1,500 m a.s.l. Lake deposits accumulated in a depression dammed by this moraine (Campo Mayor plain) provided a minimum age of 35 ka for the stabilization of the glacial front in this area (Serrano et al. 2013).

All these chronological data support a local glacial maximum for the PENP older than the global Last Glacial Maximum (ca. 21–18 ka) of the Last Glacial Cycle (late Pleistocene). The Duje valley also displays evidence of previous glaciations. Thus, the till deposits of Llomba del Toro overlie a unit of cemented breccias talus deposits that cover a previous glacially eroded surface. U-Th dating of the calcite cement of the breccias provides ages of ca 276 and 394 ka. This numerical age suggest that the erosional surface covered by the breccias could be related to a glacial episode older than 394 ka (Villa et al. 2013).

An outstanding feature in the PENP is the occurrence of elongated closed depressions reaching more than 0.2 km^2 and located up to 1,200 m a.s.l., popularly named as hou, hoyo or jou (Fig. 13.5). The term is locally used to designate a glaciokarstic depression bounded by very steep walls, where freeze-thaw cycles and snow avalanches are highly active. They have been interpreted as the result of glacial and karstic erosion controlled by E–W trending thrusts (Smart 1986). The bottom of the hollows often shows glacially polished surfaces, locally covered by till and scree deposits, where collapse dolines less than 2 m in diameter, locally called Boches, are developed. As an example, the 80-m-deep Jou Santu is formed by two coalescent hollows covering about 2 km^2 and surrounded by the highest peaks of the Picos de Europa Western Massif (Peña Santa, 2,596 m a.s.l). At the present time, the geomorphological evolution of these closed depressions is controlled by nival, karstic and rockfall processes (Fig. 13.5).

Solution dolines and karren are widespread in the Carboniferous limestone outcrops. Examples of pinnacles and solution runnels are common. In the Western Massif of the PENP, some examples of blind valleys can be recognized, as the Orandi blind valley, where a >2-km-long stream sinks into a ponor at 515 m a.s.l. The water, after flowing several hundred metres in the subsurface, emerges in a waterfall spring located at the bottom of the Covadonga Cave, which hosts the Covadonga Sanctuary, a pilgrimage site of Asturias with high historical significance.

The fluvial network comprises the rivers flowing to the Cantabrian Sea, in the Sella, Cares and Deva watersheds, and those flowing to the south, which belong to the Esla river basin, a tributary of the Duero River. The fluvial landscape is largely controlled by bedrock lithology. Extensive limestone outcrops of the PENP are crossed by deep S–N trending canyons like the dramatic 11-km-long Cares Gorge, with walls more than 1,500 m high (Fig. 13.6). A long-term fluvial incision rate of 0.3 mm/year has been estimated on the basis of speleothem age data (Smart 1986). By contrast, wide open alluvial valleys with farming activity develop on siliciclastic rocks in some areas such as the southern sector of the PENP or the Esla River basin.

13.3.2 The Underground Landscape. The Geomorphological Secret of Picos de Europa

One of the geological singularities of the PENP is the presence of limestone successions more than 3 km thick, related to the stacking of thrust sheets. The geological configuration, together with the regional incision of the fluvial network, favoured the extraordinary development of caves within an

Fig. 13.6 View of the Cares Gorge from the Los Collaos. A person has been *circled* on the trail for scale

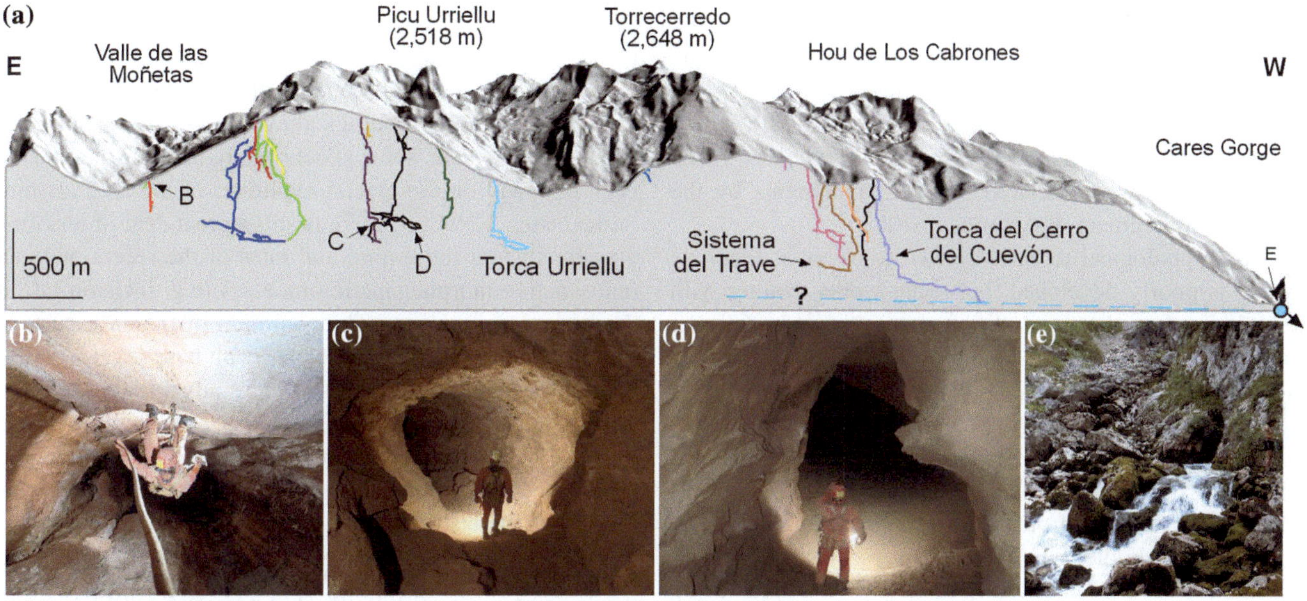

Fig. 13.7 **a** 3D model of the northern sector of the Central Massif (Picos de Europa National Park) showing the distribution of some of the deepest shafts (cave survey data courtesy of IE Valenciano, Secció d'Investigacions Subterrànies del Centre Excursionista de Terrassa, V. Ferrer and 1987–2012 Expeditions to Castil-Urriellu). Pictures **b–d** show some images of the caves depicted in (**a**). The picture **e** shows the Farfáu de la Viña spring (0.6–3 m^3/s in average discharge), most probably the main discharge point of the northern sector of the karst massif

Fig. 13.8 Examples of speleothems in the Western Massif of Picos de Europa. **a** Pool deposits covering flowstones and dripstones older than 300 ka. **b** Stalagmites, stalactites and flowstones covering a conduit developed under phreatic and vadose conditions

altitude range of 2.3 km and with the main karst springs located at altitudes between 200 to 800 m a.s.l. Caves constitute an extensive underground landscape within the park, which is being progressively inventoried and mapped since the 1960s thanks to speleological explorations carried out by national and international groups. Currently, almost five thousand caves with a total length close to 355 km are known. These include 14 out of the 96 deepest shafts of the world. The northern sector of the Central Massif of the Picos de Europa (Fig. 13.7) has several deep shafts, including five caves more than 1 km deep. One of them, Torca del Cerro del Cuevón (1,589 m deep), is the deepest cave in Spain and the seventh deepest of the world. The longest known cave in Picos de Europa is Red del Toneyu (18.7 km long), which together with Pozu'l Hitu, Sistema del Trave and Torca Urriellu shafts, were declared Natural Monuments by the Regional Government of Asturias in 2003.

Geomorphological research related to caves of Picos de Europa is poorly developed, but some works dealing with the relationships between the fracture network and the spatial pattern of the caves have been published (Fernández-Gibert et al. 2000; Ballesteros et al. 2011). These studies, together with the ongoing research in several systems, indicate that documented caves consist of conduits of phreatic origin (41 %), vadose canyons (46 %), shafts (11 %) and conduits of other types (2 %) developed since, at least, the Middle Pleistocene. Vadose canyons up to 150 m deep are present, usually associated with pitches up to 350 m high, both of them originating in relation to the incision of the fluvial network. Horizontal cave passages in the western and eastern sectors of the PENP usually show phreatic and vadose features, as well as fluvial deposits and speleothems; dripstones and flowstones (Fig. 13.8).

13.4 Conclusions

The two parks described in this chapter display landscapes representative of the high-mountain areas situated to the north and south of the main water divide of the Cantabrian Mountains. Moreover, the exceptional development of caves in PENP constitutes an added geomorphological value, although hardly accessible for most of the people. Both surface and underground landforms provide a valuable record of the geomorphological evolution of the Cantabrian Mountains. To date, some geochronological information has been obtained on the glacial evolution of the area at some critical sites. However, there is still a great deal of unsolved questions about the timing and rates of the recent and currently active morphogenetic processes (e.g. incision rate in the limestone canyons). The evolution of the cave systems in the karst massifs must be related with base-level lowering by fluvial incision. Hopefully, the study of caves, where some geomorphic and stratigraphic records have a higher preservation potential than at the surface, will provide data on the geomorphological history of the Cantabrian Mountains over the last hundreds of thousand years.

Acknowledgments We are indebted to PC-10-14 project (FICYT—Rioglass S.A., EC FEDER, PCTI Asturias), CANDELA project CGL2012-31938 (Spanish MINECO-PGE-FEDER), GEOCAVE project (MAGRAMA 580/12, Organismo Autónomo de Parques Nacionales—OAPN). D. Ballesteros and L. Rodríguez-Rodríguez are, respectively, granted by the Severo Ochoa (FICYT—Asturias) and FPU (Spanish Ministry of Education, Culture and Sport, MECD) programs. We are also grateful to cavers (especially V. Ferrer, Expeditions to Castil-Urriellu, IE Valenciano and SIS CE Terrassa).

References

Adrados-González L, Alonso-Alonso V, Bahamonde Rionda JR, Fernández-González LP, Gutiérrez-Claverol M, Heredia-Carballo N, Jiménez-Sánchez M, Meléndez-Asensio M, Merino-Tomé O, Villa-Otero E (2012) Guía Geológica del Parque Nacional de Los Picos de Europa, 2nd edn. Adrados Ediciones, IGME, Parques Nacionales. Oviedo, 333 p

Alonso V (1998) Covadonga National Park (Western Massif of Picos de Europa, NW Spain): a calcareous deglaciated area. Trabajos de Geología 20:167–181

Alonso JA, Pulgar JA, García-Ramos JC, Barba P (1996) Tertiary basins and Alpine tectonics in the Cantabrian Mountains (NW Spain). In: Friend PF, Dabrio CJ (eds) Tertiary basins of Spain. Cambridge University Press, New York, pp 214–227

Alonso JL, Marcos A, Suárez A (2009) Paleogeographic inversion resulting from large out of sequence breaching thrusts: The León Fault (Cantabrian Zone, NW Iberia). A new picture of the external Variscan Thrust Belt in the Ibero-Armorican Arc. Geologica Acta 7(4):451–473

Ballesteros D, Jiménez-Sánchez M, García-Sansegundo J, Giralt S (2011) Geological methods applied to speleological research in vertical caves: the example of Torca Teyera shaft (Picos de Europa, N Spain). Carbonates and Evaporites 26:29–40

Fernández-Gibert E, Calaforra JM, Rossi C (2000) Karst in the Picos de Europa massif (North Spain). In: Klimchouk B, Ford DC, Palmer AN, Dreybrodt W (eds) Speleogenesis: evolution of Karst Aquifers. National Speleological Society, Huntsville, pp 352–357

Jiménez-Sánchez M, Farias P (2002) New radiometric and geomorphologic evidences of Last Glacial Maximum older than 18 ka in SW European Mountains: the example of Redes Natural Park (Cantabrian Mountains, NW Spain). Geodin Acta 15:93–101

Jiménez-Sánchez M, Rodríguez-Rodríguez L, García-Ruiz JM, Domínguez-Cuesta MJ, Farias P, Valero-Garcés V, Moreno A, Rico M, Valcárcel M (2013) A review of glacial geomorphology and chronology in northern Spain: timing and regional variability during the last glacial cycle. Geomorphology 196:50–64

Merino-Tomé OA, Bahamonde JR, Colmenero JR, Heredia N, Farias P, Villa E (2009) Emplacement of the Cuera and Picos de Europa imbricate system at the core of Iberian-Armorican arc (Cantabrian zone, north Spain): New precisions concerning the timing of the arc clousure. Geol Soc Am 121:729–751

Moreno A, Valero-Garcés BL, Jiménez-Sánchez M, Domínguez-Cuesta MJ, Mata P, Navas A, González-Sampériz P, Stoll H, Farias P, Morellón M, Corella JP, Rico M (2010) The last deglaciation in the Picos de Europa National Park (Cantabrian Mountains, Northern Spain). J Quat Sci 25:1076–1091

Serrano E, González-Trueba JJ, González-García M (2012) Mountain glaciation and paleoclimate reconstruction in the Picos de Europa (Iberian Peninsula, SW Europe). Quatern Res 78:303–314

Serrano E, González-Trueba JJ, Pellitero R, González-García M, Gómez-Lende M (2013) Quaternary glacial evolution in the central Cantabrian Mountains (northern Spain). Geomorphology 165:65–82

Smart PL (1986) Origin and development of glacio-karst closed depressions in the Picos de Europa, Spain. Zeitschrift fur Geomorphologie 30:423–443

Villa E, Stoll H, Farias P, Adrados L, Edwards RL, Cheng H (2013) Age and significance of the Quaternary cemented deposits of the Duje Valley (Picos de Europa, Northern Spain). Quatern Res 79:1–5

14 The Ordesa and Monte Perdido National Park, Central Pyrenees

José M. García-Ruiz, Blas L. Valero-Garcés, Santiago Beguería, Juan I. López-Moreno, Carlos Martí-Bono, Pilar Serrano-Muela, and Yasmina Sanjuan

Abstract

The Ordesa and Monte Perdido National Park was created in 1918 and enlarged in 1982 to highlight and protect spectacular high mountain relief dominated by limestone. Alpine tectonics resulted in the piling-up of south-verging thrust sheets leading to the thick sedimentary successions exposed in impressive vertical cliffs. The presence of massive limestones has favoured the development of deep canyons and karst landforms, including karren, dolines, and caves with large shafts. Quaternary glaciations contributed to increase the geomorphic diversity, forming cirques and stunning U-shaped valleys. Small glaciers from the Little Ice Age still remain on the north-facing slopes of the Monte Perdido. Periglacial processes in the most elevated areas of the National Park, as well as erosion in thick soils developed on marly limestone have produced unique geomorphological features.

Keywords

National Park • Limestone • Canyon • Karst relief • Glaciers

14.1 Introduction

The Ordesa and Monte Perdido National Park (OMPNP) is an impressive mountain area in the central Pyrenees mostly underlain by limestone bedrock. The elevation of the mountains commonly exceeds 3,000 m a.s.l., and the landscape is largely controlled by the geological structure, reflecting also the impact of Quaternary glaciations, and the influence of past human activity. Its relief, characterised by dramatic vertical cliffs, deep canyons, U-shaped valleys, waterfalls, avalanche tracks and active glaciers, was the most important factor for the declaration of National Park in 1918. The protected area was originally 2,100 ha, but the enlargement of the National Park in 1982 provided protection for a total area of 15,608 ha. The OMPNP represents a typical high mountain landscape dominated by calcareous formations and affected by intense glacial erosion. The relief of the OMPNP provides the opportunity to study the altitudinal zonation of different landforms and geomorphic processes, as well as the role played by snow and ice in the past and present morphogenesis. Other environmental values (biodiversity, endemic species) reinforce the exceptional importance of the area.

Geomorphology has been one of the most intensively studied topics in the Pyrenees, mainly in relation to the glacial landforms and landscape evolution. Nevertheless, the OMPNP has been the subject of a limited number of geomorphological studies, beyond general descriptions in guides and popular books. A small part of the study area, the Marboré Cirque, has been the focus of various studies on periglacial and glacial landforms, including classical publications by Hernández-Pacheco and Vidal-Box (1946) and

J. M. García-Ruiz (✉) · B. L. Valero-Garcés · J. I. López-Moreno · C. Martí-Bono · P. Serrano-Muela · Y. Sanjuan
Instituto Pirenaico de Ecología, CSIC, Campus de Aula Dei, Apartado 13.034, 50080 Zaragoza, Spain
e-mail: Humberto@ipe.csic.es

S. Beguería
Estación Experimental de Aula Dei, CSIC, Campus de Aula Dei, Apartado 13.034, 50080 Zaragoza, Spain

F. Gutiérrez and M. Gutiérrez (eds.), *Landscapes and Landforms of Spain*, World Geomorphological Landscapes,
DOI: 10.1007/978-94-017-8628-7_14, © Springer Science+Business Media Dordrecht 2014

Barrère (1952). Barrère (1971) also published a geomorphological map of the Broto area in the context of a regional geomorphological map of an extensive sector of the Central Pyrenees. A part of the OMPNP was included in the Broto map, particularly the canyons of Ordesa, Añisclo, and Escuaín. Nicolás-Martínez (1981) provided more detailed information on the Marboré Cirque. García-Ruiz and Arbella (1981) studied soil erosion and shallow landsliding in the deep soils that cover the mid-altitude areas of the OMPNP. Martí-Bono and García-Ruiz (1993) produced a short report on the extent of the Quaternary glaciers, and García-Ruiz and Martí-Bono (2001) published a geomorphological map of the park, accompanied by a description of the most significant landforms. Various reports have referred to the present-day glaciers (e.g. Chueca et al. 2002) and the environmental evolution of the OMPNP during the last thousands years based on sediment cores from high mountain lakes (Oliva-Urcia et al. 2013; Salazar-Rincón et al. 2013).

This chapter synthesises historical and recent information on the geomorphology of the OMPNP, overviews the main features of its evolution, including structural, glacial, periglacial, and karst landforms, and highlights soil degradation processes in this unique protected area.

14.2 Geographical and Geological Setting

The OMPNP is located in the Central Spanish Pyrenees and comprises several valleys (Fig. 14.1): Arazas (also known Ordesa Valley), Bellos (Añisclo Valley), the Upper Cinca Valley (Pineta Valley), and Yaga (Escuaín Valley). Many peaks exceed 3,000 m a.s.l.: Monte Perdido (3,355 m), Cilindro (3,322 m), Marboré (3,247 m), Soum de Ramond (3,263 m). The Monte Perdido Massif, located on the main drainage divide, dominates the landscape. The Cinca River flows at the foot of the northern side of the massif, whereas the other rivers (the Arazas, Bellos, and Yaga rivers) drain the southern slopes. The Monte Perdido Massif is a part of the Inner Sierras, which is one of the main structural units of the southern Pyrenees. The formations exposed in the Inner Sierras correspond to Cretaceous–Eocene limestones, marly limestones, and sandstones. The compressional structural style is characterised by stacking of south-verging thrust sheets, locally affected by overturned folds (e.g. Cilindro Peak). The nappes were emplaced during the Alpine development of the Pyrenean orogeny, resulting in the development of limestone and sandstone piles thousands of metres thick. South of the Inner Sierras, the bedrock mostly consists of Eocene flysch facies, with the typical alternation of thin sandstone and marl beds. Here, the topography is gentler, with rounded divides and hillslopes displaying continuous profiles.

The climate shows predominantly Mediterranean influences: Atlantic effects barely reach the Central Pyrenees. The Góriz weather station (2,220 m a.s.l.) has recorded precipitation averaging 1,850 mm annually, the majority of which occurs in autumn and spring. Autumn rainstorms can be very intense, with more than 650 mm recorded in three days. The mean annual temperature at Góriz is approximately 4 °C, with three months (January, February, and March) having temperatures below 0 °C. The average temperature exceeds 10 °C between June and September. The 0 °C isotherm during the coldest period (November–May) is located at 1,603–1,670 m a.s.l. and thus includes most of the study area. This explains why, in spite of the low average winter precipitation, snowfall and snowmelt have a marked hydrological and geomorphological influence. Nevertheless, studies have confirmed a decline in the influence of snow in recent decades, related to a decrease in winter precipitation (López-Moreno 2005).

Altitude and topography were major limitations for agricultural use in the OMPNP. Beech tree (*Fagus sylvatica*) forests dominate in the Ordesa Valley, while mixed forests tend to prevail in the Añisclo and Escuaín valleys. Mixed stands of pine (*Pinus sylvestris*) and beech form dense forests below 1,800 m on the flysch slopes. The upper forest level is colonised by *Pinus uncinata*, which is well adapted to low winter temperatures, snow accumulation, stony soils, and steep slopes. Nevertheless, the upper forest limit is in most cases artificial, due to the removal of forest communities to enable the expansion of summer grasslands. Above 2,400–2,500 m vegetation is sparse because of the extreme climatic conditions.

14.3 Landforms

The complex and impressive relief of the OMPNP records a long geomorphological history mainly controlled by structural and climatic factors. The most outstanding landforms are largely determined by the distribution of the different lithologies and their structure, whereas fluvial and glacial processes have generated some of the best known geomorphic features of the OMPNP.

14.3.1 Structural Landforms: Cliffs and Canyons

The geology of the OMPNP and surrounding areas is characterised by the superposition of various thrust sheets, and the consequent piling-up of thick successions consisting mainly of limestone and, secondarily, sandstone and marly limestone (Ríos et al. 1982, 1989). Four nappes bounded by thrusts largely concordant with the strata have been identified. Some of these thrust structures display recumbent

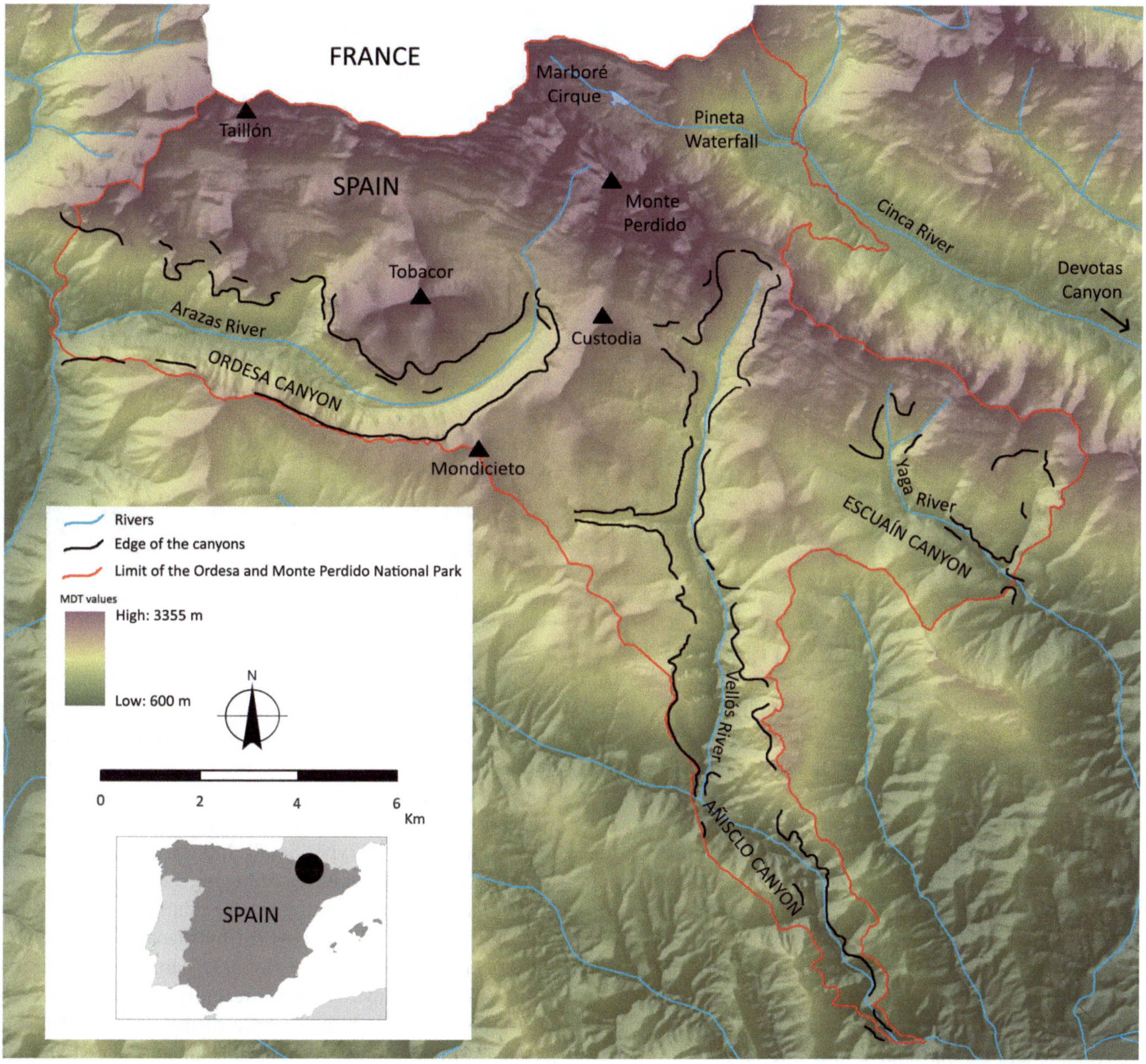

Fig. 14.1 The Ordesa and Monte Perdido National Park

anticlines at their front typically associated with nearly vertical limestone cliffs. Some sections display stacked overturned folds forming a peculiar cascade-like structure of folds. The sub-horizontal attitude of the thrusts and the limbs of the associated recumbent anticlines provide a false impression of a very simple horizontal structure, consisting of massive limestone with intercalations of sandstone beds. Structural landforms are common in the landscape. High and laterally extensive cliffs are abundant, reflecting the high strength of the bedrock; particularly good examples include the huge limestone escarpment of the Monte Perdido Massif (Fig. 14.2) and the precipitous walls of the canyons of Añisclo (Bellos River), Ordesa (Arazas River), and Escuaín (Yaga River).

The Añisclo, Ordesa, and Escuaín canyons have developed due to a long process of fluvial downcutting into a pre-existing relatively horizontal topography currently situated at 2,000–2,200 m a.s.l. The slopes of the canyons faithfully reproduce the resistance of the various lithological units, with successive vertical cliffs separated by slopes covered by scree (Fig. 14.2). In the case of the Ordesa Canyon, the valley has been carved through the pile of thrust sheets creating a tectonic window, like in the Añisclo Canyon. A number of cuestas have developed on each side of the

Fig. 14.2 The Monte Perdido Massif and the Añisclo canyon. *Photograph* Fernando Biarge

Fig. 14.3 Alpine karst landscape near the Góriz Shelter. *Photograph* Fernando Biarge

Añisclo Canyon, corresponding to the limbs of a N–S valley-centred anticline. Some of the tributaries developed short and steep hanging valleys with waterfalls associated with knick points (e.g. the Cotatuero and La Pardina valleys).

On the north-facing slopes of the Monte Perdido Massif, the Marboré Cirque is underlain by synclinorium-like structure, where alternating limestone and sandstone outcrops are evident, with small *cuestas* and depressions in the bottom of the cirque. The Pineta or Cinca waterfall, of approximately 1,000 m in height, falls into the Pineta Valley. To the southeast, the canyon of the Escuaín Valley is relatively modest in size, although the glacial cirques of Gurrundué and La Sarra have dramatic cliffs.

14.3.2 Karst Landforms

Surface karst landforms are common, including extensive karren, dolines, shafts, and sinks mainly formed by surface water. The Sierra de las Cutas, the flat areas near the Góriz Shelter, the sub-horizontal structural surfaces in the vicinity of the Añisclo Canyon, the reliefs immediately to the south of the Taillón and Casco peaks, and the right margin of the Escuaín Valley are among the many sites that display outstanding examples of karst topography and hydrology. In general, the most widespread landform corresponds to the structural karren (*kluftkarren*), mainly related to the solutional widening of stratification planes and joints. The result is a complex network of grikes that may reach more than 1 m in depth. In some places, particularly on sandy limestone, extreme dissolution controlled by discontinuity planes has resulted in isolated limestone boulders of capricious shapes (e.g. close to the Llano del Descargador; Fig. 14.3). On steep slopes underlain by massive limestone with a low density of joints, solution flutes (*rillenkarren*) is the dominant karren type. Small dolines are also relatively common and tend to cluster following fault lines.

The karren and small dolines are relatively recent, since they are located in areas covered by glaciers in the Upper Pleistocene. Some collapse dolines, particularly near the Góriz Shelter, have formed recently or are currently enlarging as the fresh appearance of their steep sides suggests. A small polje is recognisable in the La Estiva platform. Large depressions of mixed glacial and karstic origin, locally termed *llanos* (Descargador, Millaris, and Salarons *Llanos*) have developed on marly limestones and are partially filled by torrential deposits (Barrère 1964; García-Ruiz and Martí-Bono 2001). Downstream of each *llano*, a large doline has developed on grey, massive limestone, probably favoured by structural discontinuities associated with the underlying recumbent fold.

Surface water flow is generally absent, except in the main rivers and in some springs (the Fuen Blanca and Cola de Caballo wells). Throughout the OMPNP, most of the snowmelt and rainfall waters infiltrate almost immediately into the karstified bedrock. Underground drainage is favoured by a well-developed network of caves. Approximately 40 caves at altitudes >2,500 m remain iced for most of the year (e.g. the Casteret Cave).

14.3.3 Glacial Landforms

The high altitude of the OMPNP explains the importance of Quaternary glaciers in the evolution of the relief. Large ice masses developed in the cirques near the divides in the Monte Perdido Massif and descended towards the main valleys. The cirques on the southern slopes of the massif are well developed, with steep walls and frequent structural scarps. The large Marboré cirque, on the northern face of the Monte Perdido Peak, displays a structurally controlled overdeepened basin currently occupied by the Marboré Lake. Small-scale erosional glacial landforms (polished surfaces, striations, *roches mutonnées*) are absent due to limestone dissolution, except in the recently deglaciated surfaces at the front of the present-day glaciers.

The U-shaped Arazas and Añisclo valleys show the typical effects of glacial erosion at a large scale, formerly occupied by large ice tongues >400 m thick. The more rapid glacial erosion in these trunk valleys led to the development of hanging glacial valleys in the lateral tributaries, particularly the Salarons, Carriata, and Cotatuero valleys. The Arazas Valley also constitutes a hanging valley, with the Molineto waterfall at the confluence with the Ara Valley. In the Arazas glacial valley, the landscape locally has a conspicuous structural imprint, like in the Gradas de Soaso, a series of waterfalls developed on subhorizontal Maastrichtian sandstone. Other changes in the longitudinal profile reflect the presence of hard limestone outcrops, resulting in dramatic waterfalls, including the Cola de Caballo, Estrecho, Abanico, and Torrombotera in the Arazas Valley.

Glacial sediments are relatively scarce in the OMPNP. Remnants of old glacial deposits are visible on the left margin of the Añisclo Valley, perched at approximately 300 m above the valley bottom. No clear end moraines have been found in the Añisclo glacial valley, the terminal area of which would be at approximately 900 m a.s.l., near the San Urbez bridge. In the Escuaín Valley, the end moraines belonging to the Glacial Maximum are located at 1,400 m a.s.l. In the Arazas Valley, most of the moraine deposits belong to recession glacial stages and are located near the cirques. On the northern side of the Monte Perdido Massif, the Marboré Cirque has abundant glacial deposits from the Little Ice Age (LIA) (Fig. 14.4). A chaotic deposit composed of grey limestone blocks forms the front of a large rock avalanche, although it was considered to be a Late Glacial moraine (Nicolás-Martínez 1981). Beyond the OMPNP, the Cinca Valley has other glacial deposits at the foot of the Marboré Cirque, and on the left side at Espierba, 400 m above the valley bottom. The glacier ended at Salinas, next to the confluence with the Cinqueta Valley, upstream the Devotas Canyon (Martí-Bono and García-Ruiz 1993). Lewis et al. (2009) dated the last glacial maximum (LGM) at the front of the Cinca glacier at 62.7 ± 3.9 ka. The Ara Valley, into which the Arazas River flows, also has major lateral moraines associated with moraine-dammed lake basins (the Diazas and Buesa glacio-lacustrine deposits). The end moraines of the LGM are located several kilometres downstream, near Sarvisé. In the La Larri paleolake, which is located in a tributary of the Cinca Valley, sedimentation started prior to 35 ka (Salazar-Rincón et al. 2013), confirming that the glacial maximum in the Pyrenees occurred earlier than in mountains of central Europe. The age of the top of the lacustrine sediment of the La Larri paleolake is ca. 13 ka, suggesting that the main Cinca glacier had already retreated to the Balcón de Pineta area during the Younger Dryas and the transition to the Holocene.

The sediments retrieved from Marboré Lake reveal that the lake basin was deglaciated during the Younger Dryas. The sedimentation rate has been constant (0.45 mm $year^{-1}$) over the last 12.7 ka (Oliva-Urcia et al. 2013). With the onset of the Holocene, there was a clear increase in biological productivity in the lake. The most significant change during the Holocene occurred at approximately 5–4.4 ka and suggests a decrease in humidity. Increased run-off and sediment delivery, and higher productivity occurred during the LIA.

14.3.4 Periglacial Processes

Throughout the OMPNP scree slopes occur at the foot of the cliffs, forming steep accumulations of angular gravel- and boulder-sized particles. Above 2,000 m, the scree slopes are still active, whereas at lower altitudes (e.g. the Ordesa and Añisclo canyons) they have been colonised and stabilised by vegetation. Snow avalanches together with other mass movements form rectilinear canals that cross the forest and develop debris cones.

However, the most characteristic periglacial feature in the OMPNP is the patterned ground, particularly in the Marboré Cirque, above 2,600 m and near relict glaciers of the Monte Perdido Massif. The patterned ground results from the cryoturbation processes, involving the selective displacement of fine and coarse material due to the presence of ice and the development of cracks. Permafrost occurs throughout the Marboré Cirque, although the upper part of the soil thaws in summer, developing an active layer (*mollisol*). Examples of active polygons are evident in the central and western sectors of the Marboré Cirque, mostly on flat areas with poorly drained sandy soils (Fig. 14.5) and at more than 3,000 m on the Marboré Peak, where cryoturbation is a common phenomenon in early autumn. On moderately steep areas, the polygons are replaced by

Fig. 14.4 Little Ice Age moraines in the Marboré Cirque, at the foot of the Cilindro peak. *Photograph* J.M. García-Ruiz

Fig. 14.5 Sorted stone polygons in the Marboré Cirque. *Photograph* J.M. García-Ruiz

Fig. 14.6 The lower Monte Perdido glacier in September 2010. *Photograph* J.M. García-Ruiz

stripes. Gelifluction terracettes are particularly evident in the headwater of the Pardina ravine, where they develop micro-terraces with *Festuca gautieri* on the crest of the risers, growing on remnants of degraded thin soils.

14.3.5 Shallow Landslides and Soil Erosion on Hillslopes: Human Impacts

Deep silty soils have developed on marly limestone outcrops. Such soils are located in the Pardina and Capradiza ravines and near the Góriz Shelter and show relatively recent spatial redistribution because of water and wind erosion. At some sites, the soil is more than 4 m deep, particularly at the foot of small cliffs. The forests that used to cover these soils several centuries ago were burnt to facilitate grazing by transhumant sheep flocks. As in other Pyrenean areas, deforestation of the subalpine belt activated soil erosion and shallow landsliding processes (García-Ruiz and Arbella 1981; García-Ruiz et al. 2010).

14.3.6 Contemporary Glaciers

The north face of the Monte Perdido Massif (Marboré Cirque) still has small active glaciers inherited from the LIA. LIA glaciers developed on both the north- and south-facing slopes of the massif, which has fresh moraines close to the cirque walls. The size of the glaciers has decreased since the beginning of the nineteenth century and some have disappeared, providing evidence for the impact of global warming in the Pyrenees. At the beginning of the twentieth century, the north face of the Monte Perdido Peak had three glaciers with a stepped arrangement controlled by structural benches, the highest of which produced continuous ice avalanches. At the present time, only two thin glaciers remain (Fig. 14.6). The Cilindro–Marboré glaciers, which are located to the west of the Monte Perdido glaciers, have now become small ice patches. Melting has accelerated in the last three decades as a consequence of increasing temperature and decreasing winter snowfall (Chueca et al. 2002; López-Moreno 2005).

14.4 Conclusions

The OMPNP has an impressive relief (3,355 m at its highest point) mostly developed on massive limestone and some marly limestones and sandstones. The thick limestone succession is related to the stacking of south-verging thrust sheets, which favoured the development of (1) vertical cliffs in a series of steps controlled by the different lithological units; (2) deep canyons; and (3) karst landforms and very limited surface drainage, except in the main rivers. The high altitude of the Monte Perdido Massif favoured the development of large glaciers which carved stunning U-shaped valleys flanked by steep walls >700 m high. Most of the preserved glacial deposits belong to recent glacial stages, the LIA being particularly well represented. The high altitude of the Monte Perdido Massif has also favoured: (1) the presence of small glaciers that have significantly receded since the end of the LIA; and (2) the occurrence of active periglacial features, including scree slopes and patterned ground like sorted stone polygons and stripes. Deep soils developed on marly outcrops have been intensively eroded after deforestation.

References

Barrère P (1952) Evolution mécanique et nivation sur les versants calcaires de la haute montagne pyrenéenne. Pirineos 24:201–213

Barrère P (1964) Le relief karstique dans l'Ouest des Pyrénées Centrales. Revue Belge de Géographie 88:9–62

Barrère P (1971) Le relief des Pyrénées Centrales franco-espagnoles. Institut de Géographie, Université de Bordeaux, Bordeaux

Chueca J, Julián-Andrés A, Peña-Monné JL (2002) Comparación de la situación de los glaciares del Pirineo español entre el final de la Pequeña Edad del Hielo y la actualidad. Boletín Glaciológico Aragonés 3:13–41

García-Ruiz JM, Arbella M (1981) Modelos de erosión en el piso subalpino: la degradación de los loess del macizo de Monte Perdido (Pirineo central español). Pirineos 114:35–58

García-Ruiz JM, Martí-Bono C (2001) Mapa geomorfológico del Parque Nacional de Ordena y Monte Perdido. Organismo Autónomo de Parques Nacionales

García-Ruiz JM, Beguería S, Alatorre LC, Puigdefábregas J (2010) Land cover changes and shallow landsliding in the Flysch sector of the Spanish Pyrenees. Geomorphology 124:250–259

Hernández-Pacheco F, Vidal-Box C (1946) La tectónica y la morfología del macizo de Monte Perdido y de las zonas de cumbres inmediatas en el Pirineo Central. Pirineos 4:69–108

Lewis CJ, McDonald EV, Sancho C, Peña JL, Rhodes EJ (2009) Climatic implications of correlated Upper Pleistocene glacial and fluvial deposits on the Cinca and Gállego Rivers (NE Spain) based on OSL dating and soil stratigraphy. Global Planet Change 61:300–312

López-Moreno JI (2005) Recent variations of snowpack depth in the Central Spanish Pyrenees. Arct Antarct Alp Res 37:253–260

Martí-Bono C, García-Ruiz JM (1993) La extension del glaciarismo cuaternario en el Parque Nacional de Ordesa y Monte Perdido. Geographicalia 30:271–282

Nicolás-Martínez P (1981) Morfología del circo de Tucarroya (Macizo de Monte Perdido, Pirineo aragonés). Cuadernos de Investigación Geográfica 7:51–80

Oliva-Urcia B, Moreno A, Valero-Garcés B, Mata P, Grupo Horda (2013) Magnetismo y cambios ambientales en registros terrestres: el lago de Marboré, Parque Nacional de Ordesa y Monte Perdido (Huesca). Cuadernos de Investigación Geográfica 39:117–140

Ríos LM, Frutos E, Barnolas A (1982) Mapa Geológico de España, escala 1:50.000, Hoja 178, Broto. Instituto Geológico y Minero de España, Madrid

Ríos LM, Galera JM, Barettino D, Barnolas A (1989) Mapa Geológico de España, escala 1:50.000, Hoja 146, Bujaruelo. Instituto Tecnológico GeoMinero de España, Madrid

Salazar-Rincón A, Mata-Campo P, Rico-Herrera MT, Valero-Garcés BL, Oliva-Urcia B, Ibarra P, Rubio FM, Grupo Horda (2013) El paleolago de La Larri (Valle de Pineta, Pirineos): Significado en el contexto del último máximo glaciar en el Pirineo. Cuadernos de Investigación Geográfica 39:97–116

15 Glacial and Structural Geomorphology in the Maladeta Massif, Pyrenees

J. Chueca-Cía, A. Julián-Andrés, and M. Ortuño-Candela

Abstract

This work presents the general geomorphological setting of the Maladeta Massif and a more specific analysis on two outstanding aspects of this Alpine area: (1) the main lithostructural features and their relation with active deformation processes (active faults and deep-seated landslides) and (2) the glacial geomorphology of the massif, focusing on the recent recessional pattern (1981–2005) of the glacial masses (glaciers and glacierets). Historical surface and volume losses in the ice bodies are assessed, together with the climatic and topographic factors that have controlled the shrinkage of glaciers at regional and local scales.

Keywords

Glacial geomorphology • Structural geomorphology • Maladeta Massif • Pyrenees

15.1 Introduction

In recent years, several studies have analysed the geomorphology of the Maladeta Massif (Central Spanish Pyrenees), addressing issues related to geomorphological mapping (Chueca and Julián 2008), lithostructural features and their relationship with active gravitational and tectonic deformation (Gutiérrez et al. 2005, 2008; Ortuño 2008; Ortuño et al. 2008; Larrasoaña et al. 2010), or the present-day glacial processes and the evolution of the glaciers since the end of the Little Ice Age (LIA) (Chueca et al. 2003a, 2005, 2007; Chueca and Julián 2004; López-Moreno et al. 2006a, b).

One of the most significant singularities of the Maladeta Massif from the geomorphological perspective is that it hosts the largest glacial complex in the Pyrenees. Since the LIA, the high altitude and the NW–SE trend of the range have favoured the preservation of several N–NE-facing glaciers and glacierets (i.e. small ice masses derived from former glaciers with no signs of displacement). A similar situation is found in other glaciated massifs of the Pyrenees, such as Balaitús, Infiernos, Vignemale, Monte Perdido or Perdiguero. The Aneto glacier in the Maladeta Massif is the largest Pyrenean glacier (79.6 ha as measured in 2005).

This chapter provides a general setting for the Maladeta Massif and analyzes two topics of special geomorphological significance: (1) active deformation processes (active faults and deep-seated landslides) and their relationship with the lithological and structural context and (2) the glacial geomorphology of the range, specially the recent recessional patterns (1981–2005) observed in the glaciers and glacierets. The reduction in surface and volume of the ice bodies is assessed, as well as the controlling factors (climatic and topographic) of the glacial regression, both at regional and local scales.

J. Chueca-Cía (✉)
Departamento de Geografía y Ordenación del Territorio, Facultad de Filosofía y Letras, Universidad de Zaragoza, 50009, Zaragoza, Spain
e-mail: jchueca@unizar.es

A. Julián-Andrés
Departamento de Geografía y Ordenación del Territorio, Escuela Politécnica Superior, Universidad de Zaragoza, 22071, Huesca, Spain
e-mail: ajulian@unizar.es

M. Ortuño-Candela
Departament de Geodinàmica i Geofísica, Facultat de Geologia, Universitat de Barcelona, 08028, Barcelona, Spain
e-mail: maria.ortuno@ub.edu

F. Gutiérrez and M. Gutiérrez (eds.), *Landscapes and Landforms of Spain*, World Geomorphological Landscapes,
DOI: 10.1007/978-94-017-8628-7_15,

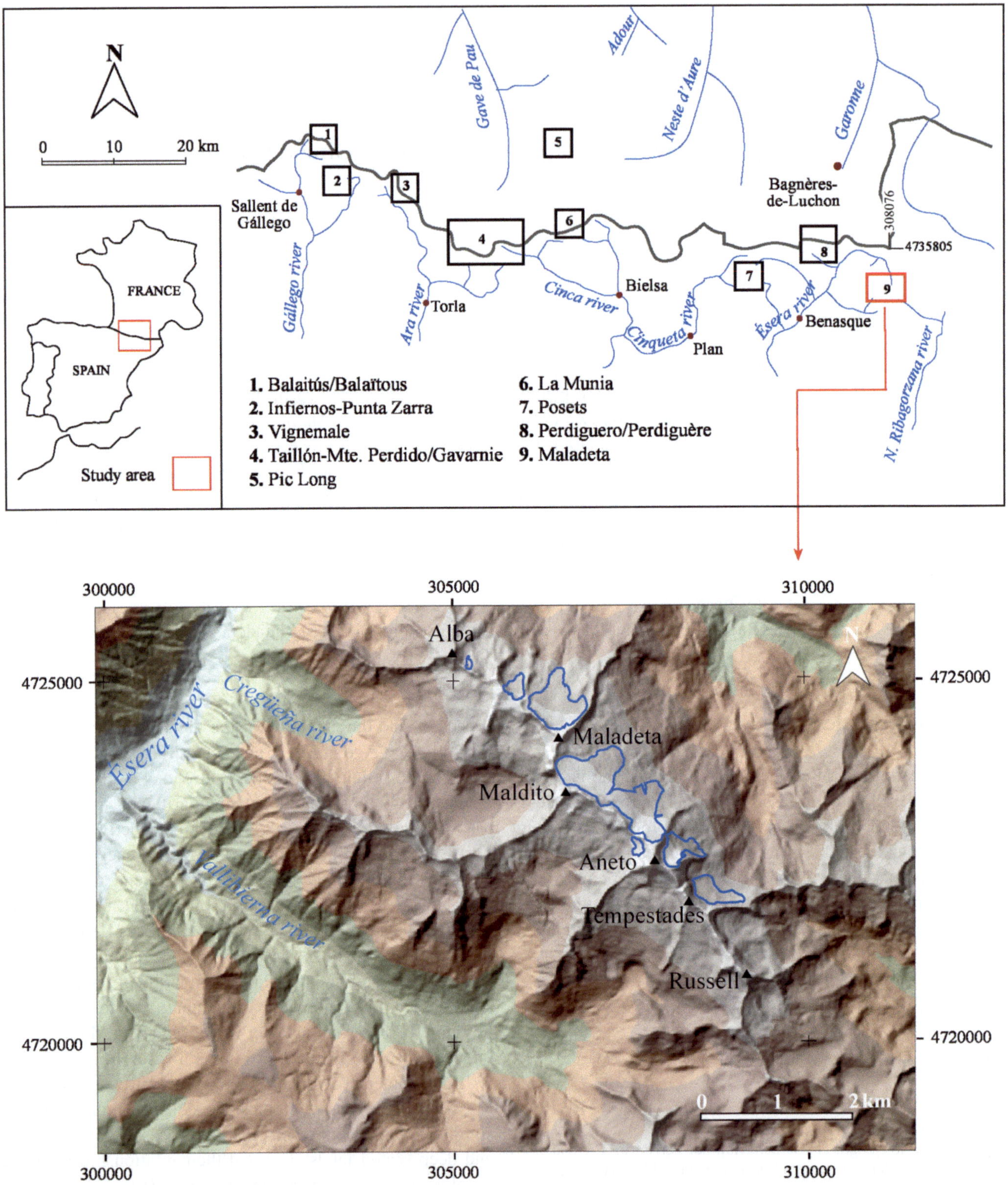

Fig. 15.1 Location of the study area (Maladeta Massif) and the present-day glaciated sectors within the Pyrenees. The lower image depicts the distribution of glaciers and glacierets in the Maladeta Massif in 2005

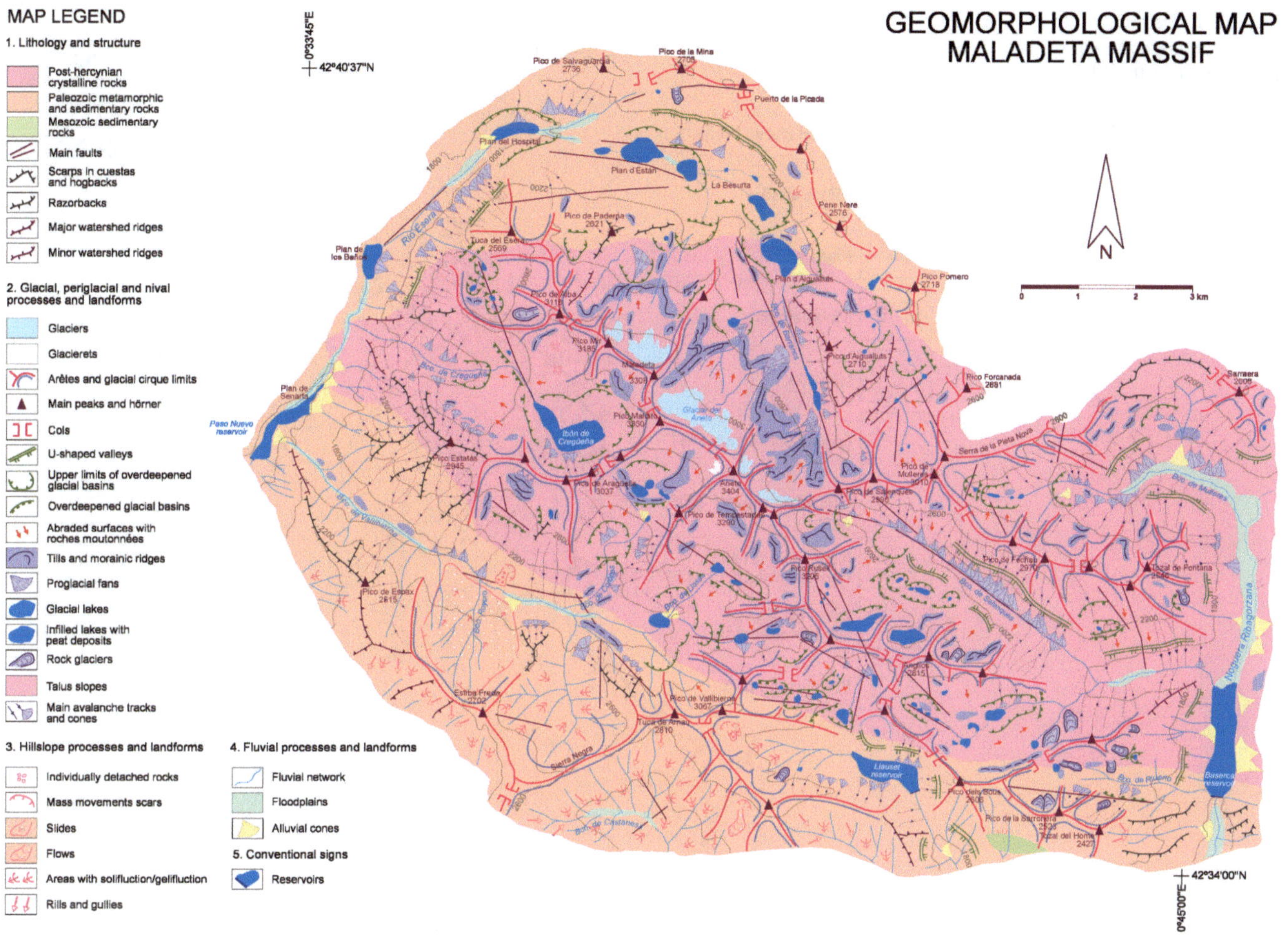

Fig. 15.2 Geomorphological map of the Maladeta Massif

15.2 Study Area

The Maladeta Massif is located in the central Spanish Pyrenees, between the Ésera and Noguera Ribagorzana valleys, and is the highest mountainous area of the Pyrenees (Fig. 15.1). It is formed of crystalline rocks of the Maladeta batholith. This igneous body extends in an E–W direction and consists mainly of homogenous masses of granodiorite and granite. The high altitudes of the massif (e.g. Aneto, 3,404 m a.s.l.; Pico Maldito, 3,350 m; Maladeta, 3,308 m; Tempestades, 3,290 m; Pico Russell, 3,205 m) are largely related to the high resistance to erosion of the massive crystalline bedrock.

The landscape in this sector of the Pyrenees is mainly the result of Pleistocene glacial activity. Limited transformation of the relict glacial landforms has occurred during the Holocene, predominantly by periglacial and mass-wasting processes. The present-day alpine relief is characterized by over-deepened glacial valleys (Alto Ésera, Vallibierna, Salenques, Alta Noguera Ribagorzana) with steep walls. The main landforms in the highest sectors include the following: (1) glacial cirques with abraded surfaces, roches moutonnées, over-deepened basins, glacial lakes and peat bogs associated with infilled lakes; (2) talus slopes; (3) moraine ridges and protalus ramparts corresponding to Neoglacial and LIA periods and (4) small rock glaciers derived from talus deposits with discontinuous permafrost, affected by flow processes (Figs. 15.2, 15.3).

Above the tree line (2,100–2,200 m), slopes are frequently covered by rock debris or till deposits. The mean annual temperature varies from −0.5 °C at 3,000 m to +6.8 °C at 1,500 m. In the coldest month, the average temperature reaches −4.3 °C at 3,000 m and +0.3 °C at 1,500 m. The annual precipitation varies from 2,600 mm at 3,000 m to 1,400 mm at 1,500 m (Chueca and Julián 2004). Discontinuous and sporadic permafrost have been reported

Fig. 15.3 Detailed geomorphological map of the central part of the Maladeta Massif; glacial circques of Aneto, Coronas and Llosás

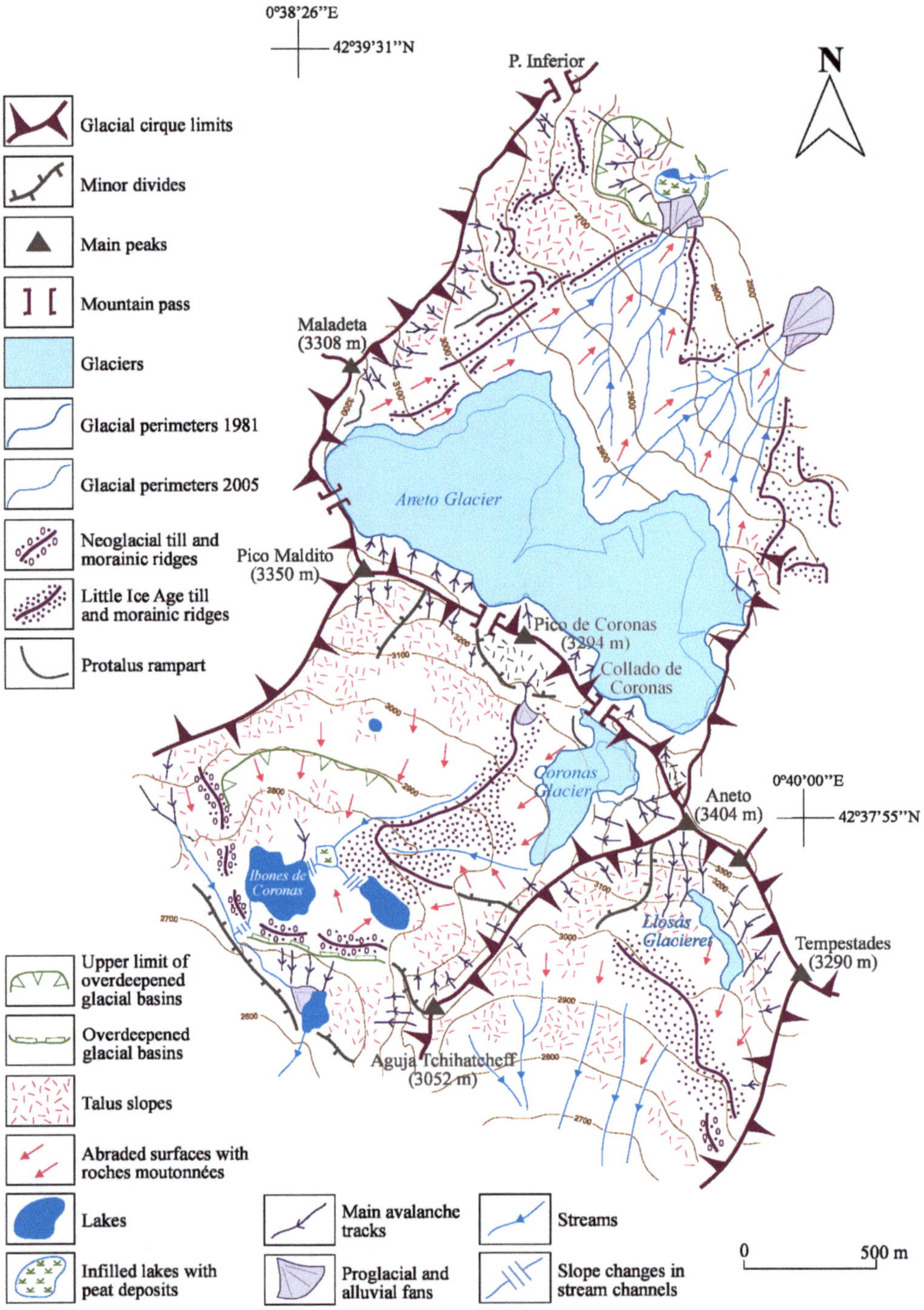

in the Maladeta Massif. Detailed studies are restricted to the Aneto glacial cirque area, where thermal characterization of the soil pointed to the probable existence of permafrost in several sectors between 2,950 and 3,050 m, including LIA moraine deposits and protalus ramparts associated with the Aneto glacier and the former Cresta de los Portillones glacier (Chueca and Julián 2010).

15.3 Geomorphology

15.3.1 Lithology and Structure

The lithology and structure of the Maladeta Massif are major factors controlling the development of landforms in the area. The massif is made up of granitoids surrounded by

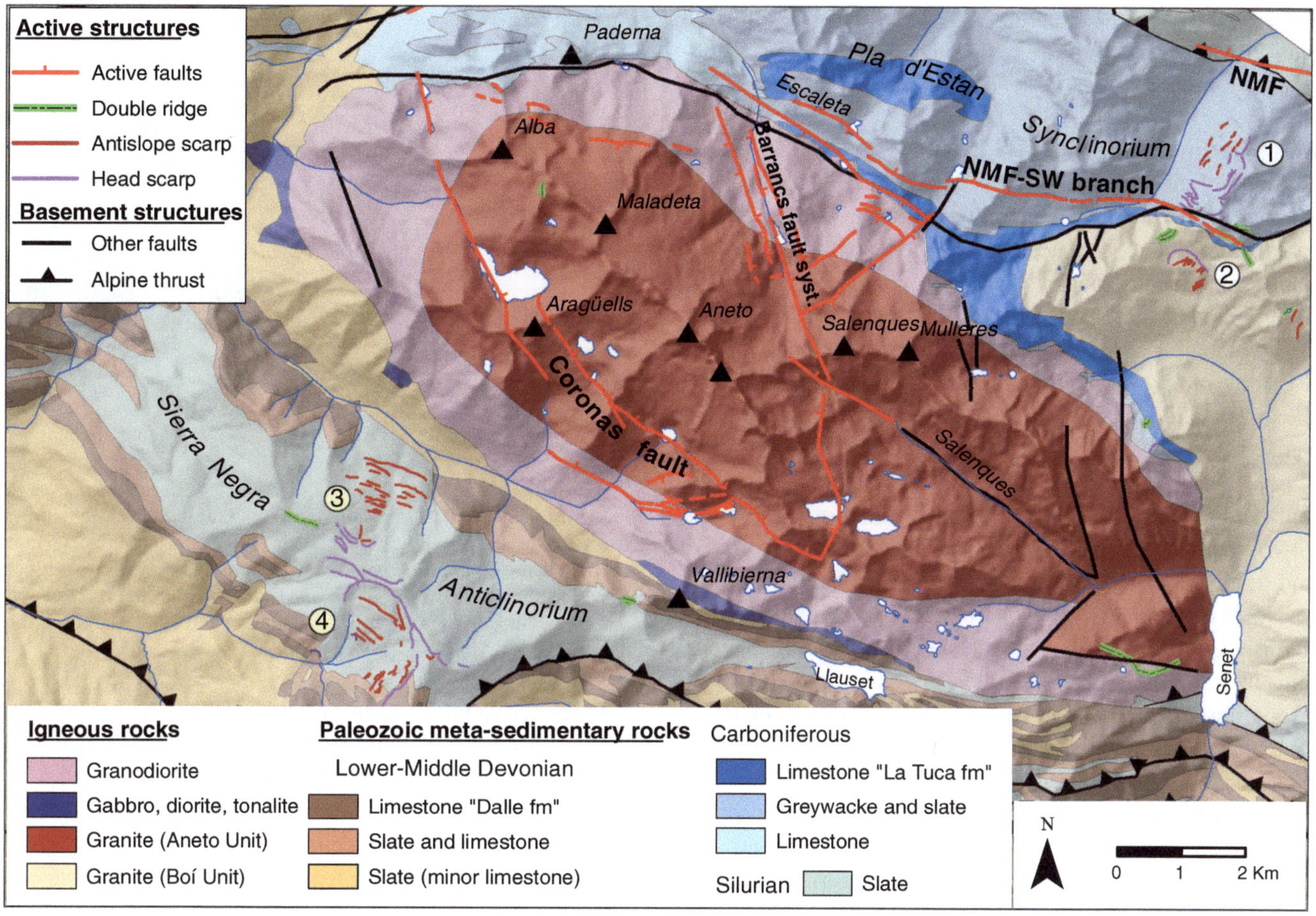

Fig. 15.4 Synthetic sketch of the main lithological units identified in the Maladeta Massif (compiled by Ortuño 2008) using data from Zwart (1979), Arranz (1997) and Ríos et al. (1997). The active faults identified in the area by Ortuño (2008) have been included. NMF: North Maladeta fault. Four conspicuous cases of gravitational slope failure occur in the metamorphic aureole: *1* the Hurno site (Bordonau and Vilaplana 1986); *2* the Mulleres sites (Ortuño 2008); *3* the Vallibierna-Fangonielles site (Lampre 1998; Gutiérrez et al. 2005) and *4* El Ubago site (Gutiérrez et al. 2008)

Paleozoic meta-sedimentary rocks. The area described in this chapter corresponds to the western sector of the Maladeta batholith, emplaced at the end of the Variscan orogeny producing a thermal metamorphic aureole (Charlet 1979; Zwart 1979; García-Sansegundo 1991, 1992; Arranz 1997; Arranz and Lago 2004). The metamorphosed sedimentary rocks surrounding the batholith were intensively folded and faulted during Variscan times. Later, during the Alpine orogeny, these basement rocks were incorporated into the Orry thrust sheet, one of the three structural units of the antiformal stack of the Pyrenean Axial Zone defined by Muñoz (1992).

The Granitic Batholith

Most of the area is dominated by the granodiorites of the Aneto unit, as defined by Charlet (1979) and Arranz (1997). The different lithofacies of this unit display a concentric zonation, with an inner part consisting of granite with two micas and cordierite and an outer zone of granodiorite. Magnetic foliations also display a concentric distribution (Leblanc et al. 1994). Some patches of mafic rocks (gabbro, diorite and tonalite) are identified in peripheral zones (Fig. 15.4). The age of the intrusion has been estimated at 298 ± 2.5 Ma by using the uranium–lead dating method (Gleizes et al. 1997; Evans et al. 1998).

In contrast to other granitic landscapes in the region, the different lithotypes do not display significant variations in rock erodibility, so that lithological changes are scarcely reflected in the erosional landforms (Ortuño 2008). The plutonic rocks are fractured and affected by Alpine thrusts. The most conspicuous Alpine fault is the Coronas fault, associated with a fault breccia up to 10 m wide (Ortuño 2008). The two main fracture sets have NNW and WNW orientations. The geomorphic expression of these discontinuities has been enhanced by glacial and periglacial erosion processes, and some of them have been reactivated as normal faults during the Neotectonic period, which, in the Central Pyrenees, starts in the Middle Miocene (Lacan and

Ortuño 2012). Furthermore, several valleys within the batholith (e.g. Salenques) are controlled by the fracture orientation or major fault zones (Fig. 15.4).

Meta-sedimentary Folded Rocks

The batholith is surrounded by meta-sedimentary rocks of Silurian to Carboniferous age. The Silurian formations are mainly detrital (e.g. slates), while the Devonian and Carboniferous successions also include calcareous units like dolomites and marble (Zwart 1979; Ríos et al. 1997).

Two major E–W structures are distinguished: the Pla d'Estan synclinorium and the Sierra Negra anticlinorium, to the north and south of the massif, respectively (García-Sansegundo 1991, 1992).They are characterized by two Variscan fold generations with E–W oriented subvertical and recumbent axial planes. A gentle foliation and several E–W-trending subvertical fault zones have been recognized and assigned to the Alpine deformational phase.

North of the batholith the carbonate rocks are affected by intense karstification, as the presence of numerous cover and bedrock collapse sinkholes reveals. The Forau de Aigualluts is a bedrock collapse sinkhole more than 100 m long that functions as a ponor. It swallows the flow of the Ésera River, which emerges more than 3 km to the north in France, after traversing the Mediterranean–Atlantic divide through an underground karst system. To the south, the Sierra Negra Anticlinorium displays a conformable topography, in which Silurian black slates underlie the crest of the ridge, and metamorphosed Devonian sandstones, lutites and limestones crop out on its northern and southern flanks (Fig. 15.4).

15.3.2 Active Faults and Deep-Seated Landslides

The slopes of the Maladeta Massif display numerous downhill- and uphill-facing scarps. Most of them are parallel to the contour lines and correspond to erosional landforms controlled by exfoliation joints (Lampre 1998; Ortuño et al. 2008). However, some scarps within the massif correspond to normal active faults attributable to a combination of neotectonics and slow landsliding. Differential uplift of the valley bottom after the glacial retreat has also been suggested by Ortuño (2008) as a probable deformational process at some particular sites (e.g. Barrancs and Vallibierna valleys).

Gravitational deformation processes in the slopes are likely to be triggered by seismic activity. The epicentres of the two most damaging historical earthquakes in the Central Pyrenees are located in this region: the 1373 Ribagorza earthquake (Mw 6.2; Olivera et al. 2006) and the 1923 Vielha earthquake (Mw 5.8; Susagna et al. 1994).

Fig. 15.5 The Escaleta antislope scarp is interpreted as the surface expression of the southwestern branch of the North Maladeta fault. The fault scarp (*black line* with teeth towards the downthrown block) is developed along the axial plane of a tight syncline (the *red line* with a cross in the *middle* shows the fold trace) affecting Carboniferous greywackes and slates

Three active fault systems have been identified in the area: (1) the 7.5-km-long southwestern branch of the North Maladeta fault (NMF); (2) the Barrancs fault system (4.5 km long) and (3) the Coronas fault, 12 km long. The NMF is the most feasible seismic source of the 1923 Vielha earthquake (Ortuño et al. 2008). The location of the 1373 Ribagorza earthquake has an uncertainty of 25 km (Olivera et al. 2006) and could have been produced by the NMF or the Coronas fault, among other faults in the area (Ortuño 2008).

The westernmost segment of the NMF southwestern branch displaces several slopes, coinciding with the axial plane of a tight syncline. This feature is especially conspicuous at the Escaleta site (Figs. 15.4, 15.5). To the south of this site, a group of scarps bounding the Barrancs lake have been interpreted as postglacial normal faults (Moya and Vilaplana 1992; Ortuño 2008). The activity of these faults, which offset polished glacial surfaces (Figs. 15.4, 15.6b), has been related to seismic shaking and could have caused the repeated sudden drainage of the Barrancs lake, as its Holocene sedimentary fill suggests (Larrasoaña et al. 2010).

The most prominent scarp in the area is that of the Coronas fault, developed in the southern flank of the massif (Fig. 15.6c). The fault displaces the slopes carved by glaciers, as attested by the systematic increase in the scarp height towards the centre. To explain the exceptional height of the scarp (more than 120 m), Ortuño (2008) has proposed that the Coronas fault is a composite fault resulting from neotectonic activity and gravitational collapse of the massif towards the north. The collapse is understood as a product by deep-seated gravitational slope deformation involving other structures such us discrete faults observed in the northern

Fig. 15.6 Images of scarps resulting from active deformation on the slopes of the Maladeta Massif and its surroundings. **a** Vallibierna-Fangonielles *sackung* in Sierra Negra range; **b** fault scarps in Barrancs lake offsetting glacially polished surfaces and **c** Coronas fault scarp next to Aragüells peak. The scarp has influenced the distribution of several over-deepening basins related to differential glacial erosion

flank of the massif or the fault system of Barrancs (Fig. 15.4). Glacial over-deepening and karstification at the base of the northern slope, together with topographic amplification of seismic shaking, may have influenced the slope failure.

The presence of *sackung* features in the area, such as double ridges and sets of antislope scarps, frequently associated with landslide head scars on the upper part of the slopes, was first reported in the slopes of the Hurno ridge by Bordonau and Vilaplana (1986). To the south of this ridge, Ortuño (2008) has identified a deep-seated gravitational slope deformation in the Mulleres valley. Two other cases of *sackung* have been reported in the Sierra Negra range: the Vallibierna-Fangonielles (Lampre 1998; Gutiérrez et al. 2005; Fig. 15.6a) and the El Ubago site (Gutiérrez et al. 2008). With the exception of the Mulleres landslide, developed on granitic rocks, the other slope failures occur on foliated metamorphic rocks (Fig. 15.4). The formation of these slow gravitational failures in the Maladeta Massif is likely influenced by the topographic amplification of the seismic waves in the over-steepened slopes. Evidence of episodic activity inferred from trenches excavated across anti-slope scarps in Vallibierna and El Ubago sites led Gutiérrez et al. (2005, 2008) to propose that the episodic activity of these surface ruptures might be related to seismic shaking.

15.3.3 Recent Glacial Evolution

General Setting

Since the end of the LIA, dated in the Pyrenees to ~1820–1830 AD (Chueca and Julián 1996), the Pyrenean glaciers, both in the Spanish and French regions, have experienced significant losses in surface area and volume. At the end of the LIA, the extent of the glaciers in the nine main mountainous massifs (Balaitús, Infiernos, Vignemale, Monte Perdido/Gavarnie, Pic Long, La Munia, Posets, Perdiguero and Maladeta) totalled slightly over 2000 ha, while nowadays, the remaining glacial ice covers 600 ha (Chueca et al. 2004) (Fig. 15.1). In the last two decades of the twentieth century and the beginning of the twenty-first century, glacial shrinkage rates were as high as those estimated for the period 1860–1900 (Chueca et al. 2003a, 2005).

The total extent of the ice masses in the Maladeta Massif reaches 155.0 ha, as measured in 2005 (Chueca et al. 2007), which represents about 25 % of the total area estimated for the final stage of the LIA (616 ha). The largest ice bodies are Aneto (79.6 ha) and eastern Maladeta (36.7 ha) glaciers, while the smallest correspond to Coronas (2.7 ha) and Alba (1.0 ha) glacierets (Figs. 15.2, 15.3). The ice thickness has been estimated by seismic reflection and ground penetrating radar in the western and eastern Maladeta and Aneto glaciers (Martínez and García 1994; Martínez et al. 1997) and in the Coronas glacieret (Chueca et al. 2003b). According to these geophysical surveys, ice thickness reaches 40–50 m and 7–9 m in the main glaciers and in the Coronas glacieret, respectively.

The glacial bodies included in the study of the recessional pattern (1981–2005) indicated above comprise all the glaciers and glacierets existing in the massif in 1981: Maladeta, Aneto, Barrancs, Tempestades, western Salenques and Coronas glaciers, as well as Alba, eastern Salenques, western and eastern Cregüeña and Llosás glacierets. The extent and volume losses were quantified with a geographical information system (GIS) integrating the results of the following procedures: (1) the analysis of 1981 and 1999 aerial photographs and global positioning system (GPS) measurements conducted in 2005 to quantify ice areal changes and (2) the comparison of digital elevation models of 1981 and 1999 to estimate changes in ice volume (Fig. 15.7). Subsequently, the results were compared with factors that control glacial recession: (1) climatic factors including the evolution of temperature (mean, maximum and minimum) and precipitation during the last decades in the Pyrenean region embracing the Maladeta Massif and (2) local factors related to topography, such as solar radiation inputs, glacier elevation or glacier initial size.

Surface and Volume Losses

The estimated magnitudes of glacier recession are shown in Tables 15.1 and 15.2 and in Fig. 15.8. All glacial bodies display a similar evolution, with clear extent and volumetric losses and increases in the mean altitude of each glacier.

During the period 1981–2005, the glaciers of the Maladeta Massif receded by 35.7 % in surface area, from 240.9 to 155.0 ha. The average surface-loss rate was 3.57 ha/year. Ice volume losses between 1981 and 1999 were estimated at 0.0137 km^3, yielding an average annual rate of 763 × 10^3 m^3/year. Considering an average density for the ice of 0.9 g/cm^3, the ice thickness reduction in the Maladeta Massif was 75.6 m w.e. (water equivalent) (4.2 m w.e./year). Data from glaciers of different extent were compared using the volume loss-to-initial surface ratio (VL/IS) recorded for the period 1981–1999. The VL/SL index value for the whole massif was 5.70 (0.31 VL/IS units/year). Finally, the mean altitude of the studied glaciers increased 43.5 m between 1981 and 2005 (1.81 m/year), from 3026.9 to 3070.4 m a.s.l. This parameter, calculated by averaging the altitudinal value of all pixels classified as glacial surface, was chosen due to its adequacy to evaluate the regression of the small cirque glaciers that characterize the massif. In such glaciers, which may have very irregular geometries, the usual measure of length reduction along one or several axes is less representative than this total value.

Fig. 15.7 **a** 3-D view of the Maladeta glacier after superimposing the geometrically corrected and georeferenced aerial photograph of the 1981 Pirineos-Sur flight on the 1981 DEM (glacier perimeter is indicated with a black line). **b** Oblique aerial photograph taken in 2005 of the eastern and western Maladeta glaciers. Major ice losses in the small tongue of the eastern Maladeta glacier are obvious

Noticeable differences in the observed shrinkage trends are directly linked to the northern or southern aspect of the studied glaciers and glacierets (Figs. 15.3, 15.8). This effect was quantified by grouping the values of extent and volume losses, and the increase in altitude into two sets: north-facing ice masses (northeast aspect in most cases; Alba, Maladeta, Aneto, Barrancs, Tempestades and western Salenques glacier and eastern Salenques glacieret) and south-facing ice masses (southwest aspect; Coronas glacier and Llosás, western and eastern; Cregüeña glacierets).

Higher volumetric and extent losses and altitudinal changes have been observed in the south-oriented glaciers and glacierets. During the period covered by the study, these glacial bodies decreased in area by 83.9 % (from 16.9 to 2.7 ha), the VL/IS index was 12.28 (0.68 VL/IS units/year) and average altitude in the case of Coronas glacier increased 75 m (3.12 m/year), from 3,081 to 3,156 m a.s.l.

Table 15.1 Volumetric and extent losses in the glaciers of the Maladeta Massif

	Vol. losses 1981–1999 (m^3)	Mean thickness losses 1981–1999 (m w.e.)	Extent 1981 (m^2)	Extent 2005 (m^2)	% Surface loss 1981–2005	VL/IS index
Alba	0.120×10^6	4.1	2.679×10^4	1.004×10^4	62.52	4.51
Maladeta	2.621×10^6	4.9	48.904×10^4	43.028×10^4 $(6.309 + 36.718) \times 10^4$	12.02	5.35
Aneto	4.736×10^6	4.1	106.770×10^4	79.699×10^4	25.36	4.43
Barrancs	1.190×10^6	4.8	22.497×10^4	13.956×10^4	37.97	5.29
Tempestades	2.051×10^6	8.4	28.792×10^4	14.564×10^4	49.42	7.12
Salenques west	0.896×10^6	6.3	12.844×10^4	–	100.00	6.98
Salenques east	0.048×10^6	2.9	1.523×10^4	–	100.00	3.17
Cregüeña west	0.202×10^6	7.9	1.994×10^4	–	100.00	10.15
Cregüeña east	0.229×10^6	10.4	2.344×10^4	–	100.00	9.77
Coronas	1.445×10^6	12.6	10.570×10^4	2.748×10^4	74.00	13.67
Llosás	0.208×10^6	9.2	2.072×10^4	–	100.00	10.08
Total	13.751×10^6	75.6	240.993×10^4	155.002×10^4	–	–

The percentage of surface loss in the Maladeta glacier was calculated after adding the 2005 extent of its two present-day fragments: western and eastern Maladeta glaciers

Table 15.2 Maximum, minimum and mean altitude (m a.s.l) of Maladeta Massif glaciers

	Max. alt. 1981	Min. alt. 1981	Mean alt. 1981	Max. alt. 2005	Min. alt. 2005	Mean alt. 2005	Increase in mean alt. 1981–2005
Alba	3,039	2,923	2,970	3,038	2,950	2,992	22
Maladeta	3,206	2,801	3,051	3,196	2,884	3,061.5	10.5
Aneto	3,325	2,855	3,101	3,316	2,911	3,126	25
Barrancs	3,332	2,899	3,080	3,323	2,938	3,121	41
Tempestades	3,085	2,792	2,928	3,070	2,890	2,966	38
Salenques west	3,122	2,854	2,988	–	–	–	–
Salenques east	2,990	2,877	2,918	–	–	–	–
Cregüeña west	3,021	2,928	2,964	–	–	–	–
Cregüeña east	3,215	3,110	3,146	–	–	–	–
Coronas	3,213	2,984	3,081	3,206	3,100	3,156	75
Llosás	3,108	3,001	3,069	–	–	–	–

The 2005 value for the Maladeta glacier was obtained averaging the altitudes of its two present-day fragments: western and eastern Maladeta glaciers

The other south-oriented ice masses underwent total degradation, and thus, their altitudinal migration is not calculable. The north-facing glaciers diminished their surfaces by 32.1 % (from 224.0 to 152.2 ha), had a VL/IS index of 5.20 (0.28 VL/IS units/year) and their average altitude increased by 48.2 m (2 m/year), from 3,005 m to 3,053 m a.s.l. Only the western Salenques glacier and eastern Salenques glacieret showed complete degradation (Fig. 15.9).

Causes of Recent Glacial Degradation

Analysis of precipitation and temperature trends in the study area reveals a deterioration of conditions favourable for the development of glaciers during the 1980s, 1990s and beginning of the 2000s (Chueca et al. 2007). Snow precipitation during the accumulation period decreased significantly, reducing the snowfall contribution to mass balance in February and March. In addition, an increase in temperature during the ablation season, particularly maximum temperatures, which exert a major influence on ice melting processes (Singh and Singh 2001), also contributed to reinforce negative glacial mass balance.

The rates of glacial degradation over the 1981–2005 period were as high as those estimated for the period 1860–1900 (Chueca et al. 2003a, 2005). Data from the beginning of the twenty-first century indicate a continuation of the negative conditions for glacier conservation. Climatic

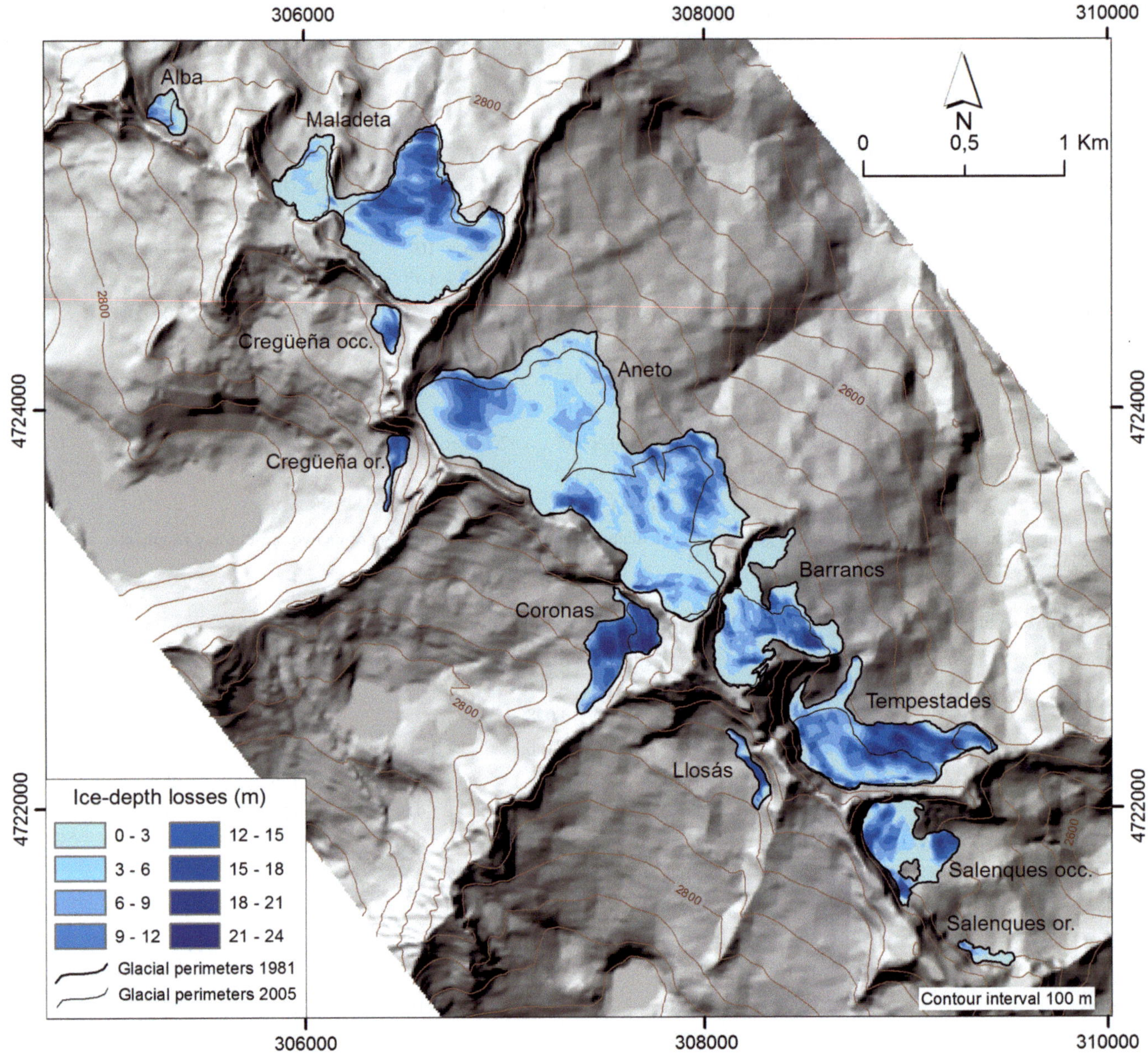

Fig. 15.8 Extent losses (glacier perimeters in 1981 and 2005 are indicated) and ice-depth losses (for the 1981–1999 period) observed in the glaciers and glacierets of the Maladeta Massif

Table 15.3 Correlation coefficients (R) between selected variables and significance level

Variables		R	Significance level
VL/IS index	Mean summer solar radiation	0.77	0.005
VL/IS index	Mean annual solar radiation	0.82	0.002
VL/IS index	Mean initial altitude (1981)	0.35	0.297
VL/IS index	Initial size (1981)	−0.38	0.248
% Surface loss	Initial size (1981)	−0.75	0.008
% Surface loss	Mean summer solar radiation	0.18	0.606
% Surface loss	Mean annual solar radiation	0.34	0.303
Increase mean alt.	Mean summer solar radiation	0.74	0.048
Increase mean alt.	Mean annual solar radiation	0.80	0.028

Fig. 15.9 Comparison of the western Salenques glacier in 1990 (**a**) and 2005 (**b**) (*Photographs* by J. Camins)

projections for the twenty-first century (Houghton et al. 2001) in the Mediterranean mountain regions predict an increase in thermal values and a diminution of precipitation that would induce a further deterioration of glacial mass balance.

According to the results of the bivariate correlation analysis (Table 15.3), the three main topographical factors that influence glacial degradation include the following: (1) glacier orientation, which controls to a great extent solar radiation input; (2) glacier altitude and (3) glacier initial size (i.e. extent in 1981). The presence of a debris cover, related either to endoglacial material brought to the ice surface or to rockfalls, can considerably reduce ablation processes on a glacial surface due to its insulating effect. This factor is so far negligible in the Maladeta ice bodies. However, debris cover will probably gain importance in the future, since some of the glaciers and glacierets are beginning to show accumulations of debris on their distal sectors (e.g. western and eastern Maladeta, Barrancs and western Tempestades glaciers, and Coronas glacieret).

Potential incoming solar radiation, in terms of both the summer and annual values, seems to be the main factor that controls the volume loss per surface unit recorded in the glaciers of the Maladeta Massif, as well as the increase in mean altitude derived from their deterioration. As the trends shown in Fig. 15.10a–d illustrate, the higher the solar radiation inputs, the higher the loss of volume per unit area and the higher the increase in the mean altitude of each glacier. This finding is consistent with the previous observations in the study area. López-Moreno et al. (2006b) studied how the influence of topography on the extent of the Maladeta cirque glaciers has changed since the LIA. They found that during the final stages of glacial degradation process, the relative influence of climatic factors linked to altitude (e.g. temperature lapse rate, amount of precipitation and snowfall versus rainfall ratio) decreases with respect to the influence of the exposure to solar radiation and other topography-related variables (e.g. terrain curvature).

Bivariate correlation analyses also confirm the important role of the initial size of glaciers in the subsequent shrinkage process. In the Maladeta Massif, higher rates of surface loss are associated with smaller glaciers and vice versa (Fig. 15.10e). This finding highlights the marked sensitivity of smaller glaciers, which usually have a reduced altitude range, to changes in the main climatic factors. The latter always exhibit a strong altitudinal gradient and thus play a significant role in the accumulation and ablation processes. Chueca et al. (2005) pointed out that there seems to be a minimum response time of a few years between temperature and precipitation changes and the shrinkage of glaciers in the Maladeta Massif. As the response time of a glacier to climatic fluctuations depends mainly on its size and latitudinal location, climatic changes, even if subtle, have a faster effect on the mass balance of small glaciers located in

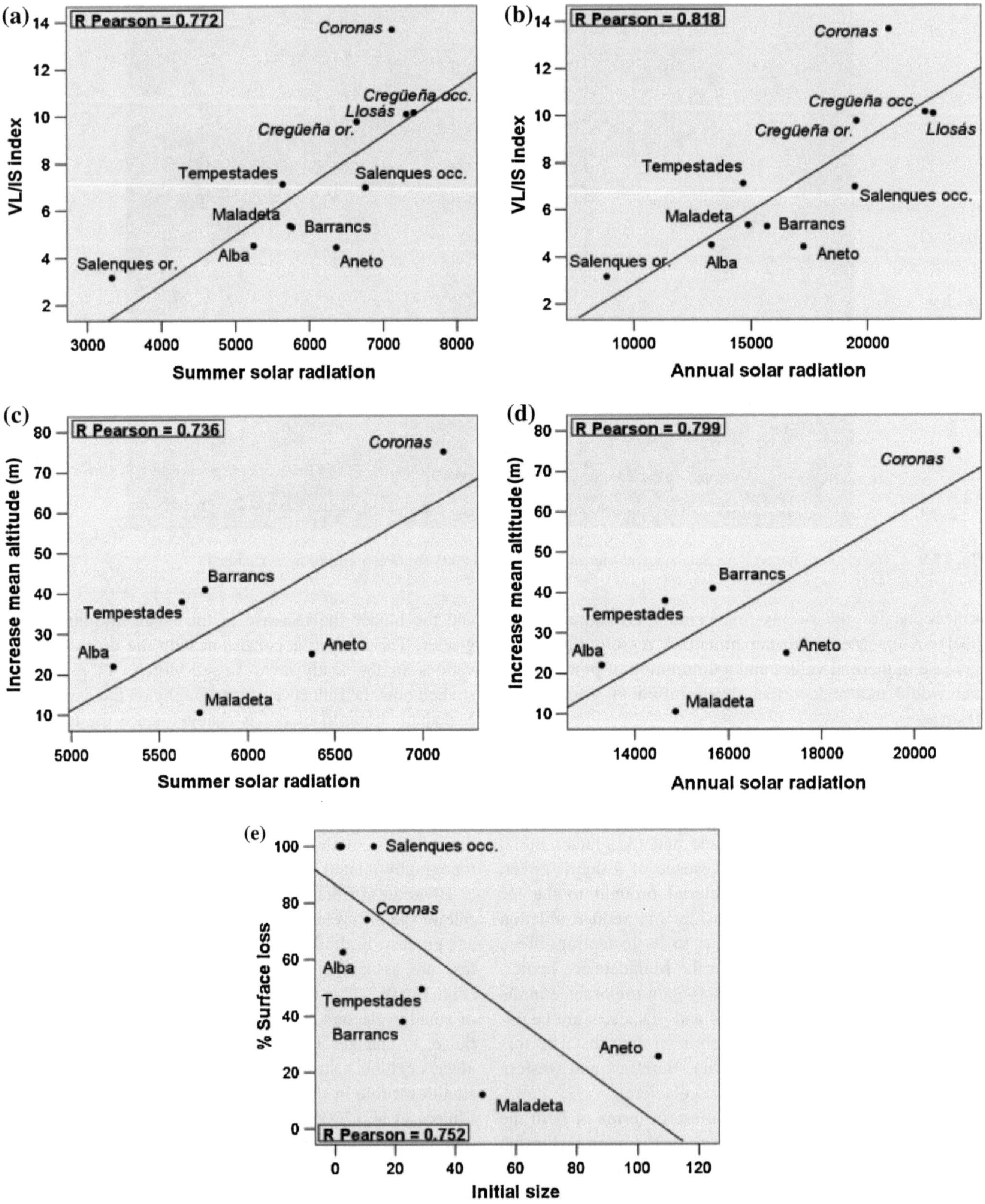

Fig. 15.10 Scatterplots of the relationships between **a** summer solar radiation (10 kJ m^{-2} d^{-1} μm^{-1}) and VL/IS index, **b** annual solar radiation (10 kJ m^{-2} d^{-1} μm^{-1}) and VL/IS index, **c** summer solar radiation (10 kJ m^{-2} d^{-1} μm^{-1}) and increase in mean altitude (1981–2005) (m), **d** annual solar radiation (10 kJ m^{-2} d^{-1} μm^{-1}) and increase in mean altitude (1981–2005) (m), and **e** glacier area in 1981 (ha) and percentage of surface loss (1981–2005). Cases of 100 % surface loss located in the *upper left* include eastern Salenques, western and eastern Cregüeña, and Llosás glaciers. South-oriented glaciers and glacierets are indicated in *italics*

warmer settings like the Pyrenees, than on larger glaciers at higher latitudes. This fact makes the Pyrenean glacial remnants useful proxies to monitor future climatic changes in mid-latitude mountains.

15.4 Conclusions

The Maladeta granitic massif, including the highest peaks of the Pyrenees, displays remarkable glacial and periglacial landforms, as well as a particularly high density of uphill- and downhill-facing fault scarps attributable to seismogenic active tectonic faults, gravitational deformation (sackung), differential erosion or a combination of these processes.

An outstanding feature of the Maladeta Massif is that it contains some of the southernmost glaciers in Europe. The high rates of volumetric and extent losses that have been observed in the glacial masses (glaciers and glacierets) from 1981 to 2005 are consistent with those detected on a global scale (Haeberli et al. 2005) and in other alpine sectors of the world with variable altitudinal and latitudinal distribution. These changes have been interpreted primarily as a result of the climatic trend observed in this Pyrenean region since the 1980s: a reduction in snowfall and an increase in maximum temperatures. Nevertheless, there are significant differences in the magnitude of these changes, which appear to be related to three factors: the exposure, which controls the input of solar radiation, the altitude, and the initial size of each glacier or glacieret.

If the climatic scenarios predicted for the twenty-first century in this region are correct, the conservation of the fragile Pyrenean glaciers is in danger, and it is reasonable to expect their accelerated degradation and complete disappearance within the next few decades.

References

Arranz E (1997) Petrología del macizo granítico de La Maladeta (Huesca-Lérida): estructura, mineralogía, geoquímica y petrogénesis. Unpublished PhD thesis. Universidad de Zaragoza, Zaragoza

Arranz E, Lago M (2004) El plutonismo sin- y tardi-varisco en los Pirineos. In: Vera JA (ed) Geología de España. SGE-IGME, Madrid, pp 263–266

Bordonau J, Vilaplana JM (1986) Géomorphologie et tectonique récente dans le Val d'Aran (Zone Axiale des Pyrénées Centrales, Espagne). Rev Géologie Dyn Géog Phys 27:303–310

Charlet JM (1979) Le massif granitique de la Maladeta (Pyrénées centrales espagnoles), synthèse des données géologiques. Ann Soc Géol Belgique 102:313–323

Chueca J, Julián A (1996) Datación de depósitos morrénicos de la Pequeña Edad del Hielo: macizo de la Maladeta. In: Pérez Alberti A, Martini P, Chesworth W, Martínez Cortizas A (eds) Dinámica y evolución de Medios Cuaternarios. Xunta de Galicia, Santiago de Compostela, pp 171–182

Chueca J, Julián A (2004) Relationship between solar radiation and the development and morphology of small cirque glaciers (Maladeta mountain massif, Central Pyrenees, Spain). Geogr Ann 86(1):81–89

Chueca J, Julián A (2008) Geomorphological map of the Alta Ribagorza (Central Pyrenees, Spain). J Maps 1:235–247

Chueca J, Julián A (2010) Caracterización térmica del suelo en el circo del Aneto (Pirineo central aragonés: cartografía de variaciones estacionales). In: Blanco JJ, de Pablo MA, Ramos M (eds) Ambientes periglaciares, permafrost y variabilidad climática. Universidad de Alcalá, Alcalá de Henares, pp 55–60

Chueca J, Julián A, López-Moreno JI (2003a) Variations of Glaciar Coronas, Pyrenees, Spain, during the 20th century. J Glaciol 49(166):449–455

Chueca J, López-Moreno JI, Julián A (2003b) Determinación de espesores en el glaciar-helero de Coronas (macizo de la Maladeta; Pirineo central español) mediante el empleo de geo-radar. Boletín Glaciológico Aragonés 4:111–124

Chueca J, Julián A, René P (2004) Estado de los glaciares en la cordillera pirenaica (vertientes española y francesa) a finales del siglo XX. In: Benito G, Díez Herrero A (eds) Contribuciones recientes sobre geomorfología (Actas VIII Reunión Nacional de Geomorfología). SEG-CSIC, Madrid, pp 91–102

Chueca J, Julián A, Saz MA, Creus J, López-Moreno JI (2005) Responses to climatic changes since the Little Ice Age on Maladeta Glacier (Central Pyrenees). Geomorphology 68:167–182

Chueca J, Julián A, López-Moreno JI (2007) Recent evolution (1981–2005) of the Maladeta glaciers, Pyrenees, Spain: extent and volume losses and their relation with climatic and topographic factors. J Glaciol 53(183):547–557

Evans NG, Gleizes G, Leblanc D, Bouchez JL (1998) Syntectonic emplacement of the Maladeta granite (Pyrenees) deduced from relationships between Hercynian deformations and contact metamorphism. J Geol Soc 155:209–216

García-Sansegundo J (1991) Estratigrafía y estructura de la Zona Axial pirenaica en la transversal del Valle de Arán y de la Alta Ribagorza (Parte I). Boletín Geológico y Minero 102:781–829

García-Sansegundo J (1992) Estratigrafía y estructura de la Zona Axial pirenaica en la transversal del Valle de Arán y de la Alta Ribagorza (Parte II). Boletín Geológico y Minero 103:42–93

Gleizes G, Leblanc D, Bouchez JL (1997) Variscan granites of the Pyrenees revisited: their role syntectonic markes of the orogen. Terra Nova 9(1):38–41

Gutiérrez F, Acosta E, Rios S, Guerrero J, Lucha P (2005) Geomorphology and geochronology of sackung features (uphill-facing scarps) in the Central Spanish Pyrenees. Geomorphology 69:298–314

Gutiérrez F, Ortuño M, Lucha P, Guerrero J, Acosta E, Coratza P, Piacentini D, Soldati M, Beguería S (2008) Late Quaternary episodic displacement on a sackung scarp in the central Spanish Pyrenees. Secondary paleoseismic evidence? Geodin Acta 21:187–202

Haeberli W, Zemp M, Frauenfelder R, Hoelzle M, Kääb A (eds) (2005) Fluctuations of glaciers 1995–2000, vol III. World Glacier Monitoring Service, Zurich

Houghton JT, Ding DJ, Griggs M, Noguer M, Van der Linden PJ, Dai X, Maskell K, Johnson CA (eds) (2001) Climate change 2001: the scientific basis. Contribution of working Group I to the Third Assessment Report of the intergovernmental panel on climate change. Cambridge University Press, New York

Lacan P, Ortuño M (2012) Active Tectonics of the Pyrenees: a review. J Iberian Geol 38(1):9–30

Lampre F (1998) Estudio geomorfológico de Ballibierna (Macizo de la Maladeta, Pirineo Aragonés): modelado glacial y periglacial. Consejo de Protección de la Naturaleza, Zaragoza

Larrasoaña JC, Ortuño M, Birks H, Valero-Garcés B, Parés JM, Copons R, Camarero L, Bordonau J (2010) Paleoenvironmental and palaeoseismic implications of a 3,700-year sedimentary record from proglacial Lake Barrancs (Maladeta Massif, Central Pyrenees, Spain). Palaeogeogr Palaeoclimatol Palaeoecol (Palaeo3) 294:83–93

Leblanc D, Gleizes G, Bouchez JL (1994) The Maladeta granite polydiapir, Spanish Pyrenees: a detailed magnetostructural study. J Struct Geol 16:223–235

López-Moreno JI, Nogués D, Chueca J, Julián A (2006a) Glacier development and topographic context. Earth Surf Proc Land 31(12):1585–1594

López-Moreno JI, Nogués D, Chueca J, Julián A (2006b) Change of topographic control on the extent of cirque glaciers since the Little Ice Age. Geophys Res Lett 33:L24505

Martínez R, García F (1994) Trabajos de glaciología en el glaciar de la Maladeta. Campaña 1991–1992. In: MOPTMA (ed) La nieve en las cordilleras españolas. MOPTMA, Madrid, pp 209–236

Martínez R, García F, Macheret Y, Navarro J, Bisbal L (1997) El sustrato subglaciar y la estructura interna de los glaciares del Aneto y la Maladeta cartografiados por geo-radar de ultra-alta frecuencia (UHF). In: MMA (ed) La nieve en las cordilleras españolas. Ministerio de Medio Ambiente, Madrid, pp 227–249

Moya J, Vilaplana JM (1992) Tectónica reciente en el Macizo de la Maladeta, sector del Alto Ésera (Pirineo Central). In: Cearreta A, Ugarte FM (eds) The late Quaternary in the Western Pyrenean region. Universidad del País Vasco, Vitoria, pp 385–403

Muñoz JA (1992) Evolution of a continental collision belt: ECORS-Pyrenees crustal balanced cross-section. In: McClay KR (ed) Thrust tectonics. Chapman and Hall, London, pp 235–246

Olivera C, Redondo E, Lambert J, Riera Melis A, Roca A (2006) Els terratrèmols dels segles XIV i XV a Catalunya. Institut Cartogràfic de Catalunya, Monografies n°30

Ortuño M (2008) Deformación activa en el Pirineo Central: la falla Norte de la Maladeta y otras fallas activas. Unpublished PhD thesis, Universitat de Barcelona, Barcelona

Ortuño M, Queralt P, Martí A, Ledo J, Masana E, Perea H, Santanach P (2008) The North Maladeta Fault (Spanish Central Pyrenees) as the Vielha 1923 earthquake seismic source: recent activity revealed by geomorphological and geophysical research. Tectonophysics 453:246–262

Ríos LM, Galera JM, Barettino D, Charlet JM (1997) Mapa Geológico de España E. 1:50.000, Hoja n° 180 (Benasque). Instituto Geológico y Minero de España, Madrid

Singh P, Singh VP (2001) Snow and glacier hydrology. Kluwer, Dordrecht

Susagna T, Roca A, Goula X, Batlló J (1994) Analysis of macroseismic and instrumental data for the study of the 19 November 1923 earthquake in the Aran Valley (Central Pyrenees). Nat Hazards 10:7–17

Zwart HJ (1979) The geology of the central Pyrenees. Leid Geol Meded 50:1–74

Block Streams in the Tremedal Massif, Central Iberian Chain

16

Mateo Gutiérrez and Francisco Gutiérrez

Abstract

The Tremedal Massif is an inlier of Paleozoic rocks that protrudes around 300 m over an extensive planation surface cut across Mesozoic formations. It reaches 1,920 m a.s.l. at Caimodorro and has a continental climate characterised by a large number of days with minimum temperatures below freezing point (80–100 days per year). The massif displays a concordant topography controlled by nearly cylindrical folds mainly developed on quartzites and shales. Differential erosion has produced synclinal valleys underlain by shales and steep quartzite ridges on the intervening anticlines. Frost shattering acting on the quartzites, affected by widely spaced jointing, has produced laterally continuous block slopes on the lower part of the hillslopes. The block slopes developed on both valley sides grade into striking block streams up to 2.5 km long along the valley bottoms. The block slopes, underlain by soft shales, display solifluction lobes and benches, suggesting that the formation of the streams is related to the progressive downhill displacement of the boulder deposits in the slopes. The block slopes and streams are relict landforms as the lichens covering the weathered surface of the boulders and the development of tree cover reveal. These periglacial deposits record a period in the Late Quaternary during which frost shattering processes were much more intense, and the slopes were essentially devoid of tree vegetation, probably due to colder and drier conditions.

Keywords

Periglacial geomorphology • Frost shattering • Block slopes • Solifluction

16.1 Introduction to the Geology and Geomorphology of the Albarracín Ranges

The Tremedal Massif is a Paleozoic outcrop located in the Albarracín Ranges, central sector of the Iberian Chain (Fig. 16.1). The Iberian Chain, with a dominant NW–SE structural and topographic grain, is one of the main mountain belts of the Iberian Peninsula. It stretches for about 400 km from its northwestern end to the Mediterranean Sea. From the geotectonic point of view, the Iberian Chain constitutes an intraplate Alpine orogene resulting from the tectonic inversion of Mesozoic sedimentary basins related to the convergence and collision between the Iberian, Euroasiatic and African plates (Sopeña et al. 2004). The rocks that crop out in this mountain belt and in the Albarracín Ranges record two large tectosedimentary cycles: Variscan (Hercynian) and Alpine. The Variscan cycle is recorded by rocks ranging in age from Precambrian to Permian and dominated by silicilastic formations. The inliers of these typically resistant rocks commonly crop out

M. Gutiérrez (✉) · F. Gutiérrez
Department of Earth Sciences, University of Zaragoza, Zaragoza, Spain
e-mail: mgelorza@unizar.es

F. Gutiérrez and M. Gutiérrez (eds.), *Landscapes and Landforms of Spain*, World Geomorphological Landscapes,
DOI: 10.1007/978-94-017-8628-7_16,

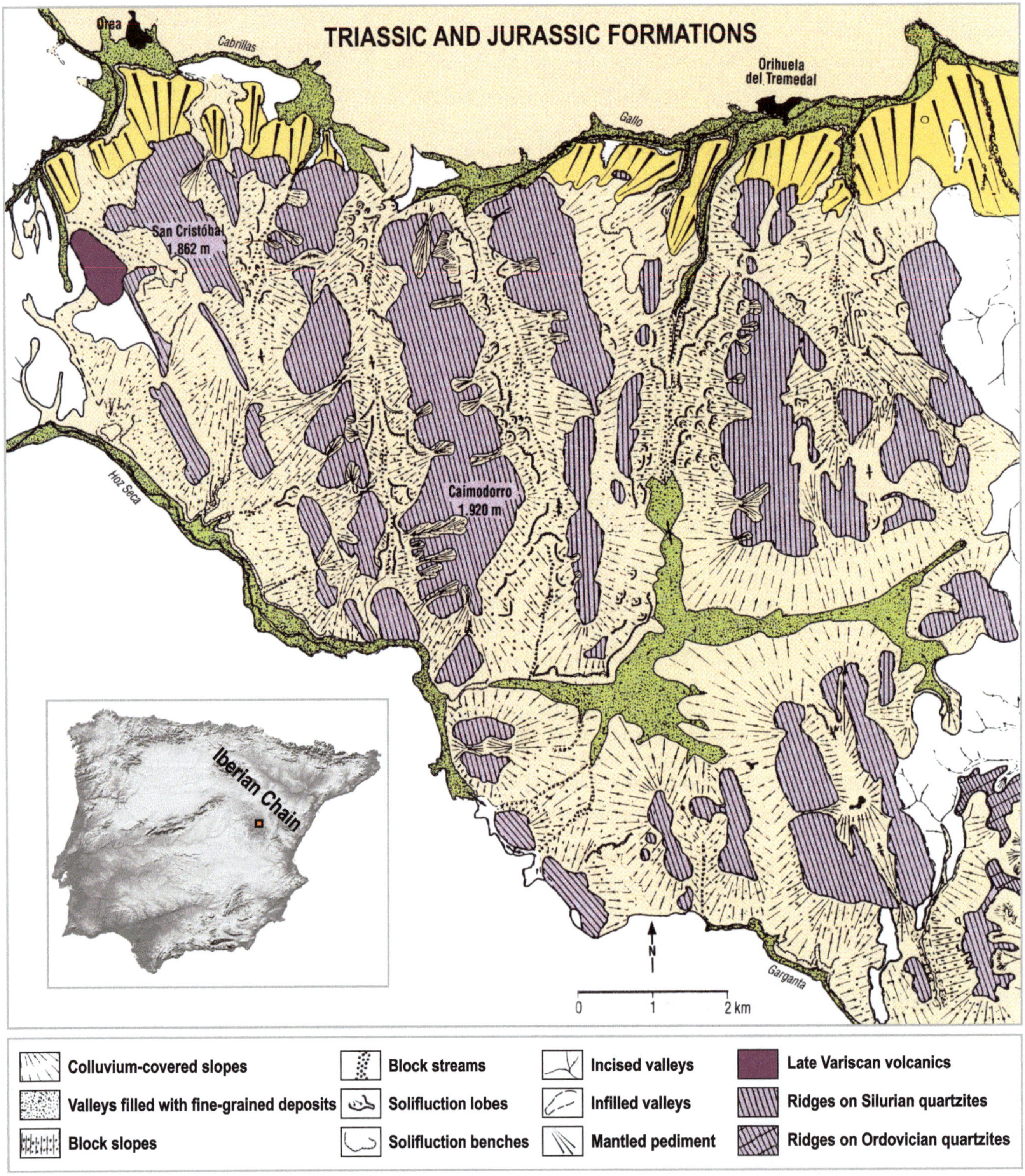

Fig. 16.1 Geomorphological map of the western sector of the Tremedal Massif, Iberian Chain (modified from Gutiérrez and Peña 1977). *Green* valley fill and fluvial deposits. *Yellow* pediments

associated with Alpine compressional structures and uplifted neotectonic blocks, and form high elevation reliefs like the Tremedal Massif (Fig. 16.2). The Alpine cycle developed from the Upper Permian to the Lower–Middle Miocene and comprises two major stages. The initial sedimentary stage is recorded by thick Late Permian to Late

Fig. 16.2 Panoramic view of the northern margin of the Tremedal Massif, a residual relief underlain by siliciclastic rocks protruding over an extensive planation surface cut across folded Mesozoic formations. The Orihuela Polje has developed on Jurassic carbonate rocks at the northern edge of the Tremedal Massif

Cretaceous successions deposited in extensional basins, including a high proportion of shallow marine carbonate rocks. In the subsequent orogenic stage, compressional tectonics caused the inversion of the sedimentary basins and the development of contractional structures affecting synorogenic Tertiary sediments and older rocks. The Alpine compressional structures in the central sector of the Iberian Chain and in the Albarracín Ranges show a dominantly thick-skinned tectonic style, with folds and thrusts involving the basement (Capote et al. 2002). Since the Middle Miocene, extensional neotectonics has produced horsts and grabens, the latter filled with Neogene continental sediments up to several hundred metres thick, like the Teruel Graben to the east of the Albarracín Ranges (Gutiérrez et al. 2008). During this postorogenic phase, the topography of the Iberian Chain was largely transformed into an elevated plateau with planation surfaces cut across folded Paleozoic and Mesozoic rocks, and structural surfaces underlain by Mio-Pliocene lacustrine limestones in the Neogene basins (Gutiérrez and Gracia 1997). The Albarracín Ranges are dominated by outcrops of folded carbonate Jurassic formations bevelled by extensive planation surfaces typically at 1,500–1,600 m a.s.l. (Gutiérrez et al. 2005; Fig. 16.2). Older planation surfaces at ca. 1,800 m are also found on Jurassic and Upper Cretaceous rocks in the southwestern sector of the area (Muela de San Juan, El Portillo de Guadalaviar). The flat topography underlain by soluble bedrock has favoured the generation of spectacular sinkhole fields (Gutiérrez and Peña 1979a, b) and structurally controlled poljes. The most striking poljes have been carved at the contact between impervious Paleozoic rocks and soluble Jurassic limestones, like the Orihuela Polje on the northern side of the Tremedal Massif (Fig. 16.2). In Pliocene and Quaternary times, the Neogene postorogenic basins were captured by the external drainage network, changing progressively from endorheic to exhoreic conditions (Gutiérrez et al. 2008). The consequent rapid capture-induced incision has propagated headwards through the drainage network, deeply entrenched into the planation surfaces. Some drainages in the Albarracín Ranges have excavated dramatic canyons cut into Mesozoic carbonate sequences (e.g. Guadalaviar River, Fuente del Berro Arroyo). Moreover, this sector of the Iberian Chain includes the main water divide between the Mediterranean and Atlantic catchments and the source of important fluvial systems like the Tajo, Guadalaviar and Cabriel rivers (Gutiérrez and Peña 1994).

16.2 Geology and Climate of the Tremedal Massif

The formation of the unique block streams of the Tremedal Massif is related to the concurrence of a set of specific geological and climatic factors. The Tremedal Massif, located to the south of Orihuela del Tremedal village, constitutes a WNW-ESE trending inlier of Paleozoic rocks around 22 km long and 8 km wide, surrounded by outcrops of Jurassic and Permo-Triassic formations. It mainly consists of an Ordovician and Silurian siliciclastic succession and some outcrops of volcanic rocks of probable Permian age (Riba 1959; Fig. 16.1). The geology of the massif was initially studied by Riba (1959) in his pioneering PhD thesis. More detailed cartographic information was provided by Lendínez et al. (1981) and Portero et al. (1983) in the 1:50,000 scale geological sheets of Checa and Tragacete, respectively. Two main lithostratigraphic units with contrasting resistance to erosion may be differentiated in the western sector of the massif, where most of the block streams have developed (Figs. 16.1, 16.3): (1) A 280-m-thick succession mainly consisting of thickly bedded quartzites ascribed to the Early Silurian (Llandovery). (2) A 150-m-thick Middle Silurian (Wenlockian) unit made up of fossil-rich dark grey shales.

The Paleozoic formations are affected by both Variscan and Alpine tectonic structures. The Variscan orogeny is recorded by two folding phases with cleavage development (Gutiérrez and Peña 1977). The second phase, which is the most conspicuous one, produced nearly cylindrical folds with a dominant NNW-SSE trend (Figs. 16.1, 16.3). The

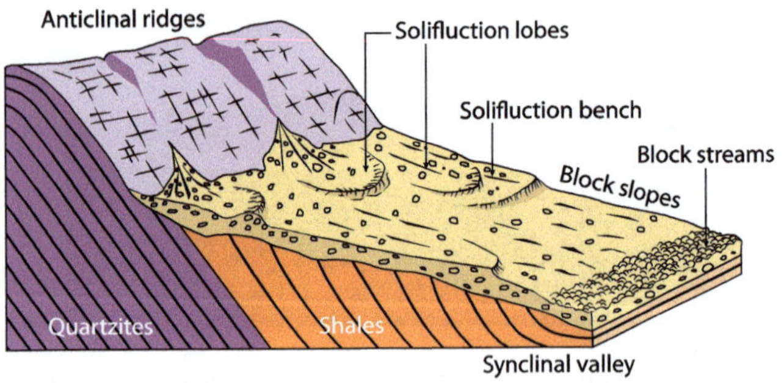

Fig. 16.3 Block diagram showing the spatial relationships between the block slopes and streams, and solifluction lobes and benches. These features have developed in synclinal valleys carved in shales and flanked by steep dip slopes underlain by quartzites (modified from Gutiérrez and Peña 1977)

Alpine structures mainly correspond to WNW-ESE and E-W reverse faults superimposed on the Variscan folding. These structures bound the Paleozoic inlier on its northern edge (Fig. 16.2), whereas the southern edge essentially corresponds to an unconformable contact between the Paleozoic basement and the Mesozoic cover. Therefore, the Tremedal Massif is a morpho-structure with an overall WNW-ESE orientation related to the development of Alpine north-verging compressional structures, whereas Variscan folding dominates the tectonic style within the massif.

The area has an average annual precipitation of 800–900 mm and a mean annual temperature of 8–9 °C. The monthly temperatures of July and January are 17–18 and 1 °C, respectively, indicating considerable thermal amplitude throughout the year (Peña et al. 2002). The number of days with temperatures below the freezing point is around 80–100 (Font 1983).

16.3 Geomorphology of the Tremadal Massif

16.3.1 General Features

The Tremedal Massif forms a large residual relief that protrudes more than 300 m above an extensive planation surface (Fig. 16.2). The highest elevation corresponds to Caimodorro at 1,920 m a.s.l. (Fig. 16.1). The initial development of the relief is most probably related to the formation of an uplifted area during the Paleogene, associated with the activity of thick-skinned Alpine contractional structures. The crests of the quartzitic ridges tend to be situated at a similar elevation, suggesting that they may correspond to an old dissected planation surface of probable Miocene age (Gutiérrez and Peña 1977; Peña et al. 1984; Fig. 16.2). The folded Mesozoic formations surrounding the massif, dominated by Jurassic limestones, display an extensive planation surface ascribed to the late Neogene (Gutiérrez and Gracia 1997; Gracia et al. 1998; Fig. 16.2). This surface locally truncates the Paleozoic rocks displaying an Appalachian relief (Gutiérrez and Peña 1977, 1994).

The topography within the massif is controlled by the presence of cylindrical folds developed in two lithological units with contrasting erodibility. Differential erosion of the shales has produced a concordant landscape consisting of NNW-SSE valleys along the synclines, flanked by anticlinal quartzite ridges (Gutiérrez and Peña 1977, 1990; Figs. 16.1, 16.3, 16.4). The local relief in the valley sides reaches more than 200 m, and gradients of around 20° are common (Fig. 16.4). This structural landscape was mainly formed by fluvial erosion before the cold Quaternary phases recorded by the block slopes and streams (Gutiérrez and Peña 1977, 1990). This sector of the Iberian Chain was never glaciated. The southernmost evidence of glacial activity (i.e. glacial cirques and moraines) in the Iberian Chain has been reported around 150 km to the north, in the Moncayo Massif (2,314 m a.s.l.).

Periglacial deposits are widely distributed in the central sector of the Iberian Chain. The most outstanding examples are found in the Tremedal Massif. Here, the accumulations mainly consist of angular decimetre- and metre-sized quartzite boulders produced by freeze–thaw action (congelifraction). The size of the clasts is determined by the relatively wide spacing of the discontinuity planes (jointing, stratification) affecting the thickly bedded quartzites (Gutiérrez and Peña 1977; Rea 2007). The quartzite boulder deposits, typically with an open-work texture, may also include some finer-grained materials, especially in the lower part. The loose and angular bouldery deposits found in the slopes and the valley bottoms originate by rockfalls from the upper part of the steep quartzite slopes (Fig. 16.4). These deposits cover most of the slopes underlain by Silurian shales, significantly restricting their exposure (Figs. 16.1, 16.3). No consistent differences in the morphology of the slopes related to their aspect have been identified, and the local asymmetry of the valleys may be attributed to structural factors.

The bouldery slope deposits developed below the steep quartzite outcrops correspond to block slopes, and those extending along the valley bottoms or in gullies are designated as block streams (Washburn 1979). In the Tremedal Massif, the block slopes and the block streams form a continuous deposit with a gradational geomorphic transition (Figs. 16.1 and 16.3). This peculiar landform association was first described by Darwin (1846) in the Falkland Islands and investigated by Andersson (1907), Clark (1972) and André et al. (2008). According to Washburn (1979), block streams may be also designated as rubble streams. However, Potter and Moss (1968) indicate that rubble deposits are covered by trees and contain interstitial fine material. They have also been described as stone runs, coulées de blocaille

Fig. 16.4 General view of a quartzite anticlinal ridge. *Dashed lines* indicate the general attitude of the bedding. In the foreground, block slope developed at the foot of a steep slope consisting of quartzites with widely spaced jointing. In the background, Orihuela del Tremedal village and an extensive planation surface developed on folded Mesozoic formations

(Tricart 1967), felsenmeer (German) and kurum (Russian). Some authors (Clark 1972; Schroder 1987) propose that in some cases, it is not possible to conclusively differentiate between rock glaciers and other bouldery slope deposits like block slopes and streams and suggest a morphologic convergence or equifinality. Relatively well-sorted boulder-sized periglacial deposits, including block streams, are not common and have been mainly reported in alpine and polar environments. They have been investigated in Tasmania, Sweden, the Carpathians, the Alps, the Appalachian Mountains, Sierra Nevada, Southern Africa, etc. (Wilson 2007; Rixhon and Demoulin 2013 and references therein).

16.3.2 Block Slopes

These are the most widespread accumulations. They occur in the lower part of the valley sides below the steep quartzite slopes of the ridges (Figs. 16.1, 16.3). On the surface, they consist of relatively well-sorted, matrix-free, angular boulders with a dominant size of 25–50 cm, although some particles may exceed 1 m (Figs. 16.4, 16.5). Some exposures reveal that the clast-supported bouldery slope deposits have a matrix made up of granules and fines in the lower part. Numerous authors attribute the textural difference between the upper and lower parts of the deposit to downward migration of primary and secondary fine particles through the large interstitial voids by water percolation and gravity (Andersson 1907; Smith and Smith 1945; Smith 1953; Tricart 1967; Potter and Moss 1968). This stratigraphic feature explains why the block slopes are widely colonised by trees rooted in an apparently highly porous blocky surface. Some sections around 2 m thick associated with artificial excavations display an obvious coarsening-upward grading. This reverse sorting is most probably related to frost heaving processes that favour the differential downward percolation of the finer clasts, with the consequent rearrangement of the particle distribution (Richmond 1962; Corte 1966; Caine 1968,1983; Washburn 1973,1979; Bennett et al. 1996). There is almost no information on the thickness of these slope deposits, but it might be highly variable, as in some sectors they likely fill pre-existing gullies.

The lower sectors of the block slopes, especially in the northern part of the massif, display solifluction lobes and benches (Washburn 1979; Figs. 16.1, 16.3) which in some cases resemble lobate talus rock glaciers with arcuate ridges and furrows (Degenhardt et al. 2007). The lobes can be identified across the whole block slopes, whereas benches mostly occur in lower and less steep sectors. The lobes have a steep downhill-facing convex side up to several metres high and are generally around 150 m wide. The coalescence of several lobes produces sinuous benches. The solifluction benches have a variable length, reaching up to 1.5 km, and the steep downhill-facing sides may be several metres high. A number of benches and lobes show closed depressions on their upper part, with small shallow lakes and peat bogs, locally known as tremedales or gotiales. An area covering 1,845 ha and including some of the best examples was declared RAMSAR site in 2011. The palynological record of one of these basins has been analysed by Menéndez and Esteras (1965).

The boulders of the block slopes were generated by frost shattering and rockfalls acting on the quartzitic ridges on

Fig. 16.5 Block slope with open-work texture composed of Silurian quartzite boulders covered with lichens. Person 1.4 m high for scale. Photograph taken next to the road to the Nuestra Señora del Tremedal Sanctuary. Orihuela del Tremedal village above the trees and outcrops of Jurassic limestone in the background

the valley flanks (Fig. 16.4). The jointing in the quartzites has a decimetre- to metre-scale spacing and the rock has a sufficient mechanical strength to prevent its disintegration as it falls and bounces while moving downslope. In most Paleozoic massifs of the Iberian Chain the periglacial talus deposits are dominated by gravel-sized particles due to the narrower spacing of the discontinuity planes in the source rocks and their lower resistance (e.g. Sierra Menera Massif situated 20 km to the N).

Frost shattering experimental studies on quartzites with a porosity of around 0.5 % using temperature-controlled cabinets reveal that these rocks appear to respond to fatigue after 300–550 freeze–thaw cycles (Evin 1987). Initially, the blocks are detached from the scarps by rockfalls and transported downslope by free fall, bouncing and rolling mechanisms. By the time the block slopes formed, the hillslopes must have been essentially forest-free. The presence of snow fields probably contributed to increase the run-out of the clasts. Subsequently, solifluction and creep, most probably favoured by the presence of weathered shale bedrock, produced a progressive and long-sustained movement of the blocks towards the valley bottom as it has been suggested in other regions (Dahl 1966; Potter and Moss 1968; Clapperton 1975; Washburn 1979; French 2007). Currently, the block slopes are relict deposits as revealed by (1) the dense lichen cover; (2) well-developed weathering rinds and patina and (3) the presence of abundant tree vegetation with no deflection in the trunks (Fig. 16.5). The limited fresh scars and boulders indicate that rockfalls are an infrequent process at the present time.

16.3.3 Block Streams

The block slopes merge in the valley axes forming block streams that cover broad low-gradient (<5°) valley bottoms (Figs. 16.1, 16.3, 16.6). The block streams, up to 2.6 km long and 0.25 km wide, are larger than the longest ones described in the Appalachian Mountains at Hickory Run (Smith 1953). Block streams up to 5 km long have been documented in other regions of the world (Clark 1972; White 1976; Harris et al. 1998; Barrows et al. 2004). These deposits also show reverse grading, and their maximum thickness is unknown. Wilson (2007) indicates frequent thicknesses of 1–3 m, and maximum values of 9 m have been reported by Boelhouwers (1999) in South Africa. The major axes of the surface blocks do not seem to have preferred fabrics (Fig. 16.7). Relatively flat areas with peat bogs are common at the divides separating north- and south-flowing valleys. The block streams are typically devoid of vegetation and flanked by trees (Fig. 16.6). This vegetation distribution pattern may be related to the presence of a higher proportion of fines in the lower portions of the slopes and the frequent and audible throughflow at the contact between the impervious bedrock and the porous rubble deposit.

Caine (1968), in his work on the block fields and block streams of Tasmania, differentiates three main layers: (1) blocks with open-work texture at the surface and to depths of over 3 m in places; (2) an intermediate layer 10–30 cm thick of boulders with interstitial humic mud and (3) a basal layer lying on bedrock consisting of silty sand or, in a few

Fig. 16.6 Flat-bottom and low-gradient block stream flanked by pines at Majada de las Vacas, next to A-1511 road. Person 1.4 m high for scale

Fig. 16.7 Close-up view of the block stream shown in Fig. 16.6. Note the chaotic fabric of the boulders covered by lichens and with subrounded edges indicative of prolonged inactivity and weathering

places, of silty sand with some blocks. The fine-grained sediment identified in the lower part of the deposit may derive from mechanical and chemical weathering (Dahl 1966), as well as aeolian input. The fine particles readily percolate downwards through the large interstitial spaces of the bouldery deposit. Moreover, frost heaving also contributes to the reverse sorting (Clark 1972; Hamson et al. 2008) and the development of an armour of large boulders on the surface. According to Smith and Smith (1945) and Caine (1968), the block stream deposits are accumulated with a fine matrix, which is subsequently flushed by percolating and throughflow water.

The block streams show bowl-shaped closed depressions up to several metres across and less than 1 m deep (Smith 1953). These hollows could be related to subsidence induced by melting of interstitial ice in the past (cryokarst), although internal erosion of fine particles by throughflow at the bedrock–cover interface seems to be the most feasible explanation. The longitudinal profile of the block streams shows gentle steps that may be related to inhomogeneities in the bedrock and irregularities in the pre-existing topography. Frequently, throughflow water emerges at these sections for a short distance, percolating again in the talus. Flow-related lobate features like those described in the block streams studied by Potter and Moss (1968), Harris et al. (1998) and Degenhardt et al. (2007) have not been observed in the Tremedal Massif.

The formation of the block streams is linked to the development of the block slopes by rockfall processes on the quartzite ridges and the progressive displacement of these deposits towards the valley bottom by solifluction and frost creep, favoured by the presence of plastic shale bedrock (e.g. Czudek and Demek 1972). The fact that the block streams do not occur downstream beyond the quartzitic slopes indicates that they have not undergone any significant longitudinal displacement along the low-gradient valleys (Fig. 16.1). Such transport would have produced block fans at the valley mouths like those described by Caine (1968). Conversely, the valley floors grade into mantled pediments underlain by fine-grained deposits with some gravels. A significant proportion of these fines may derive from the flushing out of the block streams. These pediments display numerous cover subsidence sinkholes in the areas underlain by Jurassic limestones (Gutiérrez and Peña 1977).

Currently, the block streams, covered by lichens and partially colonised by trees, do not show evidence of activity. The subrounded edges provide evidence of prolonged weathering (Fig. 16.7). Caine (1972) suggests that the stabilisation of the block streams in Tasmania is related to the removal of the fine matrix. The block streams of the Tremedal Massif are consequently relict landforms developed under different environmental conditions in the past (e.g. Potter and Moss 1968; Washburn 1979; French 2007). They record time intervals in the Late Quaternary during which frost shattering processes were much more intense and the slopes were essentially devoid of tree vegetation, most probably due to colder and drier conditions. The block streams of other regions in the world have been dated by several methods. Exposure ages from ^{36}Cl concentration measurements obtained in various places of Australia (Barrows et al. 2004) indicate a peak of periglacial activity, notably block stream development, between 23 and 16 ka, remarkably coincident with the Last Glacial Maximum (LGM). OSL dating in Fakland Islands also revealed that a major phase of block stream activity at 32–27 ka occurred under severe periglacial conditions, almost coeval with the LGM (Hamson et al. 2008). However, both studies also document other phases of block stream development that predates the LGM. In the Falkland Islands, Wilson et al. (2008) obtained by ^{10}Be and ^{26}Al ages between 42 and 731 ka, suggesting that some block streams may be very old relict deposits.

Acknowledgments This work has been supported by the Spanish national project CGL2010-16775 (Ministerio de Ciencia e Innovación and FEDER) and the Regional project 2012/GA-LC-021 (DGA-La Caixa).

References

Andersson JC (1907) Contributions to the geology of the Fakland Islands. Wiss. Ergebnisse der Schwadischen Südpolar-Expedition, 1901–1903. Stockholm, Band 8, Lief 2, pp 16–29

André M-F, Hall K, Bertran P, Arocena J (2008) Stone runs in the Falkland Islands: periglacial or tropical? Geomorphology 95:524–543

Barrows TT, Stone JO, Fifield LK (2004) Exposure ages of Pleistocene periglacial deposits in Australia. Quatern Sci Rev 23:697–708

Bennett MR, Mather AE, Glasser NF (1996) Earth hummocks and boulder runs at Merrivale, Datmoor. In: Charman OJ, Newnhom RM, Croot DG (eds) The Quaternary of Devon and East Cornwall: field guide. Quaternary Research Association, London, pp 81–96

Boelhouwers JC (1999) Relict periglacial slope deposits in the Hex River Mountains, South Africa: observations and paleoenvironmental implications. Geomorphology 30:245–258

Caine N (1968) The blockfields of northeastern Tasmania. National Department of Geography, Publication G/6, 127 p

Caine N (1972) Air photo analysis of blockfield fabrics in Talus Valley, Tasmania. J Sediment Petrol 42:33–48

Caine N (1983) The mountains of northeastern Tasmania. A study of Alpine geomorphology. Balkema, Rotterdam 200 p

Capote R, Muñoz JA, Simón JL, Liesa CL, Arlegui LE (2002) Alpine tectonics I: the Alpine system north of the Betic Cordillera. In: Gibbons G, Moreno T (eds) The Geology of Spain. The Geological Society. London, pp 367–400

Clapperton CM (1975) Further observations on the stone runs in the Fakland Islands. Biul Peryglac 24:211–217

Clark R (1972) Periglacial landforms and landscapes in the Falkland Islands. Biul Peryglac 21:33–50

Corte AE (1966) Particle sorting by repeated freezing and thawing. Biul Peryglac 15:175–240

Czudek T, Demek T (1972) Present-day cryogenic processes in the mountains of eastern Siberia. Geographia Polonica 23:5–20

Dahl R (1966) Blockfields and other weathering forms in the Narvik Mountains. Geogr Ann 48A:55–85

Darwin C (1846) Journal of Researches into the Natural History and Geology of the Countries Visited During the Voyage of It. S.M. Beagle. Harper and Bross, New York

Degenhardt J, Giardino J, Marston R, Pitty A (2007) Possible origin of the block stream in Tom Mays Canyon, Franklin Mountains, TX. Zeitschrift für Geomorphologie 51:377–396

Evin M (1987) Lithology and fracturing control of rock glaciers in Southwestern Alps of France and Italy. In: Giardino JR, Shroeder JF, Vitek JD (eds) Rock glaciers. George Allen and Unwin, London, pp 83–106

Font I (1983) Climatología de España y Portugal. Instituto Nacional de Meteorología, Madrid 296 p

French HM (2007) The periglacial environment. Wiley, Chichester 458 p

Gracia FJ, Gutiérrez M, Leránoz B (1998) Las superficies de erosión neógenas en el sector central de la Cordillera Ibérica. Revista de la Sociedad Geológica de España 1:135–142

Gutiérrez F, Gutiérrez M, Gracia FJ (2005) Karst, neotectonics and periglacial features in the Iberian Ranges. Field Trip C-5. In: Sixth international conference on geomorphology, Zaragoza, 58 p

Gutiérrez F, Gutiérrez M, Gracia FJ, McCalpin JP, Lucha P, Guerrero J (2008) Plio-Quaternary extensional seismotectonics and drainage network development in the central sector of the Iberian Range (NE Spain). Geomorphology 102:21–42

Gutiérrez M, Gracia FJ (1997) Environmental interpretation and evolution of the Tertiary erosion surfaces in the Iberian Range (Spain). In: Widdowson M (ed) Paleosurfaces: recognition, reconstruction and palaeoenvironmental interpretation, vol 120. The Geological Society, London, pp 147–158 (Special Publication)

Gutiérrez M, Peña JL (1977) Las acumulaciones periglaciares del Macizo del Tremedal (Sierra de Albarracín). Boletín Geológico y Minero 88:109–115

Gutiérrez M, Peña JL (1979a) El karst de los Llanos de Pozondón (Sierra de Albarracín). Teruel 61–62:39–46

Gutiérrez M, Peña JL (1979b) El karst de Villar del Cobo (Sierra de Albarracín). Estud Geol 35:651–654

Gutiérrez M, Peña JL (1990) Las Formas del Relieve de la Provincia de Teruel. Cartillas Turolenses, vol 7. Instituto de Estudios Turolenses. Teruel, 65 p

Gutiérrez M, Peña JL (1994) Cordillera Ibérica. Gutiérrez M (coord) Geomorfología de España. Rueda, Madrid, pp 251–286

Hamson JD, Evans DJA, Sanderson DCW, Bingham RG, Bentley MJ (2008) Constraining the age and formation of stone runs in the Falkland Islands using optically stimulated luminescence. Geomorphology 94:117–130

Harris SA, Cheng C, Zhao X, Yanggin D (1998) Nature and dynamics of an active block stream, Kunlun Pass, Qinghai Province, People's Republic of China. Geogr Ann 80A:123–133

Lendínez A et al (1981) Mapa Geológico de España 1:50.000. Checa (540). Instituto Geológico y Minero de España. Madrid, 63 p

Menéndez J, Esteras M (1965) Análisis polínico de la turbera de "Los Ojos del Tremedal" (Orihuela del Tremedal, Teruel). Revista Teruel 34:191–198

Peña JL, Gutiérrez M, Ibañez MJ, Lozano MV, Rodríguez J, Sánchez M, Simón JL, Soriano MA, Yetano LM (1984) Geomorfología de la Provincia de Teruel. Instituto de Estudios Turolenses, Teruel 149 p

Peña JL, Cuadrat JM, Sánchez M (2002) El Clima de la Provincia de Teruel. Cartillas Turolenses, no 20. Instituto de Estudios Turolenses. Teruel, 89 p

Portero JM et al (1983) Mapa Geológico de España 1:50.000. Tragacete (565). Instituto Geológico y Minero de España. Madrid, 89 p

Potter N, Moss JH (1968) Origin of the Blue Rocks block field and adjacent deposits, Berks County, Pennsylvania. Bull Geol Soc Am 79:255–262

Rea B (2007) Periglacial landforms, rock forms. Blockfields (Felsenmeer). In: Elias SA (ed) Encyclopedia of Quaternary Science. Elsevier, Amsterdam, pp 2225–2236

Riba O (1959) Estudio Geológico de la Sierra de Albarracín. Monografías Instituto Lucas Mallada, Consejo Superior de Investigaciones Científicas no 16. Madrid, 283 p

Richmond GM (1962) Quaternary stratigraphy of the La Sal Mountains, Utah. United States Geological Survey Professional Paper 324, 135 p

Rixhon G, Demoulin A (2013) Evolution of slopes in a cold climate, In: Giardino, R, Harbor J (eds) Glacial and periglacial geomorphology. Treatise of geomorphology, vol 8. Academic Press, London, pp 392–415

Schroder JF (1987) Rock glaciers and slope failures: high Plateaus and La Sal Mountains Colorado Plateau, Utah, USA. In: Giardino JR, Schroder JF, Vitek JD (eds) Rock glaciers. Allen and Unwin, London, pp 193–238

Smith HTU (1953) The Hickory Run boulderfield, Carbon County, Pennsylvania. Am J Sci 25:625–642

Smith HTU, Smith AP (1945) Periglacial rock streams in the Blue Ridge area. Bull Geol Soc Am 56:1198

Sopeña A, Gutiérrez-Marco JC, Sánchez-Moya Y, Gómez JJ, Mas R, García A, Lago M (2004) Cordilleras Ibérica y Costero Catalana. In: Vera JA (ed) IGME, Madrid, pp 467–470

Tricart J (1967) Le Modelé des Régions Périglaciaires. SEDES, Paris 512 p

Washburn AL (1973) Periglacial processes and environments. Arnold, London 320 p

Washburn AL (1979) Geocryology. A survey of periglacial processes and environments. Arnold, London 406 p

White SE (1976) Rock glaciers and block fields, review and new data. Quatern Res 6:77–97

Wilson P (2007) Periglacial landforms, rock forms. Block/rock streams. In: Elias SA (ed) Encyclopedia of Quaternary Science. Elsevier, Amsterdam, pp 2217–2235

Wilson P, Bentley MJ, Schnabel C, Clark R, Xu S (2008) Stone run (block stream) formation in the Fakland Islands over several colds stages, deduced from cosmogenic isotope (^{10}Be and ^{26}Al) surface exposure dating. J Quat Sci 23:461–473

Badlands in the Tabernas Basin, Betic Chain 17

Adolfo Calvo-Cases, Adrian M. Harvey, Roy W. Alexander, Yolanda Cantón, Roberto Lázaro, Albert Solé-Benet, and Juan Puigdefábregas

Abstract

The complex badland landscape at Tabernas results from a combination of relief amplitude generated by tectonic uplift since the Pliocene and reactivated several times during the Pleistocene, the properties of the Tortonian sedimentary rocks and a predominantly arid climate. The landscape is dominated by deep incision of the main river systems, which continues in part of the headwater tributaries, and characterized by contrasting slope morphologies and a variety of microecosystems. The Tabernas badlands exhibit a diversity of landforms resulting from the combination of multi-age soil surface components that allow a variety of processes to operate at different rates. These are dominated by rilling and shallow mass movements on south-facing hillslopes. On old surfaces and north-facing hillslopes, where biological components are present, overland flow with variable infiltration capacity and low erosion rates prevail. Incision in the gully bottoms occurs in the most active areas.

Keywords

Badlands • Landform evolution • Biological soil crusts • Erosion processes

17.1 Introduction

Tabernas badlands, in Almería, constitute one of the most extensive badland landscapes in semi-arid south-east Spain and one of the driest areas in Europe (Fig. 17.1). The badland area occupies most of the centre of a tectonic basin filled during the Upper Miocene by marine marls, shales and turbidites, with occasional interbedded calcareous sandstones. The basin is bounded by faults that delimit three important Betic mountain ranges; Sierra de los Filabres, Sierra de Gádor and Sierra Alhamilla (Fig. 17.1). The basin is elongated in an E-W direction, from the Canjáyar Corridor in the west to the Sorbas basin at the east (see Harvey et al. 2014 in this volume).

The dissection of the Upper Miocene sedimentary rocks, mainly since the Pliocene period, has created a singular erosional landscape dominated by badlands, but with a wide variety of landforms resulting from the sequence of dissection and stability phases controlled by environmental changes and pulses of uplift and tectonic deformation in the

A. Calvo-Cases (✉)
Department of Geography, University of Valencia, Valencia, Spain
e-mail: adolfo.calvo@uv.es

A. M. Harvey
School of Environmental Sciences/Geography, University of Liverpool, Liverpool, UK

R. W. Alexander
Department of Geography and Development Studies, University of Chester, Chester, UK

Y. Cantón
Dpto. de Agronomía, University of Almería, Almería, Spain

R. Lázaro · A. Solé-Benet · J. Puigdefábregas
Dpto. Desertificación y Geoecología, Estación Experimental de Zonas Áridas, CSIC, Almería, Spain

F. Gutiérrez and M. Gutiérrez (eds.), *Landscapes and Landforms of Spain*, World Geomorphological Landscapes,
DOI: 10.1007/978-94-017-8628-7_17,

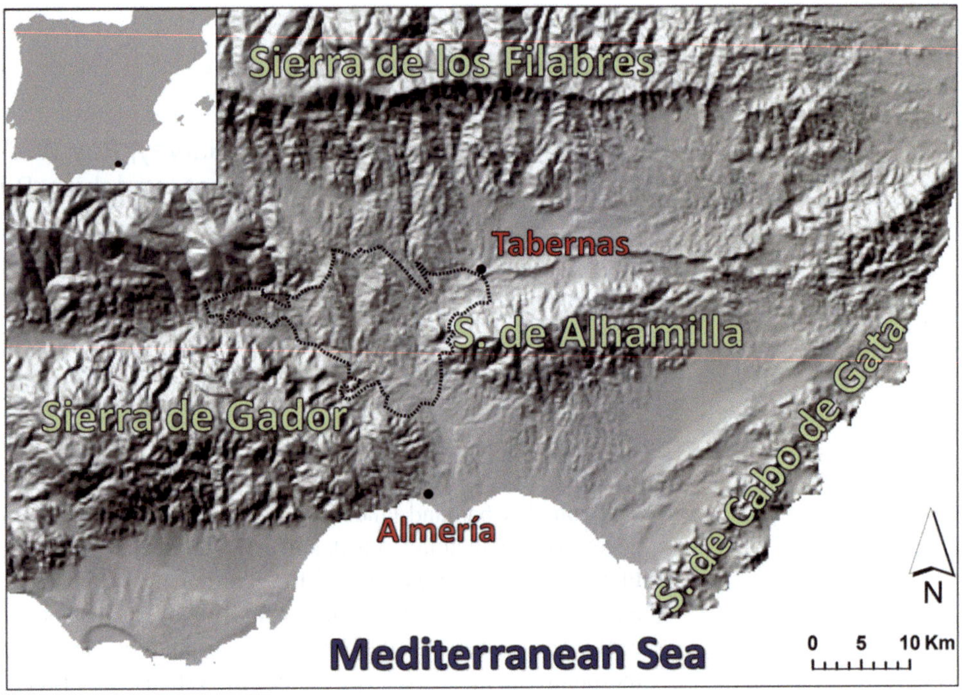

Fig. 17.1 Location map of the Tabernas Basin. Dashed line is the limit of the badlands area

basin. The role of land-use changes, that is relevant in many other badlands areas around the Mediterranean, seems to be less important here although they can be considered effective in some parts of the basin.

17.2 Geological and Geographical Setting

The Tabernas basin is a tectonic depression that was created ca. 10 Ma ago by the folding and faulting phases that have occurred since the Burdigalian–Serravallian stages of the Miocene, affecting the Alpujárride–Nevado–Filabride basement. After a continental erosion–sedimentation sequence during the first 2–3 Ma of this period, the environment became marine during the Tortonian and Messinian, resulting in the sedimentation of a sequence of more than 700 m of deep water (up to 800 m of sea depth) turbidites and gypsiferous calcareous mudstones, alternating with layers of coarser sediments and submarine fan deposits. The sedimentary sequence includes the Chozas, Turre and Yesares formations, characteristic of the region as a whole. Later, in the Pliocene, marine conditions were restricted to the west and south-west of the basin, depositing marine fan deltas and mudstones (Abrioja formation), followed by a thick cover of Plio-Pleistocene continental alluvial fan deposits (Gador formation) (see Mather et al. 2001).

Since the Pliocene, uplift of the basin switched the conditions from net deposition to net erosion (Harvey 2001). The main drainage network developed during the Pliocene and dissected the easily erodible sediments of the Tortonian basin fill. The drainage exited the basin to the south, through the Rioja corridor, which was filled with large volumes of fan-delta deposits (Postma 1984), and was later dissected during the Pleistocene, as the Andarax river system prograded towards the coast at Almería (Harvey et al. 2012, 2014 in this volume).

Some of the faults bounding and within the basin remain active today (Mather 2009), and tectonic activity during the Quaternary has affected the landscape evolution. During the Pleistocene, the central part of the basin, in the lower section of the Rambla de Tabernas, has been affected by uplift that obstructed the drainage creating two shallow water lakes. The upper lake was placed near Tabernas town (see Harvey et al. 2003, 2014 in this volume) and the lower lake upstream of the confluence with the Andarax River (Fig. 17.2). During the latest Pleistocene and through the Holocene, dissection has been dominant, resulting in the erosional modern landscape.

The climate of the area is related to its low latitude, but is also very much influenced by local factors related to the position of the basin, surrounded by high mountain ranges and especially by the fact that the circulation of the air masses coming from the west is shaded by the highest elevations (over 3000 m a.s.l.) of the Iberian Peninsula in the Sierra Nevada, a factor also to be considered in relation to past climates. The climate is classified as semi-arid thermo-Mediterranean, with a mean annual precipitation of 235 mm (Lázaro et al. 2001). The average number of days of rainfall per year is 37, although this is very variable (Solé-Benet et al. 1997). The rainfall intensity can be high during rainstorms (maximum recorded 5- and 10-minute

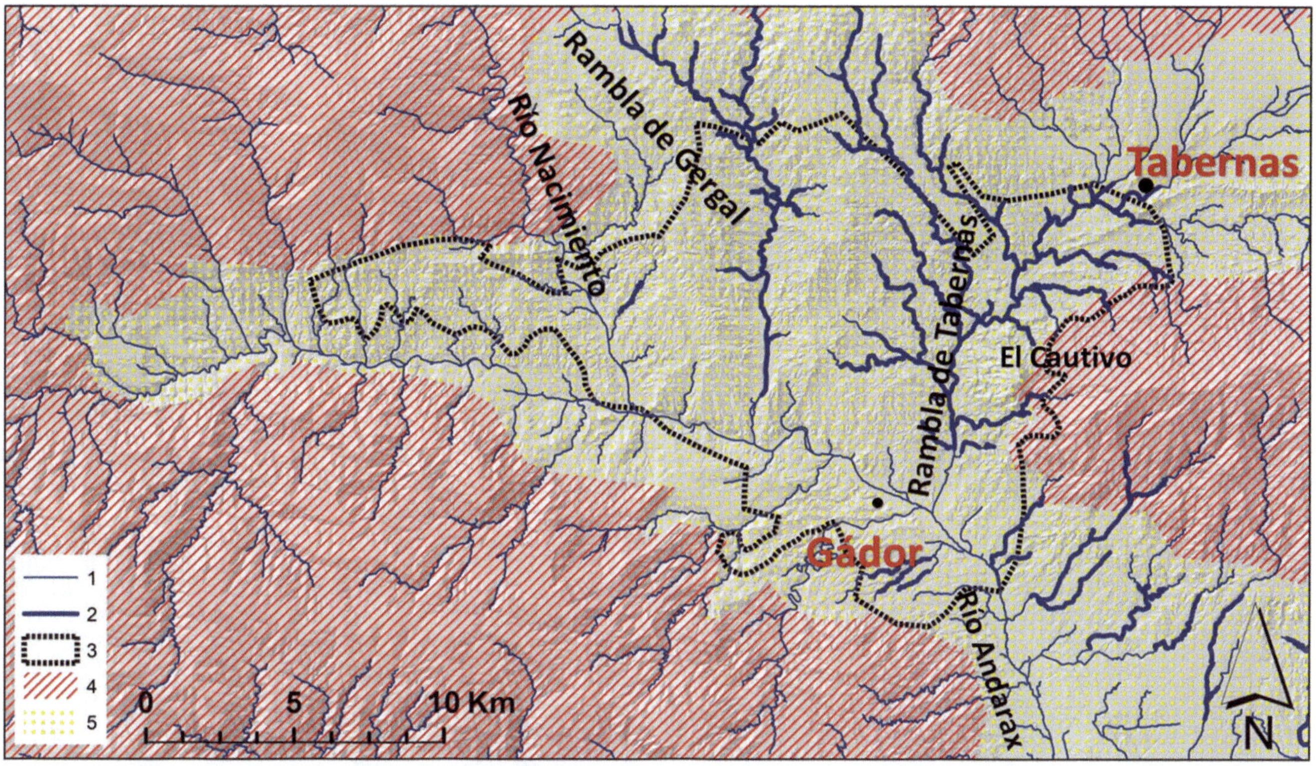

Fig. 17.2 Simplified geological–geomorphic sketch of the Tabernas Basin. *1* Main river channels. *2* Deep canyons. *3* Limit of the badlands area. *4* Surrounding reliefs on non-badlands-prone rocks. *5* Tortonian mudstones

intensities at El Cautivo experimental station are 85.2 and 41.6 mm h^{-1}, respectively). Mean annual temperature is 17.9 °C, with a minimum monthly average in January of 4.1 °C, and a maximum of 34.7 °C in August. Mean annual potential evapotranspiration reaches 1,500 mm, indicating a considerable annual water deficit (Cantón et al. 2003); between 6 and 9 months per year can be considered as dry at El Cautivo. The climatological conditions and the sparsely vegetated landscape justify the local name of "Desierto de Tabernas" despite its Mediterranean climate (Lázaro et al. 2001).

Natural vegetation is dominated by species adapted to severe drought, and although the Mediterranean element accounts for the main part of the flora, the Iberian–North African element often accounts for the majority of individuals and of the biomass. Richness of species and endemic plant rates are high. Vegetation is patchy, and both physiognomy and life-form spectrum are more similar to those of North Africa than Europe (Lázaro et al. 2001). Chamaephytes, terophytes and cryptogams are the main life-form types. Sparse dwarf shrubs or tussock grasses physiognomically dominate the landscape, whereas annual plants account for more than 50 % of the total flora, particularly in wet years. Biological soil crusts can cover a high proportion of the ground surface (more than 50 % in El Cautivo, the best studied part of the area). Trees are absent and medium and large shrubs are very sparse (Lázaro et al. 2008).

17.3 Geomorphology

The Andarax River with its three main tributaries crossing the badland area (Río Nacimiento, Rambla de Gergal and Rambla de Tabernas) has a large catchment area (ca. 2,000 km^2), with its headwaters in the Sierra Nevada, Sierra de Gádor and Sierra de los Filabres (Fig. 17.1). Here, annual rainfall is much higher and storms are more frequent, giving a high energy to the system.

In most of the central and eastern part of the Andarax catchment, drained mainly by the Rambla de Tabernas, lithology, climate, relief and rapid incision have combined to create a typical but also distinctive badlands landscape (Fig. 17.3a).

As in many other areas with badlands, the landscape comprises a series of entrenched valleys with high drainage density and short and steep hillslopes, usually dissected by rills and gullies. This is the characteristic landscape of most of the tributaries of the main rivers, where incision is more recent. In the areas where the incision is older, the drainage is characterized by wide braided channels flanked by a flood plain or lower terrace, largely disconnected from the

Fig. 17.3 General view of the lower Rambla de Tabernas valley (**a**) and two perspectives of El Cautivo area (**b**) and (**c**) where surfaces from different stages can be identified as hanging pediments, valley floors and talus flatirons. Note the contrast in the vegetation cover between the old surfaces and the more degraded hillslopes affected by incision

hillslopes. Changes in local base level have been critical for the development of these badlands. In the western part of the basin (Andarax and Nacimiento catchments), pure badlands are restricted to specific sites and most of the rest of the hillslopes are more regular, covered by sparse vegetation and only locally affected by gullying. In the central part of the basin (mid- and lower Rambla de Tabernas catchment), the coupling (sensu Harvey 2002) between hillslopes, tributaries and main channels is complete. In contrast, the north-facing hillslopes, the remains of old pediments and old hanging hillslope parts, have more stable surfaces still disconnected from the local base level.

These contrasting landscapes have been created primarily by the spatially differential response to tectonic activity that has favoured incision in some parts of the basin, while at certain times, it has impeded this process in other parts. The landscape not only shows a variety of landforms, but also landforms of different ages within the basin.

17.3.1 General Landforms

The slope map shown in Fig. 17.4 contains six slope classes that illustrate the morphology of the basin. Harvey (1987) described the erosional parts as composed of canyon landscapes, deeply dissected soft rock areas and scarplands. Half of the area consists of hillslopes <30°, and 20 % of the surface includes steeper areas, both recently dissected valley walls and scarps associated with the caprocks in the higher parts of the relief. The remaining 30 % corresponds to flat and gentle sloping areas such as valley bottoms, terraces and pediments.

The valleys of the main rivers have wide braided channels similar to those described by Thornes (1976) in the neighbouring area of the Alpujarras. They are characterized by high transport capacity during flood events and active bank erosion maintaining very steep sidewalls, as in the Rambla de Tabernas and some parts of the Rambla de

Fig. 17.4 Slope map of the badlands area and surrounding reliefs. Legend shows the upper limits of the slope intervals in degrees

Gergal. Both ramblas have excavated deep canyons in most of their mid-catchment (Fig. 17.2 and 17.3a). The rapid incision of these channels is related to the uplift episodes during the Pleistocene and especially with several interruptions of the drainage that produced at least two palustrine areas in the catchment of the Rambla de Tabernas (Harvey and Mather, 1996; Harvey et al. 2003). The easily erodible palustrine sequence around 20 m thick (Harvey 2001) has been dissected during the last ca. 15 ka (Alexander et al. 2008).

Some parts of the landscape have less steep landforms inherited from pedimentation periods during the Pleistocene, hanging above the present-day base level. These features are more frequent in the Rambla de Tabernas area due to its more recent evolution. Also remains of the two main sets of fluvial terraces (Harvey 1987) contribute to the lower gradient areas on the map (Fig. 17.4). These terraces correlate in part with the undissected alluvial fans that characterize the eastern part of the basin (Harvey and Mather 1996; Harvey et al. 2003).

The older terraces exhibit calcrete development (Nash and Smith 1998) (Fig. 17.5a), including pedogenic calcretes at the mesa surfaces and groundwater calcretes at the contact between the terrace gravels and the underlying Tortonian mudstones. Locally, like at Las Salinas site (Fig. 17.5b), $CaCO_3$ precipitation occurs as travertine curtains (Mather et al. 1997; Mather and Stokes 1999).

17.3.2 The Role of Biological Activity

Geomorphological processes affect, and are affected by, biotic components such as shrubs, annual plants or biological soil crusts (BSC) (Fig. 17.5d), and these interactions act across different spatial and temporal scales. According to Lázaro et al. (2000), the rich variety of microhabitats existing in Tabernas badlands is a consequence of the landform diversity, which in turn is related to the complex geomorphology of the system. The microhabitats show different topographical, erosional and edaphic features, explaining the spatial patterns exhibited by the main physiognomic types of vegetation. The cover of BSCs, annual and perennial vegetation is largely controlled by the landform type, and consequently, landform distribution is a proxy for the vegetation map (Alexander et al. 1994). Cantón et al. (2004) demonstrated the significant correlation

Fig. 17.5 Dissected hillslopes on the Tortonian mudstones capped by calcrete (**a**). Travertine curtain at Las Salinas site (**b**). Rock fall on the walls of the Rambla de Tabernas main channel (**c**). North-facing hillslope covered by a continuous biological soil crust (BSC) in an advanced development stage (**d**)

between the landforms resulting from a non-hierarchical classification of the terrain attributes, and the ground-cover pattern. This association shows the influence of geomorphology in determining the type and characteristics of the biotic components.

Two levels of spatial segregation can be observed in the area: (1) between eroded and vegetated areas. The former mainly correspond to steep south- or south-west-facing hillslopes, almost completely devoid of vegetation and practically without soil development, which account for one-third of the total area. The latter are mainly associated with pediments and less steep north- or north-east-facing hillslopes; (2) inside the vegetated areas, spatial segregation often exists among the communities dominated by different life forms: grasses, dwarf shrubs, annual herbs and BSCs. This segregation is associated with landforms and, in addition, to the nested topographical levels (see Alexander et al. 1994, 2008 and below for a description of these nested levels). The older the level, the longer the time elapsed since the stabilization of the surface and consequently the more developed the vegetation succession and soils, particularly leaching of the topsoil.

The community dominated by perennial grasses, mainly *Macrochloa tenacissima* (=*Stipa tennacissima*, alpha grass), which is a late-successional stage, is located mainly on the oldest remains of hillslopes and pediments, in the higher topographical positions, often disconnected from the current drainage network. Typical semi-arid dwarf shrubs (often endemic or Ibero-North African species, e.g. *Helianthemum almeriense, Euzomodendron bourgeanum, Hammada articulata, Anabasis articulata* and others) and annual plants (e.g. *Stipa capensis, Plantago ovata, Bromus rubens, Linum strictum, Asteriscus aquaticus, Brachypodium distachyon* and many others) occur mainly in the intermediate nested levels (on the older surfaces they have been progressively replaced by grassland). Shrubs and annuals often occupy the same habitats although they are not stratified, but form a non-random mosaic, the shrubs being slightly more water demanding than the annuals, which require more stable areas (Cantón et al. 2004). BSCs are distributed on the driest sites, on recent surfaces and in the upper half of the stabilized slopes of the medium levels (Lázaro et al. 2000; Cantón et al. 2004). On the oldest levels, they also appear, but are less frequent, as they have already been replaced in the great part of these sites. BSCs show an important cover on more than two-thirds of the total area and in one-third constitute the main or even the only cover. BSCs are composed of three main communities, dominated, respectively, by Cyanobacteria and diverse small colonizing lichen species; *Diploschistes diacapsis* and *Squamarina lentigera*, and *Lepraria isidiata* and *Squamarina cartilaginea*. This abundance of BSC is characteristic of these badlands and is due to the high proportion of the rainy days with only small total rainfall amounts and the frequency of the dew. This favours poikilohydric organisms constituting the BSC in comparison with vascular plants, as the main part of the low-magnitude rainfall events is intercepted by the canopies and does not reach the roots (Lázaro 2004).

The features and activity of biotic components also control geomorphological processes at a detailed scale. These controls, which have implications at the landform scale, operate through several clusters of mechanisms:

1. By stabilizing the soil surface, decreasing erosion and allowing soil development (Alexander et al. 1994). In addition to the known stabilizing role of the vascular vegetation, BSCs have been suggested to be the first colonizers and often constitute the initial stabilization mechanism (Lázaro et al. 2008). The role of BSCs in reducing soil erosion at Tabernas badlands has been widely demonstrated (Alexander and Calvo 1990; Calvo-Cases et al. 1991a; Solé-Benet et al. 1997; Cantón et al. 2001b; Lázaro et al. 2008; Chamizo et al. 2012).
2. By modifying infiltration, run-off paths and amount, as well as the amount and spatial distribution of sediment deposition (Puigdefabregas et al. 1999). This effect of vegetation is dynamic and species-specific, and particularly significant in tussock grasses because they are able to intercept an important percentage of run-off and sediment flow; about 50 % for *Macrochloa tenacissima*, the dominant grass at Tabernas, according to Puigdefábregas and Sánchez (1996). Very local biotic components affect feedback processes at landform scale such as the asymmetry present in many of the small catchments within these badlands. This asymmetry is always linked to the vegetation pattern: north- to east-facing hillslopes are vegetated, larger, more stable and less steep. The opposite slopes are affected by erosion; they are shorter, steeper and bare.
3. By introducing feedbacks in the surface processes and in the plant-soil relationships tending towards vegetation development (Eviner and Chapin 2003; Ehrenfeld et al. 2005). Many of the soil–plant interactions are based on the effects of vegetation on the hydric fluxes (Pugnaire et al. 1996a; Horton and Hart 1998), gas fluxes (Raich and Tufekcioglu 2000), energy fluxes (i.e. temperature modifications; Hillel 1998), or on the soil enrichment by organic matter (Mora and Lázaro 2013). The most labile fractions of organic matter, coming from the root exudates and from the litter, favour the activity and diversity of the soil microbial communities (Kourtev et al. 2003; Ravit et al. 2003), which play a key role in the formation of soil aggregates (Jastrow et al. 1998; Rillig et al. 2002) and in the nutrient availability for plants (Northup et al. 1998; Hinsinger 2001). The sum of these processes gives rise to "islands of fertility" under the main vegetation patches, in which the facilitation relationships among plants dominate (Pugnaire et al. 1996b).

In the Tabernas badlands, the vegetation development, during which the mechanisms detailed above operate, can be synthesized as follows:

1. The eroded and unstable sites are colonized only by a few vascular species (mainly the shrub *Salsola genistoides* and the annual *Moricandia foetida*), creating the typical appearance of these badlands, particularly in wet years, during which *Moricandia* grows abundantly and flowers in a short period. However, none of them is able to have significant stabilization or pedogenesis effects on these sites (Lázaro 1995; Gallart et al. 2002), and no vegetation succession exists.
2. Constructive colonization able to feedback towards stabilization and soil and vegetation development requires a certain degree of previous stabilization caused by abiotic factors (disconnection from the drainage network, obstruction of the channel by sediment deposits, etc.), together with reduced land use (as currently in the Tabernas badlands). Constructive colonization often begins with BSC, which are able to resist splash erosion but unable to resist either trampling, or some erosional processes frequent in the area, such as mass movement, rill/gully development and collapse produced by previous piping (Lázaro et al. 2000). BSCs enrich bare soils with organic matter (Miralles et al. 2012a), initiate biogeochemical cycling of elements (Miralles et al. 2012b), modify local water balance and strongly decrease erosion (Chamizo et al. 2012; Solé-Benet et al. 1997), and thus provide additional stability and enable the ground to be colonized by vascular plants. Vegetation and erosion compete for soil fertility and depth as proposed by Thornes (1990) and beyond a certain threshold; vegetation (and soil) receives a positive feedback. Below that threshold, abiotic stabilization is necessary; otherwise, erosion receives positive feedback and colonization fails. In the Tabernas badlands, vegetation and soil development are usually successful on pediments and in the north- to east-facing slopes of the old and medium nested levels, whereas in the rest of landforms, they often fail.

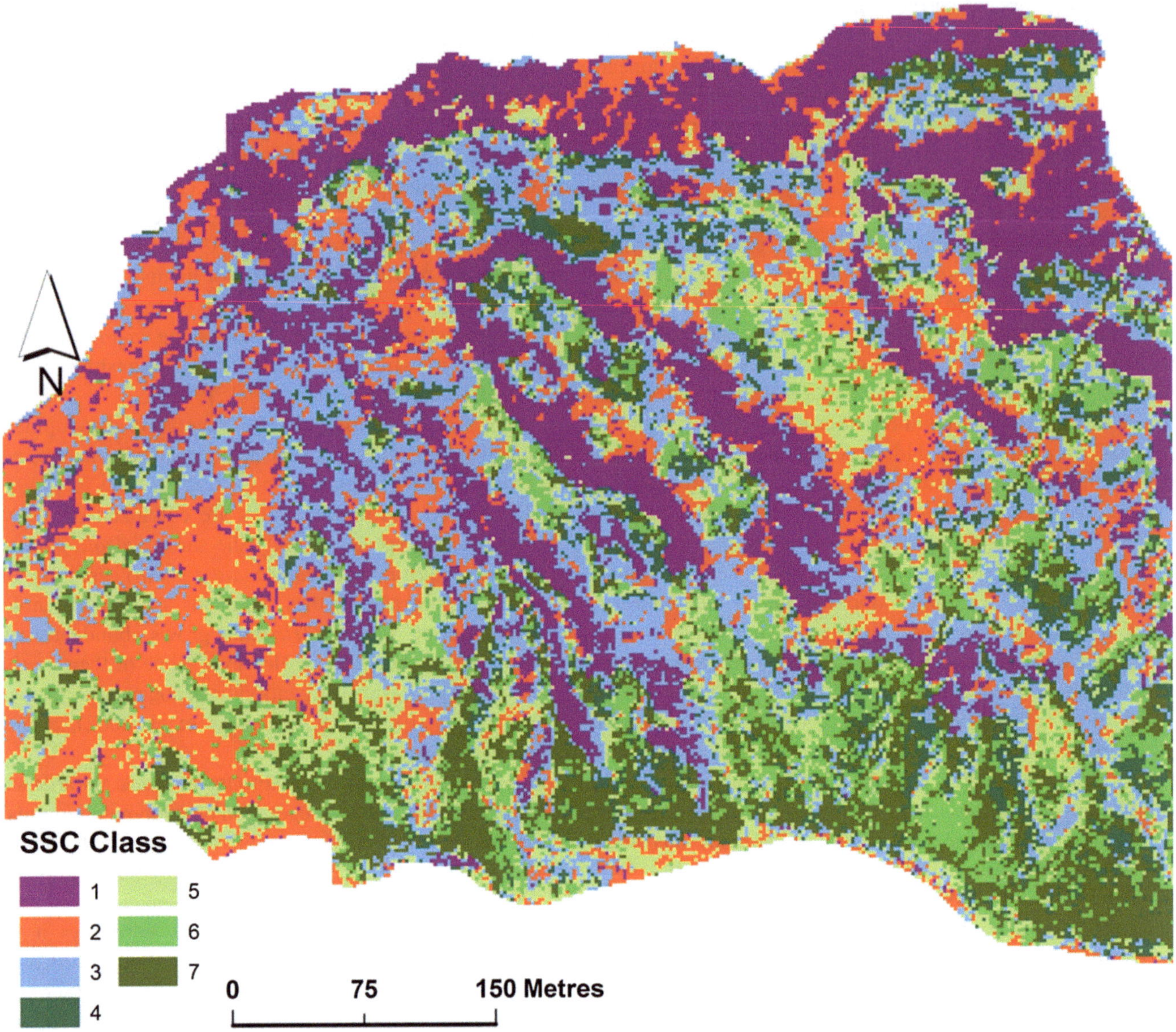

Fig. 17.6 Soil surface components (SSC) map of El Cautivo area by ATM classification (Modified from Calvo-Cases et al. 2009). *1* Actual bare badland hillslopes; (*2*) Rock fragment cover; *3* BSC of lichens; *4* Low-density bushes and BSC; *5* Annual plants cover; *6* Bushes; *7* *Macrochloa tenacissima*. See location in Fig. 17.2

17.3.3 Present-day Geomorphic Processes

One of the best studied parts of the Tabernas badlands is the El Cautivo catchment, located in the eastern part of the area. The catchment is operated as an experimental station for process monitoring by the EEZA-CSIC institute (http://www.eeza.csic.es/eeza/g_geo.aspx).

At El Cautivo, the landscape comprises three main components: (1) bare hillslopes representing 40 % of the surface; (2) remains of Pleistocene and early Holocene pediments that occupy 10 % of the landscape and constitute stable areas with a well-developed plant cover; (3) the remaining 50 % is composed of hillslopes with a BSC cover (33 %) and/or vascular plants (Fig. 17.6).

The Tortonian series, with alternating mudstones and thin calcareous sandstone layers, are easily weathered by wetting and drying sequences (Cantón et al. 2001a), in which the role of the frequent small rainfall events (more than 84 % of the total) is essential. These do not yield run-off, but weather the regolith providing easy-to-transport sediments. Cantón et al. (2001b) showed that annual erosion rates, in most years, are lower than weathering and regolith production rates. The climate is arid enough to restrict the development of plant cover on the south-facing slopes, but provides enough rainfall to occasionally allow run-off generation with sufficient energy to remove most of the previously weathered material and keep the surface clear of regolith.

Mudstone Weathering

The main rock forming the badlands is a marl or more precisely a gypsum–calcareous mudstone, essentially formed by very homogeneous silts: the D_{60} of its mineral grains ranges between 25 and 20 μm, D_{10} between 2 μm and 3.7 μm, and the uniformity coefficient (D_{60}/D_{10}) is 6.5 (Solé-Benet et al. 2009). The main minerals are quartz, muscovite, paragonite, calcite (up to 30 %) and gypsum (up to 30 %), and minor amounts of clay (5 %) of which small amounts are smectites (Solé-Benet et al. 1997). According to Cantón et al. (2001a), a combination of three factors seems to be responsible for mudstone weathering: repeated wetting–drying cycles, the presence of geologically induced cracks and fissures, and dissolution–crystallization of relatively soluble minerals like gypsum. In the laboratory, a few wetting–drying cycles applied to the fresh mudstone are sufficient to enhance a network of cracks and fissures favouring mineral dissolution. In the field, surface weathering rates from 0.7 mm $year^{-1}$ (average after three years of natural rainfall without removing the produced regolith) to 4.7 mm $year^{-1}$ (average when removing the regolith after the main rainfall events, like natural bed scouring) were measured on weathering plots set in 1995 over fresh rock, after removing the original regolith cover (Cantón et al. 2001a). Weathering rates were found to be proportional to the number of rainfall events during the sampling periods, confirming the results of laboratory experiments; the number of wetting–drying cycles has the greatest influence on weathering. These findings confirm: (a) the long-term observations about available material and erosion on bare, regolith-covered slopes: the thickness of weathered, prone to erosion material, is proportional to the number of low-magnitude rains before an erosive high-intensity or high-magnitude event, which occurs on average once or twice a year; (b) a long-term erosion of 3.72 mm in 16 years (equivalent to 3 t ha^{-1} $year^{-1}$) was measured by erosion pins, laser scanner and profiles in a 100 m^2 sloping area on bare regolith under south-west exposure (Solé-Benet et al. 2012), though a single event caused a surface lowering of 86 mm in a rill.

These weathering rates might be considered as the probable range of incision under present semi-arid conditions. Moreover, the quite homogeneous particle-size distribution of this mudstone indicates susceptibility for piping processes, as observed, in the area.

Rainfall-runoff Processes

The main processes identified by the resulting forms are rainfall-runoff related, rill erosion being the most evident (see Calvo-Cases and Harvey 1996), especially following moderate rainstorms. In contrast, after periods of low rainfall intensity, the regolith becomes thicker (Cantón et al. 2001a) and the rills may be partially destroyed by surface weathering. When subsequent intense rainfall does occur, shallow mudflows can be identified in the concavities usually occupied by the rill network. On south-facing slopes, the rill networks form part of a hierarchy of rills, gullies and small catchments.

Available data on erosion rates, provided by rainfall simulation experiments, measurements in plots and catchments with different sizes, show figures that are lower than could be expected for badland areas. In fact, most of the data analysed in Calvo-Cases et al. (1991a) comparing badland areas in south-east Spain show the Tabernas data in a lower range, for similar soil surface components (SSC), than many other badland areas. Figure 17.6 shows the soil surface components (SSC) at El Cautivo, which are associated with soils of different degrees of evolution (Alexander et al. 1994), controlled by aspect and the distribution of surfaces with different ages.

Experimental data obtained by Calvo-Cases et al. (1991b), Solé-Benet et al. (1997) and Cantón (2001b) indicate that the BSC composed of lichens and the mineral soil crust (MSC) characteristic of the bare areas are the soil surface components with higher runoff coefficients. Erodibility is low when BSC and MSC remain, but increases significantly when they are destroyed (Chamizo et al. 2012). Soils with well-developed vegetation cover have higher infiltration capacity and lower erosion rates.

During ten years of monitoring at El Cautivo, average ground lowering ranges from 2.15 to 0.08 mm $year^{-1}$ for a bare microcatchment (ca. 200 m^2) and 0.35 mm $year^{-1}$ for a small catchment (2 ha) (Solé-Benet et al. 2003). The differential hydrological response is controlled by the distribution of different SSC in the catchment (Cantón et al. 2002).

In comparison with other badland areas in south-east Spain, the lower erosion values of the Tabernas badlands are related partly to the drier climatological conditions with less erosive rainfall Lázaro et al. 2001), and the rock properties yield less erodible soil. The preservation of a complex mosaic of SSC associated with soils with different degree of development is one of the main factors that explain the variability of the response.

Piping

In comparison with other badland areas, the dispersiveness of the Tabernas badlands bedrock is relatively low (Faulkner et al. 2000 and Faulkner 2013; Cantón et al. 2003) and

Fig. 17.7 Gully dissection in a north-facing hillslope and pediment covered by some vascular plants and a well-developed BSC made of lichens; white spots (**a**). Gullies developed in the lower terrace of the Rambla de Tabernas (**b**) and along tracks (**c**). Some human activities damaging the BSC and vegetation cover on the hillslopes of the natural park (**d**)

piping processes are very restricted. The only evidence of forms related to piping appears in collapses of the lower part of hanging pediments where the water flows through macropores produced by roots and salt solution cracks (Solé-Benet et al. 2009), developing galleries in the contact zone between Quaternary soils and regolith and the underlying mudstones. A higher concentration of sodium was found by Alexander et al. (1994) at certain depths (c. 60 cm) as a consequence of the leaching of bases.

Gullies

Some relatively shallow arroyo-like gullies with flat floors and vertical walls dissect the flat terrace surfaces (Calvo-Cases et al. 1991b). They are decoupled from the main rivers by agricultural practices. Since such activities have ceased, episodes of rapid gully development have occurred (Fig. 17.7b). On some north-facing hillslopes with thick regolith cover, longitudinal gullies (Fig. 17.7a), some decimetres wide and initially as deep as the regolith, appear associated with tracks (Fig. 17.7c) and other areas where the BSC has been degraded by human activities (Fig 17.7d).

Mass Movements

Together with the shallow mudslides and mudflows (described above) that occasionally occur on badland hillslopes, areas where fast incision has created scarps show frequent earth falls consisting of mudstone block accumulations on the gully channel floors, released by the development of unloading cracks. These deposits are easily removed by floods. On the main canyon walls, similar

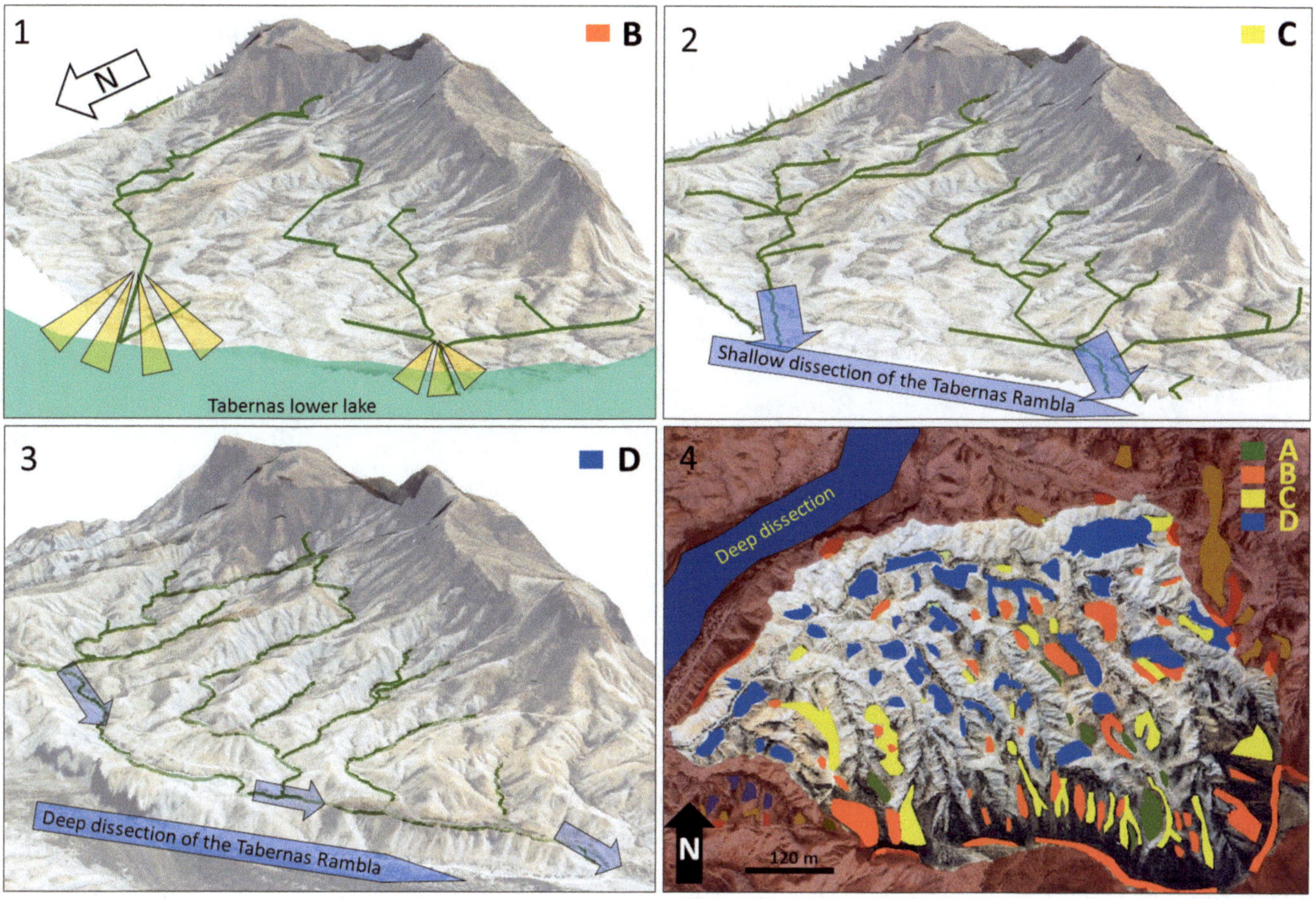

Fig. 17.8 Topographical reconstruction of El Cautivo catchment at three times: from 14 ka ago *1* to the present *3*, and map of the remains of landforms of the different ages *4* (modified from Alexander et al. 2008)

mechanisms produce rock falls of bigger blocks (Fig. 17.5c). Far from the main channels, on the upper part of the relief, rock falls derived from sandstones and Quaternary cemented gravel caprocks underlain by mudstones are also frequent (Fig. 17.5a).

17.4 Evolution

El Cautivo (Fig. 17.3b and c) as part of the lower Rambla de Tabernas catchment has been affected by the above-mentioned differential tectonic uplift and lake formation sequences. The landscape records stabilization and dissection sequences that occurred during the Pleistocene and perhaps over the last few thousand years of the Holocene. Alexander et al. (1994) identified five stages of aggradation separated by incision sequences.

The earlier stages (A, B and C in Fig. 17.8) represent successive badland and pediment stages carved in the marls. They involve long-term hillslope evolution through the retreat of the northern side of a sandstone scarp that forms a caprock over the Tortonian marls (Fig. 17.3c).

Stage A is represented by some talus flatirons (sensu Gutiérrez-Elorza and Sesé-Martínez 2001) remaining in the upper part of the relief, characterized by regular concave hillslopes armoured by a detrital cover and with a distinctive vegetation cover dominated by *Macrochloa tenacissima* tussocks.

Stage B appears in the upper parts of graded hillslopes below the scarp, between the remnants of stage A. It is also recorded by debris cones and fan deposits, small remnants of pediments in the divides and some valley sections hanging over the contemporary relief. Most of these stage B remnants are also associated with a distinctive regolith-soil and vegetation cover similar to stage A sites. The more widespread presence of these remnants allowed Alexander et al. (2008) to create a reconstruction of the stage B topography (Fig. 17.8-1) and to correlate these forms with tributary catchments draining to the north and north-west towards the lower Tabernas Lake. The terminal age of these forms can be considered as contemporary with the top of the lake sediments (OSL date of 14 ± 1 ka BP, Alexander et al. 2008), although they developed during the previous ca. 150 Ka when the lakes were present, U/Th date of

Fig. 17.9 Three-dimensional view of a recent ortophoto with a 5-m resolution DEM (www.cnig.es) covering most of the badlands of the Tabernas Basin

150 ± 50 ka from the lower part of the sediments of the Tabernas upper lake (Delgado-Castilla et al. 1993).

After a limited incision (ca. 3 m) along the main channels, a new period of pediment formation and hillslope stabilization occurred (stage C). The reconstructed topography does not show significant changes in the drainage network from stage B (Fig. 17.8-2). The incision appears to have been driven by a small base-level drop as the palaeo-Rambla de Tabernas began to cut down through the soft lake sediments.

After this stage C, a major change occurred in the El Cautivo catchment (Fig. 17.8-3). All the drainage directed to the north-western side was captured (following the arrows in Fig. 17.8-3) by the western tributary to the Rambla de Tabernas, inducing deep incision of the entire drainage network. The local base level dropped about 30 m due to the entrenchment of the modern Rambla de Tabernas in Tortonian marls, creating the modern canyons. In El Cautivo catchment, the main channel incision (ca. 50 m) and the consequent badlands landscape development during this stage (C–D), that supposed a total volume of denudation of c. 1.4 hm^3 of rock within El Cautivo catchment (Alexander et al. 2008), possibly linking to the dry–warm period around 5–4 ka BP (Nogueras et al. 2000). Small pediments of stages D and E occur within the incised drainage network that were formed sometime during the late Holocene, prior to the present-day active incision.

The detailed study of the landform evolution of the lower Rambla de Tabernas catchment suggests an average ground lowering of 0.16 mm $year^{-1}$ in a period of 14 ka although most of the erosion work was done between stages C and D, as a consequence of the incision of the Rambla de Tabernas. The Quaternary stages of stabilization and incision of the hillslope-pediments-channel systems identified in El Cautivo and other parts of the lower part of the Rambla de Tabernas (stages A, B, C, D) relate to changes in the main drainage axis induced by tectonics, which seem to be primarily confined to the Rambla de Tabernas drainage and seem to apply less to the western part of the basin, drained by the Andarax River main valley, and the lower catchments of the Río Nacimiento and Rambla de Gergal, although Garcia et al. (2003) have suggested some Quaternary fault activity there. The western areas appear as darker tones on the aerial photographs (Fig. 17.9), reflecting a greater vegetation cover and less active hillslope erosion. The geology is also different; this is the main outcrop area of the Plio-Pleistocene Gador conglomerates. Furthermore, late Quaternary tectonically induced incision has perhaps been less active there. In contrast, the Rambla de Tabernas area presents lighter colours on the aerial photograph (Fig. 17.9). The late

Pleistocene reactivation following the post-lake incision of the Rambla de Tabernas has led to the formation of the well-coupled canyon and badland landscapes. Within that area, at a more local scale, coupling is weaker in parts of the north-facing hillslopes, where local microclimatic and dynamic factors decouple the hillslopes from the channels. Also decoupled are places like the upper part of the hillslopes of El Cautivo and many other tributaries, where the wave of aggression has had insufficient time to arrive. Similarly, the upper catchment of the Rambla de Tabernas, upstream of the incision headcuts, preserves a totally different landscape of coalescent alluvial fans where base-level-induced dissection is only just beginning (Harvey 2006).

The development of the Tabernas badlands can be explained, according to Faulkner (2008), by the interaction between a succession of stages of decoupling in the system driven by climatic and/or intrinsic controls under local base-level stability, and a succession of stages in which the reactivation of incision in the main fluvial systems propagates upstream creating a wave of aggression (see Brunsden 2001) increasing the coupling and connectivity.

17.5 Conclusions

The whole Tabernas basin has three morphologically different areas: (1) The eastern part, upstream of Tabernas town, has not been dissected and retains a low-gradient morphology dominated by alluvial fans. (2) The western part has the morphology of an old badlands area, today graded to the main base level with more regular and vegetated hillslopes. (3) The most active badlands appear in the central part, and especially in the lower Rambla de Tabernas catchment, as the consequence of the recent (c. 14 Ka) dissection of the lower Tabernas Lake. Also the upper catchments of the Rambla de Gergal and some parts of the upper Río Nacimiento, both inside the basin sedimentation area, have active badlands.

Tectonic chronology seems to be the main factor differentiating these parts as there has been enough relief created to allow the dissection in some areas while others remain unaffected. Landforms show complex multi-stage development and contain a variety of microecosystems. This variety is enhanced by the control of aspect and the special erosive conditions introduced by climate and bedrock that allow a singular diversity and extension of biological soil crusts and especially of lichens species.

Although erosion rates are high in bare hillslopes oriented to the south, for the area as a whole, they are low in relation to badlands elsewhere in south-east Spain. The low overall rates are due to the scarcity of rainfall events that generate runoff and transport sediments, as well as to the mosaic of soil surface components that results from differences in lithology, landform ages and microclimate, and more than 50 % coverage of the catchments by plants or biological soil crusts. The role of the crusts has been found to be critical in maintaining the stability of the hillslopes when a lack of disturbance allows positive feedbacks that are conducive to runoff rate reduction as vegetation colonizes.

References

Alexander RW, Calvo A (1990) The influence of lichens on slope processes in some Spanish badlands. In: Thornes JB (ed) Vegetation and Erosion. Wiley, Chichester, pp 385–398

Alexander RW, Calvo A, Arnau E, Mather AE, Lázaro R (2008) Erosion and stabilization sequences in relation to base level changes in the El Cautivo badlands, SE Spain. Geomorphology 100:83–90

Alexander RW, Harvey AM, Calvo A, James PA, Cerdá A (1994) Natural stabilisation mechanisms on Badland Slopes: Tabernas, Almería, Spain. In: Millington AC, Pye K (eds) Environmental change in Drylands. Wiley, pp 85–111

Brunsden D (2001) A critical assessment of the sensitivity concept in geomorphology. Catena 42:99–123

Calvo-Cases A, Harvey AM (1996) Morphology and development of selected badlands in southeast Spain: Implications of climatic change. Earth Surf Proc Land 21:725–735

Calvo-Cases A, Alexander RW, Arnau-Rosalén E, Bevan J, Cantón Y, Lázaro R, Puigdefábregas J, Solé-Benet A (2009) Interacción de procesos geomórficos y distribución de componentes de la superficie del suelo en relación a la evolución de los abarrancamientos de Tabernas (Almería). Cuadernos de Investigación Geográfica 35:43–62

Calvo-Cases A, Harvey AM, Paya-Serrano J, Alexander RW (1991a) Response of badland surfaces in southeast Spain to simulated rainfall. Cuaternario y Geomorfologia 5:3–14

Calvo-Cases A, Harvey AM, Paya-Serrano J (1991b) Process interactions and badland development in SE Spain. In: Sala M, Rubio JL, García-Ruiz JM (eds) Soil erosion studies in Spain. Geoforma Ediciones, pp 75–90

Cantón Y, Del Barrio G, Solé-Benet A, Lázaro R (2004) Topographic controls on the spatial distribution of ground cover in the Tabernas badlands of SE Spain. Catena 55:341–365

Cantón Y, Solé-Benet A, Queralt I, Pini R (2001a) Weathering of a gypsum-calcareous mudstone under semi-arid environment in SE Spain: laboratory and field-based experimental approaches. Catena 44:111–132

Cantón Y, Domingo F, Solé-Benet A, Puigdefábregas J (2001b) Hydrological and erosion response of a badlands system in semiarid SE Spain. J Hydrol 252:65–84

Cantón Y, Domingo F, Solé-Benet A, Puigdefabregas J (2002) Influence of soil surface types on the overall runoff of the Tabernas badlands (SE Spain). Field data and model approaches. Hydrol Process 16:2621–2643

Cantón Y, Solé-Benet A, Lázaro R (2003) Soil-geomorphology relations in gypsiferous materials of the Tabernas desert (Almeria, SE Spain). Geoderma 115:193–222

Chamizo S, Cantón Y, Lázaro R, Solé-Benet A, Domingo F (2012) Crust composition and disturbance drive infiltration through biological soil crusts in semiarid ecosystems. Ecosystems 15:148–161

Delgado-Castilla L, Pascual-Molina A, Ruiz-Bustos A (1993) Geology and micromammals of the Serra-1 site (Tabernas Basin, Betic Cordilera). Estud Geol 49:361–366

Ehrenfeld JG, Ravit B, Elgersma K (2005) Feedback in the plant-soil system. Ann Rev Environ Resour 30:75–115
Eviner VT, Chapin FS III (2003) Functional matrix: a conceptual framework for predicting multiple plant effects on ecosystem processes. Annu Rev Ecol Evol Syst 34:455–485
Faulkner H, Spivey D, Alexander R (2000) The role of some site geochemical processes in the development and stabilisation of three badland sites in Almería, Southern Spain. Geomorphology 35:87–99
Faulkner H (2008) Connectivity as a crucial determinant of badland morphology and evolution. Geomorphology 100:91–103
Faulkner H (2013) Badlands in marl lithologies: a field guide to soil dispersion, subsurface erosion and piping-origin gullies. Catena 106:42–53
Gallart F, Solé A, Puigdefàbregas J, Lázaro R (2002) Badland systems in the Mediterranean. In: Kirkby MJ, Bull NLJ (eds) Dryland rivers: Hydrology and geomorphology of semi-arid channels. Wiley, Chichester, pp 299–326
Garcia AF, Zhu Z, Ku TL, Sanz de Galdeano C, Chadwick OA, Chacon Montero J (2003) Tectonically driven landscape development within the Eastern alpujarran Corridor, Betic Cordillera, SE Spain (Almeria). Geomorphology 50:83–110
Gutiérrez-Elorza M, Sesé-Martínez VH (2001) Multiple talus flatirons, variations of scarp retreat rates and the evolution of slopes in Almazán Basin (semi-arid central Spain). Geomorphology 38:19–29
Harvey AM (1987) Patterns of Quaternary aggradational and dissectional landform development in the Almeria region, southeast Spain: a dry-region tectonically active landscape. Erde 118:193–215
Harvey AM (2002) Effective timescales of coupling within fluvial systems. Geomorphology 44:175–201
Harvey AM (2006) Interactions between tectonics, climate and base level: Quaternary fluvial systems of Almeria, southeast Spain. In: Alberti AP, Bedoya JP (eds) Geomorfologia y Territorio. Actas de la IX Reunion Nacional de Geomorfología, Universidade de Santiago de Compostella, Santiago de Compostella, pp 25–48
Harvey AM, Mather AE (1996) Tabernas Quaternary alluvial fans and 'lake' system. In: Mather AE, Stokes, M (ed) 2nd Cortijo Urra field meeting, Southeast Spain, Field Guide. University of Plymouth, pp 39–42
Harvey AM (2001) Uplift, dissection and landform evolution: the Quaternary. In: Mather AE, Martin JM, Harvey AM, Braga JC (eds) A field guide to the Neogene sedimentary basins of the Almeria Province, South–East Spain. Blackwell Science, Oxford, pp 225–322
Harvey AM, Foster GC, Hannam J, Mather AE (2003) The Tabernas alluvial fan and lake system, southeast Spain: applications of mineral magnetic and pedogenic iron oxide analyses towards clarifying the Quaternary sediment sequences. Geomorphology 50:151–171
Harvey AM, Whitfield (nee Maher) E, Mather AE, Stokes M (2012) The transformation of alluvial fans into continuous fluvial systems: the Quaternary evolution of the Neogene sedimentary basins of the eastern Betic Cordillera, Almeria. In: González-Diaz A (coord). Avances de la Geomorfología en España 2010-2012. PUbliCan, Santander, pp 13–17
Harvey A et al (2014) in this volume (chapter 3)
Hillel D (1998) Environmental soil physics. Academic Press, San Diego
Hinsinger P (2001) Bioavailability of soil inorganic P in the rhizosphere as affected by root-induced chemical changes: a review. Plant Soil 237:173–195
Horton JL, Hart SC (1998) Hydraulic lift: a potentially important ecosystem process. Trends Ecol Evol 13:232–235
Jastrow JD, Miller RM, Lussenhop J (1998) Contributions of interacting biological mechanisms to soil aggregate stabilization in restored prairie. Soil Biol Biochem 30:905–916
Kourtev PS, Ehrenfeld JG, Häggblom MM (2003) Experimental analysis of the effect of exotic and native plant species on the structure and function of soil microbial communities. Soil Biol Biochem 35:895–905
Lázaro R (1995) Relaciones entre vegetación y geomorfología en el área acarcavada del Desierto de Tabernas. Universidad de Valencia, Tesis doctoral
Lázaro R (2004) Implications of precipitation on vegetation of water-limited lands. In: Pandalai SG (ed) Recent research development in environmental biology. Research Signpost, Kerala, vol I. pp 553–591
Lázaro R, Alexander RW, Puigdefábregas J (2000) Cover distribution patterns of lichens, annuals and shrubs in the Tabernas Desert, Almería, Spain. In: Alexander RW, Millington AC (eds) Vegetation mapping: from patch to planet. Wiley, Chichester, pp 19–40
Lázaro R, Rodrigo FS, Gutiérrez L, Domingo F, Puigdefábregas J (2001) Analysis of a thirty year rainfall record (1967–1997) from semi-arid SE Spain for implications on vegetation. J Arid Environ 48:373–395
Lázaro R, Cantón Y, Solé-Benet A, Bevan J, Alexander R, Sancho LG, Puigdefábregas J (2008) The influence of competition between lichen colonization and erosion on the evolution of soil surfaces in the Tabernas badlands (SE Spain) and its landscape effects. Geomorphology 102:252–266
Mather AE (2009) Tectonic setting and landscape development. In: Woodward J (ed) Physical geography of the Mediterranean region. Oxford University Press, Oxford
Mather AE, Stokes M (1999) Pleistocene travertines and lakes of Tabernas: evidence for a wetter climate? In: Mather AE, Stokes M (eds) BSRG/BGRG SE Spain field meeting. University of Plymouth, Field Guide, pp 63–71
Mather AE, Martin JM, Harvey AM, Braga JC (2001) A field guide to the Neogene sedimentary basins of the Almeria Province, South–East Spain. Blackwell Science, Oxford
Mather AE, Stokes M, Pirrie D (1997) Marl braid bars and their significance: an example from SE Spain. In: 6th International conference Fluvial Sedimentol., University of Cape Town, South Africa, Abstracts Volume
Miralles I, van Wesemael B, Cantón Y, Chamizo S, Ortega R, Domingo F, Almendros G (2012a) Surrogate descriptors of C-storage processes on crusted semiarid ecosystems. Geoderma 189:227–235
Miralles I, Domingo F, Cantón Y, Trasar-Cepeda C, Leirós MC, Gil-Sotres F (2012b) Hydrolase enzyme activities in a successional gradient of biological soil crusts in arid and semi-arid zones. Soil Biol Biochem 53:124–132
Mora-Hernández JL, Lázaro-Suau R (2013) Evidence of a threshold in soil erodibility generating differences in vegetation development and resilience between two semiarid grasslands. J Arid Environ 89:57–66
Nash DJ, Smith RF (1998) Multiple calcrete profiles in the Tabernas basin, southeast Spain: their origins and geomorphic implications. Earth Surf Process Land 23:1009–1029
Nogueras P, Burjachs F, Gallart F, Puigdefàbregas J (2000) Recent gully erosion in the El Cautivo badlands (Tabernas, SE Spain). Catena 40:203–215
Northup R, Dahlgren R, McColl J (1998) Polyphenols as regulators of plant-litter-soil interactions in northern California's pygmy forest: a positive feedback? Biogeochemistry 42:189–220
Postma G (1984) Mass-flow conglomerates in a submarine canyon: Abrioja fan-delta, Pliocene Southeast Spain. In: Koster KH, Steel R (eds) Sedimentology of gravels and conglomerates. (Mem. Can. Soc. Petrol. Geol. 10:237–258)

Pugnaire FI, Haase P, Puigdefabregas J (1996a) Facilitation between higher plant species in a semiarid environment. Ecology 77:1420–1426

Pugnaire FI, Haase P, Puigdefabregas J, Cueto M, Clark SC, Incoll LD (1996b) Facilitation and succession under the canopy of a leguminous shrub, Retama sphaerocarpa, in a semi-arid environment in south-east Spain. Oikos 76:455–464

Puigdefabregas J, Solé A, Gutiérrez L, del Barrio G, Boer M (1999) Scales and processes of water and sediment redistribution in drylands: results from the Rambla Honda field site in Southeast Spain. Earth-Sci Rev 48:39–70

Puigdefábregas J, Sánchez G (1996) Geomorphological implications of vegetation patchiness on semiarid slopes. In: Anderson MG, Brooks SM (eds) Advances in hillslope processes, vol II. Wiley, Chichester, pp 1027–1060

Raich JW, Tufekcioglu A (2000) Vegetation and soil respiration: correlations and controls. Biogeochemistry 48:71–90

Ravit E, Ehrenfeld JG, Häggblom MM (2003) A comparison of sediment microbial communities associated with Phragmites australis and Spartina alterniflora in brackish wetlands of New Jersey. Estuaries 26(2B):465–474

Rillig M, Wright S, Eviner VT (2002) The role of arbuscular mycorrhizal fungi and glomalin in soil aggregation: comparing effects of five plant species. Plant Soil 238:325–333

Solé-Benet A, Calvo A, Cerdà A, Lázaro R, Pini R, Barbero J (1997) Influences of micro-relief patterns and plant cover on runoff related processes in badlands from Tabernas (SE Spain). Catena 31:23–38

Solé-Benet A, Cantón Y, Domingo F, Del Barrio G, Lázaro R, Domene MA, Vidal S, Puigdefábregas J (2003) Long term hydrological monitoring of two micro-catchments in semi-arid SE Spain. In: Holko L, Miklánek P (eds) Interdisciplinary approaches in small catchment hydrology: Monitoring and research IHP-VI. UNESCO, Paris, Technical Documents in Hydrology 67, pp 183–188

Solé-Benet A, Cantón Y, Lázaro R, Puigdefabregas J (2009) Meteorización y erosión en el sub-desierto de Tabernas. Almería. Cuadernos de Investigación Geográfica 35:151–163

Solé-Benet A, Afana A, Cantón Y (2012) Erosion pins, profile and laser scanners for soil erosion monitoring in active hillslopes in badlands of SE Spain. In: González-Díez A (coord) Avances de la Geomorfología en España 2010-2012. Ediciones de la Universidad de Cantabria, pp 575–578

Thornes JB (1976) Semi-arid erosional systems: Case studies from Spain. Geographical Papers, London School of Economics 7, p 104

Thornes JB (1990) The interaction of erosional and vegetational dynamics in land degradation: spatial outcomes. In: Thornes JB (ed) Vegetation and erosion. Wiley, Chichester

Geology and Geomorphological Evolution of the Ebro River Delta

18

Luis Somoza and Inmaculada Rodríguez-Santalla

Abstract

From a geological perspective, deltas are ephemeral geomorphic systems whose development is controlled by the sensitive equilibrium between fluvial sediment supply, global sea level, wave-induced energy and regional subsidence. Once this equilibrium is lost, the delta may start to retreat and eventually disappear. The ephemeral character of deltas is a serious threat to their lands, as well as to the economic activities developed in these areas for centuries. This is the case of the Ebro Delta, the third largest delta in the Mediterranean area, which is extensively used for agriculture and includes a Natural Park with special significance for migrating birds. Historically, the geomorphological features of the Ebro Delta have been controlled by the discharge of the Ebro River that used to supply large amounts of sediment during flood events. This sediment is redistributed along the coast in response to wave-induced sedimentary dynamics. However, the construction of dams during the twentieth century has drastically changed the equilibrium responsible for the previous morphological evolution of the delta. Nowadays, after the drastic reduction in sediment supply to the delta, the wave-induced processes cause strong erosion during storm events and rapid littoral drift affecting the shape of coastal landforms. This chapter shows the morphological evolution of the Ebro delta over the last 8000 years in response to the deceleration of Holocene sea-level rise, along with its most recent evolution and the predicted future morphology of delta according to the climate change scenario of the IPCC.

Keywords

Ebro Delta • Holocene • Morphology • Sea-level changes • Mediterranean Sea

18.1 Introduction

Deltas are depositional landforms built at the mouth of rivers. The study of morpho-sedimentary changes in these systems has a special relevance due to their high dependence on global climate. Their evolution is mainly controlled by changes in the river water discharge, type and amount of sediment supply, regional and global sea-level fluctuations and variations in wave energy. Morphology and sedimentary sequences of deltas depend on the relative magnitude of tides, waves and currents (Wright and Coleman 1973; Galloway 1975).

L. Somoza (✉)
Marine Geology Division, Geological Survey of Spain (IGME), Madrid, Spain
e-mail: l.somoza@igme.es

I. Rodríguez-Santalla
University Rey Juan Carlos, Madrid, Spain
e-mail: inmaculada.rodriguez@urjc.es

F. Gutiérrez and M. Gutiérrez (eds.), *Landscapes and Landforms of Spain*, World Geomorphological Landscapes,
DOI: 10.1007/978-94-017-8628-7_18,

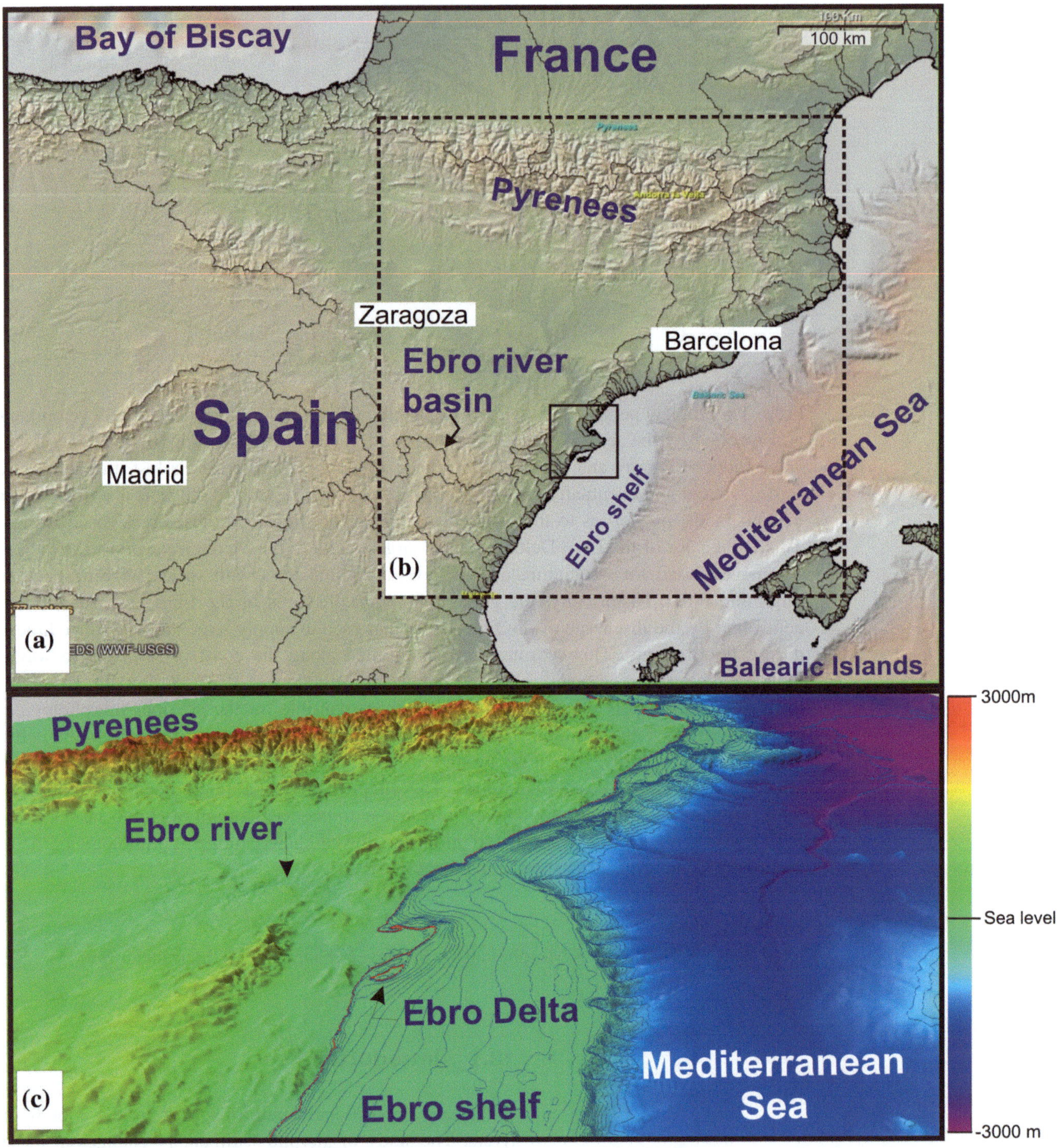

Fig. 18.1 **a** North-eastern sector of the Iberian Peninsula showing the location of the Ebro Delta. The boundaries of the Ebro River basin are also shown. **b** Shaded relief DTM of the emerged and submerged region of the Ebro Delta

The Ebro River Delta is the third largest delta in the Mediterranean Sea, after those of the Nile and Rhone rivers. Its emerged surface covers approximately 325 km^2, whereas the submerged area (pro-delta) extends over approximately 2,172 km^2 onto the continental shelf (Rodríguez 1999) (Fig. 18.1). The emerged part has a coastal length of approximately 50 km, being essentially a plain that reaches a maximum altitude of 4–5 m above the sea level (Serra 1997).

Like other large deltas of the world (e.g. Mississippi, Nile, Yangtze, Rhone), the construction of the Ebro Delta is related the deceleration of the sea-level rise over the last

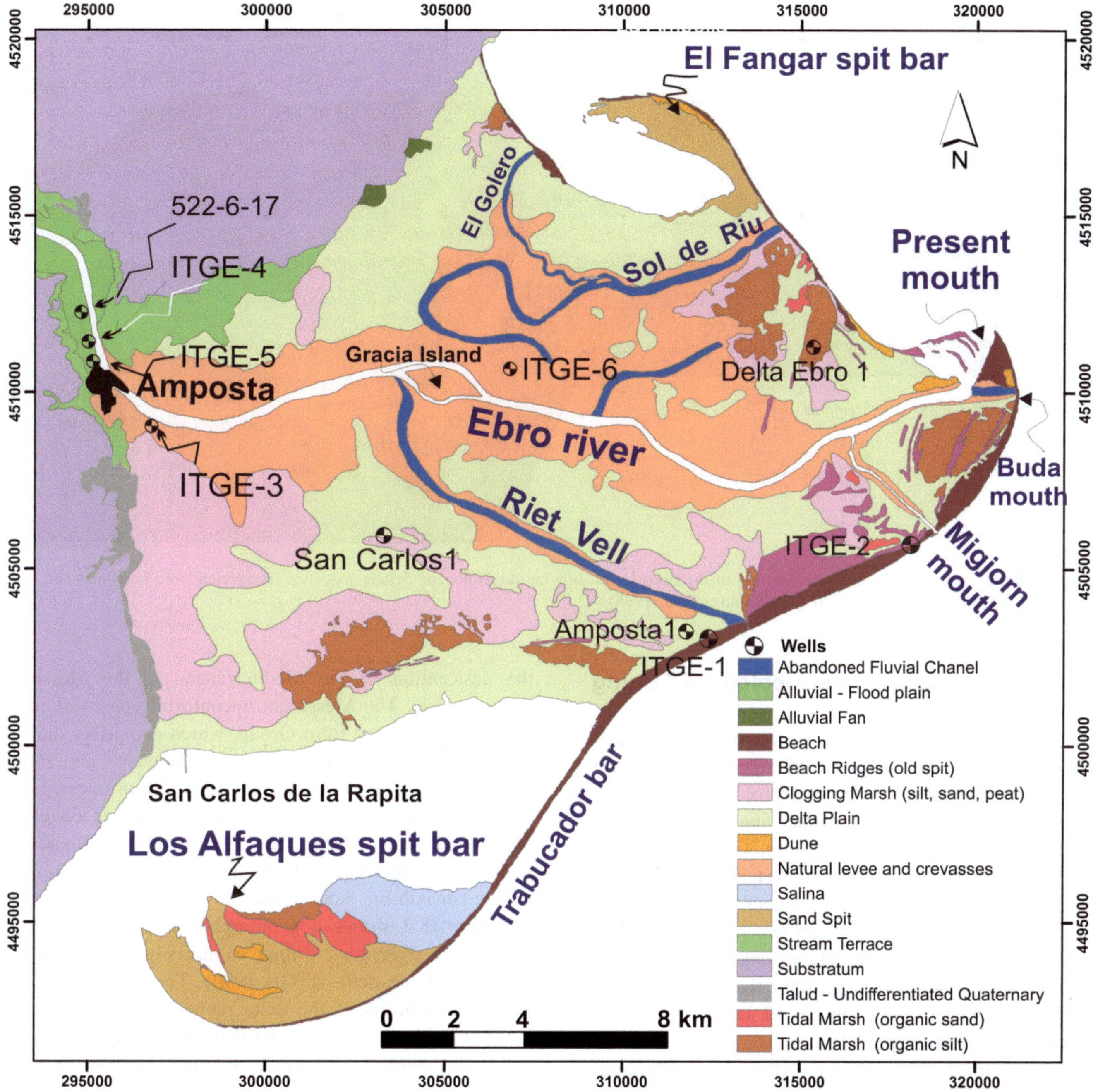

Fig. 18.2 Geomorphological map of the Ebro Delta. Location of boreholes used to construct the cross section shown in Fig. 18.3

8,000 years (e.g. Stanley and Warne 1994), involving the accumulation of a total volume of Holocene sediments higher than 28 km^3 (Guillén 1992). The Ebro Delta is considered to be a micro-tidal delta dominated by the river regime and wave action. However, the drastic reduction in sediment flux in the Ebro River due to the construction of dams (e.g. Mequinenza, Flix, Ribarroja) during the first half of the twentieth century has greatly increased the role of wave storms in the origin of coastal landforms.

From an economic perspective, the Ebro Delta is one of the main areas of rice crops; 98 % of the production in Catalonia and the third in the European market (Fatoric and Chelleri 2012). Besides, the relevance of its ecosystems (salt marshes, lakes, springs, etc.) makes it an area of great ecological value. This has allowed the inclusion of the Ebro Delta in different national and international frameworks for environment conservation (Rodríguez et al. 2010).

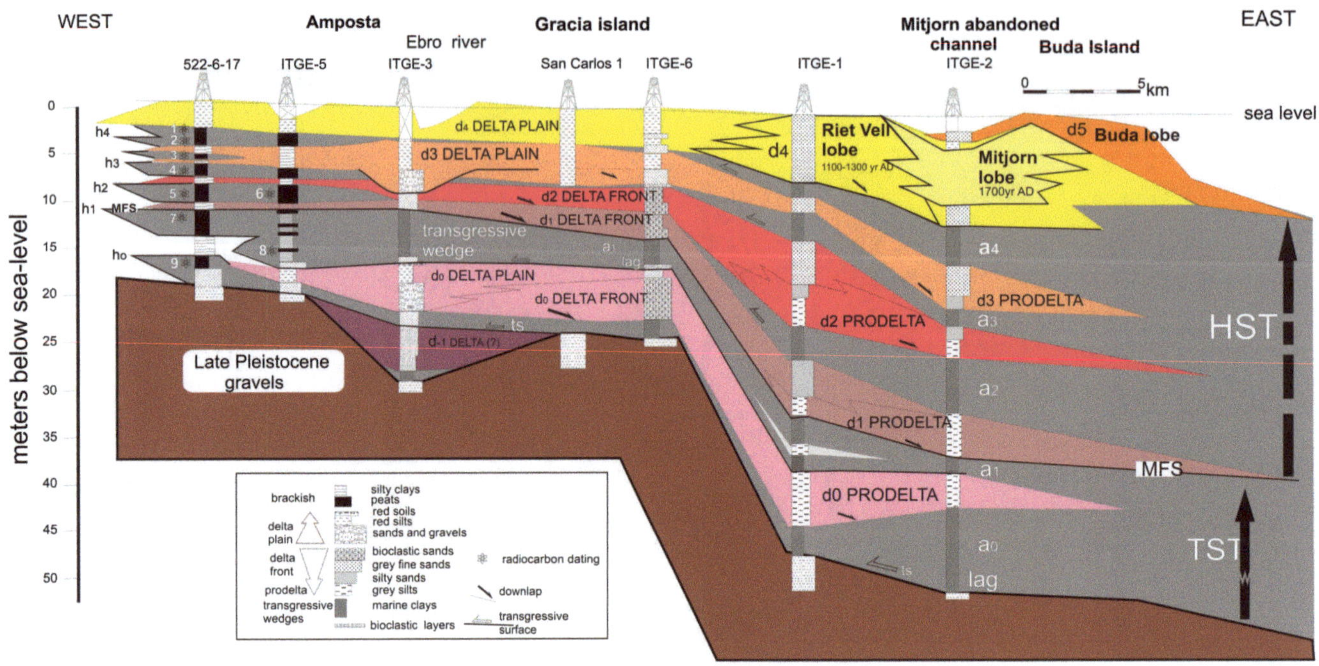

Fig. 18.3 Interpreted geological cross section of the Ebro Delta showing the Holocene deltaic sequence interpreted from boreholes and high-resolution seismic profiles (modified from Somoza et al. 1998)

18.2 Geographical and Geological Setting

The Ebro Delta is developed in an area of the Mediterranean Sea (Fig. 18.1b) with a micro-tidal regime, characterized by maximum astronomical and meteorological tides of 0.25 and 1 m, respectively. The average annual significant wave height in deep water is around 0.7 m, with a mean period of the order of 4 s (Sánchez-Arcilla et al. 1997). The main longshore transport is towards the south, but the east waves break against the present mouth generating sediment drift towards the NW and SW. These longshore currents have generated two spits which constitute the lateral limits of the delta: El Fangar to the north and El Trabucador-Los Alfaques to the south (Fig. 18.2).

The sedimentary succession of the Ebro Delta is underlain by 500–2,500 m of Plio-Quaternary deposits, which in turn unconformably overlie Miocene deposits (IGME 1987). The basement consists mostly of Lower Jurassic limestones and dolomites and Upper Cretaceous marls and bioclastic limestones (e.g. Maldonado 1975). These Mesozoic formations are cross-cut by a dense network of normal faults with prevailing N70E and N110E strikes, corresponding to the structural grains of the Catalan Coastal Chain and the Iberian Chain, respectively. These faults bound concealed structural highs and lows parallel to the present coast. The top of the Miocene sediments beneath the delta is marked by a strong reflection that corresponds to a regional erosional surface (unconformity) developed during the desiccation of the Mediterranean in the Messinian salinity crisis. The Messinian unconformity is overlain by the Plio-Quaternary Ebro Group, which comprises the Pliocene Lower Ebro Clays that grade upwards and shorewards into the Plio-Pleistocene Ebro Sandstones. These sandstones contain interbedded shallow marine conglomerates and marly clays. The Ebro Sandstones and the underlying Ebro Clays form a huge landward-thinning wedge beneath the delta plain.

Figure 18.2 shows a geomorphological map with the location of boreholes drilled to investigate the internal structure of the delta (ITGE 1996). The thickness of the Holocene deposits of the delta ranges from 18 m on the landward side, to 51 m at the delta front (Fig. 18.3). The maximum thickness of delta sediments occurs near the mouth of the present-day Ebro River (Maldonado 1972). Offshore radiocarbon ages (Díaz et al. 1990) indicate that deposition of the pro-delta on the shelf began between about 10000 and 11000 yrs BP.

18.3 Geomorphology

18.3.1 Morphology of the Coast

The coastal morphology of the Ebro Delta corresponds to a delta front (Fig. 18.4) composed of the present river mouth and two large spit bars that partially close two adjacent

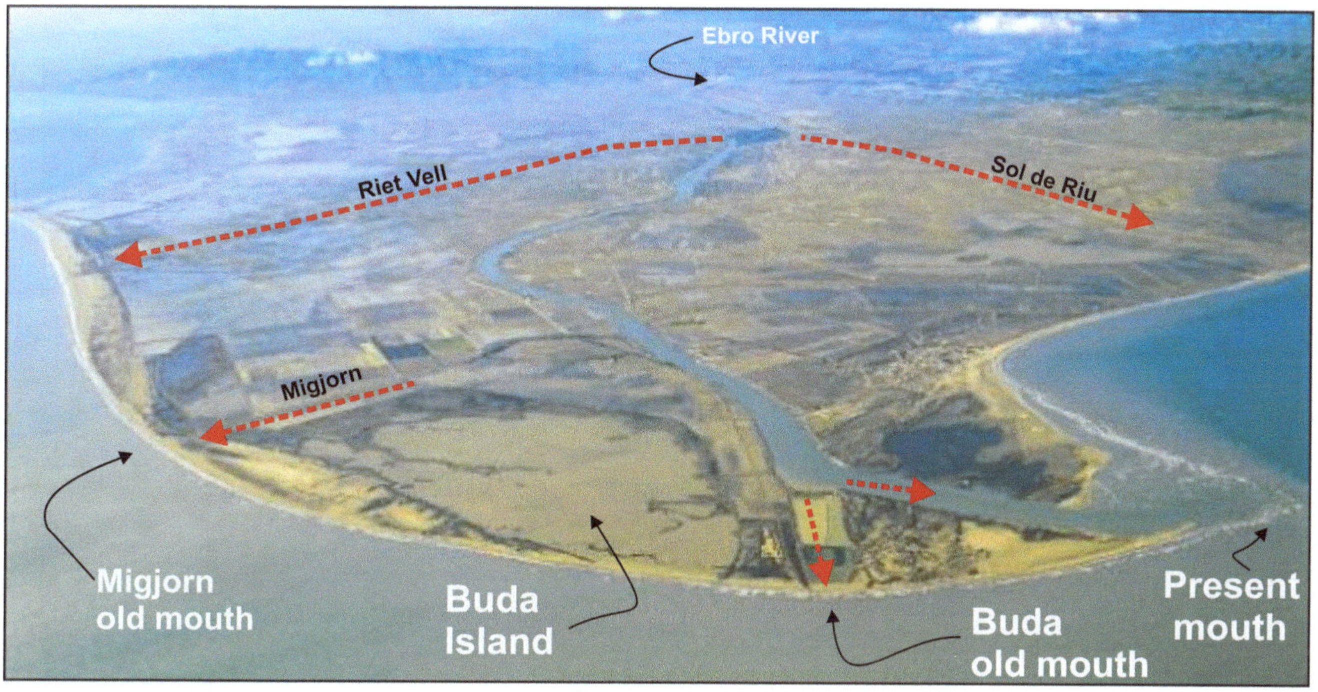

Fig. 18.4 Present-day morphology of the Ebro Delta mouth. To the *left*, the abandoned Buda delta. To the *right*, the present deltaic lobe at the mouth of the Ebro River. *Source* Aeroguía del Litoral. Cataluña (1998)

lagoons: El Fangar spit bar to the NW (Fig. 18.5a) and Los Alfaques spit bar to the SW (Fig. 18.5b). The latter spit is joined to the rest of the delta by the Trabucador bar, approximately 250 m wide and 6 km long (Fig. 18.5b). The mouth shows a very flat submerged profile with a low slope, which favours the development of bars. The current geometry of the Ebro Delta is related to various processes acting directly on it which include, firstly, the retention of the sediment flux by dams built along the river and the increase in water consumption, basically for irrigation and, secondly, marine processes that continuously reshape the delta and alter its configuration (Serra et al. 2012).

The mechanism of accretion of the spits is the annexation of sand bars terminated in a hook which are deposited in the contour of the coast (Fig. 18.6a). The longshore bars have been studied by Guillén and Díaz (1990) and followed and mapped from satellite images (Fig. 18.6b) by Rodríguez and Rodríguez (2012). El Fangar is a 2-km-long sand spit bar with a maximum width of 1.4 km at its centre, where an active dune field is developed (Fig. 18.7a). Information on the morphological characteristics and evolution of this dune field may be found in Rodríguez et al. (2009) and Serra et al. (2012). Los Alfaques spit bar is bigger than El Fangar, with foredunes developed on the coast (Fig. 18.7b).

The mouth has changed its geometry since the second half of twentieth century due to strong erosion. Ramírez et al. (2011) analysed the changes in the different environments that develop in the mouth, showing that an extensive active aeolian mantle has accumulated over the old beach ridges in the mouth. This demonstrates the importance of aeolian processes in the Ebro Delta, related mainly to northerly winds.

18.3.2 Holocene Changes in the Morphology of the Delta

The morphological evolution of the delta during the Holocene is the result of successive accumulation of prograding units at the mouth of the river. These morpho-stratigraphic units advance radially seawards from an avulsion point on the lowest lateral zones of the delta plain (Fig. 18.8). Periods of slight fall in sea level and/or major river floods seem to have an important role in switching the location of deltaic constructions (Fairbridge 1988). In the case of the Ebro Delta, the main periods of delta-lobe switching seem to have a recurrence of thousands of years, whereas the frequency of the morphological modifications of the mouth is of the order of hundreds of years.

Based on borehole logs and radiocarbon ages from peat deposits, a tentative chronology for the formation of the present morphology of the Holocene Ebro Delta has been proposed (Somoza et al. 1998, Fig. 18.8). This chronology is based on the assumption that the main deltaic

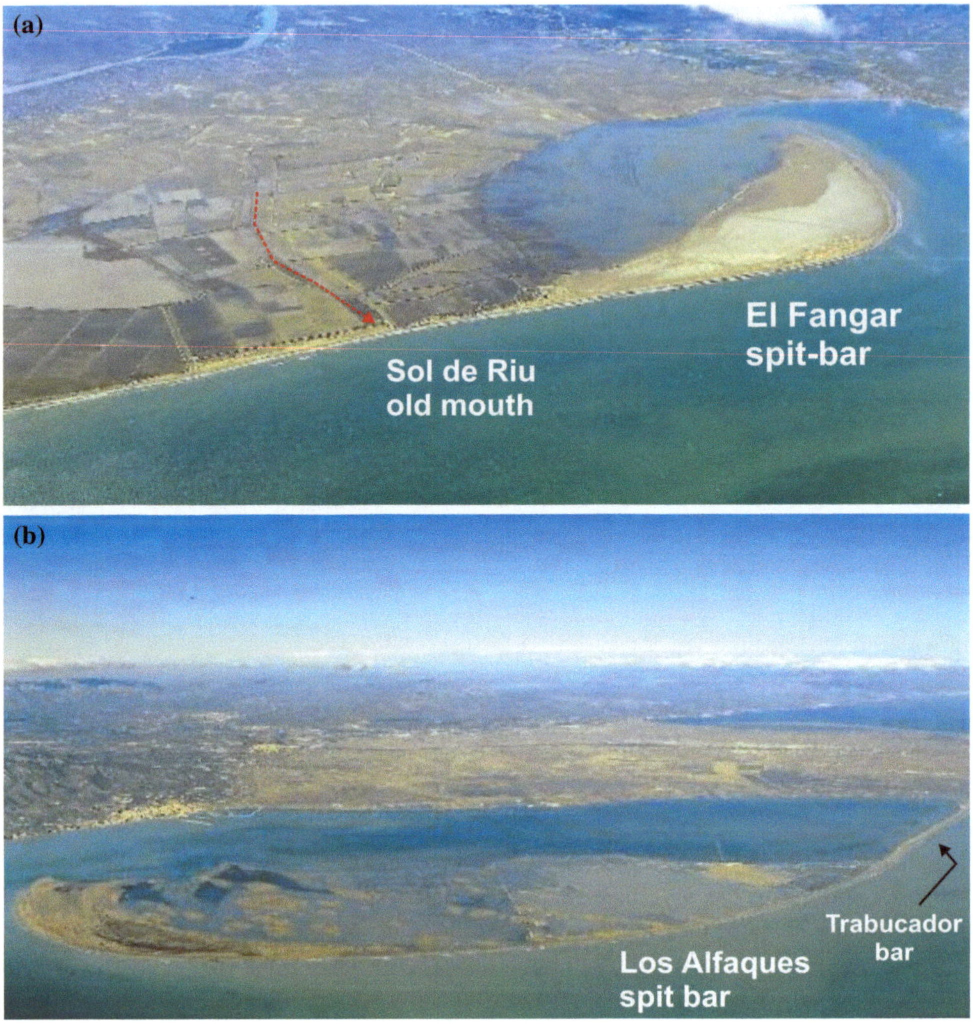

Fig. 18.5 Morphology of the large spit bars of the Ebro Delta: **a** El Fangar spit bar located at the north. **b** Los Alfaques spit bar, linked to the main delta by the El Trabucador bar. *Source* Aeroguía del Litoral. Cataluña (1998)

constructions (d1, d2, d3, d4 and d5) were formed during periods of sea-level stability (stillstands) or slight fall, whereas peat units recorded in drillholes and deposited in inland lagoons are related to flooding during periods of slight sea-level rise. Five major periods of deltaic construction occurred over the last 7,000 years and have been differentiated in the following way (dates in yrs BP correspond to conventional ages): d1 (6150–5350 yrs BP), d2 (4400–3600 yrs BP), d3 (2910 and 2700 yrs BP), d4 and d5. The deltaic construction d4 consists of three historical lobes: the Riet Vell lobe in the south (ca. 1149–1362 AD), the Sol-de-Riu lobe in the north (ca. 1350–1700 AD) and the Migjorn lobe in the centre (ca. 1700 AD). The deltaic construction d5 corresponds to the Buda lobe (active till 1950 AD) and the present active mouth. Subsidence related to sediment compaction allows the stacking and preservation of successive progradational deltaic deposits formed during periods of relative sea-level lowering.

18.3.3 Historical Evolution

The first historical reference to the Ebro Delta is from Posidonio of Apamea (born in 135 BC), who described the delta as a “zone with numerous floods apparently not related to discharges of the river”. Plinio (23–79 yr AD) reported navigation upstream along the Ebro River for more than 320 km in the Roman period. Thus, during Roman times, the coast probably retreated as far as Tortosa, possibly related to a period of sea-level rise that occurred from ca. 2700 yrs BP to 1100 yrs BP (Fig. 18.8, Somoza et al. 1998). In the twelfth century, the coast advanced to Amposta. In the fifteenth century, during the d4 period, the delta prograded significantly, extending over 130 km^2 due to the formation of two main deltaic constructions: the Riet de Vell and Sol-de-Riu lobes (Fig. 18.8). The southern Riet Vell lobe was the main active mouth in 1149 AD, which, according to the conquerors of Tortosa, was abandoned in

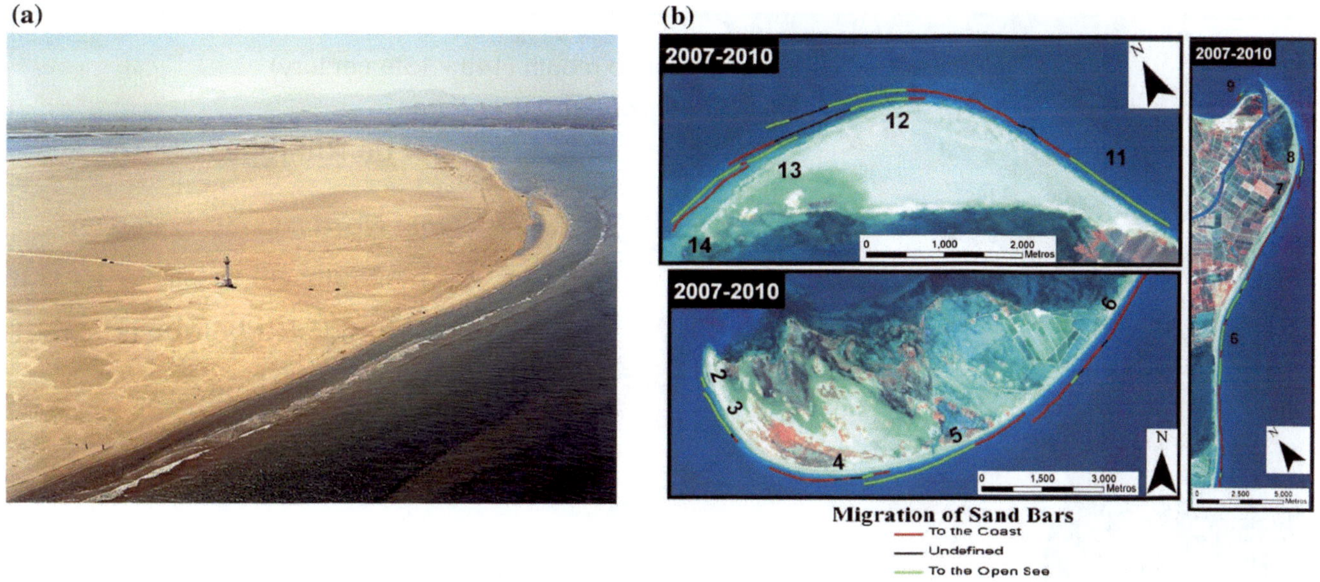

Fig. 18.6 Longshore bars in Ebro Delta. **a** Image of the tip of El Fangar [*Source* Guía de playas, MARM. (MAGRAMA 2013)] showing the process of bars annexation of the emerged area. **b** Dynamic of the longshore *bars* in the deltaic coast (from Rodríguez-Martín and Rodríguez-Santalla 2012)

Fig. 18.7 Dune field in Ebro Delta. **a** Mobile dune field in El Fangar spit. **b** Foredunes in Los Alfaques spits

1362 AD. After this period, it seems that the main mouth of the Ebro Delta moved northward to the Sol-de-Riu lobe. Thus, this mouth was active at least in 1575 AD, as reported during the construction of the now disappeared "Tower of the Sol-de Riu", built up to prevent the entrance of pirates to the Amposta harbour.

Between the beginning of the sixteenth century and the middle of the seventeenth century, the sharp increase in the construction of ships, following the discovery of America, led to a great deforestation in this sector of the Iberian Peninsula. This is considered to be one of the main factors responsible for a new expansion of the delta, as it induced a great increase in the sediment supply feeding the delta mouth. The development of the Migjorn lobe, active around 1700 AD (Fig. 18.8), coincides with this period of deforestation. Another possible explanation for the advance of this lobe is a small sea-level fall between 0.5 and 1 m below the present mean sea level (MSL), closely related to the neoglacial "Little Ice Age" events. It has been proposed that a relative sea-level rise event following this period might have been the cause for the growth of major spit bars during the eighteenth and nineteenth centuries that are observed in the current morphology of the Ebro Delta giving rise to the development of the Los Alfaques spit bar

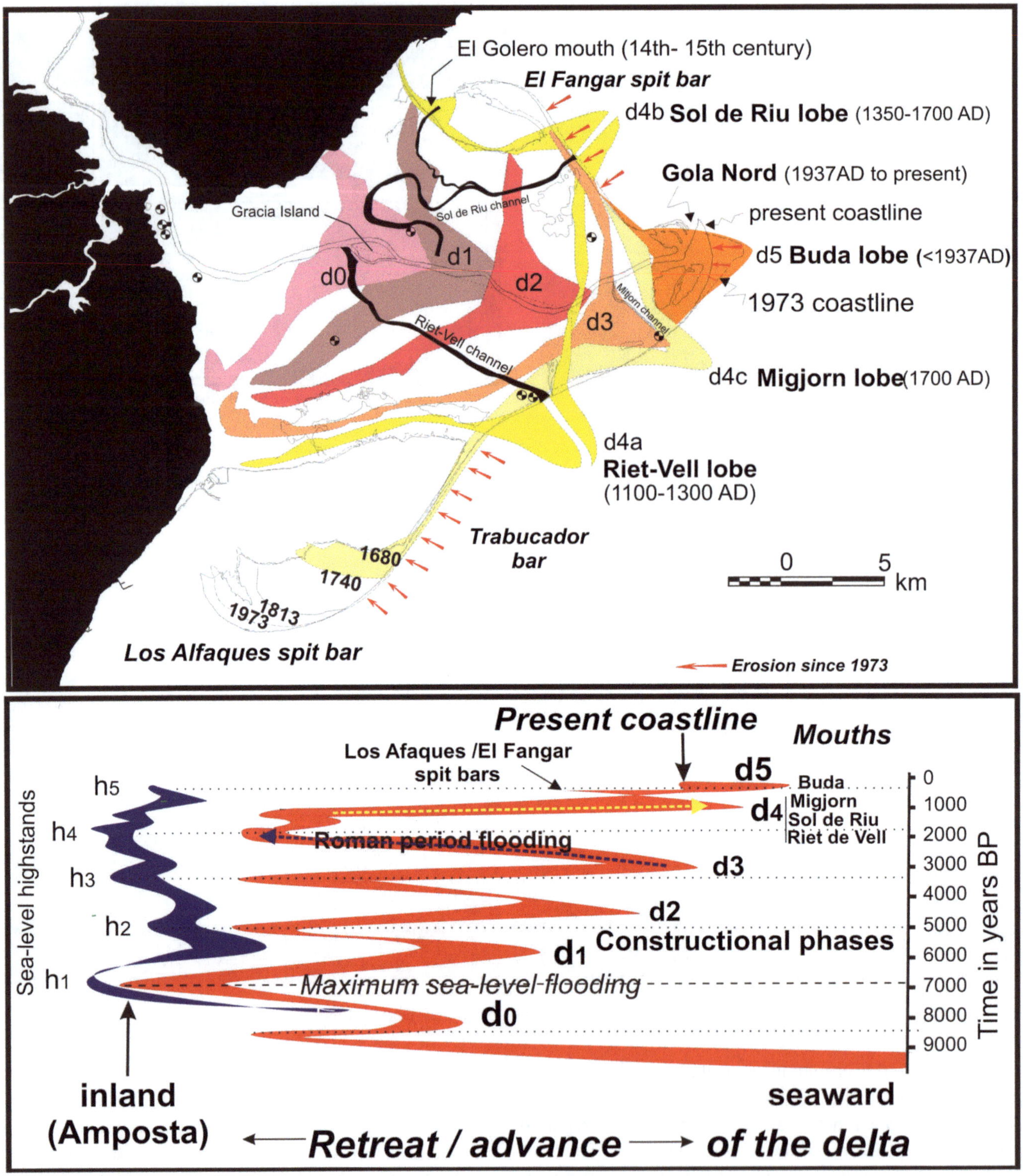

Fig. 18.8 Proposed schematic evolution of the Holocene Ebro Delta. The progradation of delta lobes is associated with sea-level falls, whilst periods of relative sea-level rise induce the retreat of the coastline and flooding of the delta plain (modified from Somoza et al. 1998)

(ca. 1680–1813 AD, Fig. 18.8). On the other hand, the northern El Fangar spit bar was constructed with sediments derived from the destruction of the Sol-de-Riu lobe (Fig. 18.8). A linkage between sea-level rise, erosion of deltaic lobes and formation of spit bars has also been suggested. Thus, warm periods in the eighteenth century,

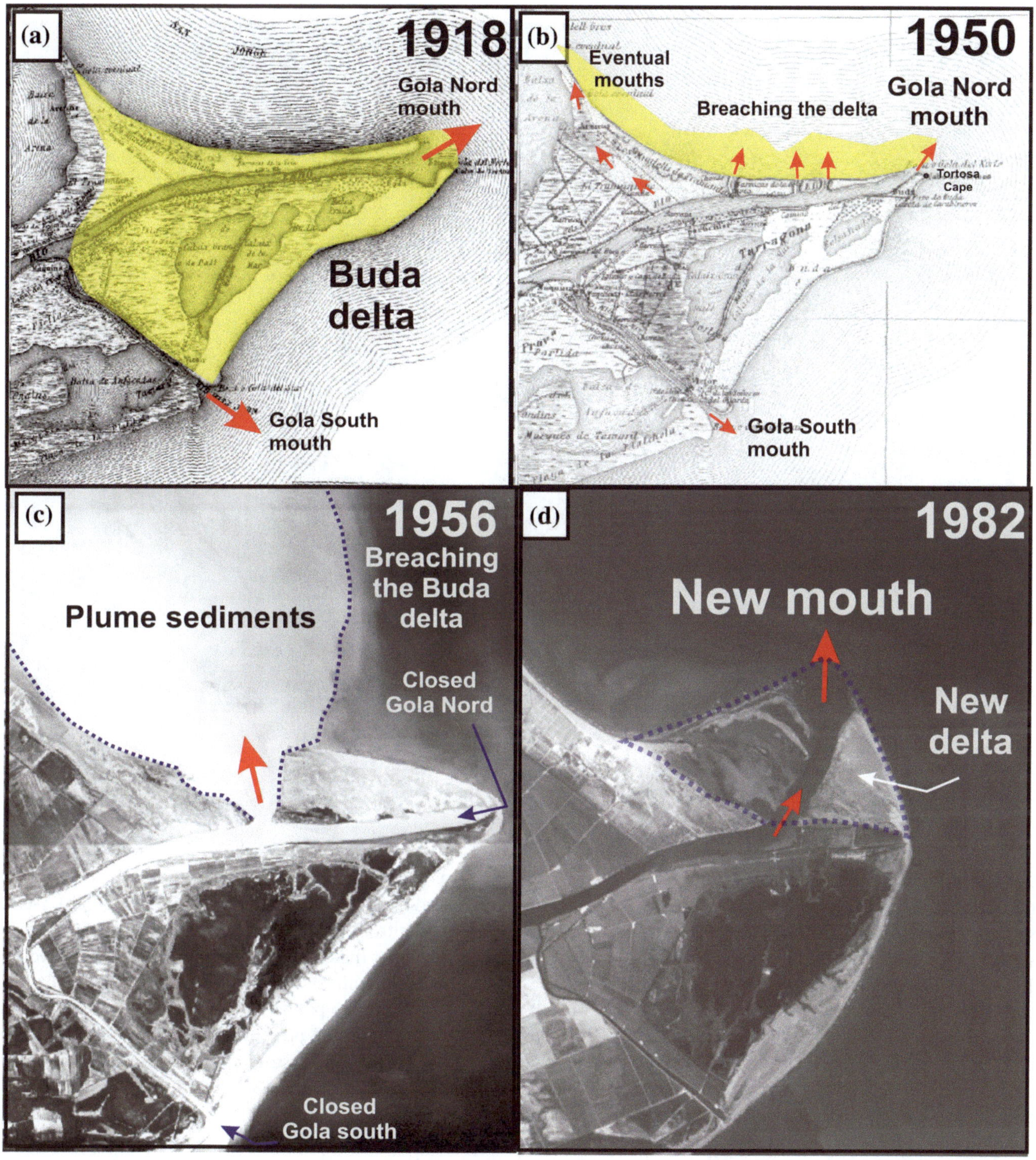

Fig. 18.9 Maps and aerial photographs from different dates illustrating the historical changes in the morphology of the mouth of the Ebro Delta over the last 90 years

related to sunspot maxima of $R_M = 158$ in 1778 AD and $R_M = 141$ in 1788 AD (R_M = maximum amplitude of solar cycle) (Fairbridge 1988), coincide with the main growth period of the Los Alfaques spit bar, which took place between about 1740 and 1813 AD (Fig. 18.8).

18.3.4 Morphological Changes in the Twentieth Century

The evolution of the Ebro Delta over past centuries has shown a relative equilibrium between river-dominated and

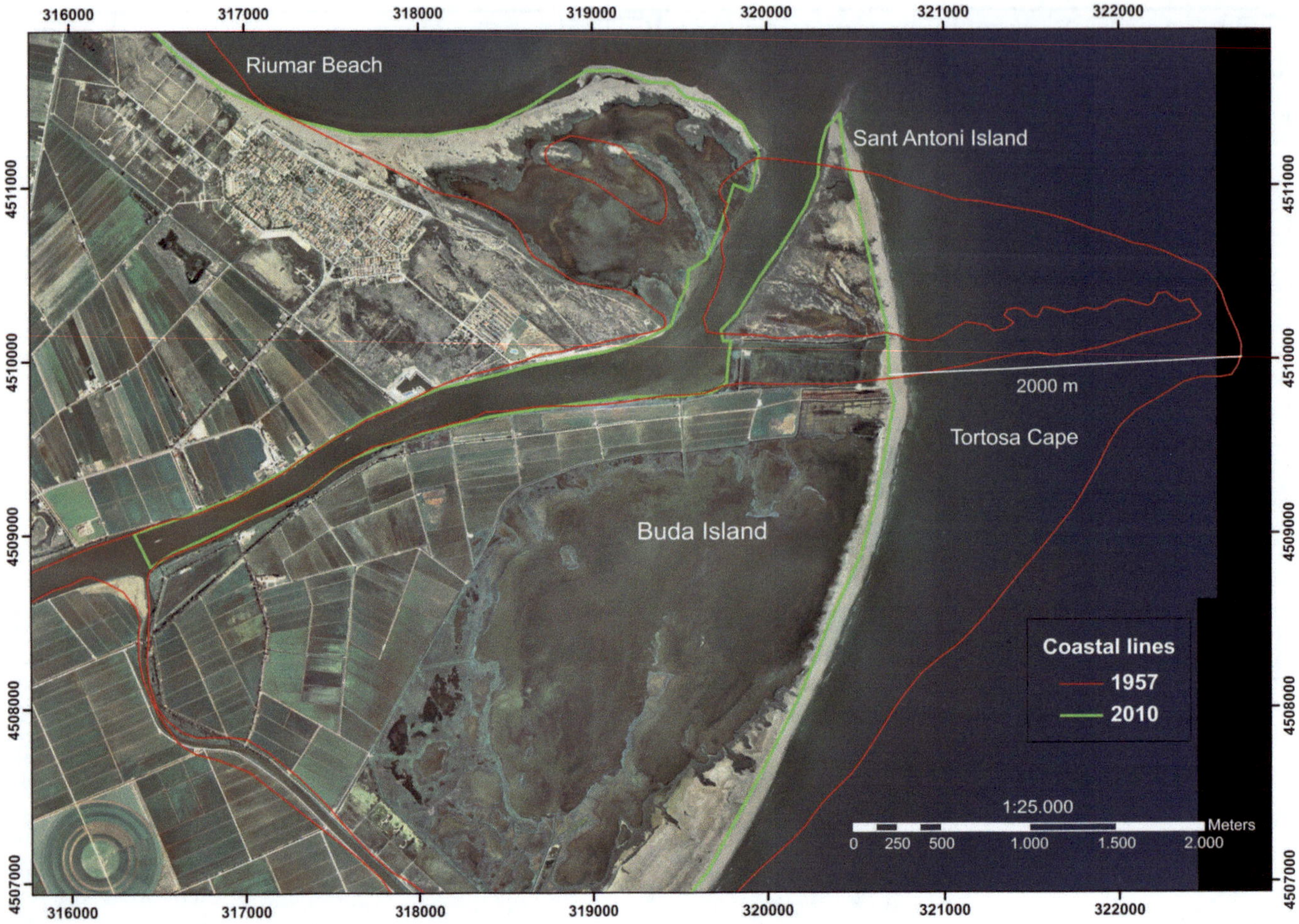

Fig. 18.10 Changes in the coastline of the Ebro Delta over the last 50 years

wave-dominated processes. Before massive dam construction, maximum floods reached 20,000 m^3 s^{-1}, whereas minimum flows of about 50 m^3 s^{-1} were common in summer (Maldonado 1972). The construction of dams started at the beginning of the twentieth century (the first in 1913), but most were built in the period 1940–1975. The construction of the two largest reservoirs in the lower reach of the Ebro River, Mequinenza and Ribarroja, was finished in 1964 and 1969, respectively. Dams have almost entailed the suppression of river flooding over the delta plain and a drastic reduction in sediment feeding the mouth. The last major floods in the lower Ebro took place in 1907 and 1937, with peaks of around 23,000 m^3 s^{-1} in Tortosa (40 km upstream from the mouth). During the recent decades, only small floods (2,000–3,000 m^3 s^{-1}) have occurred from time to time. These last floods were probably the cause of the present configuration of the Ebro Delta mouth. The National Topographic Map from 1918 depicts the mouth of the Ebro River on the eastern side of the pro-delta, named as the "Gola Nord" (Fig. 18.9a). The topographic map from the national series made in 1950 shows several wave washover breaches located on the north side of the Buda lobe, but the Gola Nord remained as the main mouth (Fig. 18.9b). Aerial photographs from 1956 show a huge breach at the north of the Buda lobe, linked to a large sediment plume, and the Gola Nord mouth is clearly closed (Fig. 18.9c). This sequence of images points out to that the switch from the Buda delta lobe to the present-day one took place between 1950 and 1956. The construction of dams along the lower course of the river in the 1960s could be the main factor conditioning the change in the morphology of the delta front. Thus, erosion by wave washovers without the counterbalancing effect of fluvial sediment supply was probably the cause of multiple breaches that facilitated the northward shift of the river mouth. In images taken in 1982, the new lobe displays a characteristic "delta" morphology feed by the present mouth (Figs. 18.9d, 18.10).

Figure 18.11 illustrates the areas of the coast affected by net retreat and accretion processes. The mouth area of the delta front is the zone with greater erosion; 2 km of retreat

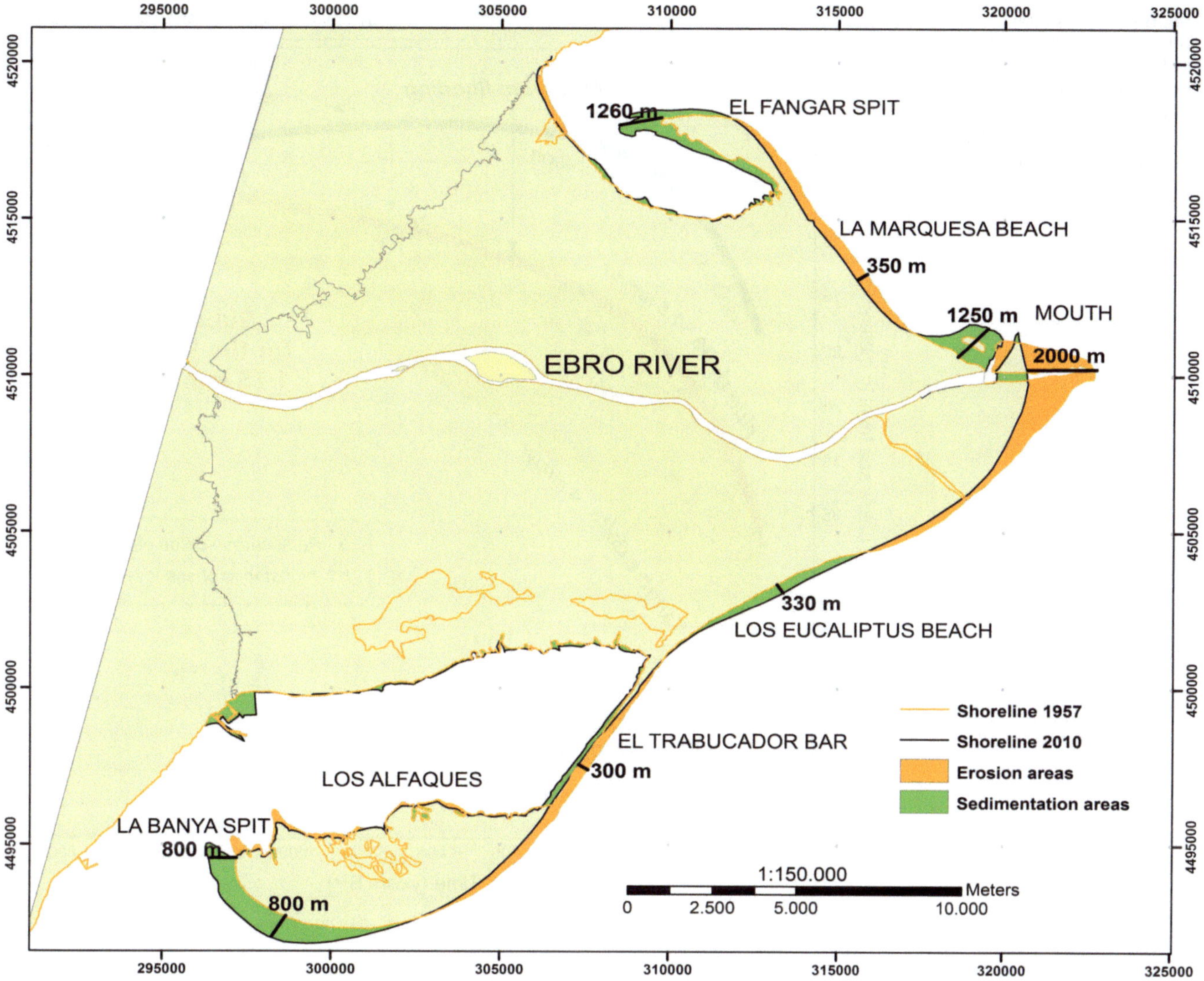

Fig. 18.11 Distribution of erosion and sedimentation areas for the period 1957–2010 (modified from Rodríguez 1999; Rodríguez et al. 2010)

during the last 50 years yields a rate of approximately 40 m/yr (Fig. 18.10). This high erosion rate is related to its position, facing easterly waves, and to the drastic reduction in fluvial sediment transport since the second half of the last century. Conversely, the northern and southern spits record high deposition and rapid progradation of the shoreline. This spatial pattern results from the combined effect of wave and wind action. The longitudinal sediment drift from the old eastern mouth has NW and SW components along the flanks of the delta, and the main winds, coming from the NW, influence the current coastal design and the distinctive configuration and evolutionary behaviour of the two hemi-deltas (Rodríguez et al. 2010).

18.4 The Future of the Delta

Two main factors control the morphological evolution of a delta (Galloway 1975): erosion/progradation of the coast and the relative sea-level changes (eustatic rise due to global warming and local subsidence). For decades, the Ebro Delta has been considered as a closed morpho-sedimentary system (e.g. Maldonado 1972; Jiménez 1996; Rodríguez 1999). However, where sediment cannot be replaced, net erosion and regression of the shoreline occur. In terms of sediment budget, the system should not be considered as stable because factors such as subsidence, submerged areas and cross-shore sediment transport were not taken into account

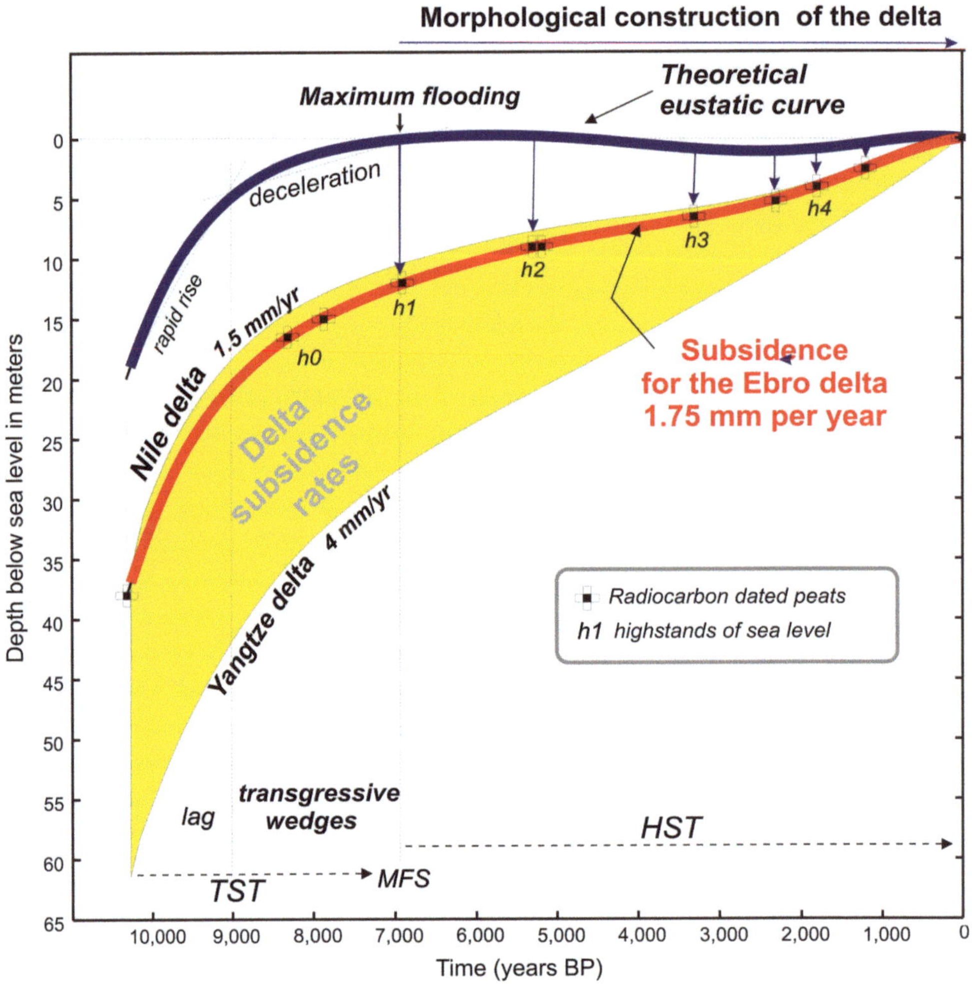

Fig. 18.12 An average subsidence rate of 1.75 mm/yr has been calculated for the Ebro Delta on the basis of the depth of dated buried peat deposits. The rates estimated for other deltas in the world are also shown

(Rodríguez et al. 2010). Several proposals have been made as to how to recover the delta ecosystem and its natural dynamics. A review on the studies and measures may be found in Rodríguez et al. (2010, 2012).

According to Rodríguez (1999), between 1957 and 1998, there was a slight increase in the emerged surface of the delta by about 5 km^2, which represents approximately 1.5 % of the total delta area (~325 km^2). This enlargement of the delta may be related to the increase in the influence of wave transport along the coast as the main factor capable of modifying the coastal morphologies of the delta. Strong erosion of the coastal barriers that protect the delta against the consequences of a relative sea-level rise (global rise and/or local subsidence) is observed today. This threat is illustrated by the breaching of the Trabucador bar in October 1990 due to a storm, which resulted in flooding of around 85 % of the spit bar and erosion of about 70,000 m^3 of sediment (Sánchez-Arcilla et al. 1997). This event led to the beginning of emergency works in January 1991, with the construction of a 1-km-long dune. The dune, 1.5 m high and 12 and 24 m wide at the crest and the base, respectively, was fixed using cane stakes and dune vegetation (*Amophila arenaria*, *Othanthus marítima* and *Elymus factus*).

The other threat is the relative sea-level rise taking into account two components: a possible global warming-related sea-level rise and the progressive subsidence of the delta. Average subsidence rate in the Ebro Delta has been estimated at 1.75 mm/yr during the last 7,000 years (Fig. 18.12, Somoza et al. 1998). This average subsidence rate that falls within the range of other deltas (1–4 mm/yr) is mainly caused by compaction of deltaic sediments, degassing of peats and growth faults developed at the base of deltaic sediments. In the case of deltas, this progressive subsidence must be added to the sea-level rise in a scenario of global warming.

A potential sea-level rise of the Ebro would modify drastically the present morphology of the delta (Fig. 18.13), which will begin with the erosion of the outer sand barriers. Flooding of areas that presently are only 0.75 m above sea level would reduce dramatically the extent of the delta,

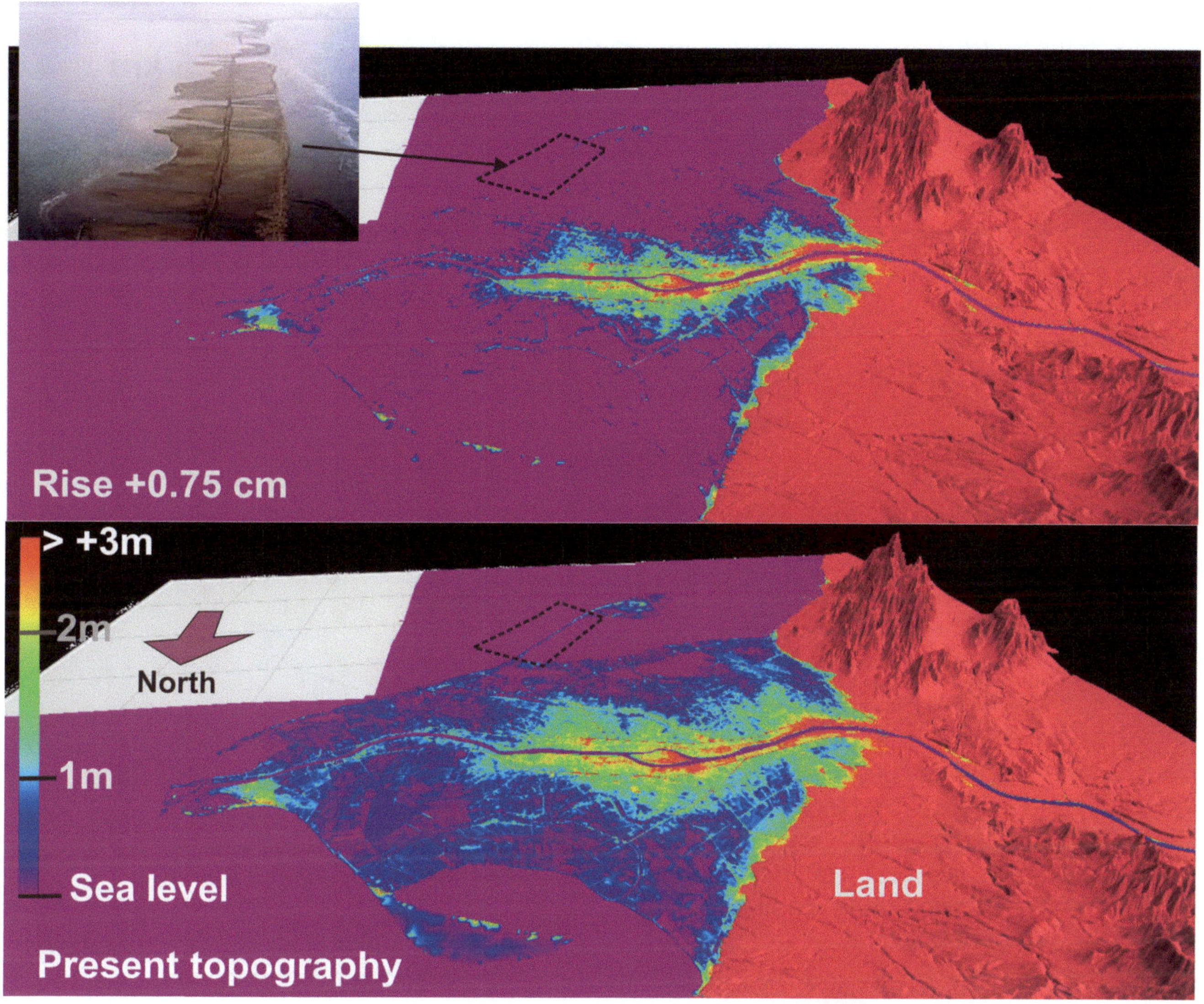

Fig. 18.13 Model of the morphology Ebro Delta after a potential relative sea-level rise of 0.75 m associated with global warming and local subsidence without taking in account coastal adaptation to the marine dynamics. The digital model, with a spatial resolution of 5 m, has been constructed with data from the IGN (Instituto Geográfico Nacional)

producing significant shoreline retreat and damaging the rice fields. The only areas in the delta that would remain emerged would be those formed by the growth of natural levees created by major historical floods of Ebro River (Fig. 18.13), which do not occur nowadays due to the construction of dams. Figure 18.14 shows an estimation of the changes in coastal morphology with various scenarios if a sea-level rise of half a metre occurs (Sánchez-Arcilla et al. 2008).

The potential changes in the Ebro Delta have been analysed considering different scenarios (Sánchez-Arcilla et al. 1996, 2008; Jiménez and Sánchez-Arcilla 1997; Ministerio de Medioambiente 2004; Generalitat de Catalunya 2010; Ibañez et al. 2010; Serra and Roca 2010; Alvarado et al. 2012; Fatoric and Chelleri 2012). All of them agree regarding the devastating effect that an increase in sea level would have on the delta.

The IPCC (2001) recommends several management strategies to reduce the impact of climatic changes in coastal areas: retreat, protection and adaptation. In the case of the Ebro Delta, Galofré (2011) suggests retreat strategies by incorporation of land to public coastal property through acquisition, granting, donation, transfer or by any other means, in order to guarantee and protect the environmental value of the coast, its use and recreation. Alvarado et al. (2012) propose adaptation as a plausible option for managing the delta under an RSLR (relative sea-level rise)

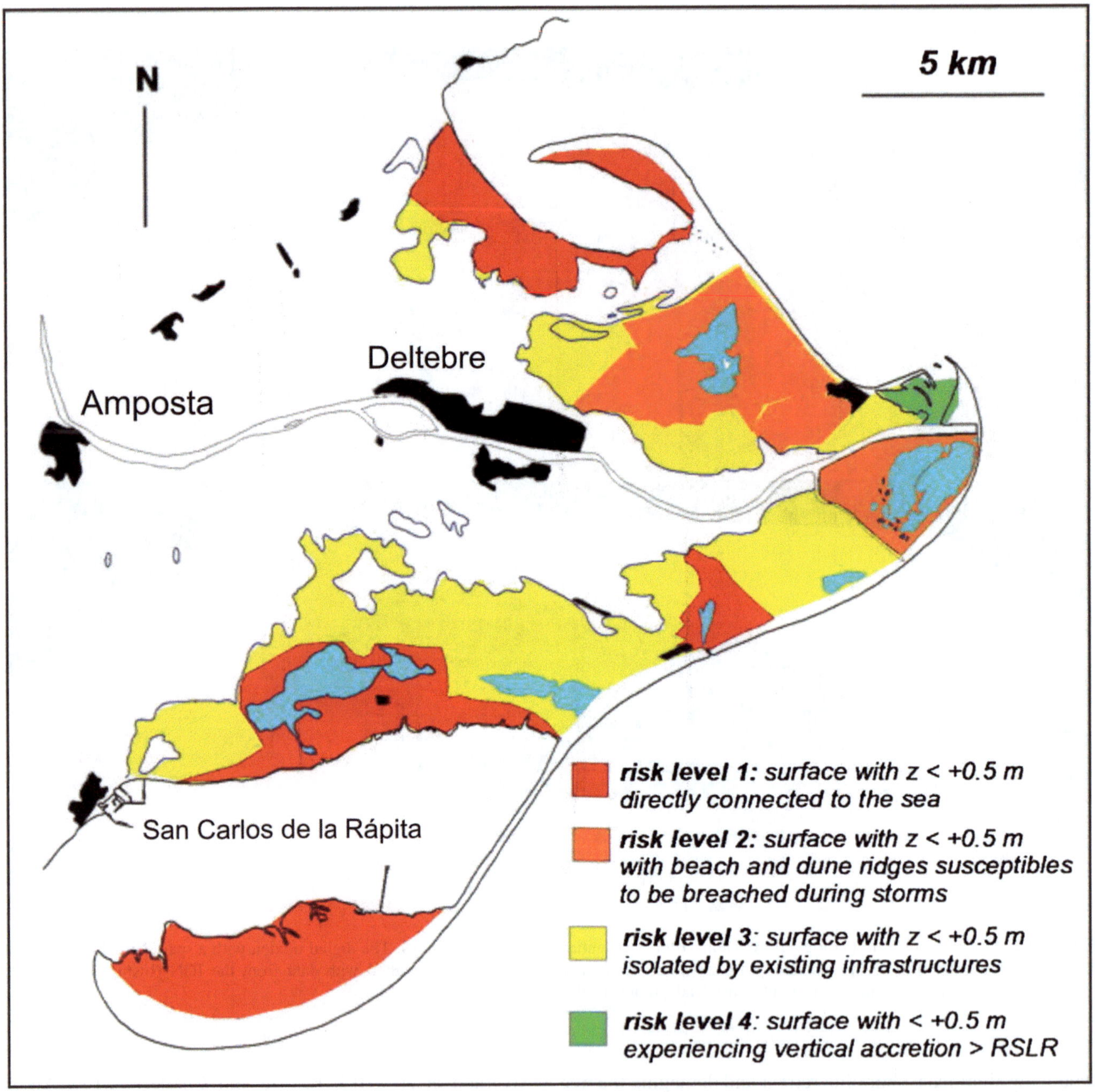

Fig. 18.14 Areas in the Ebro Delta below 0.5 m and vulnerable areas that would be flooded by a sea-level rise of 0.5 m (modified from Sánchez-Arcilla et al. 2008)

scenario, whilst the Ebro River is not playing a significant role in supplying sediment to the deltaic plain. Fatoric and Chelleri (2012) suggest that the most feasible option for adaptation in the Ebro Delta would be preserving the existing dunes and creating new dunes and wetlands. A wide range of prevention and adaptation measures against climate change applicable to the Ebro Delta can be found in Generalitat de Catalunya (2010).

The Ebro Delta represents a special ecosystem from a geomorphological and environmental point of view. Since climate change will produce the reduction in the deltaic surface, it is necessary to monitor the coastal processes, both in the short and medium term, in order to design adequate measures aimed at reducing the detrimental effects on the natural system, as well as on the human activities carried out in the Ebro Delta.

References

Aeroguía del Litoral. Cataluña (1998) Aeroguías Mini. Ed. Geoplaneta, pp 270

Alvarado-Aguilar D, Jimenez JÁ, Nicholls RJ (2012) Flood hazard and damage assessment in the Ebro Delta (NW Mediterranean) to relative sea level rise. Nat Hazards 62:1301–1321

Díaz JI, Nelson CH, Barber JH, Giro S (1990) Late Pleistocene and Holocene sedimentary facies on the Ebro continental shelf. Mar Geol 95:333–352

Fairbridge RW (1988) Mississipi delta-lobe switching during Holocene eustatic fluctuations. Am Assoc Pet Geol Bull 72:183–184

Fatoric S, Chelleri L (2012) Vulnerability to the effects of climate change and adaptation: the case of the Spanish Ebro Delta. Ocean Coast Manage 60:1–10

Galloway WE (1975) Process framework for describing the morphologic and stratigraphic evolution of deltaic depositional systems. In: Broussard ML (ed) Deltas, models for exploration. Houston Geological Society, Houston, pp 87–98

Galofré J (2011) Aplicación de medidas de retranqueo en la costa, mediante criterios geomorfológicos, como herramienta para la gestión integral de zonas costeras. In: Montoya-Montes I, Rodríguez-Santalla I, Sánchez-García MJ (eds) Avances en geomorfología litoral. VI Jornadas de Geomorfología Litoral, Tarragona, pp 7–17

Generalitat de Catalunya (2010) Framework studies for preventing and adapting to climate change in Catalonia. Study N1, Ebro Delta: summary document, 206 pp

Guillén J (1992) Dinámica y balance sedimentario en los ambientes fluviales y litoral del Delta del Ebro. PhD, Instituto de Ciencias del Mar de Barcelona

Guillén J, Díaz JI (1990) Elementos morfológicos en la zona litoral: ejemplos en el delta del Ebro. Sci Mar 54(4):359–373

Ibañez C, Sharpe PJ, Day JW, Day JN, Prat N (2010) Vertical accretion and relative sea level rise in the Ebro Delta Wetlands (Catalonia, Spain). Wetlands 30:979–988

IGME (1987) Contribución de la exploración petrolífera al conocimiento de la geología de España. Publicaciones Instituto Geológico y Minero de España, Madrid

IPCC (2001) IPCC third assessment report: climate change 2001. Working Group II: impacts, adaptation and vulnerability. Cambridge University Press, Cambridge

ITGE (1996) Estudio geológico del Delta del Ebro. Proyecto para la evaluación de la tasa de subsidencia actual. Informes técnicos Instituto Tecnológico y Geominero de España, Madrid

Jiménez J (1996) Evolución costera en el Delta del Ebro. Un proceso a diferentes escalas de tiempo y espacio. PhD, Universidad Politécnica de Cataluña

Jiménez JA, Sánchez-Arcilla A (1997) Physical impacts of climatic change on deltaic coastal systems (II): driving terms. Clim Change 35(1):95–118

MAGRAMA (2013) Guía de playas. Ministerio de Agricultura, Alimentación y Medio Ambiente. http://www.magrama.gob.es/es/costas/servicios/guia-playas/

Maldonado A (1972) El delta del Ebro: estudio sedimentológico y estratigráfico. Boletín de Estratigrafía 1, Universidad de Barcelona

Maldonado A (1975) Sedimentation, stratigraphy, and development of the Ebro Delta, Spain. In: Broussard ML (ed) Delta Models For Exploration. Houston Geological Society, Houston, pp 311–338

Ministerio de Medioambiente (2004) Impactos en la costa española por efecto del cambio climático. Fase II. Evaluación de efectos en la costa española. Ministerio de Medioambiente, Madrid

Ramírez-Cuesta JM, Rodríguez-Santalla I, Sánchez-García MJ, Montoya-Montes I, Romero-Calcerrada R, Gracia-Ramírez FJ (2011) Análisis de las variaciones geomorfológicas ocurridas en la desembocadura del delta del Ebro mediante el empleo de técnicas de detección de cambios (período 1957–2009). In: Montoya-Montes I, Rodríguez-Santalla I, Sánchez-García, MJ (eds) Avances en Geomorfología Litoral. VI Jornadas de Geomorfología Litoral, Tarragona, pp 95–98

Rodríguez I (1999) Evolución geomorfológica del Delta del Ebro y prognosis de su Evolución. PhD thesis, Departamento de Geografía, Universidad de Alcalá de Henares

Rodríguez I, Sánchez MJ, Montoya I, Gómez D, Martín T, Serra J (2009) Internal structure of the aeolian sand dunes of El Fangar spit, Ebro Delta (Tarragona, Spain). Geomorphology 104(3–4):238–252

Rodríguez I, Serra J, Montoya I, Sánchez MJ (2010) The Ebro Delta: from its origin to present uncertainty. In: River deltas: types structures & ecology. Nova Science Publishers Inc., New York, pp 161–171

Rodríguez I, Serra J, Montoya I, Sánchez MJ (2012) El Delta del Ebro. Características dinámicas y ambientales y propuestas para su protección. In: Montoya I, Sánchez MJ, Rodríguez I (eds) El litoral Tarraconense. JMC Ofimática S.L, Barcelona

Rodríguez-Martín R, Rodríguez-Santalla I (2012) Detection of Submerged Sand Bars in the Ebro Delta using ASTER Images. In: Huang Y, Wu F (eds) New Frontiers in Engineering Geology and the Environment. Springer, Berlin, pp 103–106

Sánchez-Arcilla A, Jiménez JA, Stive MJF, Ibañez C, Pratt N, Day JW, Capobianco M (1996) Impacts of sea-level rise on the Ebro Delta: a first approach. Ocean Coast Manage 30(2–3):197–216

Sánchez-Arcilla A, Jiménez JA, Gelonch G, Nieto J (1997) El problema erosivo en el Delta del Ebro. Revista de Obras Públicas 3(368):23–32

Sánchez-Arcilla A, Jiménez JA, Valdemoro H, Gracia V (2008) Implications of climatic change on Spanish Mediterranean low-lying coasts: the Ebro Delta case. J Coast Res 24(2):306–316

Serra J (1997) El sistema sedimentario del Delta del Ebro. Revista de Obras Públicas 3:15–22

Serra J, Roca E (2010) L'equilibri dels sistemes deltaics i els efectes del canvi climàtic. Noves oportunitats de gestió. In "El sistema litoral, un equilibri feble amenaçat". Diputación de Barcelona

Serra J, Rodríguez I, Sánchez MJ, Montoya I (2012) Delta del Ebro: papel del sistema dunar frente a la regresión deltaica (actuaciones y medidas paliativas). In: Rodríguez-Perea A, Roig X, Pons GX, Martín JA (eds) La gestión integrada de playas y dunas: experiencias en Latinoamérica y Europa. Mon Soc Hist Nat Balears 19:237–242

Somoza L, Barnolas A, Arasa A, Maestro A, Rees JG, Hernandez-Molina FJ (1998) Architectural stacking patterns of the Ebro delta controlled by Holocene high-frequency eustatic fluctuations, delta-lobe switching and subsidence processes. Sed Geol 117:11–32

Stanley DJ, Warne AG (1994) Worlwide initiation of Holocene marine deltas by deceleration of sea-level rise. Science 265:228–231

Wright LD, Coleman JM (1973) Variations in morphology of major river deltas as functions of ocean wave and river discharge regimes. AAPG Bull 57:370–398

Coastal Dunes and Marshes in Doñana National Park

19

Joaquín Rodríguez-Vidal, Teresa Bardají, Cari Zazo, José L. Goy, Francisco Borja, Cristino J. Dabrio, Javier Lario, Luis M. Cáceres, Francisco Ruiz, and Manuel Abad

Abstract

Doñana Natural Park is a good global example of the sedimentary filling of a broad tidal estuary during the Mid-Late Holocene, after the last postglacial sea-level rise. The timing of this rise is not well defined yet in the Gulf of Cádiz, since the oldest evidence of coastal sedimentation, located at the right bank of the mouth of the old Guadalquivir Estuary, dates back to ca. 5,000 years ago. The first evolutionary stages of the embayment indicate an obvious marine influence, dominated by waves and storms from the SW. Since ca. 4,000 years ago, protection provided by the growing coastal spit barrier of Doñana favored the development of a sheltered marsh dominated by tides and fluvial currents. About 2,200 years ago, since the time Romans controlled the area, the estuary was dominated by marshlands with a wide lagoon at its mouth (Lacus Ligustinus), and the current landscape of Doñana started to form. The evolution of the last 2,000 years includes the quick and continuous growth of coastal barriers by longshore drift, the origin of the present-day marshland landscape and the development of dune fields migrating inland towards the wetlands.

Keywords

Coastal dunes • Spit bar • Marshland • Holocene • Guadalquivir Basin

J. Rodríguez-Vidal (✉) · L. M. Cáceres · F. Ruiz · M. Abad
Departamento de Geodinámica y Paleontología, Universidad de Huelva, 21071, Huelva, Spain
e-mail: jrvidal@dgeo.uhu.es

L. M. Cáceres
e-mail: mcaceres@uhu.es

F. Ruiz
e-mail: ruizmu@uhu.es

M. Abad
e-mail: manuel.abad@dgyp.uhu.es

T. Bardají
Unidad Docente de Geología, Universidad de Alcalá, 28871, Alcalá de Henares, Spain
e-mail: teresa.bardaji@uah.es

C. Zazo
Departamento de Geología, Museo Nacional de Ciencias Naturales (CSIC), 28006, Madrid, Spain
e-mail: mncnczc65@mncn.csic.es

J. L. Goy
Departamento de Geología, Universidad de Salamanca, 37008, Salamanca, Spain
e-mail: joselgoy@usal.es

F. Borja
Área de Geografía Física, Universidad de Huelva, 21007, Huelva, Spain
e-mail: fborja@dgf.uhu.es

C. J. Dabrio
Departamento de Estratigrafía, Universidad Complutense, 28040, Madrid, Spain
e-mail: dabrio@geo.ucm.es

J. Lario
Departamento de Ciencias Analíticas, UNED, 28040, Madrid, Spain
e-mail: javier.lario@ccia.uned.es

F. Gutiérrez and M. Gutiérrez (eds.), *Landscapes and Landforms of Spain*, World Geomorphological Landscapes,
DOI: 10.1007/978-94-017-8628-7_19, © Springer Science+Business Media Dordrecht 2014

19.1 Introduction

Doñana Natural and National Parks constitute one of the most outstanding natural protected areas in the Iberian Peninsula, as well as one of the widest lowlands. Wetlands, marshlands and dune fields cover around 1,100 km^2 within Huelva, Seville and Cádiz provinces in Andalucía. The Doñana National Park covers more than 54,720 ha, and its buffer protection area, which constitutes the Natural Park, around 54,250 ha. The natural values of this privileged landscape have not been ignored by national or international bodies. It was declared Nature Reserve by the WWF in 1963, National Park in 1969, Biosphere Reserve by the UNESCO in 1980, and UNESCO World Heritage Site in 1994.

The present-day Doñana landscape is a clear outcome of the geological evolution since Late Miocene times (Fig. 19.1), and mainly since the Last Glacial Maximum (LGM). The postglacial sea-level rise, which reached its maximum at ca. 6,500 years BP, caused the drowning of the lower Guadalquivir River valley. Since then, littoral dynamics has largely controlled the development of the present landscape. The growth of littoral spit bar systems at the mouth of the Guadalquivir Estuary induced the creation of protected marshland areas and the formation of dune systems on top of them. Detailed analyses of spit bars, dune systems and marshlands allow the reconstruction of the geomorphological evolution of this particular landscape in south-western Iberian Peninsula.

19.2 Geographical and Geological Setting

Doñana National Park is located in the SW littoral zone of the Iberian Peninsula, including the lowermost reach of the Guadalquivir River. The high biodiversity found in different ecosystems make this landscape an exceptional natural environment.

The present-day climatic conditions, and particularly the strong local winds, play an instrumental role in the morphogenetic processes. Climate can be defined as Mediterranean with Atlantic influence, with a mean annual rainfall below 600 mm (two maxima in late autumn and spring) and an average annual temperature of around 18 °C (summer maximum can reach 35–40 °C). Prevailing winds blow from the SW. Winds from the SE and E are less common but also important since they are commonly associated with storm events (Rodríguez-Ramírez 1998).

Regarding the geology, Doñana is located at the seaward edge of the Guadalquivir Cenozoic Basin (Fig. 19.1), which constitutes the foreland basin of the Betic Cordillera (Sanz de Galdeano and Vera 1992). The northern passive margin of this basin corresponds to the Variscan Massif, against which the collisional Alpine orogenic belt of the Betic Cordillera develops (Fig. 19.1a).

The sedimentary marine fill of the basin has been dated on the basis of microfaunal assemblages (Sierro et al. 1996) as Late Miocene–Early Pliocene. The palaeogeographic reconstruction (Fig. 19.1b, c, and d) based on sedimentary and palaeontological analyses shows a progressive shallowing of the basin, recorded by a stratigraphic succession ranging from Late Miocene inner platform bioclastic sandstones to Pliocene wave- and tide-dominated delta deposits. The Pleistocene evolution is characterized by a progressive estuarine fill and major changes in the course of the main rivers (Gutiérrez-Elorza 2002).

The ca. 100–120 m sea-level drop during the LGM induced the incision of wide valleys by the main rivers, which were subsequently flooded during the Holocene sea-level rise, becoming wide estuaries (Zazo et al. 1999; Dabrio et al. 2000). At present, most estuaries are silted up and spit bars and marshes have developed at the mouth of the main rivers (Fig. 19.2).

19.3 Geomorphology of Doñana National Park

Doñana National Park exhibits a series of morphosedimentary units related to the progressive silting-up of the Guadalquivir Estuary after the maximum postglacial sea-level rise at ca. 6,500 cal year BP (Benavente et al. 2005). Initial vertical accretion in the estuaries changed after ca. 2,700 cal year BP into growing progradational spit bars and associated dune systems, resulting in the present-day morphology dominated by low-lying marshlands at the mouth of the main river valleys (Rodríguez-Ramírez et al. 2005). However, the understanding of this recent evolution requires a good knowledge of the Late Pleistocene history of the lowest Guadalquivir Basin, which constitutes the initial evolutionary stage of the geomorphological configuration of Doñana.

19.3.1 El Abalario Dune Field

El Abalario area (Fig. 19.2) displays a NW–SE trending elongated domal morphology that separates the lower Cenozoic Guadalquivir Basin from the Atlantic Ocean, and outlines the headland of the spit bar system whose growth promoted the progressive closure of the Guadalquivir Estuary.

This dome structure was caused by a progressive upwarping of underlying Pliocene–Pleistocene prograding delta sediments, produced by the large NW–SE Torre del

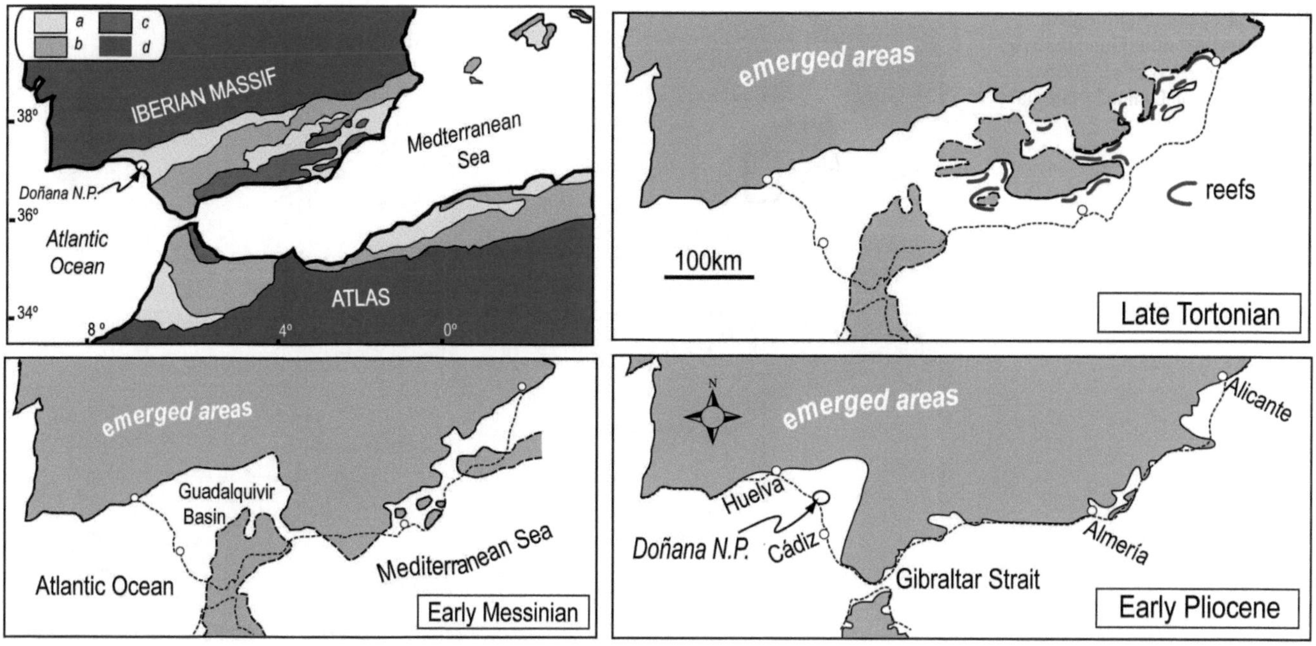

Fig. 19.1 **a** Location of the Doñana National Park within the Betic Cordillera (modified from Azañón et al. 2002). Palaeogeography of the area, **b** Late Tortonian, **c** Early Messinian, and **d** Early Pliocene (modified from Alonso-Zarza et al. 2002)

Loro gravitational fault (Fig. 19.2) (Zazo et al. 2005a). Sedimentary units cropping out along the Asperillo cliff (Fig. 19.2) record the complex evolution of the area (Zazo et al. 2005a, 2008), controlled by the interaction of littoral processes and sea-level changes in an emergent coastal plain. Upwarping at the end of Marine Isotope Stage (MIS)-5 produced a WNW–ESE normal fault (Torre del Loro Fault, Fig. 19.2), along which gravitational displacement created an upthrown block to the north and a downthrown block to the south, generating a faulted sea-cliff. Three aeolian units separated by widespread weathering surfaces and palaeosoils (U1 to U3 in Zazo et al. 1999, 2005a) accumulated in the downthrown block against the fault scarp, between the late MIS-5, under a falling sea-level scenario, and the Last Deglaciation (chronology based on radiocarbon and OSL dating, Zazo et al. 2005a). The onset of moist and temperate climate during the Holocene Climatic Optimum (9.0–6.5 cal ky BP) induced the development of a wide flat erosional surface with an iron-rich crust, at present partially destroyed. This surface is easily recognizable along the Asperillo cliff (Fig. 19.3) (Zazo et al. 2005a, b, 2008). The development of at least four younger semi-mobile and mobile dune systems on top of this surface (U4–U7, Zazo et al. 1999, 2005a) by W-SW winds has been interpreted as the result of a general aridity trend since ca. 5 cal ky BP, also described in other localities from the Southern Iberian Peninsula (Zazo et al. 1994; Santos et al. 2003). The lithic workshop level (Late Neolithic–Calcolithic; ca. 5,000 cal ky BP after Martín de la Cruz et al. 2000) found on top of this surface and coeval to the first aeolian unit (U4 in Zazo et al. 2005a) supports this chronology (Borja et al. 1999; Zazo et al. 2005b).

Radiocarbon dating on charcoal found with the U5 aeolian unit indicates an age of ca. 2.7 cal ky BP for this unit, coinciding with an arid period described in the region and an increase in coastal progradation (Zazo et al. 1999; Goy et al. 2003). Archaeological remains associated with this dune system allow one to infer that sedimentation continued until Roman and Medieval times. The more recent aeolian unit U6 seems to be coeval with the 16th–17th centuries coastal watch-towers (Borja et al. 1999), and the accumulation of the dunes corresponding to the youngest unit described in this area (U7 in Zazo et al. 2005a) began in the 17th century, but they are still active under winds blowing from the SW.

19.3.2 The Doñana Dune Systems

Although the dune systems reach the maximum development at El Abalario, the Doñana National Park includes the largest active dune field in Europe. Five dune sequences (Fig. 19.4), that can be partially correlated with those described in El Abalario area, have been differentiated by means of geomorphological studies (Rodríguez-Vidal et al. 1993; Rodríguez-Ramírez et al. 1996, 2005) (Fig. 19.4).

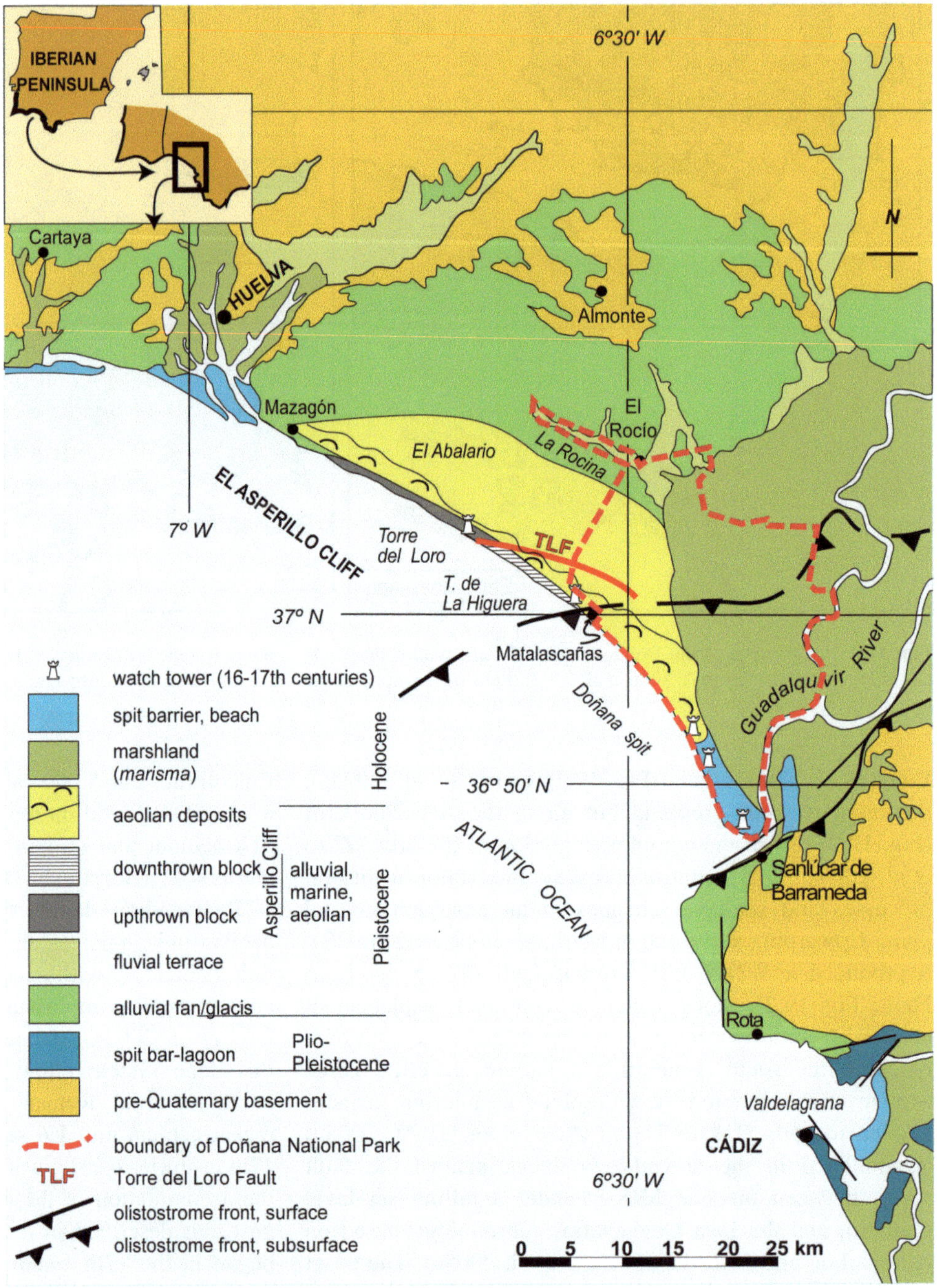

Fig. 19.2 Geomorphological map of the Atlantic Guadalquivir Basin (after Zazo et al. 2005a) and location of Doñana National Park

Stabilized Dune Systems

The first dune system described in the Doñana National Park (System I, after Rodríguez-Ramírez et al. 1996) covers the previous topography and includes some scattered and scarcely developed parabolic dunes generated by WSW winds. The correlation of this system with U2 dunes (after Zazo et al. 1999, 2005a) supports an age between 31 and 18 ky BP.

System II is represented by extensive dunes, mainly with long parabolic shapes that in some places grade into transverse dunes with undulating crests (Rodríguez-Ramírez et al. 2005; Rodríguez-Ramírez 2011). These dunes are usually less than 100 m wide, with lengths that can reach several kilometres, and heights up to 80 m. Their formation is mainly related to westerly winds and their age has been estimated for between 14 and 11.5 ky BP by correlation with the U3 system of Zazo et al. (1999).

System III marks a change in prevailing wind direction that turns again into WSW. Dunes show a parabolic shape, whose lateral coalescence results in an undulating

Fig. 19.3 Aeolian units *U2* and *U3* in El Asperillo cliff (photo J.L. Goy). Organic-rich beds within *U3* are related to humid conditions during the last deglaciation. Upper surface corresponds to an iron-rich palaeosoil generated during the Holocene climatic optimum (Zazo et al. 2005a)

transverse morphology. Their age, considered to be correlative to U4 dunes after Zazo et al. (1999), is constrained by the ages of the immediately older (11 ky BP) and younger (5–4 ky BP) systems.

Active Dune Systems

Two active dune systems (Systems IV and V, Fig. 19.4) overlapping the previous ones and migrating across the Doñana marshlands have been described (Rodríguez-Vidal et al. 1993; Rodríguez-Ramírez et al. 2005; Rodríguez-Ramírez 2011).

System IV comprises sparse and small parabolic dunes, partially trapped by vegetation, moving towards the NE. The migration of these parabolic dunes leaves a kind of elongated sand tracks in the interdune depressions, locally called "worms" (Fig. 19.5). The trapping effect of vegetation, together with a high water-table during periods of limited dune advance, have been suggested as the primary causes of these "worms" or sand tracks (Rodríguez-Ramírez et al. 2005). Archaeological remains from Late Neolithic/Calcolithic and Roman times are covered by these dunes, which also include post-Medieval ceramics (Borja et al. 1999), thus suggesting an age between the 14th and 17th centuries, like the U6 unit described by Zazo et al. (1999, 2005a).

System V is the most recent and active aeolian system in Doñana. It comprises large NW-migrating transverse dunes (Rodríguez-Ramírez 2011) which can be correlated with the U7 unit of Zazo et al. (2005a). These dunes partially cover the 17th century coastal watch-towers, and they occur associated with the beach ridges that started to prograde immediately after the construction of these towers (Borja et al. 1999). These relationships suggest that they started to form around the end of 17th century, continuing their activity under prevailing SW winds until the present time (Fig. 19.6).

19.3.3 Littoral Spit Barrier

The Holocene geological record of this littoral Atlantic region has received increasing attention in recent years. Five morphosedimentary units separated by erosional surfaces have been distinguished and dated (Zazo et al. 1994; Rodríguez-Ramírez et al. 1996): H1: 6,900–4,500 cal year BP, H2: 4,200–2,600 cal year BP, H3: 2,300–1,100 cal year BP, and H4: 1,000 cal year BP–Present. They represent the subaerial record of the Mid-Late Holocene coastal wave action and the dominant southeastward drift current.

The great sand barrier of Doñana grew from northwest to southeast, starting at the western bank of the Guadalquivir Estuary outlet (Fig. 19.2). The landward growing dune field covers a wide area of this spit barrier and its oldest outcropping beach ridge has been AMS-dated for about 5,000 cal year BP (Carrizosa-Vetalarena). Its morphological arrangement, perpendicular to the WNW–ESE direction of Doñana spit barrier, together with the small size, are evidence of a short-lived but highly intense erosional event (Ruiz et al. 2005) that destroyed an earlier spit formation (H1 of Zazo et al. 1994).

Fig. 19.4 Dune systems (I–V, after Rodríguez-Ramírez 1998) and spit bars (*H3* and *H4*, La Marismilla and San Jacinto) in Doñana National Park (Google Earth image: 2013 Digital Globe and 2013 Instituto de Cartografía de Andalucía)

The second progradation unit (4,200–2,600 cal year BP) isolated the Guadalquivir Estuary from the sea and several short rivers filled the inner shore with digitated deltas. This is the time when marshes were first protected by the early Doñana spit barrier. The prevalence of fluvial activity caused a marked biological crisis on the previous marine fauna, and frequent accumulation of organic remains (shell beds) with ridge morphology.

During the H2 to H3 transition the mouth of the Guadalquivir Estuary—the ancient Roman *Lacus Ligustinus*—suffered the action of a great tsunami event (218–209 BC) that substantially changed the coastal landscape and produced new geomorphic features (Rodríguez-Vidal et al. 2011). The earthquake had probably a similar or greater magnitude (Mw $\approx$ 8.5) than the AD 1755 event (Great Lisbon quake), also recorded in this Atlantic coast (Ruiz et al. 2013). The erosion of the tsunamigenic waves focused in the littoral spits, in which morphological evidence remains, including cliffs and incisions. The spit barrier of La Algaida became an island, and the pre-Roman human settlements were abandoned. After that, the eroded foredunes migrated inland, in the form of transgressive blown sand sheets. The marine bioclastic sand, brought into the estuary, was accumulated in the lagoon margins, leading the development of the estuarine sandy ridges of Vetalengua and Las Nuevas.

The late progradation units (H3 and H4) produced a considerable growth of beach and dune reaches, the retreat of cliffs, and the gradual fill of the Guadalquivir Estuary. In the Doñana spit barrier these phases are recorded by the

Fig. 19.5 Active dune systems in Doñana National Park. System V (foreground) is overlapping system IV (photo by L. Menanteau)

Fig. 19.6 Active aeolian dunes surrounding the 17th century coastal watch-tower in Doñana coast (photo by M. Abad)

strands of La Marismilla (H3) and San Jacinto (H4) (Fig. 19.4).

Today, the longshore drift is becoming stronger, and erosion is increasing at the end of San Jacinto strand. Hooks with NE orientation are being formed, encroaching into the Guadalquivir channel (Rodríguez-Ramírez et al. 1996).

19.3.4 Marshland

The Doñana marshlands are located in the area once occupied by the former Guadalquivir Estuary, forming a vast silt–clay plain of approximately 2000 km^2. The National Park also includes the entire fluvio-tidal complex. Although it is essentially a natural marsh, anthropogenic actions have modified the dynamics of some fluvial channels that supply it with fresh water. This environment is characterized by a monotonous plain at less than 3 m a.s.l. in the entire area, with small topographic irregularities (Fig. 19.7). Old levees stand out at the banks of the main tidal and fluvial channels, which isolate them from the rest of the plain and constitute the most elevated sectors of the marsh. These levees retain the fresh water of the rain, which floods the lowest zones, isolating the marshland from the sea. The degraded levees form rounded and elongated ridges locally known as *vetas*. The low-lying zones that remain flooded for longer periods throughout the year are called *lucios*, which constitute, together with the semi-blocked fluvio-tidal channels (*caños*), the low marsh (Rodríguez-Ramírez 1998).

Furthermore, within the silt–clay plain, there are some rather continuous long ridges, made up of sandy or shelly deposits, corresponding to relict beaches (Rodríguez-Ramírez and Yáñez-Camacho 2008). These are thin and narrow accumulations generated either by intense erosional events such as storms or tsunamis, or as a result of the

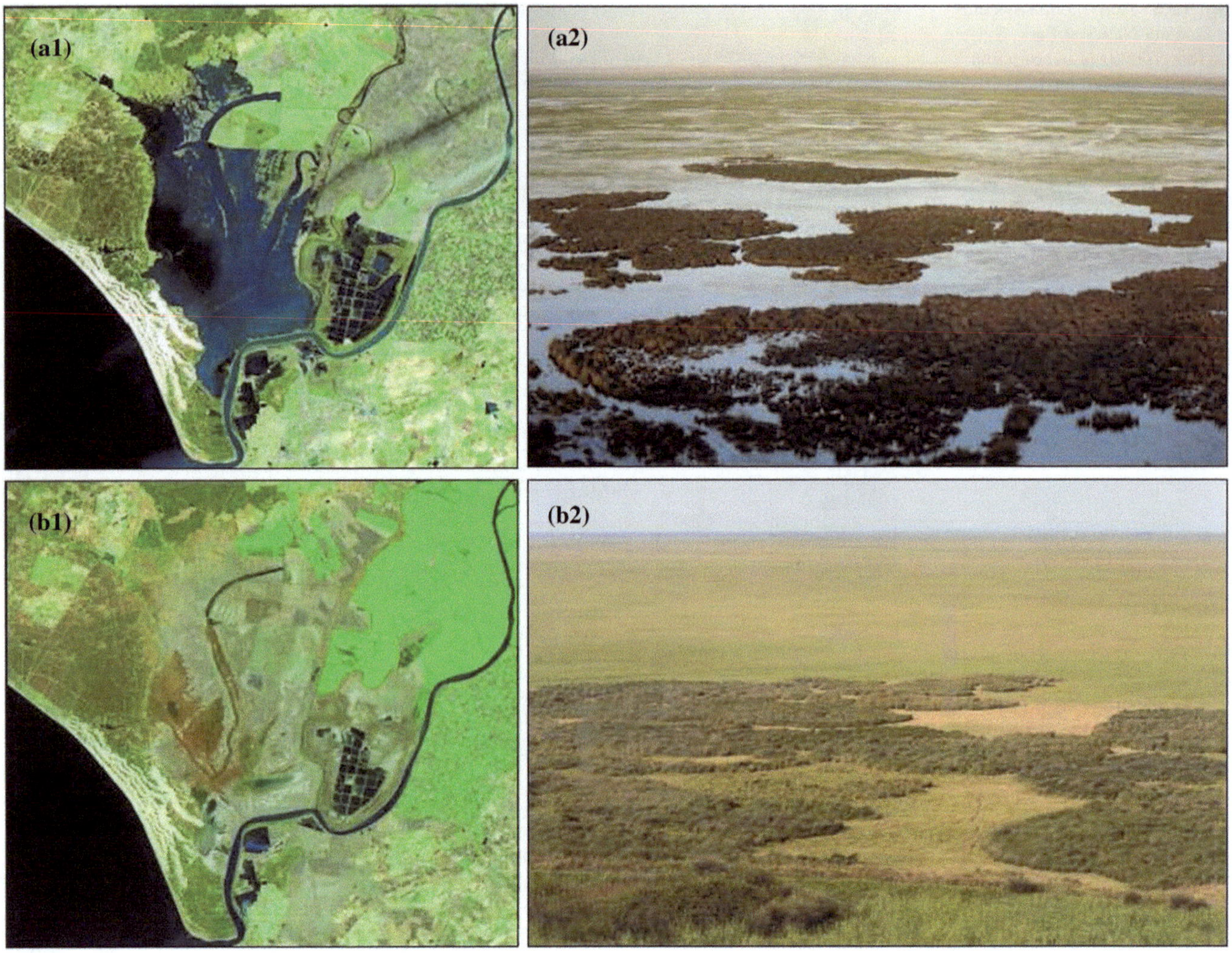

Fig. 19.7 General (**a1**, **b1**) and local (**a2**, **b2**) views of Doñana marshland during wet and dry seasons. Landsat images elaborated by LAST-EBD (CSIC) (photos by C. Finlayson)

reworking of the deposits accumulated by these events (Rodríguez-Vidal et al. 2011). The oldest of these ridges are those of Carrizosa and Vetalarena, with ages between 5,000 and 4,000 years BP (Ruiz et al. 2005), and the most recent ones are Vetalengua and Las Nuevas (2200–1900 year BP).

19.4 Evolution and Conclusions

The lower Guadalquivir River valley became a large estuary as a result of the postglacial sea-level rise, which was progressively silted up by fluvial sediment supply. Aggradation was favoured by the growth of littoral spit bars during Holocene times that eventually blocked the river flow. Changes in littoral dynamics and sea-level variations which occurred since the LGM are the main factors that have controlled the geomorphological evolution of the area.

During the LGM the coastline in the SW Iberian Peninsula was located at 120 m below the present sea-level, with wide and deep valleys excavated by the main rivers. As the sea advanced over the coast during the subsequent postglacial rise, the coastline acquired an uneven morphology with large inlets, estuaries and headlands. The estuarine deposits accumulated in the main valleys of the area indicate that the succeeding sea-level rise occurred in two phases: an initial rapid rise (7.0–5.7 mm/year) between 13,000 and 6,500 cal year BP, and a subsequent deceleration phase (2.6–0.9 mm/year) (Dabrio et al. 2000; Boski et al. 2001; Lario et al. 2002; Zazo et al. 2008). This twofold sea-level history caused a general change from vertical aggradation to lateral progradation recorded in several estuaries of the area (Dabrio et al. 2000; Zazo et al. 2008). Climatically induced sea-level changes during the Holocene highstand marked the alternation between progradational and erosional phases (Rodríguez-Ramírez et al. 1996; Goy et al. 2003; Zazo et al. 2008).

Six prograding spit bar systems have been identified in the southern Iberian Peninsula (H1 to H6, Goy et al. 2003; Zazo et al. 2008), punctuated by short periods of limited progradation or erosion, that have been correlated with an increased aridity during Bond events (Zazo et al. 2008). Nevertheless, in the Guadalquivir Estuary area, only four phases have been described (Zazo et al. 1994; Rodríguez-Ramírez et al. 1996, 2005), and only the two most recent ones crop out in Doñana area (Dabrio et al. 2000; Goy et al. 2003; Rodríguez-Ramírez et al. 2005; Zazo et al. 2008).

Identification of beach ridges on spit bars in Doñana is a hard task due to both the fine grain size that favours aeolian reworking, and the occurrence of superimposed dunes. Archaeological remains (Menanteau 1979; Borja et al. 1999), the scarce available outcrops of former shorelines, as well as data from drill cores obtained in the Guadalquivir marshlands (Ruiz et al. 2005) and radiocarbon dating (Lario et al. 1995; Rodríguez-Ramírez et al. 1996; Rodríguez-Ramírez and Yáñez-Camacho 2008; etc.), allow the reconstruction of the geomorphological evolution in several phases:

- Between MIS-5 and MIS-4 (El Asperillo Cliff). Upwarping of El Abalario area induced, on one hand, the southward deviation of Guadalquivir River, and on the other hand, vertical displacement on the Torre del Loro Fault, favouring the development of the first dune systems.
- Deglaciation: Rapid sea-level rise (7.0–5.7 mm/year) between 13,000 and 6,500 cal year BP.
- Holocene Climatic Optimum (ca. 6,500 cal year BP). The sea-level reached the maximum highstand, resulting in the development of wide estuaries at the mouth of the main rivers. Wet and temperate climate favoured the formation of an iron-rich palaeosoil on an erosional surface in El Abalario area.
- First spit bar progradation phase (6,500–4,500 cal year BP). Although there are no exposed remains, data from drill cores reveal that the Doñana littoral barrier was already growing between 5,500 and 5,400 cal year BP. The chenier of Carrizosa-Vetalarena (ca. 5,000 cal year BP) may be the result of the catastrophic destruction of this early spit bar. Prevailing winds form the WSW caused the growth of U4 dune system.
- Second spit bar progradation phase (4,200–2,600 cal year BP). Fluvial dominance in the Guadalquivir Estuary resulted in biological crisis and the development of the Roman *Lacus Ligustinus*. Pre-Roman settlement at La Algaida (Guadalquivir Estuary left bank) indicates that H2 spit bar already existed at that time.
- Third (2,300–1,100 cal year BP) and Fourth (1,000–Present) spit bar progradation phases. By these two phases the estuary was definitively closed from the open sea. Acceleration in progradation of H3 and H4 spit bars was coeval to cliff retreat upstream the longshore current at El Asperillo. Dune systems IV and V are correlative to H3 and H4 spit bars, as well as San Jacinto-La Marismilla strands, respectively.

Acknowledgments This work has been supported by MICINN-FEDER Projects CGL2010-15810, HAR2011-23798, CGL2012-33430, CGL2012-378, HAR2012-36008, and Excellence Project of the Andalousia Board (SEJ-4770) funded by the EU. It is a contribution to INQUA-CMP and Working Group UCM 910198 (Palaeoclimatology and Global Change).

References

Alonso-Zarza A, Armenteros I, Braga JC, Muñoz A, Pujalte V, Ramos E (2002) Tertiary. In: Gibbons W, Moreno T (eds) The Geology of Spain. The Geological Society, London, pp 293–334

Azañón JM, Galindo-Zaldívar J, García-Dueñas V, Jabaloy A (2002) Alpine tectonics II: Betic Cordillera and Balearic Islands. In: Gibbons W, Moreno T (eds) The Geology of Spain. The Geological Society, London, pp 401–416

Benavente J, Borja F, Gracia FJ, Rodríguez A (2005) Introduction to the Gulf of Cadiz coast. In: Desir G, Gutiérrez F, Gutiérrez M (eds) Sixth international conference on Geomorphology, field trip guides, Vol 2. 357–368

Borja F, Zazo C, Dabrio CJ, Díaz del Olmo F, Goy JL, Lario J (1999) Holocene aeolian phases and human settlements along the Atlantic coast of southern Spain. The Holocene 9:333–339

Boski T, Moura D, Veiga-Pires C, Camacho S, Duarte D, Scott DB, Fernández SG (2001) Postglacial sea-level rise and sedimentary response in the Guadiana Estuary. Sed Geol 150:103–121

Dabrio C, Zazo C, Goy JL, Sierro FJ, Borja F, Lario J, González JA, Flores JA (2000) Depositional history of estuarine infill during the last postglacial transgression (Gulf of Cadiz, southern Spain). Mar Geol 1162:381–404

Goy JL, Zazo C, Dabrio CJ (2003) A beach-ridge progradation complex reflecting periodical sea-level and climate variability during the Holocene (Gulf of Almería, Western Mediterranean). Geomorphology 50:251–268

Gutiérrez-Elorza M (2002) Quaternay. In: Gibbons W, Moreno T (eds) The Geology of Spain. The Geological Society, London, pp 335–336

Lario J, Zazo C, Goy JL, Dabrio C, Somoza L, Goy JL, Bardají T, Silva PG (1995) Record of recent Holocene sediment input on spit bars and deltas of South Spain. In: Core B (ed) Holocene cyclic pulses and sedimentation. J Coast Res Spec. (17):241–245

Lario J, Zazo C, Goy JL, Dabrio CJ, Borja F, Silva PG, Sierro F, González F, Soler V, Yll R (2002) Changes in sedimentation trends in SW Iberia Holocene estuaries. Quatern Int 93–94:171–176

Martín de la Cruz JC, Delgado MR, Sanz MP, Vera JC (2000) Novedades en el conocimiento sobre el Neolítico y Calcolítico en Andalucía: panorámica de una década de investigaciones. Trabalhos de Arqueología 16:215–241

Menanteau L (1979) Les Marismas du Guadalquivir. Exemple de transformation d'un paysage alluvial au cours du Quaternaire récent. These 3e cycle, Univ. Paris-Sorbonne, pp 154

Rodríguez-Ramírez A (1998) Geomorfología del Parque Nacional de Doñana y su entorno. Organismo Autónomo Parques Nacionales, Huelva

Rodríguez-Ramírez A (2011) Las dunas del litoral onubense. In: Sanjaume E, Gracia J (eds) Las dunas en España, pp 387–406

Rodríguez-Ramírez A, Yáñez-Camacho CM (2008) Formation of chenier plain of the Doñana marshland (SW Spain): observations and geomorphic model. Mar Geol 254:187–196

Rodríguez-Ramírez A, Rodríguez-Vidal J, Cáceres L, Clemente L, Belluomini G, Manfra I, Improta S, de Andrés JR (1996) Recent coastal evolution of the Doñana national park (SW Spain). Quatern Sci Rev 15:803–809

Rodríguez-Ramírez A, Cáceres L, Rodríguez-Vidal J (2005) Geomorphology of the Doñana National Park. In: Desir G, Gutiérrez F, Gutiérrez M (eds) Sixth international conference on Geomorphology, field trip guides, vol 2. pp 388–396

Rodríguez-Vidal J, Rodríguez-Ramírez A, Cáceres LM, Clemente L (1993) Coastal dunes and post-Flandrian shoreline changes. Gulf of Cádiz (SW Spain). INQUA Mediterr Black Sea Shoreline Subcomm Newsl 15:12–15

Rodríguez-Vidal J, Ruiz F, Cáceres LM, Abad M, González-Regalado ML, Pozo M, Carretero MI, Monge Soares AM, Gómez-Toscano F (2011) Geomarkers of the 218–209 BC Atlantic tsunami in the Roman *Lacus Ligustinus* (SW Spain): a palaeogeograhical approach. Quatern Int 242:201–212

Ruiz F, Rodríguez-Ramírez A, Cáceres LM, Rodríguez-Vidal J, Carretero MI, Abad M, Olías M, Pozo M (2005) Evidence of high-energy events in the geological record: mid-Holocene evolution of the southwestern Doñana National Park (SW Spain). Palaeogeogr Palaeoclimatol Palaeoecol 229:212–229

Ruiz F, Rodríguez-Vidal J, Abad M, Cáceres LM, Carretero MI, Pozo M, Rodríguez-Llanes JM, Gómez-Toscano, F, Izquierdo T, Font E, Toscano A (2013) Sedimentological and geomorphological imprints of Holocene tsunamis in southwestern Spain: an approach to establish the recurrence period. Geomorphology 203:97–104

Santos L, Sánchez-Goñi MF, Freitas MC, Andrade, C (2003) Climatic and environmental changes in the Santo Andre coastal area (SW Portugal) Turing the last 15000 years. In: Ruiz Zapata B, Dorado M, Valdeolmillos A, Gil MJ (eds) Quaternary climatic changes and environmental crises in the mediterranean region. Publ. Universidad de Alcalá de Henares, Madrid (Spain), pp 175–179

Sanz de Galdeano C, Vera JA (1992) Stratigraphic record and palaeogeographical context of the Neogene basins in the Betic Cordillera, Spain. Basin Res 4:21–36

Sierro FJ, González-Delgado JA, Dabrio CJ, Flores JA, Civis J (1996) Late Neogene depositional sequences in the foreland basin of Guadalquivir (SW Spain). In: Friend P, Dabrio C (eds) Tertiary basins of Spain: the stratigraphic record of crustal kinematics. Cambridge University Press, Cambridge, pp 339–345

Zazo C, Goy JL, Somoza L, Dabrio CJ, Belluomini G, Improta S, Lario J, Bardají T, Silva PG (1994) Holocene sequence of sea-level fluctuations in relation to climatic trends in the Atlantic-Mediterranean linkage coast. J Coastal Res 10:933–945

Zazo C, Dabrio CJ, Borja J, Goy JL, Lezine AM, Lario J, Polo MD, Hoyos M, Boersma JR (1999) Pleistocene and Holocene aeolian facies along the Huelva coast (southern Spain): climatic and neotectonic implications. Geol Mijnbouw 77:209–224

Zazo C, Mercier N, Silva PG, Dabrio CJ, Goy JL, Roquero E, Soler V, Borja F, Lario J, Polo D, Luque L (2005a) Landscape evolution and geodynamic controls in the Gulf of Cádiz (Huelva coast, SW Spain) during the late Quaternary. Geomorphology 68:269–290

Zazo C, Silva PG, Dabrio CJ, Roquero E, Borja F, Mercier J, Lario J, Goy JL (2005b) El Asperillo Cliffs near Torre del Loro fault. In: Desir G, Gutiérrez F, Gutiérrez M (eds) Sixth international conference on Geomorphology, field trip guides, vol 2, pp 375–379

Zazo C, Dabrio CJ, Goy JL, Lario J, Cabero A, Silva PG, Bardají T, Mercier N, Borja F, Roquero E (2008) The coastal archives of the last 15 ka in the Atlantic-Mediterranean Spanish linkage area: sea level and climate changes. Quatern Int 181:72–87

Raised Beaches in the Cantabrian Coast

20

Germán Flor and Germán Flor-Blanco

Abstract

Raised beaches along the Cantabrian coast are related to erosion surfaces, locally known as rasas, affected by tectonic uplift. The higher surfaces, probably of Pliocene age, have a maximum relative height of 285 m above mean sea level (MSL). Generally, they have gentle seaward slopes and highly variable lateral and longitudinal distribution. They reach 20 km in maximum width in central Asturias, and the lower levels cover smaller areas. Two W-E-oriented zones may be differentiated along the coast: Burela-Nalón (107 km) and Nalón-France (365 km). Two or three levels of rasas were generated in the first sector, which gradually merge into one towards the west. In the Nalón-France zone, 12 levels have been recognised. These geomorphic surfaces are apparently slightly deformed, although some authors proposed that they are strongly faulted. Some aggradation and fluvial terraces, abrasion surfaces, as well aeolian sand deposits, can be correlated with rasa levels. Other old deposits disconnected from the rasas were generated associated with exposed and estuarine beaches and may include slope deposits less than 1.0 m thick. Most beach and aeolian dune deposits are siliciclastic, with a limited proportion of bioclastic sands and debris. Relevant pending issues to be resolved are the numerical age of the rasas and their possible correlation with erosion surfaces, eustatic changes and uplift, considering that neotectonic activity is thought to be low and localised. The most recent marine terraces have not been uplifted and are affected by a general recession due to sea-level rise.

Keywords

Rasas • Erosion surfaces • Tectonic uplift • Marine terraces

20.1 Introduction

The 605-km long and E-W trending Cantabrian coast stretches from Estaca de Bares cape to the Txingudi estuary, at the border between Spain and France. It is characterised by steep cliffs locally interrupted by embayed beaches with aeolian dune fields and small estuaries. The coastal fringe displays numerous plains gently sloping towards the sea, usually forming stepped chronosequences at different relative heights (Fig. 20.1).

These flat surfaces are locally known as rasas and have been attributed to marine processes, although some may have a continental origin. They have been investigated since the nineteenth century (Schulz 1858; Barrois 1882), although most studies have been conducted since the second half of the twentieth century (Nonn 1966; Flor 1983;

G. Flor (✉) · G. Flor-Blanco
Department of Geology, University of Oviedo, Oviedo, Spain
e-mail: gflor@geol.uniovi.es

G. Flor-Blanco
e-mail: gfb@geol.uniovi.es

F. Gutiérrez and M. Gutiérrez (eds.), *Landscapes and Landforms of Spain*, World Geomorphological Landscapes,
DOI: 10.1007/978-94-017-8628-7_20,

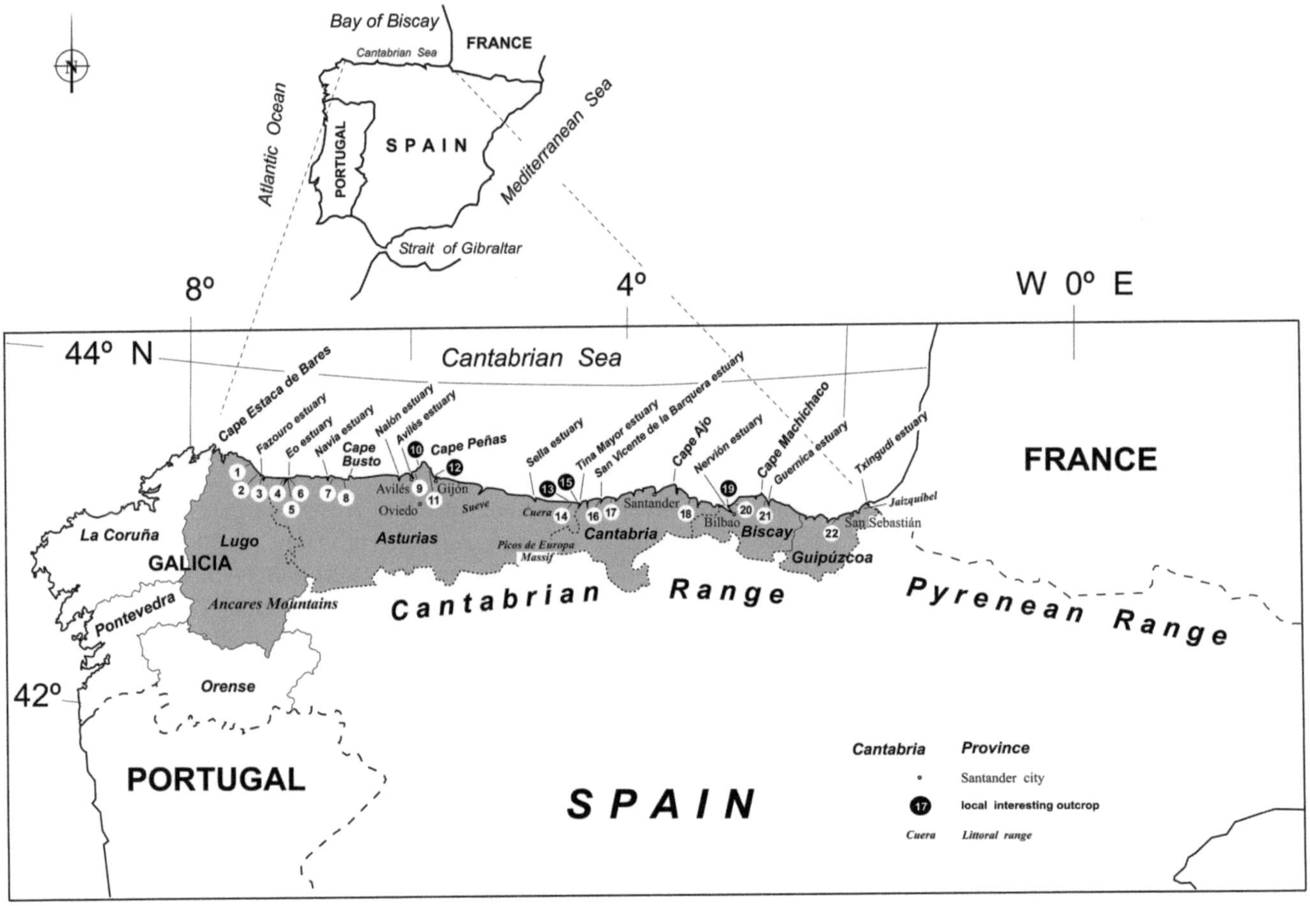

Fig. 20.1 Location of the raised beaches and relevant outcrops of coastal sediments

Mary 1983; Moñino 1986; Rivas 2000; González-Amuchástegui 2000; Moreno et al. 2009).

There are also small outcrops of deposits up to several metres thick related to raised beaches. These marine sediments are frequently overlain by aeolian sands, colluvium and periglacial talus. Some erosional landforms on cliffs and slopes, such as wave-cut notches and honeycomb features, constitute evidence of old sea levels, which may be related to marine erosion surfaces. Recent beach deposits, such as those associated with terraces, remain in contact with active supratidal areas, subject to recession since the 1970s. The correlations between the different surfaces are based on the relative height (Flor 1983).

20.2 Geological and Geographical Setting

The Cantabrian coast is a mesotidal environment where the main waves come from the NW. The main wind components are NW, SW and NE, and there is a weak persistent eastwards drift current. The steep cliffs are related to widespread uplift processes associated with the construction of the Cantabrian Mountains throughout the Cenozoic (Flor and Peón 2004).

The main morphostructural feature is the Cantabrian Range, a large monoclinal flexure that in Lugo province and part of western Asturias is replaced by the Ancares Mountains. It is the western prolongation of the Pyrenees, generated in the Cenozoic by the collision of the Euro-Asiatic plate and the Iberian microplate (Alonso et al. 2007) with partial subduction (ECORS-Pyrenean Team 1988). The maximum elevations (2,648 m a.s.l.) are reached in the Picos de Europa Massif, Asturias and Cantabria. The strongest tectonic uplift occurred from the Upper Eocene to the Middle Miocene. Remains of a faulted planation surface of probable Cretaceous–Eocene age (fundamental surface in Table 20.1) have been identified at a wide range of elevations Llopis-Lladó (1954). To the north, there is a stepped sequence of six continental planation surfaces generated under dry and warm climates, some of them underlain by slightly deformed alluvial sediments. The oldest surface is ascribed to the Oligocene (P in Table 20.1), whereas the remaining ones (Q, A, B, C and D in Table 20.1) are considered to be Miocene in age.

Table 20.1 Continental and marine erosion surfaces and their absolute and relative heights

	EROSION UPLIFTED PALAEOSURFACES	HEIGHTS (m) ABSOLUTE above the present mean sea-level	HEIGHTS (m) DIFFERENCES	MAIN GEODYNAMIC PROCESSES		CLIMATIC TRENDS ZONES	CLIMATIC TRENDS WINDS	AGES		
CONTINENTAL	(O) FUNDAMENTAL SURFACE	variable	>250 to <1,350	sea-level fall and continentalization; preorogenic cortical uplift; block faulting; synorogenic sedimentation	uplifting and faulting Cantabrian Range	tropical rainy	equatorial	Upper Eocene	PALAEOGENE	65.0 Ma; 33.7 Ma
	(P) CONTINENTAL PLANATION	1,450	250			subarid and warm		Oligocene		23.8 Ma
	(Q) CONTINENTAL PLANATION	1,200	450	compressive faults	postorogenic continental deposits			Lower Miocene	NEOGENE	11.0 Ma
	(A) CONTINENTAL PLANATION	750	200	cortical uplift			trade winds	Miocene		
	(B) CONTINENTAL PLANATION	550	100		general uplifting					
	(C) CONTINENTAL PLANATION	450	125	cortical uplift		subaride and warm				
	(D) LOWER CONTINENTAL PLANATION	325	40	strong cortical uplift and fluvial incision				?		5.3 Ma
	(I) UPPER RASA	285	55	cortical uplift				Pliocene		
	(II) RASA	230		cortical tilting from the Nalón estuary to the west						
MARINE	(VII) RASA	95	30		general uplifting					2.6 Ma
	(VIII) RASA	65	30	uniform cortical uplift		alternating humid temperate and cold	westerlies + subpolar	Pleistocene		
	(IX) RASA	35	15	eustatic sedimentary outcrops						
	(X) RASA	20	13							
	(XI) RASA	7	3							
	(XII) LOWER RASA	4								10.0 ka
MIXED	WÜRMIAN-FLANDRIAN SEDIMENTARY TERRACE	5-7		ongoing transgression / anthropogenic transformations		humid temperate	westerlies	Holocene		

The main geodynamic processes and climates throughout the Palaeogene and Neogene in the western and central Cantabrian Range are indicated (modified from Flor and Peón 2004)

Pérez-Alberti (1993) reported at least 11 levels of flat continental surfaces in north-western Spain. Eight surfaces in the inner Galician region were referred to by Vidal-Romaní et al. (1998) and Pagés and Vidal-Romaní (1998). These oldest planation surfaces were formed under a prevailing dry and warm climate, similar to that attributed to the surfaces mentioned above.

20.3 Littoral Platforms and Raised Beaches

There is consensus on the existence of one or more erosional surfaces (rasas) in the Cantabrian coast, although there is no agreement as to whether they have a continental or marine origin. They can be traced from the French border to the central coast of Lugo province. In the Burela area, Lugo (point 1 in Fig. 20.1), they are situated at only +5.0 m and cover a strip less than 5 km wide (Nonn 1966), called "rasa field" by Flor and Peón (2004).

There are flat erosional surfaces with a gentle seawards slope (<5°). The uplifted wave-cut platforms were considered to be true rasas (Mary 1983), including those situated as high as 285 m above sea level (Flor and Flor-Blanco 2009), but this idea must be reviewed. They are better preserved on quartzite and sandstone bedrock. The flat surfaces cut across Carboniferous, Jurassic and Cretaceous limestones are very small due to dissolution and erosion processes, especially in the coasts of eastern Asturias and western Cantabria.

Some authors have reported only one marine surface (rasa) of Plio-Pleistocene age at +100–120 m. This surface is tilted towards the W, reducing its relative height to 60 m in the area of the Eo estuary (Hoyos-Gómez 1989), as previously suggested by Birot and Solé-Sabarís (1954). In Galicia, neotectonics seems to be governed by regional uplift, rather than differential movement of blocks (Alonso and Pagés 2007).

Two broad segments can be differentiated west and east of the Nalón estuary. In the western area, erosion surfaces reach low heights and are represented by a smaller number (five levels in Asturias). From the Eo estuary, there is just one rasa that progressively reduces its height and disappears 13 km to the NW in Burela (E Galicia). The rest of the Cantabrian coast contains 12 surfaces, which can be correlated where they occur separated by tens of kilometres. In the Basque coast, they are poorly preserved and scarcely studied.

In central Asturias, continental and rasa erosion levels form extensive stepped morphosequences. Between the Nalón estuary and Gijón, including the Cape Peñas, upper rasas and lower erosion surfaces, as well as Würmian-

Fig. 20.2 **a** Aerial view of the Asturian eastern coast showing rasa levels II and III developed on quartzites and the rasa VIII on Carboniferous limestone. **b** The Jaizquíbel coastal range in Guipúzcoa, with upper rasa levels strongly dissected by streams

Flandrian, sedimentary terraces were generated. An aeolian sand sheet 1.1 m thick is the oldest marine deposit studied that was associated with rasa VIII in El Otero (Fig. 20.1, point 9). The sands are overlain by periglacial gravels several decimetres thick, in which a hand axe ascribed to the Older Palaeolithic Acheulean tradition and affected by aeolian abrasion was found (Rodríguez-Asensio and Flor 1983).

In eastern Asturias and western Cantabria numerous upper levels occur. In eastern Cantabria and western Biscay, they form noteworthy surfaces discontinuously distributed, largely due to the high density of streams and rivers and the rugged coast. In most of the Biscay and Guipuzcoan coasts, approximately from the Guernica to the Txingudi estuaries, the rasas are barely developed. Twelve erosion surfaces considered as rasas were defined as stepped flat surfaces from 285 to 5 m, which are best preserved on sandstones and quartzites.

20.3.1 The Upper Surfaces

The upper rasas (I–VIII), similar to the continental ones (Table 20.1; Fig. 20.2a, b), are broad surfaces a few kilometres wide covered by siliciclastic sediments several

Table 20.2 Rasa levels, their heights and altitude differences

Set	RASA LEVEL	FOOT / NICK POINT HEIGHT (m)	HEIGHT DIFFERENCES (m)	MAIN DEPOSITS	EUSTATIC SEDIMENTARY RECORDS	AGES
UPPER SET	I	285		*alluvial/peats*		
			55			
	II	230		*alluvial/peats*		
			45			
	III	185		*alluvial/slope*		
			25			
	IV	160		*alluvial/slope*		PLIOCENE
			25			
INTERMEDIATE SET	V	145		*alluvial/slope*		
			30			
	VI	115		*alluvial/slope*		
			20			
	VII	95		*alluvial/slope*		?
			30			
	VIII	65		*beach/aeolian dunes*		
			30			
LOWER SET	IX	35		*beach*		
			15			
	X	20		*beach/slope*	■	PLEISTOCENE
			13			
	XI	7		*beach/aeolian dunes* *slope/peats*	■	
			3		■	
	XII	4		*beach*	■	
Würmian-Flandrian sedimentary terrace		5-7		*sand and/or gravel beaches*	■	HOLOCENE
				solifluction / periglacial / peats	■	

The rasas are grouped into three sets, each with four levels. The presence of continental and coastal deposits is indicated, as well as eustatic records together with a tentative chronology in the central and eastern Asturias coast, Cantabria, western Biscay and eastern Guipúzcoa (modified from Flor and Flor-Blanco 2009)

metres thick, including torrential gravel facies and slope deposits. Only the lower levels (IX–XII) contain coastal sediments (Table 20.2), mainly beach deposits and aeolian sands, usually intercalated with colluvium, solifluction lobes, debris flows and periglacial stratified scree (Mary 1983). Those coastal deposits record eustatic changes, while the upper surfaces contain thin siliceous conglomerates of continental origin. Many subrounded quartzite gravel and sand deposits arranged in thick sequences of metric scale underlie different rasa levels in western Asturias. These are generally interpreted as marine facies, as indicated in Villapedre at 60 m (Fig. 20.1, point 7; Hernández-Pacheco and Asensio-Amor 1961; Álvarez-Marrón et al. 2008), but they can be considered as torrential fans.

In western Bilbao and eastern Cantabria, and between San Sebastián and the French border, there are also flat surfaces. In western Biscay, Hernández-Pacheco et al. (1957) mentioned several rasas, such as VII (60–65 m) in Bermeo and Mundaca (Fig. 20.1, point 21) and II (220–230 m) in the hills in the Asúa valley (N Bilbao), as well as the VII rasa, which is dissected by several coastal streams. In this work, levels I and II are first cited in the northern slope of Jaizquíbel range (Figs. 20.1, 20.2b).

Hazera (1968) refers to only one continental surface but located at different heights due to faulting and considerable degradation. He finds the surfaces at 80 m in Algorta and Punta Galea (Fig. 20.1, point 19) in the eastern mouth of the Nervión estuary, which belong to rasa VIII. In eastern Guipúzcoa, along the northern side of the Jaizquíbel range, levels I, II (Figs. 20.1, 20.2b), IV, V, VI and VIII can be identified (Edeso and Ugarte 1990).

20.3.2 The Lower Surfaces

The area covered by the lower rasas (IX–XII; Table 20.2) decreases as they become younger. The best examples of rasa XI are found in eastern Galicia and western Cantabria. In the Galician coast, several flat surfaces at heights close to the present-day sea level were interpreted as rasas by Pagés (2000).

The lower set of rasas attributed to the Pleistocene (Mary 1983) comprises three levels at 35–40, 15–18 and 5–6 m, well developed in eastern Asturias. In the Basque Country, the upper surface reaches relative heights of 40–45 m. In western and central Asturias, these rasa levels are situated at

Fig. 20.3 **a** Sedimentary sequence of the rasa XI in Oyambre beach (courtesy of M. Moñino). *1* Marine Oligocene blue grey silts and marls truncated by strath surface, *2* Imbricated calcareous boulders and cobbles with sands 1.7 m thick containing scarce mollusc debris, *3* Terrestrial grey silts and mud flow deposits (4 m thick), *4*. Aeolian sands 7 m thick. **b** Rasas XI (*upper level* containing cemented gravels inside the cave (*right*) and the lowest rasa XII developed in the supratidal belt of the La Arena beach (western end of Cantabria, near the Tina Mayor estuary)

lower heights. According to Mary (1983), this is related to a faster uplift in the eastern sector and towards the eastern end of the Bay of Biscay. The lower levels remain at a constant height along the coast.

In the western Cantabrian coast, deposits of rasa XI situated between 2.0 and 3.0 m above the mean sea level (MSL) were studied by Alonso and Pagés (2007). The basal beach gravels of Castro de Fazouro (Fig. 20.1, point 2) were dated by thermoluminescence (TL) to 75.8 ± 5.9 ka, corresponding to oxygen isotope stage (OIS) 5, while the overlying continental unit (solifluction deposits) in Paizás (Fig. 20.1, point 3) was dated by means of optically stimulated luminescence at 34.1 ± 2.7 ka, OIS stage 3.

The best sedimentary record is preserved in Oyambre (Fig. 20.1, point 16; Fig. 20.3a) and Bendía-La Franca in rasa XI (Fig. 20.1, point 13). The first outcrop has been studied by Hernández-Pacheco and Asensio-Amor (1966), Flor (1980), Mary (1983), Moñino (1986), Garzón et al. (1996), and Flor and Martínez-Cedrún (2004). The 1.7 m thick lower unit consists of imbricated calcareous boulders and cobble-pebble gravels with a matrix of granules and medium-fine sands including bioclastic debris and molluscs (e.g. *Acanthocardia echinata*, *Mytilus edulis*, *Patella* sp., *Murex* sp.). They were dated by aminoacid racemisation at 71,570 ± 13,400 year BP (Garzón et al. 1996), but an age of 100–73 ka would be more realistic according to Alonso and Pagés (2007). The uppermost 1.3 m thick unit consists of subrounded and polished quartz-rich coarse sands that contain iron and manganese oxides. Both deposits are covered by grey and bluish clays and silts and an uppermost unit of siliceous sands up to 7.0 m thick.

Another sedimentary sequence (Fig. 20.1, point 15) has been found within a cavity excavated at 2.5–3.0 m in calcareous bedrock west of the mouth of San Vicente de la Barquera estuary. The cavity is partially filled with rounded gravels and sands containing the following molluscs: *Ocenebra erinaceus*, *Nassa reticulata*, *Bittium reticulatum*, *Littorina saxatilis*, *Patella vulgata*, *P. lusitanica*, *Barnea candida* and *Pholas dactylus* (Mary 1983).

Inside the cave of Gesa, eastern Asturias (Fig. 20.1, point 12), remains of *Elephas (Paleoxodon) antiquus*, dated at 23,575 ± 1125 year BP (OIS stage 5) have been found (Pinto-Llona and Aguirre (1999). The skeletal remains are relatively well preserved in beach sand (Flor 1999).

In eastern Biscay, rasa XI was identified in the surroundings of the Guernica estuary, and levels X and IX in Bermeo (Hernández-Pacheco et al. 1957; Fig. 20.1, point 20). In eastern Guipúzcoa, Edeso and Ugarte (1990) refer to numerous patches of rasa IX located at a slightly higher elevation (40–45 m) than the equivalent surfaces in Asturias. In eastern Barrika (Fig. 20.1, point 19), there is a 15-m-thick deposit of aeolian sand related to a climbing dune that reaches a height of 50 m (Cruz-Sanjulián et al. 1984). These deposits are older than 40 ka (Cearreta et al. 1990) and are very probably linked to rasa XI.

Rasa XII is not frequent, and the best outcrop is found in the La Arena beach, right at the mouth of the Tina Mayor estuary, where rasa XI includes marine-cemented gravels in a small cave (Fig. 20.3b). Outside the study area, in coastal areas of Galicia, rasas and associated deposits have been studied by Martínez-Graña et al. (2007): R-0 (+1 m: Holocene), R-1 (+5–6 m: Upper Pleistocene), R-2 (+12–15 m: Middle Pleistocene), R-3 (+30 m: Lower Pleistocene) and R-4 (+60 m: Plio-Pleistocene). Two levels of marine deposits were found at +2 m (OIS 5) and +0.5 m (OIS 1).

20.3.3 The Age of the Rasas

Several hypotheses have been proposed about the age and origin of these surfaces. Mary (1983) suggested that their development began either during the Aquitanian–Langhian great transgression (Early Miocene) or in the Lower Pliocene. However, the upper rasas have a clear continental origin, whereas the lower ones are more probably marine, as suggest their lower extent and composition, including coastal deposits (Table 20.2). According to the first alternative, rasa I would be older than 22 Ma, indicating a long-term uplift rate of 0.013 mm/year. If the age of rasa I were 5.3 Ma (Lower Pliocene), the uplift rate would be 0.05 mm/year.

Jiménez-Sánchez et al. (2006), based on U/Th dates from speleothems sampled in El Pindal cave (eastern Asturias; Fig. 20.1, point 14), calculated an uplift rate of 0.19 mm/year for the last 300 kyr. In western and central Asturias, Álvarez-Marrón et al. (2008) dated rasa IV by cosmogenic nuclides. They obtained a minimum age of 1–2 Ma and estimated an uplift rate of 0.07–0.15 mm/year, with local changes related to faulting. Applying those rates and assuming a constant uplift rate, rasa I at 285 m would be 4.07–1.9 Ma old. The last numerical age of 1–2 Ma agrees with the age estimated by Mary (1983). The height differences between the lower rasas (Table 20.2) decrease gradually towards rasa XII, indicating that their formation required progressively shorter time intervals.

20.4 Würm-Flandrian Sedimentary Terraces

In the Würm-Flandrian terraces, gravel and sand beach deposits locally overlie other continental sediments, such as mud flow facies, silts derived from weathering processes and slope deposits, including talus related to periglacial conditions. Additionally, peat layers and trees, some in upright position, can be found within the beach deposits in Asturias and Cantabria (Mary 1983; Garzón et al. 1996; Salas et al. 1996). In Oyambre and Trengandín (Fig. 20.1, points 16 and 17, respectively), these sequences can be observed overlying a planar rock-cut surface.

These deposits are not abundant and cover small areas (<1,500 m^2), but can be found along the entire coastline, regardless of the distribution of the lower rasas. They are more common in small bays, cropping out at about 2.0–3.0 m above the average spring tides. They record a period from the end of the Pleistocene–Holocene transgression until the Flandrian maximum and are designated as the Würm-Flandrian terraces (Rodríguez-Asensio and Flor 1979).

In Gallín, eastern Galicia (Fig. 20.1, point 4), Feal-Pérez et al. (2009) and Feal-Pérez and Blanco-Chao (2012) dated a marine deposit at 5,580–5,530 cal BP ascribable to the Flandrian transgression and assigned to the Late Flandrian a higher gravel deposit that have yielded a numerical age of 1,735–1,590 cal BP.

One of the best outcrops of the Würm-Flandrian terrace is found at Portizuelo beach (Fig. 20.1, point 8; Fig. 20.4a). Mary (1983) inferred relatively dry periglacial conditions from a basal unit composed of angular pebbles and cobbles with mud–clay matrix, and attributed the upper beach deposits, consisting of rounded quartzite gravels, to the Flandrian sea-level rise (Fig. 20.4a). The upper part of the sedimentary sequence corresponds to fluvial gravels and brown sands. The terrace of Bañugues (Fig. 20.1, point 10) is composed by yellowish mud flow facies and a few debris flow units, including Lower Palaeolithic (Acheulean tradition) quartzite lithic objects (Rodríguez-Asensio and Flor 1979), or according to Álvarez-Alonso (2011), to the Old-Medium Palaeolithic (between OIS 5 and 6). At the top of the sequence, light brown fine sands contain Asturian picks of Mesolithic age.

In Oyambre beach, a sequence including clays, peats and tree debris was dated by Mary (1983) between 5,880 ± 30 and 5,250 ± 90 year BP. The basal peat has been dated at 6,210 ± 85 year BP by Garzón et al. (1996). The deposits recording the Flandrian transgression are capped by fine brown sands.

The transgressive Flandrian beach sediments are overlain by aeolianites in Tenrero, Górliz (Biscay) and Zarauz (Guipúzcoa). The aeolianites of Górliz, including Cervidae footprints (Flor 1989), were dated by Cearreta et al. (1990) at 5,710 ± 50 and 6,020 ± 50 year BP. In Zarauz (Fig. 20.1, point 21), this transgressive event occurred before 5,810 ± 170 BP (1–15,352), while the late Flandrian, located at 2.15 m, took place after 4,920 ± 100 BP (Edeso 1994).

Some scarce deposits, such as marine gravels in Las Fontías (Fig. 20.1, point 6), were dated by Mary (1983) at 1,920 ± 110 year BP, which record a small sea-level rise

Fig. 20.4 Würm-Flandrian deposits at the beaches of Portizuelo (**a**) and Bañugues (**b**). In the latter, the darker clayey basal unit (lower half of the outcrop) corresponds to highly weathered Devonian limestone. The sea level at spring high tides reaches the foot of both terraces

named Tardi-Flandrian, beginning with the Roman era. An estuarine beach deposit from this event, made up of sands and bivalves, was discovered by the authors in La Linera in the Eo estuary (Fig. 20.1, point 5).

Fragments of the gastropod *Purpura (= Thais) hoemastoma* were related to a sea-level rise dated at 2,150 + 110 year BP by Mary (1983) in Xivares (western Gijón; Fig. 20.1, point 12). These deposits are situated at 2.0–2.5 m, but the presence of man-made charcoal indicates an anthropogenic origin.

Outside the study area, the Corrubedo beachrock (Fig. 20.1), which have yielded dates of 1,045 ± 125 and 2,280 ± 60 year BP, can be interpreted as the result of a relative sea-level rise in the Galician coast during the last 2 kyr (Vilas et al. 1991).

20.5 Conclusions

In the Cantabrian coast, many erosional surfaces, locally known as rasas, have developed along the foot of the mountain ranges. They are distributed in two sectors: the western area exhibits only one low surface, which gradually splits into five rasas towards the Nalón estuary (Asturias). From that area to the French border, up to twelve stepped levels can be differentiated. The five lower rasa levels contain coastal deposits. They may be interpreted as raised beaches that record eustatic changes.

The development of the rasa sequences is related to epeirogenic uplift, active since the Miocene, which has generated broad erosional surfaces with poor sedimentary covers. Tectonic uplift was low in Galicia, increasing gradually towards the Nalón estuary, and further east to the French border. This deformation is rather homogeneous, as reveals the fact that the upper rasas are at approximately the same heights in Cape Peñas and in the Jaizquíbel Range.

Many small marine terrace deposits were generated during the last eustatic cycle, including the Flandrian transgression in the Holocene. Other sedimentary outcrops are attributed to the last sea-level rise occurred during the Tardi-Flandrian.

References

Alonso A, Pagés JL (2007) Stratigraphy of Late Pleistocene coastal deposits in northern Spain. J Iber Geol 33:207–220

Alonso JL, Álvarez-Pulgar J, Pedreira V (2007) El relieve de la Cordillera Cantábrica. Ens Cien Tierra 15(2):151–163

Álvarez-Alonso D (2011) El Paleolítico antiguo en la región cantábrica: un estado de la cuestión. Férvedes 7:29–37

Álvarez-Marrón J, Hetzel R, Niedermann S, Menéndez R, Marquínez J (2008) Origin, structure and exposure history of a wave-cut platform more than 1 Ma in age at the coast of northern Spain: a multiple cosmogenic nuclide approach. Geomorphology 93:316–334

Barrois Ch (1882) Recherches sur les terrains anciens des Asturies et de la Galicia. Mém Soc Géol Nord 2. Lille

Birot P, Solé-Sabarís L (1954) Recherches morphologiques dans le Nord-Ouest de la Péninsule Ibérique. Mém et Doc C D C G 4:11–61

Cearreta A, Edeso JM, Merino A, Ugalde T, Ugarte FM (1990) Las dunas litorales de Barrika (costa occidental de Bizkaia). Kobie 19:77–83

Cruz-Sanjulián JJ, García-Mondéjar J, Granda JM, Pujalte V (1984) Características y evolución de unos depósitos de "cliff-top" localizados sobre la rasa costera vizcaína. Thalassas 2:31–34

ECORS-Pyrenean Team (1988) The ECORS deep reflection seismic survey across the Pyrenees. Nature 331:508–511

Edeso JM (1994) El relleno holoceno de la depresión de Zarauz. Lurralde 17:115–152

Edeso JM, Ugarte FM (1990) Algunos datos sobre la paleo-geografía litoral cuaternaria de la costa: Jaizkibel-Bahía de Txingudi (Golfo de Vizcaya). Cuad Sec Eusko Ikaskuntza 16:27–76

Feal-Pérez A, Blanco-Chao R (2012) Evidencias de una pulsación positia del mar durante el Holoceno en la costa NE de Galicia. Implicaciones morfodinámicas. XII Reunión Nacional de Geomorfología, PUbliCan Ed. University of Cantabria, pp 307–310

Feal-Pérez A, Blanco-Chao R, Valcárcel-Díaz M (2009) Influencia de formas y procesos heredados en la evolución reciente y en los procesos morfodinámicos actuales en un sector de la costa rocosa: Punta Gallín, costa Cantábrica gallega. Rev Soc Geol España 22:67–78

Flor G (1980) Las playas levantadas de Merón y Oyambre (Santander). Proc. IX Congreso Nacional de Sedimentología. Pub. Univ. Salamanca. I:357–371

Flor G (1983) Las rasas asturianas: ensayos de correlación y emplazamiento. Trab Geol 13:65–81

Flor G (1989) Estructuras de deformación por pisadas de cérvidos en la duna cementada de Górliz (Vizcaya, N de España). Rev Soc Geol España 2:23–29

Flor G (1999) Estudio geológico de la cueva de la Silluca (Concejo de Llanes). Excavaciones Arqueológicas en Asturias 1995–98. Principado de Asturias, pp 213–224

Flor G, Flor-Blanco G (2009) Guía de Campo. Aspectos morfológicos, dinámicos y sedimentarios del sector costero: desembocadura del Nalón-playa de Bañugues. Problemática ambiental. 6° Simposio sobre el Margen Ibérico Atlántico (MIA09), pp 1–61

Flor G, Martínez-Cedrún P (2004) Excursión costera Liencres-San Vicente de la Barquera. Geocantabria II. In: FJ Barba (coord) Cantabria. Itinerarios Geológicos. Consejería de Cultura. Gobierno de Cantabria, pp 26–47

Flor G, Peón A (2004) Rasas y superficies de erosión continental en el relieve alpídico del noroeste peninsular y los depósitos terciarios. In: Araújo MA, Gomes A (eds) Faculdade de Letras. University of Porto, Geomor NW Penín Ibérica, pp 13–31

Garzón G, Alonso A, Torres J, Llamas J (1996) Edad de las playas colgadas y de las turberas de Oyambre y Merón (Cantabria). Geogaceta 20:498–501

González-Amuchástegui M (2000) Evolución morfoclimática del País Vasco durante el Cuaternario: estado de la cuestión. Rev C&G 14:79–99

Hazera J (1968) La région de Bilbao et son arrière-pays. Étude géomophologique. Thèse de Doctorat. Sociedad Ciencias Naturales de Aranzadi. San Sebastián. Munibe, XX (1/4)

Hernández-Pacheco F, Asensio-Amor I (1961) Materiales sedimentarios sobre la rasa cantábrica III. Tramo asturiano comprendido entre Santiago de Villapedre (Navia) y Cadavedo (Luarca). Bol R Soc Esp Hist Nat (Geol) 59:207–223

Hernández-Pacheco F, Asensio-Amor I (1966) Fisiografía y sedimentología en la playa y ría de San Vicente de la Barquera (Santander). Estud Geol 22:1–23

Hernández-Pacheco F, Llopis-Lladó N, Jordá-Cerdá J, Martínez JA (1954) Guía de la excursión N2. El Cuaternario de la Región Cantábrica. V Congreso Internacional INQUA. Diputación Provincial de Asturias. Oviedo, pp 7–41

Hoyos-Gómez M (1989). La Cornisa Cantábrica. In: Pérez-González A, Cabra-Gil P, Martín-Serrano A, Heras JA (coords) Mapa Cuat España (escala 1:1.000.000). Instituto Tecnológico Geominero de España. Madrid, pp 105–118

Jiménez-Sánchez M, Bischoff JI, Stoll H, Aranburu A (2006) A geochronological approach for cave evolution in the Cantabrian Coast (Pindal Cave, NW Spain). Z Geomorph N F 147:129–141

Martínez-Graña AM, Goy JL, Zazo C (2007) Cartografía geomorfológica y patrimonio geológico cuaternario en la ría de Arosa (Pontevedra-La Coruña, Galicia España). Proc XII Reunión Nacional de Cuaternario, pp 229–230

Mary G (1983) Evolución del margen costero de la Cordillera Cantábrica de Asturias desde el Mioceno. Trab Geol 13:3–35

Moñino M (1986) Establecimiento y cartografía de los niveles de rasa existentes en Cantabria. MSc Thesis (unpublished). Facultad de Geografía. University of Cantabria

Moreno F, Mediato JF, Canas V (2009) Terrazas marinas en el litoral de Cantabria. Controles litológicos del sustrato. In: Proc. 6° Simposio sobre el Margen Ibérico Atlántico MIA09, pp 181–184

Nonn H (1966) Les régions côtières de la Galice (Espagne). Étude géomorphologique. Pub. Faculté de Lettres. University of Strasbourg, Fondation Baulig

Pagés JL (2000) Origen y evolución geomorfológica de las rías atlánticas de Galicia. Rev Soc Geol España 13:393–403

Pagés JL, Vidal-Romaní JR (1998) Síntesis de la evolución geomorfológica de Galicia Occidental. Geogaceta 23:119–122

Pérez-Alberti A (1993). La interacción entre procesos geomorfológicos en la génesis del relieve del sudeste de Galicia: El ejemplo del Macizo de Manzaneda y de la Depresión de Maceda. In: Pérez-Alberti A, Guitián-Rivera L, Ramil-Rego P (eds) La Evolución del Paisaje de las Montañas del Entorno de los Caminos Jacobeos. Xunta de Galicia, pp 1–24

Pinto-Llona AC, Aguirre E (1999) Presencia del elefante antiguo *Elephas (Paleoloxodon) antiquus* en la cueva de la Silluca (Buelna, Asturias). Excavaciones arqueológicas en Asturias 1995–98. Servicio de Publicaciones. Consejería de Cultura. Principado de Asturias, pp 225–232

Rivas V (2000) Clima y nivel del mar: reconstrucción de las posiciones marinas cuaternarias a través de las evidencias en el litoral cantábrico. In: García-Codrón JC (coord) La reconstrucción del clima de época preinstrumental. Pub. University of Cantabria, Santander, pp 179–212

Rodríguez-Asensio JA, Flor G (1979) Estudio del yacimiento prehistórico de Bañugues y su medio de depósito (Gozón, Asturias). Zephyrus 29:161–178

Rodríguez-Asensio JA, Flor G (1983) Industrias paleolíticas enlizadas en la región de Cabo Peñas. Cuad Lab Xeol Laxe 5:23–46

Salas L, Remondo J, Martínez-Cedrún P (1996) Cambios del nivel del mar durante el Holoceno en el Cantábrico a partir del estudio de la turbera de Trengandín. Proc. IV Reunión de Geomorfología. O Castro. A Coruña, pp 237–247

Schulz G (1858). Descripción geológica de la provincia de Oviedo. Imp. J. González. Madrid

Vidal-Romaní JR, Yepes-Temiño J, Martínez-Conde R (1998) Evolución geomorfológica del macizo Hespérico Peninsular. Estudio de un sector comprendido entre las provincias de Lugo y Ourense (Galicia, NW de España). Cuad Lab Xeol Laxe 23:165–199

Vilas F, Sopeña A, Rey L, Ramos A, Nombela MA, Arche A (1991) The Corrubedo beach–lagoon complex, Galicia, Spain; dynamics, sediments and recent evolution of a mesotidal coastal embayment. Mar Geol 97:391–404

The Olot Volcanic Field

21

Carles Roqué, Rogelio Linares, Mario Zarroca, and Lluís Pallí

Abstract

The Olot Volcanic Field (OVF) records the youngest eruptions (0.5–0.01 Ma) of the Catalan Volcanic Zone (CVZ), associated with the post-Alpine European intraplate rift. Magma reached the surface through a fault system cross-cutting the Catalan Transversal Range, producing about 50 monogenetic volcanoes built up by strombolian and/or phreatomagmatic activity. The volcanic cones have maximum and average basal widths of 1,290 and 500 m, respectively. Maximum and mean heights are 189–100 m, respectively. Given the age of the eruptions, the primary volcanic landforms are well preserved and geomorphic features related to erosional processes are rare. Lava flows emerging from most of the volcanoes flowed along river valleys, reaching up to 10 km in length and locally generating volcanic dams.

Keywords

Quaternary monogenetic volcanism • Strombolian cones • Lava flows • Phreatomagmatism • Catalan Volcanic Zone

21.1 Introduction

The Olot Volcanic Field (OVF) comprises the youngest and best preserved volcanic features of the Catalan Volcanic Zone (CVZ) (Fig. 21.1a). The CVZ is part of a set of volcanic areas developed along the intraplate rift system of western Europe. Neogene extensional tectonics generated a series of grabens and half-grabens, which extends along 2,000 km from the North Sea and the Rhine Graben to the Mediterranean coast of the Iberian Peninsula (Martí et al. 1992, 2000).

In the CVZ, volcanic activity has been continuous over the past 14.7 Ma. Rapidly ascending magmas from the asthenospheric mantle underwent minor compositional changes; the outcropping volcanic rocks are mainly alkali basalts and basanites (Cebriá et al. 2000). The volcanoes were constructed during single eruptions that often included several phases of activity. Most eruptions encompassed one or more phases of effusive (emission of lava flows) and strombolian (ejection of pyroclasts) activity. Occasionally, the interaction between the ascending magma and groundwater in bedrock, triggered phases of phreatomagmatic explosive activity (Martí and Mallarach 1987).

The CVZ is conventionally divided into three subzones on the basis of the geographical distribution and the age of the volcanic products (e.g. Araña et al. 1983; Martí et al. 2000; Pallí et al. 2005): Empordà Volcanic Field (EVF), Selva Volcanic Field (SVF) and Olot Volcanic Field (OVF), the latter also known as Garrotxa volcanic field (Fig. 21.1b).

C. Roqué (✉) · L. Pallí
Department of Environmental Sciences, University of Girona, Girona, Spain
e-mail: carles.roque@udg.edu

R. Linares · M. Zarroca
Department of Geology, Autonomous University of Barcelona, Barcelona, Spain
e-mail: rogelio.linares@uab.cat

M. Zarroca
e-mail: mario.zarroca.hernandez@uab.cat

F. Gutiérrez and M. Gutiérrez (eds.), *Landscapes and Landforms of Spain*, World Geomorphological Landscapes,
DOI: 10.1007/978-94-017-8628-7_21, © Springer Science+Business Media Dordrecht 2014

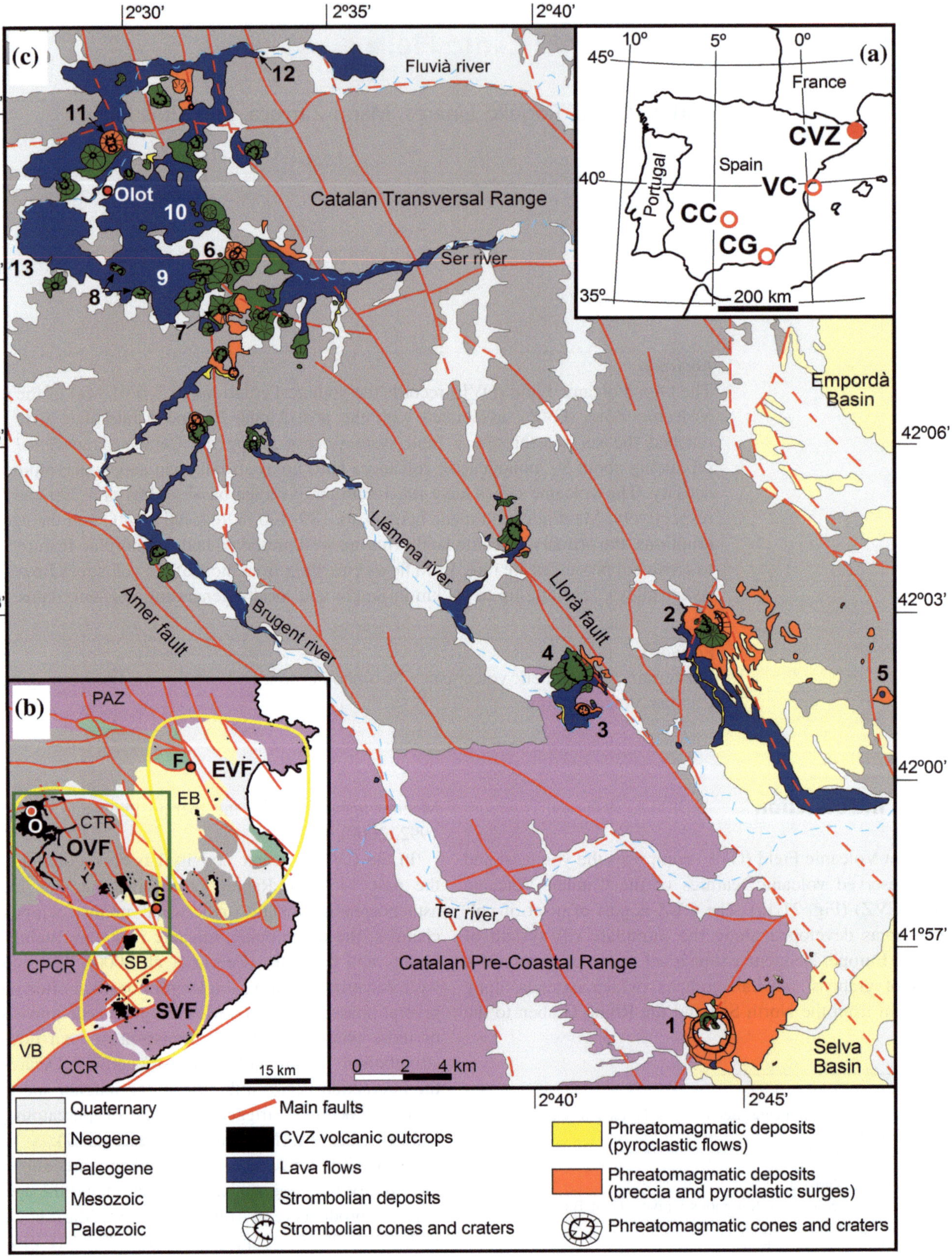
2°30'
2°35'
2°40'
2°45'
42°12'
42°09'
42°06'
42°03'
42°00'
41°57'
(c)
Fluvià river
Olot
Catalan Transversal Range
Ser river
Lléмena river
Llorà fault
Amer fault
Brugent river
Empordà Basin
Ter river
Catalan Pre-Coastal Range
Selva Basin
1
2
3
4
5
6
7
8
9
10
11
12
13
0 2 4 km
(a)
45°
40°
35°
10°
5°
0°
France
Portugal
Spain
CVZ
VC
CC
CG
200 km
(b)
PAZ
F
EVF
EB
CTR
O
OVF
G
CPCR
SB
SVF
VB
CCR
15 km
Quaternary
Neogene
Paleogene
Mesozoic
Paleozoic
Main faults
CVZ volcanic outcrops
Lava flows
Strombolian deposits
Strombolian cones and craters
Phreatomagmatic deposits (pyroclastic flows)
Phreatomagmatic deposits (breccia and pyroclastic surges)
Phreatomagmatic cones and craters

◀ **Fig. 21.1** Geological setting of the Olot Volcanic Field. **a** Location of Neogene–Quaternary volcanic areas in peninsular Spain: *CC* Campo de Calatrava, *CG* Cabo de Gata, *VC* Valencia and Columbretes, *CVZ* Catalan Volcanic Zone. **b** Geological sketch of the CVZ: Subzones: *EVF* Empordà Volcanic Field, *SVF* Selva Volcanic Field, *OVF* Olot Volcanic Field. Main morphostructural units: *EB* Empordà Basin, *SB* Selva Basin, *VB* Vallès-Penedès Basin, *PAZ* Pyrenean Axial Zone, *CTR* Catalan Transversal Range, *CCR* Catalan Coastal Range, *CPCR* Catalan Pre-Coastal Range. Main towns: *O* Olot, *G* Girona, *F* Figueres. *Green rectangle* corresponds to the area covered by the geological map 1C. **c** Simplified geological map of the OVF (after Puig et al. 1997, 2003; Cirés et al. 2002; Losantos and Planagumà 2007). Main volcanoes and localities referred to in the text: *1* Crosa de Sant Dalmai, *2* Puig d'Adri, *3* Clot de l'Omera, *4* Puig de la Banya de Boc, *5* Can Guilana, *6* Croscat, *7* Santa Margarida, *8* Cabrioler and Puig Jordà, *9* Fagueda d'en Jordà, *10* Pla de Batet, *11* Puig de la Garrinada, *12* Castellfollit de la Roca, *13* Pla de les Preses

Igneous activity in the EVF has been dated for between 14.7–6.7 Ma. It comprises about 45 volcanic outcrops associated with a normal fault system defining and cross-cutting the Empordà Basin. Most of the outcrops correspond to relict basaltic lava flows partially overlain by Miocene and Pliocene continental and marine sediments. Only the dikes and necks of the original volcanic built-ups are preserved. The trachytes of Vilacolum and Arenys d'Empordà (e.g. Díaz et al. 1996) are the most differentiated rocks of the CVZ in terms of petrology.

Volcanic activity in SVF commenced about 7.9 Ma and finished around 1.7 Ma. The ca. 100 volcanic outcrops of the area are mainly located along the south-west and south-east margins of the Selva Basin, extending locally towards the Vallès-Penedès Basin (i.e. Hostalric volcano) (Pujadas et al. 2000). The preserved volcanic products generally correspond to basaltic lava flows, interstratified in places with alluvial fan deposits (Picart et al. 2008). Alike the EVF area, the volcanic edifices have been considerably eroded and generally only necks and dikes are preserved. However, three Pliocene volcanic cones that record phases of phreatomagmatic activity are preserved: Hostalric volcano (Pujadas et al. 2000), Vidreres volcano (Picart et al. 2008) and Camp dels Ninots maar (Vehí et al. 1999). The fossil-rich lacustrine sediments deposited within the latter constitute one of the most significant Late Pliocene palaeontological sites in Europe (Gómez et al. 2012).

The Crosa de Sant Dalmai maar (Fig. 21.1c, number 1) is the best preserved volcanic cone of the SVF and the largest in the CVZ. This maar-type volcano is mainly composed of phreatomagmatic deposits (pyroclastic breccia and surge deposits) arranged as a circular tephra ring around a shallow crater 50 m deep and 1,250 m in diameter. The pyroclastic surges were directed towards the north-east, reaching a distance of over 4 km from the eruptive centre. A scoria cone and a basaltic lava flow are located within the maar, both related to the final stages of volcanic activity (Pujadas et al. 2000; Bolós et al. 2012). Picart et al. (2009) ascribe the volcano to the Pleistocene. The stratigraphic relationships observed between the distal pyroclastic surge deposits and alluvial deposits of the Ter River indicate that this eruption took place in the Late Pleistocene.

The volcanic activity in the OVF started about 0.5 Ma and lasted up to ca. 10 ka, with recurrent eruptive episodes at intervals of around 15,000 years (Martí et al. 2000). This area includes about fifty well preserved volcanic cones (Fig. 21.1c) built chiefly through phases of strombolian activity and occasional phreatomagmatic events. The lava flows emanating from these volcanoes travelled along preceding river valleys covering distances of up to 10 km.

21.2 Geological and Geographical Setting

The OVF is located within the Catalan Transversal Range, on the southern Pyrenees. Most volcanoes (36) are concentrated in the upper regions of the Fluvià and Ser river basins, in the vicinity of Olot town, in a relatively flat plain at 450–550 m a.s.l. flanked by mountains reaching altitudes of around 1,500 m. The remaining sites with evidence of eruptive activity (14) are distributed along the Llémena and Brugent river valleys (Fig. 21.1c). The area is characterized by Mediterranean mountain climate, relatively more humid than in the surrounding areas. The annual precipitation is about 1,000 mm, distributed in about 100 days of rain. The vegetation is dense. The north-facing slopes are forested with *Fagus sylvatica*, *Quercus pubescens*, *Castanea sativa* and *Betula pendula*, while the south-facing forests are dominated by *Quercus ilex* and *Pinus sylvestris*. The Olot plain and the valley lower region are occupied by farmland. In this context, geology, vegetation and human activity merge into a unique landscape, where volcanic cones and lava flows preserve their original morphology under a ground cover displaying shifting colours throughout the year. The OVF is a protected natural area since 1982; the Garrotxa Volcanic Zone Natural Park.

The Catalan Transversal Range comprises sedimentary rocks of Palaeogene age, mainly sandstones, marls, limestones and gypsum, reaching as much as 4,500 m in thickness. At the southern end of the range, these Palaeogene sediments unconformably overlie the Palaeozoic bedrock of the Catalan Pre-coastal Range. The structure of the Catalan Transversal Range consists of an E-W trending Alpine fold and thrust belt, and a superimposed system of NW–SE

Fig. 21.2 Croscat volcano. The pyroclastic deposits forming the scoria cone crop out in a restored quarry

trending normal faults. These post-orogenic faults progressively downdrop blocks to the east, towards the Empordà Basin (Fig. 21.1c). The Amer Fault, located at the westernmost region of the system, is considered to be the most active (Zarroca et al. 2012). The magma rose to the surface through these faults, producing lava flows and monogenetic volcanoes of the OVF. These single-eruption volcanoes are generally small, reaching volumes of extruded magma of the order of 0.01–0.2 km^3 (Martí et al. 2011). Most eruptions encompassed several phases of activity, generally with the alternation of: (1) one or more lava emission phases; and (2) one or more strombolian phases, during which cinder cones were built. This pattern was occasionally interrupted by phreatomagmatic explosive activity, which led to the development of more complex volcanic cone structures (e.g. Di Traglia et al. 2009; Gisbert et al. 2009; Martí et al. 2011).

21.3 Geomorphology

Given the age of the eruptions in the OVF (0.5 to 0.01 Ma), the primary volcanic landforms (volcanic cones and lava flows) are well preserved and features related to erosion processes, prevalent in other areas of the CVZ, are rare.

21.3.1 Primary Volcanic Landforms

Volcanic Cones

The occurrence of phreatomagmatic phases during the eruptions largely controls the geometry of the cones. Two categories of volcanic cones may be identified (Martí et al. 2011): (1) those corresponding to scoria-type cones, constructed during phases of strombolian activity and occasional emission of lava flows, and (2) those recording evidence of phreatomagmatic explosive phases, in addition to strombolian and effusive activity.

The scoria cones exclusively or mostly constructed by strombolian activity consist of pyroclastic deposits (lapilli and volcanic blocks and bombs) ejected ballistically from the central conduit (Fig. 21.2). The thirty-six volcanic edifices belonging to this category commonly show radial symmetry with a central crater (12 %), a side crater in horseshoe-shaped cones (72 %), or with no crater (16 %). The Croscat Volcano (Fig. 21.1c, number 6), presenting a basal cone width (Wco) of 1.290 m and a cone height (Hco) of 189 m, is the largest volcanic cone in this area (Fig. 21.2). The scoria cones of the OVF have mean Wco and Hco values of 500 and 100 m, respectively. The aspect ratio (Hco/Wco) ranges from 0.12 to 0.29.

Some among the twelve cones with evidence of phreatomagmatic activity display morphological characteristics similar to the strombolian scoria cones (e.g. Puig de la Garrinada volcano, Fig. 21.1c, number 11). They can be differentiated from the latter by the presence of lithic clasts in the pyroclastic deposits and the occurrence of pyroclastic surge and/or pyroclastic flow deposits (e.g. Gisbert et al. 2009) and by a lower aspect ratio (0.11–0.15). Six edifices in the OVF display characteristics of maar-type volcanoes: craters excavated in the pre-eruptive bedrock, encircled by a tephra ring with poor topographic expression. The Clot de l'Omera maar (Fig. 21.1c, number 3), with a 550 m diameter crater and excavated 20 m into the pre-eruptive bedrock, is the largest example (e.g. Pujadas et al. 1997; Martí

Fig. 21.3 Santa Margarida volcano. The phreatomagmatic maar is partly covered by strombolian deposits derived from the Croscat volcano. Despite its conical shape, the southern flank of the mountain (far edge of the image) is actually formed by Palaeogene sedimentary rocks. *Photograph* courtesy of David Soler

Fig. 21.4 Fageda d'en Jordà lava flow (tree-covered area in the foreground) associated with the Croscat volcano (horseshoe-shaped cone, to the *right* of the image). The volcanic primary landforms in the OVF, Pleistocene in age, are well preserved. The farmlands are located on intra- and post-volcanic lacustrine deposits. The mountains in the background correspond to Palaeogene sedimentary rocks

et al. 2011). The morphology of these edifices is partially distorted by emissions from the surrounding volcanoes. For instance, the Santa Margarita maar (Fig. 21.1c, number 7, Fig. 21.3) is partly covered by strombolian deposits of the nearby Croscat volcano. The Clot de l'Omera maar is overlain to some extent by a lava flow derived from the Puig de la Banya de Boc volcano. Four rather complex volcanoes constructed by alternating phases of strombolian, phreatomagmatic and effusive activity are composed of several overlapping volcanic cones. The best example is the Puig d'Adri volcano, a cone 152 m high and 1,200 m in basal width (Fig. 21.1c, number 2). This volcano consists of a phreatomagmatic tuff-ring and two scoria cones. Five phases of activity have been identified from the volcanostratigraphic sequence and the cartographic relationships among the different units: phreatomagmatic–strombolian–phreatomagmatic–strombolian–hawaiian (e.g. Pujadas et al. 1997; Martí et al. 2000, 2011).

Fig. 21.5 Castellfollit de la Roca cliff. Fluvial erosion has exposed the internal structure of the lava flows. Columnar jointing can be observed in the upper lava flow

Lava Flows

Most eruptions of the OVF included at least one phase of effusive activity, involving significant accumulations of basaltic and basanitic lava flows reaching in places a thickness of more than 40 m. Lava flows cover about 29.5 km^2 of the Olot plain. A notable example is the Fageda d'en Jordà lava flow, attributed to the Croscat volcano eruption, 3.2 km long and covering about 4.9 km^2 (Fig. 21.4). Its surface is riddled with blisters and hornitos up to 15 m high. This lava flow partly overlies a previous flow event associated with the Cabrioler and Puig Jordà volcanoes (Fig. 21.1c, number 8). This flow is exposed in an area of 5.3 km^2 and extends about 4.5 km north-eastwards as far as the Fluvià River bed. The lava flows of the

Olot plain led to the formation of volcanic dams in the Fluvià River valley, currently filled by lacustrine and alluvial deposits. These deposits reach over 35 m in thickness in the el Pla de les Preses (Fig. 21.1c, number 13). Another remarkable lava flow is the one associated with Puig d'Adri volcano, which flowed over a distance of 10.1 km.

21.3.2 Erosional Landforms

As stated above, erosional features in the primary volcanic landforms of the OVF are very limited due to the recent age of the latter. The four volcanic necks of the Pla de Batet (Fig. 21.1c, number 10) are unusual features, probably corresponding to remnants of the oldest volcanic cones in the area. An additional rather degraded structure is the neck at Can Guilana (Fig. 21.1c, number 5), located in the east margin of the OVF, where remains of phreatomagmatic deposits are preserved (Pujadas et al. 1997).

The lava flows that channelled along the valleys have been partly eroded by the rivers, especially along the contact with the underlying sedimentary rocks. Erosion by the Fluvià River, and its tributary the Turonell Stream, has carved a 50 m high and 1,000 m long cliff, over which the Castellfollit de la Roca village is settled (Fig. 21.1c, number 12, Fig. 21.5). The cliff exposure shows two basaltic lava flows underlain by pre-eruption alluvial sediments (e.g. Martí et al. 2000; Pallí et al. 2005).

21.4 Conclusions

The thirty-six scoria cones preserved in the OVF were mainly constructed by strombolian activity, and most of them are of the horseshoe-shaped type. These cones have maximum and mean basal widths of 1,290 and 500 m, respectively. Maximum and mean heights are 189 and 100 m, respectively. The aspect ratio of the cones (height/basal width) ranges between 0.12 and 0.29. Six maar-type volcanoes have been identified, the largest of which has a crater 550 m in diameter and 20 m deep developed by phreatomagmatic explosive eruptions. Six additional volcanoes were constructed by alternating phases of strombolian, phreatomagmatic and effusive activity and are composed of a number of superimposed (overlapping) volcanic cones. The effusive activity resulted in significant accumulations of basaltic and basanitic lava flows reaching over 40 m in thickness and covering about 29.5 km^2 in the Olot plain. The lava flowed along pre-existing river valleys, extending for up to 10 km and generating volcanic dams. Subsequent differential fluvial erosion produced cliffs up to 50 m high underlain by resistant basaltic flows.

References

Araña V, Aparicio A, Martín-Escorza C, García L, Ortiz R, Vaquer R, Barberi F, Ferrara G, Albert J, Gassiot X (1983) El volcanismo Neógeno-Cuaternario de Cataluña: caracteres estructurales, petrológicos y geodinámicos. Acta Geol Hisp 18:1–17

Bolós X, Barde-Cabusson S, Pedrazzi D, Martí J, Casas A, Himi M, Lovera R (2012) Investigation of the inner structure of La Crosa de Sant Dalmai maar (Catalan Volcanic Zone, Spain). J Volcanol Geoth Res 247–248:37–48

Cebriá JM, López-Ruiz J, Doblas M, Oyarzun R, Hertogen J, Benito R (2000) Geochemistry of the Quaternary alkali basalts of Garrotxa (NE Volcanic Province, Spain): a case of double enrichment of the mantle lithosphere. J Volcanol Geoth Res 102:217–235

Cirés J, Gimeno D, Diaz E, Mallarach JM, Solà J, Montaner J, Viñals E, Mató E (2002) Mapa Geològic de Catalunya 1:25.000. Santa Pau 295-1-1. IGC, Barcelona

Di Traglia F, Cimarelli C, de Rita D, Gimeno D (2009) Changing eruptive styles in basaltic explosive volcanism: Examples from Croscat complex scoria cone, Garrotxa Volcanic Field (NE Iberian Peninsula). J Volcanol Geoth Res 180:89–109

Díaz N, Gimeno D, Losantos M, Segura C (1996) Las traquitas de Arenys d'Empordà (Alt Empordà, NE de la Península Ibérica): características generales. Geogaceta 20:572–575

Gisbert G, Gimeno D, Fernandez-Turiel JL (2009) Eruptive mechanisms of the Puig De La Garrinada volcano (Olot, Garrotxa volcanic field, Northeastern Spain): A methodological study based on proximal pyroclastic deposits. J Volcanol Geoth Res 180:259–276

Gómez B, Campeny G, Van der Made J, Oms O, Agustí J, Sala R, Blain HA, Burjachs F, Claude J, García S, Riba D, Rosillo R (2012) A new key locality for the Pliocene vertebrate record of Europe: the Camp dels Ninots maar (NE Spain). Geologica Acta 10:1–17

Losantos M, Planagumà L (2007) Carta Vulcanològica de la Zona Volcànica de la Garrotxa 1:25.000. IGC, ICC & PN Zona Volcànica Garrotxa, Barcelona

Martí J, Mallarach JM (1987) Erupciones hidromagmáticas en el volcanismo cuaternario de Olot. Estudios Geológicos 43:31–40

Martí J, Mitjavila J, Roca E, Aparicio A (1992) Cenozoic magmatism of the Valencia trough (Western Mediterranean): Relationship between structural evolution and volcanism. Tectonophysics 203:145–166

Martí J, Planagumà L, Geyer A, Canal E, Pedrazzi D (2011) Complex interaction between strombolian and phreatomagmatic eruptions in the Quaternary monogenetic volcanism of the Catalan Volcanic Zone (NE of Spain). J Volcanol Geoth Res 201:178–193

Martí J, Pujadas A, Ferrés D, Planagumà L, Mallarach JM (2000) El Vulcanisme. Guia de camp de la Zona Volcànica de la Garrotxa. PN Zona Volcànica Garrotxa, Olot

Pallí L, Brusi D, Soler D, Roqué C, Linares R, Cebrià A (2005) Travertines and volcanic landforms in the Eastern Pyrenees margin. Field Trip Guide C-3. Sixth International Conference on Geomorphology, Zaragoza

Picart J, Roqué C, Pallí L, Linares R, Vehí M, Soler D (2009) Mapa Geològic de Catalunya. Salt 333-2-1. IGC, Barcelona

Picart J, Roqué C, Pallí L, Soler D, Linares R, Vehí M (2008) Mapa Geològic de Catalunya. Vidreres 365-2-1. IGC, Barcelona

Puig C, Badia R, Bernat E, Diaz E, Martínez A, Samsó JM, Planagumà L, Mallarach JM, Solà J, Montaner J (2003) Mapa Geològic de Catalunya 1:25.000. Olot 257-1-2. IGC, Barcelona

Puig C, Mató E, Saula E, Picart J, Solà J, Montaner J, Samsó JM, Serra J, Llenas M, Agustí J, Mallarach JM (1997) Mapa Geològic de Catalunya 1:25.000. Canet d'Adri 295-2-2. IGC, Barcelona

Pujadas A, Pallí L, Brusi D, Roqué C (1997) El vulcanisme de la Vall de Llémena. Dialogant amb les Pedres, 5. Universitat de Girona, Girona

Pujadas A, Pallí L, Roqué C, Brusi D (2000) El vulcanisme de la Selva. Dialogant amb les Pedres, 8. Universitat de Girona, Girona

Vehí M, Pujadas A, Roqué C, Pallí L (1999) Un edifici volcànic inèdit a Caldes de Malavella: el volcà del camp dels Ninots. Quaderns de la Selva 11:45–72

Zarroca M, Linares R, Bach J, Roqué C, Moreno V, Font L, Baixeras C (2012) Integrated geophysics and soil gas profiles as a tool to characterize active faults: the Amer fault example (Pyrenees, NE Spain). Environ Earth Sci 67:889–910

The Teide Volcano, Tenerife, Canary Islands

22

Juan Carlos Carracedo

Abstract

The Teide Volcano rises 3,718 m a.s.l. and 7,500 m above the seafloor and is the world's third-highest volcanic structure. The last eruptions of this active volcanic system occurred in the Early Middle Ages in the Teide stratocone summit and in the western rift zone in 1909. The explicitness of the Teide's volcanism led von Buch and von Humboldt to abandon Neptunism in favour of Plutonism, a crucial step in the progress of modern geology and volcanology. The Teide volcanic complex comprises a spectacular volcanic system that includes mafic eruptions from active rift zones and a pair of felsic stratocones encircled by peripheral lava domes. This volcanic system is nested within the depression originated by a giant landslide that occurred about 200 ka ago. The gravitational collapse favoured the emplacement of shallow felsic magma chambers under the main stratovolcanoes, interacting with the deeper mafic magmas that feed the rift zones. This led to a continuous compositional progression in a bimodal basanite–phonolite series, with the mafic terms at the distal ends of the rifts and the felsic component in the central stratocones. Compositional differences are reflected in the diversity of eruptive mechanisms and in the variety of volcanic landforms and structures and their associated landscapes. The Teide Volcano was included in the UNESCO World Heritage List in 2007 as a site of extraordinary natural beauty and exceptional geological values, which provides highly significant evidence helping to understand geological processes in the evolution of oceanic volcanic islands.

Keywords

Teide Volcano • Felsic stratocones • Lava domes • OIB felsic magmatism • World Heritage Site • Tenerife • Canary Islands

22.1 Introduction

Mount Teide is an active volcano. The first references to volcanic eruptions in Tenerife correspond to distant sightings by seafarers in the fifteenth century, who used the Teide Volcano as a natural landmark during their voyages across the Atlantic. After the arrival of the pre-Hispanic population of Tenerife (the Guanches) about 2000 years ago, several eruptions, some of them explosive, account for the name given to this volcano: Teide (Echeide), the Guanche word for Hell.

The only remarkable aspect of Mount Teide until the eighteenth century was its height. The Teide was considered the highest mountain on Earth until Mont Blanc and the Andean volcanoes were measured, although the Teide (3,718 m a.s.l., over 7,000 m above the seafloor) is among the highest volcanic structures on the planet.

J. C. Carracedo (✉)
Departamento de Física-Geología, Universidad de Las Palmas de Gran Canaria, Las Palmas de Gran Canaria, Canary Islands, Spain
e-mail: jcarracedo@proyinves.ulpgc.es

F. Gutiérrez and M. Gutiérrez (eds.), *Landscapes and Landforms of Spain*, World Geomorphological Landscapes,
DOI: 10.1007/978-94-017-8628-7_22, © Springer Science+Business Media Dordrecht 2014

The role played by Mount Teide for the scientific community changed upon the arrival of naturalists such as Leopold von Buch, Charles Lyell, Alexander von Humboldt and Georg Hartung, among many others. The volcano became a privileged site for discussing a long-lasting controversy between Neptunists (i.e. Buch, Humboldt) and Plutonists (Lyell). The explicitness and magnificence of the Teide's volcanism inspired the former to definitively abandon the obsolete ideas of Neptunism and become ardent defenders of Plutonism, thereby providing a crucial starting point to overcome religious restrictions and open the door to the advancement of geology and volcanology.

Unfortunately, the geological studies on the Teide Volcano after the eighteenth century declined until well into the twentieth century. Particularly surprising was the almost total lack of geochronological information until 2007, in contrast with other volcanoes like those of Hawaii. The accepted idea was that dating the Teide volcanic complex (TVC) was unfeasible, due to the impossibility of applying K/Ar and $^{40}Ar/^{39}Ar$ techniques to this period and the absence of suitable organic material for radiocarbon dating (e.g. Araña et al. 2000). Nevertheless, a set of radiometric ages provided for the first time precise age constraints for the recent eruptive history of the Teide Volcano (Carracedo et al. 2007), establishing a geochronological framework for the structural and volcanic evolution of the TVC and allowing a more precise and realistic assessment of potential eruptive hazards.

The Teide National Park (TNP) was included in the UNESCO's World Heritage List (WHL) in 2007 (Unesco 2007), recognizing its natural beauty and importance for providing evidence of the geological processes that underpin the evolution of oceanic islands (Carracedo 2008).

22.2 Pre-Teide Stage

The centre of Tenerife is formed by a circular volcano 3,718 m high and 40 km in diameter (Fig. 22.1), constructed from about 12 million years ago and still active. Three well-defined stages can be identified in its development: (1) a large shield volcano, the Central Shield Volcano (CSV), formed from about 11.8 to 8.8 million years (Guillou et al. 2004); (2) Las Cañadas Volcano (LCV), built from about 3.5 million years ago (Ancochea et al. 1999; Huertas et al. 2002) on top of the shield volcano after a period of eruptive repose and erosion of about 5 million years; and (3) the TVC, built nested within the depression formed by a north-directed lateral collapse at around 200–180 ka, representing the current eruptive phase in Tenerife (Fig. 22.2).

The Miocene CSV has a crescent-shaped central part with a NE extension that can only be recognized inside *galerías* (artificial adits excavated for water supply), since it is almost entirely overlain by younger volcanic products (Guillou et al. 2004; Carracedo et al. 2011). Two parasitic shields developed attached to the CSV: Teno on the western flank (6.11–5.15 Ma, Guillou et al. 2004) and Anaga (4.89–3.95 Ma, Guillou et al. 2004) on the eastern flank (Fig. 22.2).

The development of the TVC, as well as its constructive and petrological characteristics, can only be completely understood in the framework of the continued magmatism that built the central part of Tenerife, probably fed by the same source and through the same magmatic plumbing system, together with additional shallow-level magma chambers associated with the LCV and the TVC felsic volcanism.

Prior to the TVC outset, the island summit was formed by the wide dome-shaped LCV, probably over 3,500 m a.s.l. This central felsic edifice may have undergone vertical collapses (Marti et al. 1994), and several north-directed lateral collapses (Huertas et al. 2002), suggesting that the rift system, that may have already been active as early as the Miocene (Carracedo et al. 2011), was established during the LCV construction, controlling the development of large landslides and the reconstruction of successive Cañadas edifices (Ancochea et al. 1999). At around 200–180 ka, the latest giant landslide truncated the LCV and left behind the Las Cañadas Caldera and the Icod Valley (its northern seaward prolongation) collapse scar (Watts and Masson 1995; Carracedo 1999; Carracedo et al. 2007; Márquez et al. 2008). Pre-existing vertical collapse structures may have contributed to the pre-landslide lateral instability of LCV (e.g. Martí et al. 1994, 1997; Troll et al. 2002). The landslide unroofed the rift zone triple junction and initiated the construction of a proto-Teide inside the collapse embayment (see Figs. 22.1 and 22.2). The giant landslide is the event that defines the stratigraphic erosional base of the TVC.

22.3 The Teide Volcanic Complex

Although the Teide and Pico Viejo volcanoes are geographically well-defined stratovolcanoes built on the floor of Las Cañadas Caldera (Fig. 22.3), the geological distinction of these volcanoes is not clear in spatial, temporal or compositional terms. The definition of volcano-stratigraphic units is based, among others, on the petrology of lavas. However, in this geological setting, there is a progression of composition in a bimodal basanite–phonolite series, with the mafic terms in the distal ends of the rifts and the felsic component in the central stratocones (Fig. 22.4). The Montaña Reventada eruption, located near the area of influence of the Teide's felsic magma chamber (see lower inset in Fig. 22.4), emitted a composite flow with a lower

Fig. 22.1 Satellite (NASA) vertical view of Tenerife, topped by the Teide Volcano

basanite and a much thicker upper phonolite, suggesting that the two magmas erupted successively. Therefore, since the rift zones and the Teide-Pico Viejo volcanoes function as coeval and interactive volcanic systems, the best approach is to consider them to be components of the TVC (Carracedo et al. 2007), which can be clearly separated from the LCV by the lateral collapse (Fig. 22.4).

22.4 Main Components of the TVC

In this regional-scale approach, the TVC includes the Teide–Pico Viejo complex (4 in Fig. 22.4) and the northwest and northeast active rift zones (2 in Fig. 22.4). Monotonous isotopic signatures of these rift zone lavas, along with systematic spatial variations in major and trace element chemistry, support this argument (Wiesmaier et al. 2012; Troll et al. 2013). Pre-TVC features, i.e. LCV (1 in Fig. 22.4) and the giant landslide embayment (3 in Fig. 22.4), are indicated as reference elements.

22.4.1 The Teide and Pico Viejo (T-PV) Stratovolcanoes

The Teide and Pico Viejo (PV) composite volcanoes top the island of Tenerife with 3,718 m and 3,100 m a.s.l. (Fig. 22.5a). The former is the third-highest volcanic structure in the world, after Mauna Loa and Mauna Kea volcanoes on the island of Hawaii. The geometry and stratigraphy of

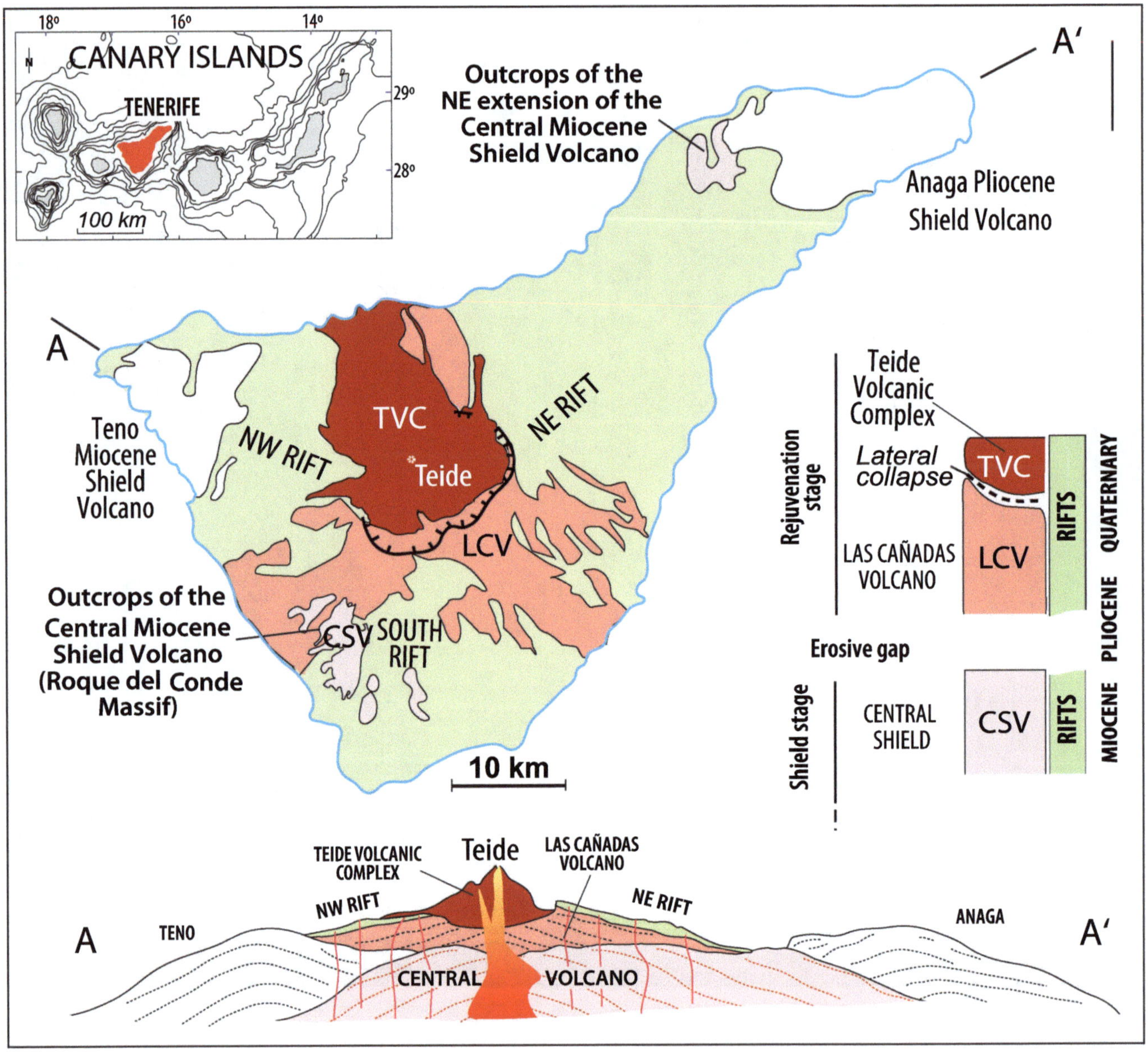

Fig. 22.2 Geological map and cross-section of the island of Tenerife showing the three Miocene–Pliocene shields that constitute the island, the post-erosional Las Cañadas Volcano and the rift zones (after Guillou et al. 2004)

overlapping stratocones are difficult to define, particularly without radiometric data; the first radiometric ages of the T-PV stratocones were published very recently (Carracedo et al. 2007). Previously, the only date reported was the 2 ka age of Montaña Blanca (Ablay et al. 1995). This lack of geochronological data led to different interpretations on how the T-PV compound volcano developed, generally regarded as twin volcanoes, with the Teide edifice overlapping Pico Viejo (Fig. 22.5b) (Fúster et al. 1968; Ablay and Marti 2000). However, 54 new radiocarbon and K/Ar ages from the T-PV volcano (Carracedo et al. 2007), together with stratigraphic relationships, reveal that Pico Viejo is a parasitic stratocone, apparently developed when the Teide stratocone was at its terminal phase of development. Therefore, it seems that Pico Viejo reaches an altitude as high as 3,100 m because its basement is perched at about 2,500 m, attached to the western flank of the Teide Volcano (Fig. 22.5c).

Two circumstances may have favoured this parasitic growth of PV: (1) the excessive altitude (3,718 m a.s.l.) reached by the Teide cone, thus probably forcing vent migration. Empirical evidence suggests that lithostatic load limits the local relief of composite volcanoes to about 3,000 m (Davidson and de Silva 2000) and (2) most PV lavas were very fluid plagioclase basalts (pahoehoe flows) and, in contrast with the Teide lavas, resulted in a low aspect ratio (height/width) volcanic edifice.

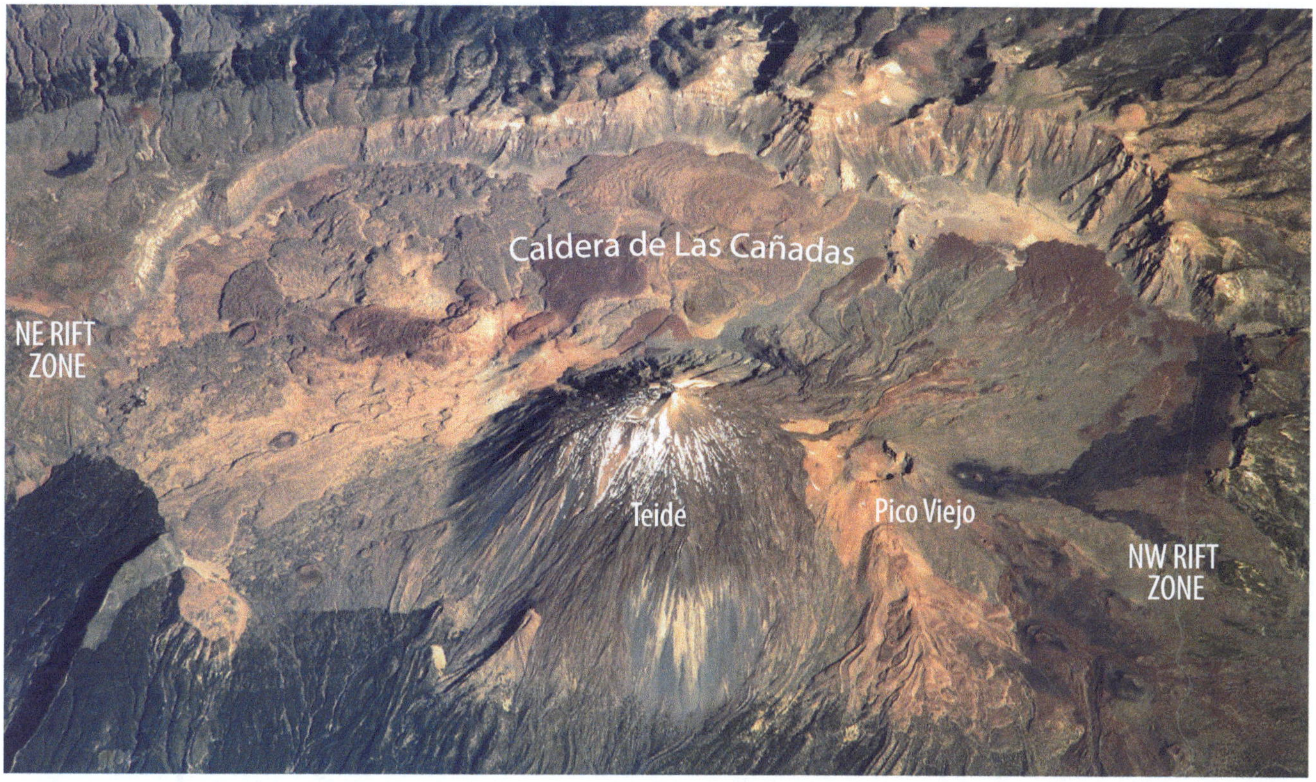

Fig. 22.3 Oblique satellite view of the Caldera de Las Cañadas and Teide Volcano. Photograph (ISS013-E-23272) taken on 8 June 2006 by an expedition 13 crew member on the International Space Station

22.4.2 The Teide Volcano

The reconstruction of the volcanic history of the Teide Volcano (TV) is based on field observations and radiometric dating. Initial eruptions at Teide produced mafic lavas indistinguishable from those of the rift zone eruptions (the proto-Teide or Old Teide) that progressively filled the bottom of the Las Cañadas Caldera and the Icod Valley, providing the basement for the development of the Teide stratocone (Carracedo et al. 2007). These early sequences can only be reached through *galerías* (Fig. 22.6). Three ages from the same flow sampled at the deepest part of the sequence in the Galería Salto del Frontón provided consistent ages of 198 ka (K/Ar), and 195 and 192 ka ($^{40}Ar/^{39}Ar$), the latter from two different laboratories (Carracedo et al. 2007). These ages constrain the timing of the pre-TVC lateral collapse.

The ages of the Old Teide lavas form the base of cliffs at the mouth of the Icod Valley range from 124 to 88 ka (see Fig. 22.6). Later on, the Teide eruptions progressed towards more differentiated lavas, yielding phonolite from about 32 ka (Carracedo et al. 2007). This variation probably resulted from the establishment of a shallow magmatic plumbing system underneath the progressively growing central complex. Meanwhile, the rift zones continued erupting mafic products, likely from deeper sub-Moho reservoirs. Phonolite lavas at the foot of the rim of the Caldera de Las Cañadas and outflowing into the Orotava Valley and down to the Icod coast (Playa de San Marcos) consistently yield radiocarbon and K/Ar ages of 33–32 ka (Fig. 22.6, Carracedo et al. 2007).

An interesting eruptive episode of the Teide likely interacted with summit ice, resulting in an explosive phreatomagmatic event that mantled its north flank with slabs of surge-type, indurated, whitish breccias (Calvas del Teide, Fig. 22.7a), interbedded within the final phonolitic lavas of the old Teide (Pérez Torrado et al. 2006, 2013).

The latest activity of the Teide stratocone appears to be the phonolitic Lavas Negras eruption (Fig. 22.7b), dated at 1240 ± 120 yr BP (1150 ± 140 cal. yr BP; Carracedo et al. 2007). After a period without further growth of the main stratocone, its activity apparently shifted to the construction of the PV parasitic volcano and the peripheral lava domes.

22.4.3 Pico Viejo Volcano

The summit of the parasitic Pico Viejo Volcano has an 800-m-wide crater at 3,100 m a.s.l. (Fig. 22.8). Field observations and radiometric dating (^{14}C and K/Ar) allowed the

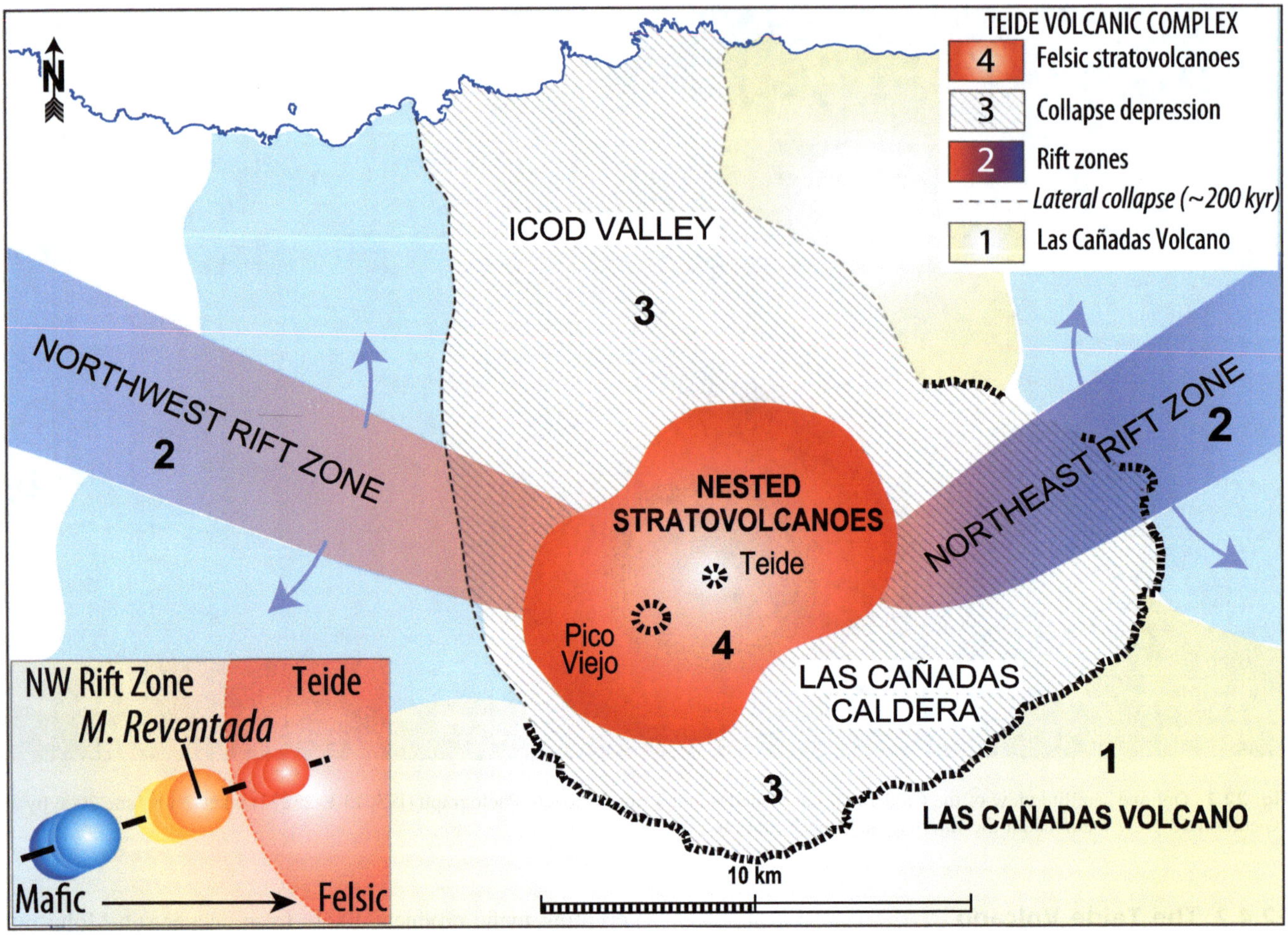

Fig. 22.4 Features forming the Teide volcanic complex: *1* The pre-collapse Las Cañadas Volcano. *2* The rift zones (NW and NE). *3* The partially filled Las Cañadas–Icod Valley landslide depression. *4* The main stratocones (Teide and Pico Viejo)

separation of three main volcano-stratigraphic units: initial (basaltic), transitional (intermediate and evolved lavas) and terminal (phonolite lavas).

The successions corresponding to the initial constructional eruptive phase of PV, perched on the flank of the main Teide stratocone, constitute the main volume of the volcano and consist of very fluid, mainly pahoehoe plagioclase basalts. These lavas flowed over an extensive area of western Tenerife (about 120 km2), reaching the north and west coasts and confined by the Miocene Teno Volcano, the Teide edifice and Las Cañadas Caldera scarp (Fig. 22.9). A radiocarbon age of about 27 ka (Carracedo et al. 2007), from a pahoehoe lava at the top of the Teide sequence at the Icod coast, probably dates the bottom of the sequence, consistently with the growth of the PV at the terminal phase of the Teide Volcano. This sequence largely modified the previous relief, providing a smooth topography for the Holocene vent lineations of the NW rift zone.

The transitional stage, comprising lavas of intermediate and evolved compositions, has been dated at 20 ka (Carracedo et al. 2007), while the more evolved terminal lavas (phonolites), capping the flanks of PV, date to about 17 ka, constraining the end of the main period of development of PV.

Contrarily to the relatively long constructional period of the TV, encompassing about 200 ka, the PV main activity only lasted about 10–15 ka. This marked difference is related to the large volume needed to fill the gigantic landslide depression before the TV could develop its present configuration, whereas the initial vents of the PV stratocone rest on the flank of the TV at an altitude of more than 2,000 m.

At the terminal stage, a lava lake filled the crater, with the magma level fluctuating as in the Nyiragongo lava lake (Tazieff 1977), producing two stepped terraces within the crater: (1) an upper terrace, formed by lavas that completely

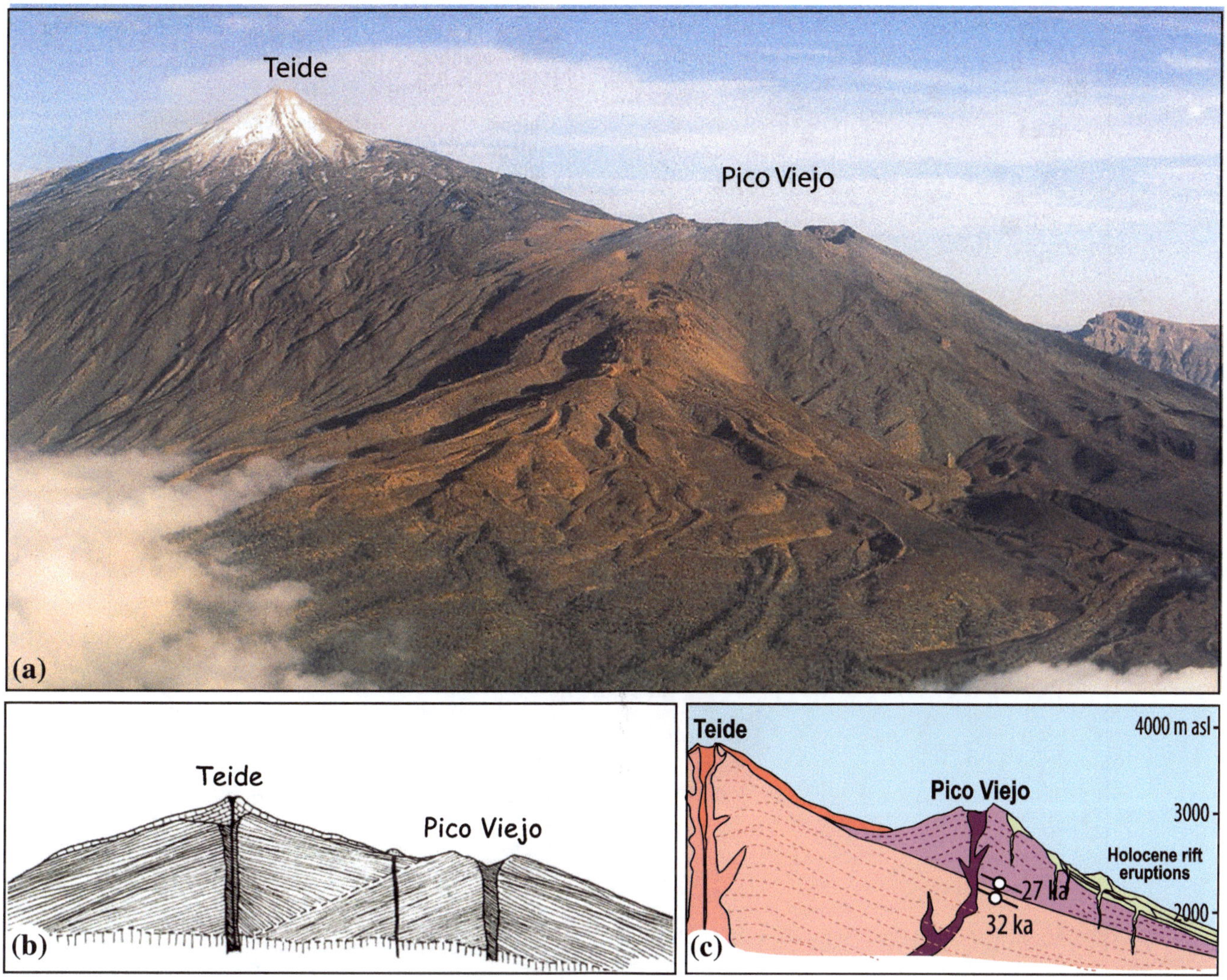

Fig. 22.5 **a** The Teide and Pico Viejo (PV) composite volcanoes topping the island of Tenerife with heights of 3,718 and 3,100 m a.s.l. In the foreground, the Roques Blancos parasitic lava dome and thick phonolitic coulees. **b** These central stratocones have been considered to be twin volcanoes, with the Teide edifice overlapping Pico Viejo (figure from Fúster et al. 1968). **c** Volcanic chronostratigraphy based on new radiocarbon and K/Ar ages from the T-PV volcano, revealing that Pico Viejo is a parasitic stratocone, apparently developed when the Teide stratocone was at its terminal phase (Carracedo et al. 2007)

filled and eventually overflowed the crater, represented by a small bench, and (2) a lower terrace, created after the magma drained back, forming the current floor of the crater (Fig. 22.8b).

Minor eruptions at the flanks and inside the main crater of the Pico Viejo Volcano persisted until relatively recent times (Holocene), but it is only a matter of definition as to whether these eruptions correspond to Pico Viejo or to the NW rift zone. One of these eruptions, with explosive phreatomagmatic episodes (Pérez Torrado et al. 2006, 2013), likely driven by snow melt and magma interaction, similar to that described in TV, opened an explosion pit in the SW part of the crater (3), spreading surge and grey breccia deposits (4) around the volcano (Fig. 22.8b).

22.4.4 Peripheral Lava Domes

An outstanding geological feature of the Teide Volcano is a set of phonolitic lava domes located around the central volcanoes (Fig. 22.10). The increasing elevation of both stratocones above the critical level likely forced intrusion of these felsic lavas through radial fissures at their basal perimeter. This late episode of volcanism of the TVC mainly occurred during the Holocene (see Fig. 22.10). As volcanoes develop, the distance the magma has to ascend to reach the summit of the edifice increases. As illustrated in Fig. 22.11, the eruption-driving overpressure (Pex) required for the lava to ascend to the summit vent progressively augments with increasing edifice altitude (1 in Fig. 22.11). At a critical

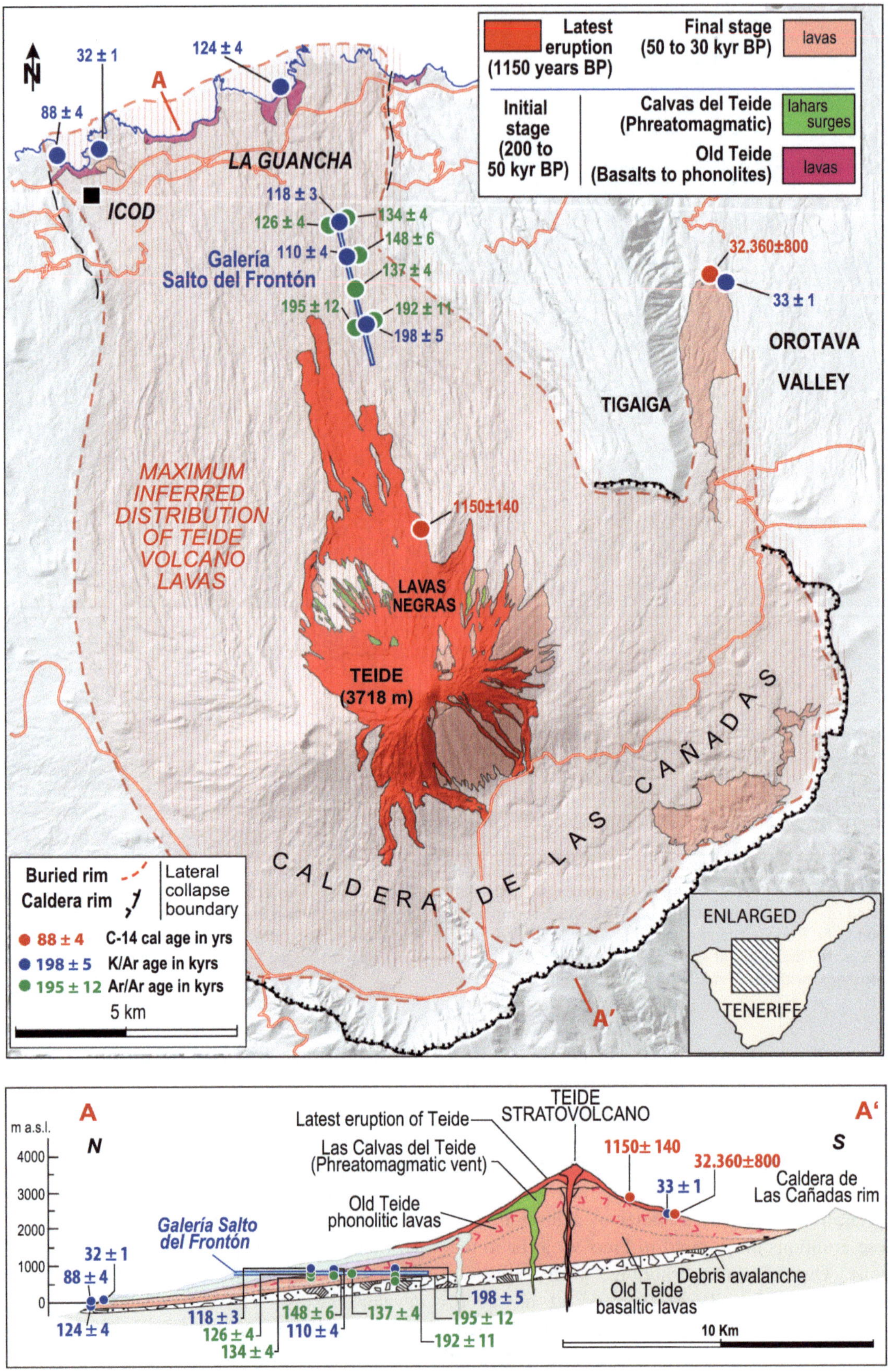

Fig. 22.6 Geological map of the TV, showing the location of radiocarbon (in years B.P.) and K/Ar (in ka) ages (*upper part*), and N–S cross-section (*lower part*) depicting the entire post-Las Cañadas lateral collapse stratigraphic sequence (modified from Carracedo et al. 2007)

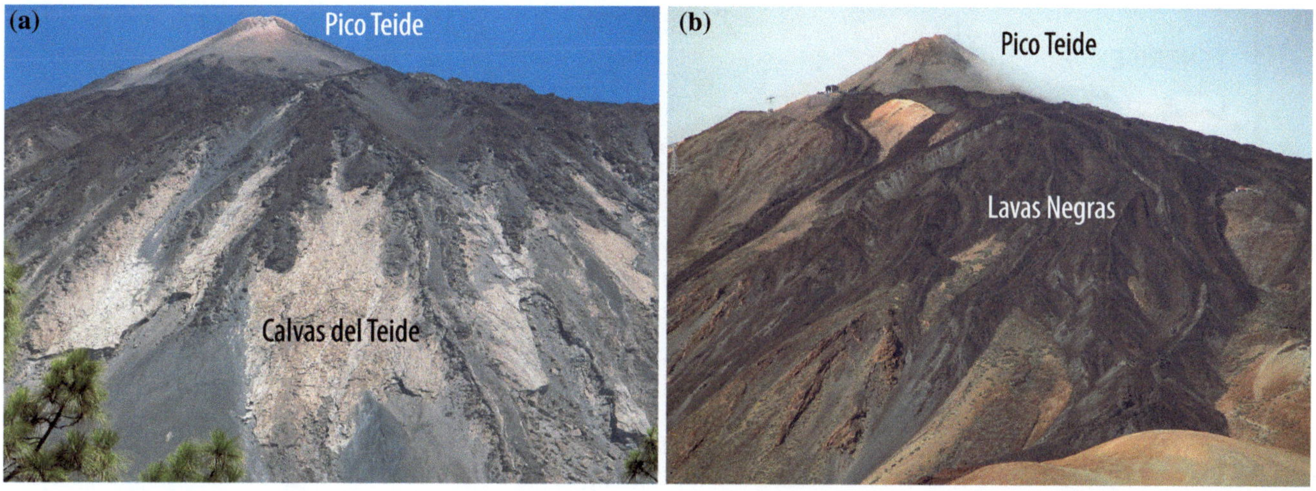

Fig. 22.7 Explosive phreatomagmatic eruption of Las Calvas del Teide (**a**), which mantled the northern flank of the stratovolcano with slabs of surge-type, indurated, whitish breccias interbedded within the final phonolitic lavas of the old Teide and topped by the latest (Middle Ages) eruption of Lavas Negras (**b**)

Fig. 22.8 **a** Pico Viejo Volcano viewed from the summit of Teide, with the south coast of Tenerife in the background. **b** Vertical view (Google Earth image) of the impressive 800-m-wide PV summit crater with a block *1* of the lava lake that filled the crater, subsequently drained to a lower level *2*. The explosion pit *3* is related to a phreatomagmatic episode that mantled with grey explosion breccias and surge deposits *4* the top and flanks of the volcano

height, lithostatic pressure finally exceeds Pex in the magmatic system and summit eruptions are strongly hindered or no longer feasible (Davidson and De Silva 2000). Lateral stresses induced by the lava at the foot of the volcano cause its bulging and the development of radial fractures. The latter evolve into open flank vents (2 in Fig. 22.11), the upper ones of which generally degas the system in explosive events, while the effusive lower vents form bulges (cryptodomes) and lava domes (3 and 4 in Fig. 22.11). Lava dome eruptions tend to be moderately explosive episodes, forming extensive air-fall deposits of pumice.

The morphology of the phonolitic domes is mostly restricted to near hemispherical low lava domes (or tortas) and coulées (Blake 1990), whose shape is strongly conditioned by the slope. Tortas are generally erupted on flat ground (see inset in Fig. 22.12a), where the lava is able to push outwards, but not far, lifting erupted flows aside from the vent, like petals in a rose, forming an onion-like internal structure and a typical external structure "in rosette" (Fig. 22.12a). Tortas and short and thick flows are mainly located at the confined part of the Caldera de Las Cañadas floor (see Fig. 22.10).

Spectacular steep-fronted blocky flows or coulees are from lava domes that develop on steep slopes (Fig. 22.12b) and produce lavas that flow for long distances. Lava oozes slowly down the slope, continuing to travel, while the vent keeps erupting, frequently forming prominent flow levees (Fig. 22.12b). The lava outpouring from the vent pushes the

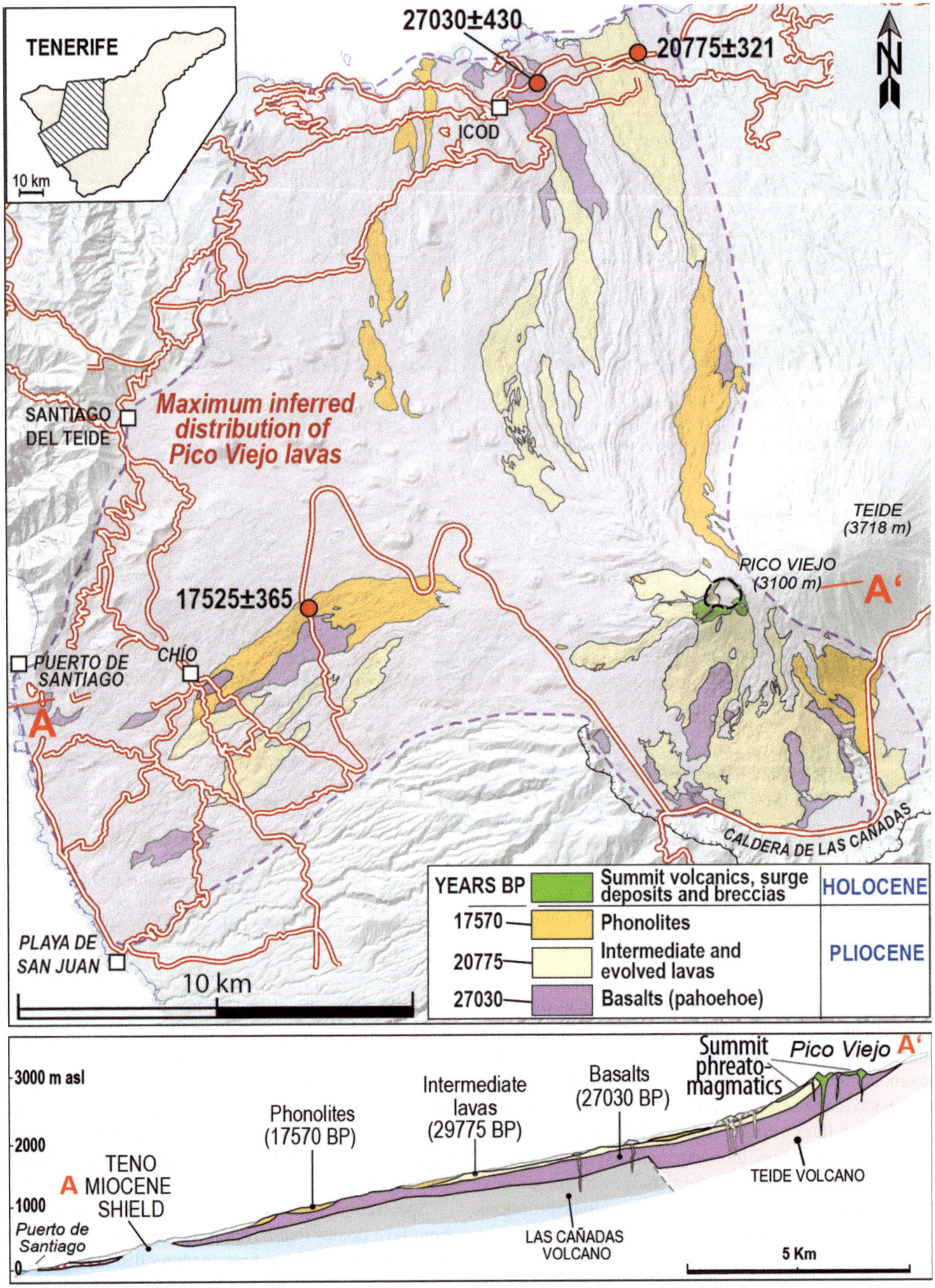

Fig. 22.9 Geological map of the PV (*upper part*), showing the location of radiocarbon ages (in yr B.P.), and W–E cross-section (*lower part*), depicting the PV stratigraphic sequence. Double red lines are roads (modified from Carracedo et al. 2007)

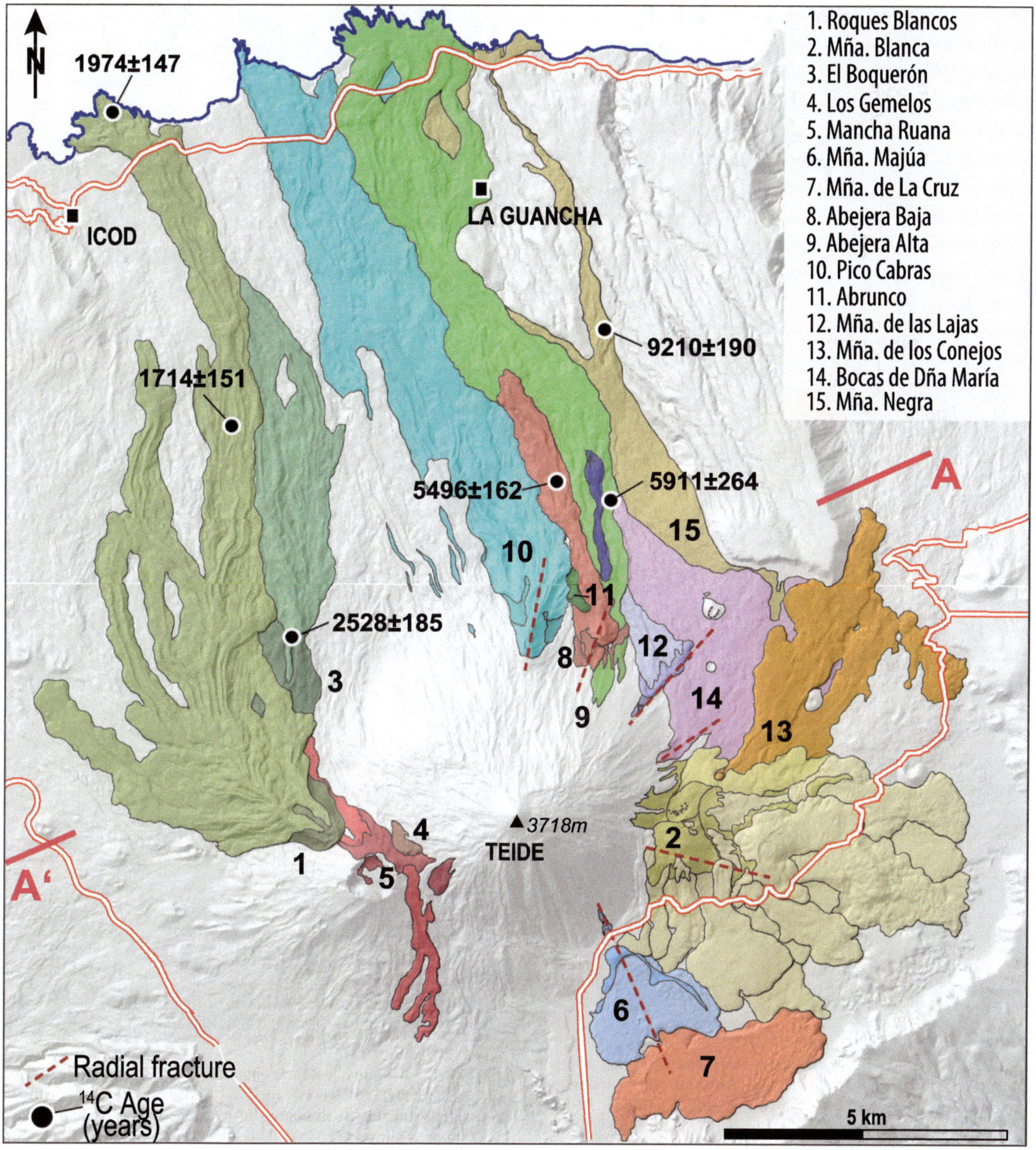

Fig. 22.10 Phonolitic lava domes and related coulées of the T-PV volcanic complex. Radiocarbon ages in yr B.P. (modified from Carracedo et al. 2007)

colder and more viscous frontal zone of the coulée and forms conspicuous pressure ridges perpendicular to the direction of flow on the distal surface of the flow. Frequently, these ridges become convex (ogives), due to the lateral velocity gradient of the lava flow inside the channel. Coulees in the T-PV complex are mainly located at the northern, unconfined flank of the volcanoes.

An interesting question is why these viscous phonolitic flows can travel for such long distances, up to 15 km in the T-PV complex (e.g. the Roques Blancos coulée, Fig. 22.12b).

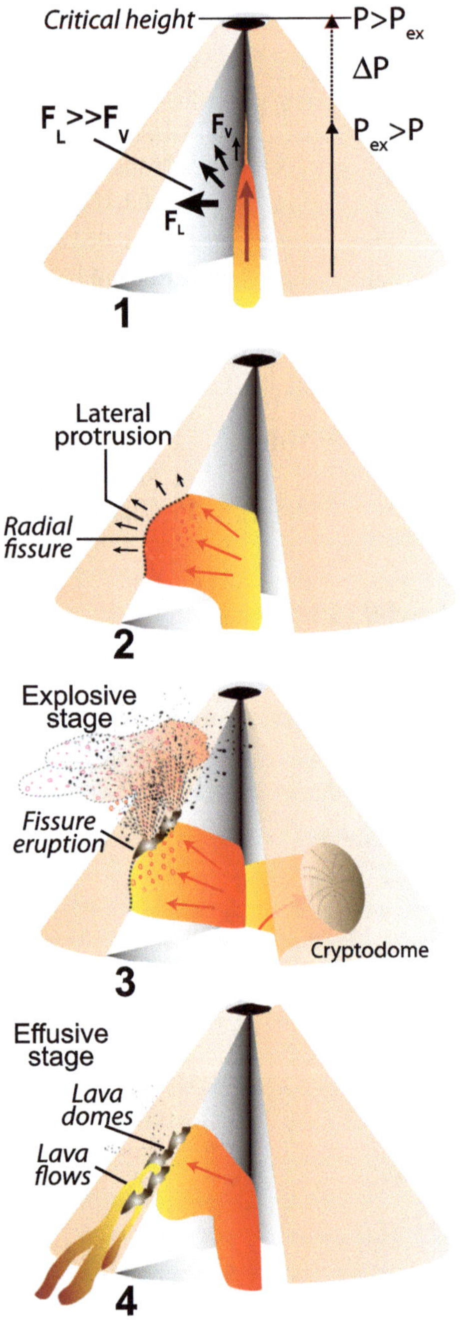

Fig. 22.11 Cartoon illustrating how the eruption-driving overpressure (Pex) required to allow lava ascent to the summit vent progressively increases with increasing edifice altitude *1*, finally leading to the opening of radial fractures *2* to form phonolitic peripheral lava domes and coulees *3*, *4*

Fig. 22.12 **a** Montaña Rajada, phonolitic lava dome (torta) formed on flat ground (the floor of Las Cañadas Caldera), with the characteristic external structure "in rosette". **b** The Roques Blancos lava dome, built on the steep flank of Pico Viejo, emitted phonolitic coulees 15 km long that reached the northern coast

The principal factors controlling lava mobility are magma discharge rate, physical properties and slope. In the case of the T-PV peripheral phonolitic lava domes, discharge rate is the most important factor, the distance travelled by the different flows being proportional to the discharge rate.

However, other factors may also help in keeping these long flows in motion, particularly those favouring thermal isolation of the hot lava. The Roques Blancos coulée (Fig. 22.12b) shows two different flow dispositions. In one case, the lava cools and obstructs the channel, dividing the main flow into many short branches that cool and halt in a typical anastomosed configuration (a in Fig. 22.12b). Other channelled flows (b) run for long distances without signs of

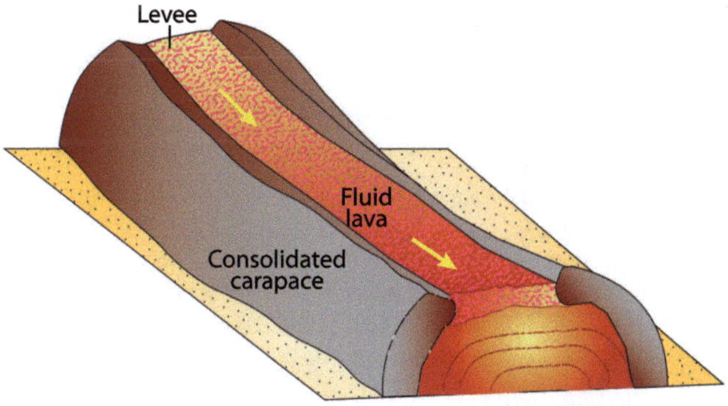

Fig. 22.13 The long run-out of the flows may be favoured by features that increase thermal insulation of the hot lava, in particular the formation of an external carapace of solid rock that isolates the lava within the lava channel and maintains the flow

obstruction, suggesting that lava was maintained hot and fluid, probably because of the formation of an external carapace of solid rock that isolated the lava and prolonged the flow (Fig. 22.13).

22.5 Rift Zone Holocene Eruptions

Eruptions have been frequent in Tenerife along the rifts, but Holocene volcanism concentrated mainly in the NW rift zone (NWRZ) (Fig. 22.14), including four of the island's five historical eruptions: Boca Cangrejo (1492), Garachico (1706), Chahorra (1798) and Chinyero (1909) (Fig. 22.15). Coexistence of basaltic rift volcanism with central phonolitic eruptions is evident in the prominent exposure on the main Dorsal Ridge road locally known as La Tarta (the cake), a sequence formed by a white Plinian phonolitic unit interbedded between two basaltic lapilli deposits from cinder cones of the NE rift (Fig. 22.16).

22.6 The Teide World Heritage

Besides its outstanding geological and geomorphological values, the TVC, the centrepiece of TNP, is a prime tourist attraction: the most heavily visited national park of any European country and the second most visited worldwide. Teide is a huge volcano that towers 3,718 m (a.s.l.) above the central part of Tenerife, reaching the highest elevation in the Canaries and Spain. Moreover, if its height is measured relative to the seafloor, the Teide is the third-tallest (7,718 m) volcanic edifice on Earth after the Hawaiian shield volcanoes Mauna Kea and Mauna Loa.

Spain presented to the United Nations Educational, Scientific and Cultural Organization (UNESCO) a proposal for the inscription of the TNP as a World Heritage Site. To be included on the WHL, sites must be of outstanding universal value and meet at least one out of ten selection criteria (six cultural and four natural). TNP represents an example of a site of extraordinary natural beauty (criterion 7) and exceptional geological values (criterion 8), which provides highly significant evidence helping us to understand geological processes important in the evolution of oceanic volcanic islands (http://whc.unesco.org/en/list/1258). This does not detract, by any means, from other natural and cultural values (flora and fauna, archaeology, history, etc.), which are also outstanding in TNP. But these alone may not have been sufficient for nomination, since UNESCO is extremely strict in demanding not only the extraordinary value of natural assets, but also that such assets be of a universal nature, and a strong initial drawback for TNP inscription was the many volcanic sites already included in the WHL, all with truly exceptional cultural values or spectacular flora and fauna (e.g. the Galapagos National Park was inscribed in 1979). In this regard, the most promising option for the TNP was to emphasize its exceptional character as an archetypal example of hot-spot-related volcanic islands, focusing the application on criteria 7 (natural beauty) and 8 (geological–volcanological values). However, even in this highly specific field, the WHL already included the Hawaii Volcanoes National Park (HVNP), nominated in 1987 for containing two of the most active volcanoes in the world, Mauna Loa (4,170 m high) and Kilauea (1,250 m high), both of which tower over the Pacific Ocean. Therefore, the HVNP apparently already represented oceanic volcanoes to the highest standard.

The basic argument demonstrating how exceptional Mount Teide is, and how the TNP could complement the HVNP in the representation of oceanic volcanoes in the WHL came from a purely geological fact, reflected in the Total Alkali Silica (TAS) diagram of Fig. 22.17. Although the Hawaiian and Canary Islands are very similar, both being excellent examples of oceanic islands originated by the volcanic activity associated with hot spots, the former are an extreme example of hotspot activity (very active and fertile), while the Canaries are the best example of the opposite situation. The Hawaiian Islands are built from a high magma-producing hot spot on the rapidly drifting Pacific Plate (about 10 cm/year) such that they are built up at a very high rate and have short active lifespans. Hawaiian volcanism is characterized by a high degree of partial melting of the mantle and very limited magma differentiation. The Canaries, in contrast, are a reference archipelago for much slower rates: the slow movement of the African Plate (some 10 mm/year), low magma production rates with slow construction and a long active life of the islands, low degrees of partial mantle melting and abundant volcanism of differentiated magmas. As shown in the TAS diagram, the HVNP comprises only mafic lavas, while the TNP

Fig. 22.14 View of the NW rift zone from the PV summit, with the Teno Miocene shield in the background, and La Gomera (*left*) and La Palma (*right*) in the far distance

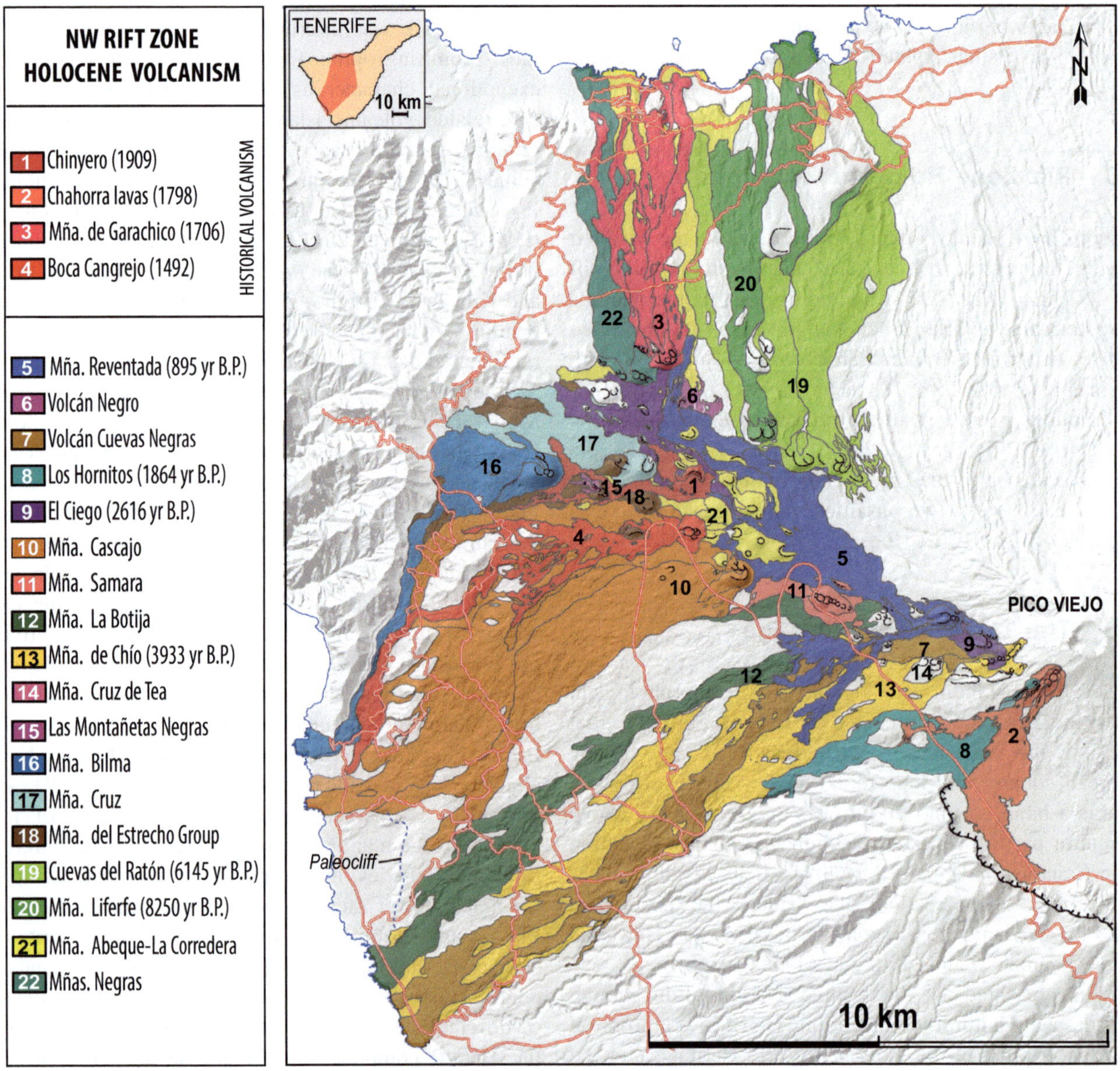

Fig. 22.15 Geological map showing the Holocene eruptions of the NW rift zone (modified from Carracedo et al. 2007)

Fig. 22.16 La Tarta (the cake), a sequence formed by a white Plinian phonolitic unit from the central Las Cañadas volcanic complex interbedded between basaltic lapilli deposits from NE rift cinder cones

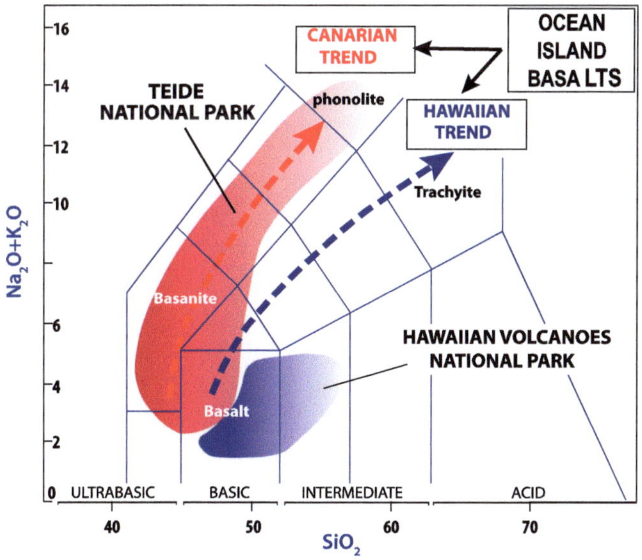

Fig. 22.17 Basic argument demonstrating that Mt Teide is an exceptional volcano, and how the Teide National Park could complement the Hawaii Volcanoes National Park in representing ocean volcanoes in the World Heritage List. As shown in the TAS diagram, because of geodynamic differences in the development of both islands, the HVNP magmas did not fully evolve, comprising only mafic lavas, while the TNP includes all terms in the magmatic variation trend, from mafic to very evolved phonolites. These differences are reflected in the types of volcanism in both archipelagos and thus in the variety of their respective volcanic forms, structures and landscapes

includes all terms in the spectrum of magma compositions, from mafic to very evolved phonolites. These differences are reflected in the types of volcanism in both archipelagos and thus in the variety of their volcanic forms and structures. Consequently, oceanic island volcanism can only be completely represented if both the HVNP and the TNP are included in the WHL, as UNESCO accepted in 2007 (Carracedo 2008).

22.7 Conclusions

The Teide Volcano, the world's third-tallest volcanic structure and one of the most impressive, played a crucial role in the early development of volcanology in the eighteenth and nineteenth centuries. The active TVC comprises a rift system and two central stratocones encircled by phonolitic peripheral lava domes. The bulk of the complex developed nested within the depression originated by a giant landslide about 200 ka ago. The lateral collapse favoured migration of magma to shallow depths and subsequent fractionation. Interaction between shallow felsic magma chambers under the main stratovolcanoes and deeper mafic magmas feeding the rift zones resulted in a continuous progression of compositions in a bimodal basanite–phonolite series, with the mafic terms at the distal ends of the rifts and the felsic component at the central stratocones. The TVC underwent numerous basaltic, intermediate and felsic eruptions, the latest of which occurred in 1909. Compositional differences explain the diversity of eruptive mechanisms and the variety of volcanic products, forms and structures, justifying the UNESCO declaration of the Teide Volcano as a World Heritage Site in 2007.

References

Ablay GJ, Ernst GGJ, Martí J, Sparks RSJ (1995) The 2 ka subplinian eruption of Mña. Blanca Tenerife. Bull Volcanol 57:337–355

Ablay GJ, Martí J (2000) Stratigraphy, structure, and volcanic evolution of the Pico Teide–Pico Viejo formation, Tenerife, Canary Islands. J Volcanol Geoth Res 103:175–208

Ancochea E, Huertas MJ, Cantagrel JM, Coello J, Fúster JM, Arnaud N, Ibarrola E (1999) Evolution of the Cañadas edifice and its implications for the origin of the Cañadas Caldera (Tenerife, Canary Islands). J Volcanol Geoth Res 88:177–199

Araña V, Felpeto A, Astiz M, García A, Ortiz R, Abella R (2000) Zonation of the main volcanic hazards (lava flows and ash falls) in Tenerife, CI. A proposal for a surveillance network. J Volcanol Geoth Res 103:377–391

Blake S (1990) Viscoplastic models of lava domes. In: Fink JH (ed) Lava flows and domes. Springer, Heidelberg, pp 88–126

Carracedo JC (1999) Growth structure instability and collapse of Canarian volcanoes and comparisons with Hawaiian volcanoes. J Volcanol Geoth Res Spec Issue 94:1–19

Carracedo JC, Rodríguez-Badiola E, Guillou H, Paterne M, Scaillet S, Pérez-Torrado FJ, Paris R, Fra-Paleo U (2007) Eruptive and structural history of Teide volcano and rift zones of Tenerife, Canary Islands. Geol Soc Am Bull 119:1027–1051

Carracedo JC (2008) Outstanding geological values: the basis of Mt Teide's World Heritage nomination. Geol Today 24(3):104–111

Carracedo JC, Guillou H, Nomade S, Rodríguez-Badiola E, Pérez-Torrado FJ, Rodríguez-González A, Paris R, Troll VR, Wiesmaier S, Delcamp A, Fernández-Turiel JL (2011) Evolution of ocean-island rifts: the northeast rift zone of Tenerife, Canary Islands. Geol Soc Am Bull 123:562–584

Davidson J, De Silva S (2000) Composite volcanoes. In: Sidgurdsson H (ed) Encyclopedia of volcanoes. Academic Press, San Diego, pp 663–681

Fúster JM, Araña V, Brandle JL, Navarro M, Alonso U, Aparicio A (1968) Geología y volcanología de las Islas Canarias: Tenerife: Madrid, Instituto "Lucas Mallada", Consejo Superior de Investigaciones Cientificas, p 218

Guillou H, Carracedo JC, Paris R, Pérez-Torrado FJ (2004) K/Ar ages and magnetic stratigraphy of the Miocene–Pliocene shield volcanoes of Tenerife, Canary Islands: implications for the early evolution of Tenerife and the Canarian hotspot age progression. Earth Planet Sci Lett 222:599–614

Huertas MJ, Arnaud NO, Ancochea E, Cantagrel JM, Fúster JM (2002) 40Ar/39Ar stratigraphy of main pyroclastic units from the Cañadas volcanic edifice (Tenerife, Canary Islands) and their bearing on structural evolution. J Volcanol Geoth Res 115:351–365

Márquez A, López I, Herrera R, Martín-González F, Izquierdo T, Carreño F (2008) Spreading and potential instability of Teide volcano, Tenerife Canary Islands. Geophys Res Lett 35:L05305

Martí J, Mitjavila J, Araña V (1994) Stratigraphy, structure and geochronology of the Las Cañadas Caldera (Tenerife, Canary Islands). Geol Mag 131:715–727

Martí J, Hurlimann M, Ablay GJ, Gudmundsson A (1997) Vertical and lateral collapses on Tenerife (Canary Islands) and other volcanic ocean islands. Geology 25:879–882

Pérez-Torrado FJ, Paris R, Cabrera MC, Schneider JL, Wassmer P, Carracedo JC, Rodríguez- Santana A, Santana F (2006) Tsunami deposits related to flank collapse in oceanic volcanoes: the Agaete Valley evidence, Gran Canaria, Canary Islands. Mar Geol 227:135–149

Pérez-Torrado FJ, Carracedo JC, Rodríguez-González A, Rodríguez-Badiola E, Paris R, Troll VR, Clarke H, Wiesmaier S (2013) Eruptive styles at the Teide volcanic complex. In: Carracedo JC, Troll V (eds) Teide volcano: Geology and eruptions of a highly differentiated oceanic stratovolcano. Active volcanoes of the World. Springer, pp 213–231

Tazieff H (1977) An exceptional eruption: Mt. Nyiragongo, Jan. 10th, 1977. Bull Volc 40:189–200

Troll VR, Walter T, Schmincke H-U (2002) Cyclic caldera collapse: Piston or piecemeal subsidence? Field and experimental evidence. Geology 30–2:135–138

Troll V, Deegan FM, Delcamp A, Carracedo JC, Harris C, van Wyk de Vries B, Petronis MS, Pérez-Torrado FJ, Chadwick JP, Barker AK, Wiesmaier S (2013) Pre-Teide volcanic activity on the Northeast volcanic rift zone. In: Carracedo and Troll (eds) Teide volcano: Geology and eruptions of a highly differentiated oceanic stratovolcano. Active volcanoes of the World. Springer, pp 75–92

UNESCO (2007) World heritage centre official site. Inscription of Teide National Park (http://whc.unesco.org/en/list/1258)

Watts AB, Masson DG (1995) A giant landslide on the north flank of Tenerife Canary Islands. J Geophys Res 100(B2):24487–24498

Wiesmaier S, Troll VR, Carracedo JC, Ellam RM, Bindeman I, Wolff JA (2012) Bimodality of lavas in the Teide–Pico Viejo succession in Tenerife: the role of crustal melting in the origin of recent phonolites. J Petrol 53–12:2465–2495

23 The 1730–1736 Eruption of Lanzarote, Canary Islands

Juan Carlos Carracedo

Abstract

Eruptions resumed in 1730 in Lanzarote Island after a prolonged period of volcanic repose, probably encompassing the entire Holocene. This historical eruption involved about 3–5 km^3 of basaltic pyroclasts and lavas, covering some 225 km^2 (one third of the island). The accumulation of volcanic products had a strong impact on the landscape of this Miocene oceanic island. This was the second largest effusive basaltic event in recorded history, surpassed only by the 1783 Lakagigar eruption in Iceland. The central part of Lanzarote was mantled by lapilli-derived soils and aeolian sands, which provided a strongly contrasting ground for the basaltic products of the 1730 eruption. After the initial phase of the eruption, the style changed and new vents were controlled by a 15-km-long volcano–tectonic zip-like eastwards-progressing fissure, with the first vents opening offshore west of the island. This abrupt modification may explain the progression of this eruption, from the average duration of historical Canarian eruptions (a few months), towards an exceptionally prolonged period of about six years. Besides duration, other outstanding features of the 1730–1736 eruption include the tholeiitic composition of lavas and the length of flows and lava tubes, particularly in the final stages. Initially, the eruption had a catastrophic impact on the resources of the island, since most of the farmland was covered by lavas and lapilli. However, agriculture significantly improved after the eruption with the introduction of dry farming, using lapilli cover as a new mulching technique.

Keywords

Fissure eruptions • Historical volcanism • Morphology of Oceanic Islands • Eruptive hazards • Lanzarote • Canary Islands

23.1 Introduction

The 1730–1736 Lanzarote basaltic eruption formed a complex fissural volcanic system with more than 30 eruptive centres, developed in at least five main, well-defined, multi-event eruptive phases. The vents aligned along a W–E trending volcano–tectonic fracture about 15 km long (Figs. 23.1 and 23.2).

This outstanding eruption, lasting almost 6 years, differs greatly from the normal pattern of historic volcanism in the Canarian Archipelago in duration (typically 3–4 months), magnitude, as well as type and evolution of magmas (Fúster et al. 1968), and has attracted scientific interest practically since the activity ceased (Hartung 1857; Brun 1908; Hernández-Pacheco 1909; Carracedo et al. 1990, 1992).

About 3–5 km^3 of basaltic pyroclasts and lavas were accumulated, covering some 225 km^2 (one third of the island) and burying the most productive farmland and

J. C. Carracedo (✉)
Departamento de Física-Geología, Universidad de Las Palmas de Gran Canaria, Las Palmas de Gran Canaria, Canary Islands, Spain
e-mail: jcarracedo@proyinves.ulpgc.es

F. Gutiérrez and M. Gutiérrez (eds.), *Landscapes and Landforms of Spain*, World Geomorphological Landscapes,
DOI: 10.1007/978-94-017-8628-7_23, © Springer Science+Business Media Dordrecht 2014

Fig. 23.1 Map on a scale of approximately 1/200,000, an oil painting from November 1730. The extent of lavas and pyroclasts emitted in the first months of the 1730 volcanic eruption and the emplacement of the villages destroyed can be estimated on the basis of the location in the map of present-day towns and geographical features, situated on the map with considerable precision. The initial vent—Caldera de los Cuervos—is indicated with an arrow (Real Audiencia de Canarias, Gracia y Justicia, Leg. 89, Archivo General de Simancas)

numerous villages (Carracedo and Rodríguez-Badiola 1991; Carracedo et al. 1992). The Lanzarote eruption is the second largest effusive basaltic event in recorded history, surpassed only by the 1783–1784 Lakagigar eruption in Iceland, with a volume of 14 km^3 (Thordarson and Self 1993).

Before the 1730–1736 eruption, the central part of the deeply eroded, 15 million years old island of Lanzarote, was dominated by a low-altitude flat topography (200 m asl), largely underlain by white aeolian sands. This landscape drastically changed after the eruption, which covered an extensive part of the central plain (about 225 km^2, one third of the island area) with black basaltic lapilli, lava flows, and abundant volcanic cones, causing a spectacular rejuvenation of the volcanic landscape of the island in some sort of "geologic make-up". Lanzarote is the oldest island of the Canaries, after Fuerteventura, and is in its late rejuvenation stage, although it was considered for a long time to be one of the youngest and most volcanically active islands of the archipelago. However, the 1730–1736 and the 1824 historical events are probably the only Holocene eruptions in the island.

The consequences of the prolonged volcanism on the island economy and population were devastating. A great part of the most fertile land was destroyed, together with 26 villages, and the subsequent famine eventually forced the majority of the inhabitants to leave the island. However, soon after the eruption, the island's agricultural resources improved with the introduction of dry farming using lapilli cover as a new mulching technique, which took advantage of effective reduction in evaporation losses and the enhanced nocturnal condensation of water vapour from the atmosphere (Graf et al. 2004).

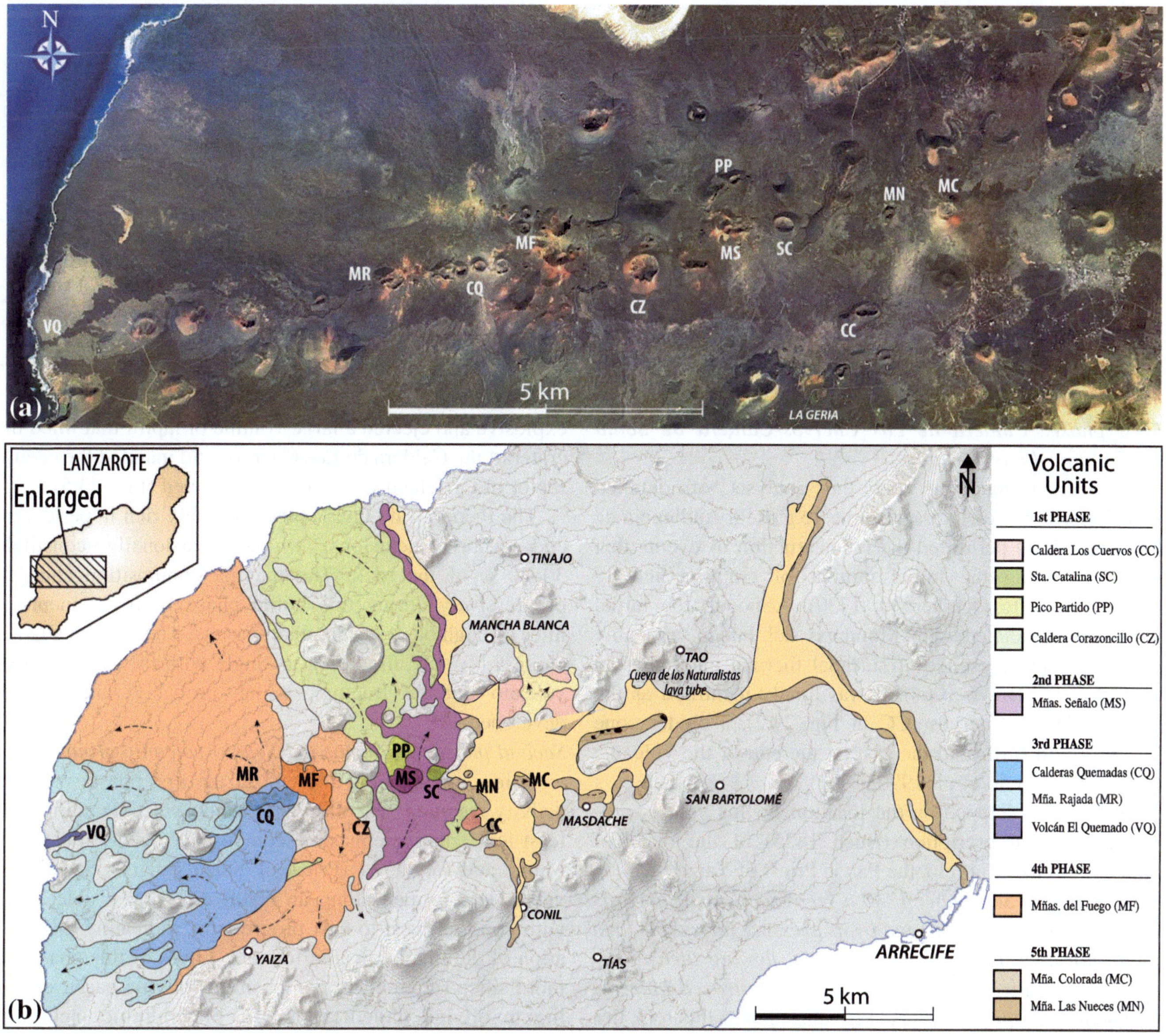

Fig. 23.2 A. Google Earth vertical view of the 1730–1736 volcanic fissure. B. Simplified geological map of the 1730–1736 eruption. CC = Caldera de los Cuervos (initial eruptive centre of the eruption); PP = Pico Partido; SC = Caldera de Santa Catalina; MS = Montañas del Señalo; VQ = Volcán de El Quemado; MR = Montaña Rajada; CQ = Calderas Quemadas; MF = Montañas del Fuego; MN = Montaña de Las Nueces; MC = Montaña Colorada, the last eruptive vent (modified from Carracedo et al. 1992)

Several eyewitness accounts contributed to the identification of this long and complex eruption and the successive eruptive vents and lava fields, particularly the detailed entries in the diary kept by the Parish Priest of Yaiza (included in the work of Buch 1825) and the official reports to the Royal Court of Justice of the Canary Islands, from the local authorities appointed to manage the crisis (Real Audiencia de Canarias 1731). The latter document, found on file in the Spanish Archivo General de Simancas, includes a map (Carracedo et al. 1990, 1992; Fig. 23.1).

23.2 The 1730–1736 Volcanic Eruption

At least ten main eruptive centres can be distinguished in the 14-km-long volcanic fissure of the 1730–1736 eruption (Fig. 23.2a). Significant changes in the eruptive dynamics, magma composition, stratigraphic relationships, and comparison with the descriptions of contemporary accounts, allowed the differentiation of five eruptive phases, some of them comprising several volcanic events and emission centres (Fig. 23.2b).

Fig. 23.3 Caldera de Los Cuervos, the initial vent of the eruption, identified by the detached and rafted block, described in contemporary eyewitness accounts (photograph JC Carracedo)

First phase: Caldera de Los Cuervos-Caldera de Santa Catalina-Pico Partido

Geological mapping and the analysis of detailed descriptions in eyewitness accounts (e.g. Real Audiencia de Canarias 1731) permitted the reconstruction of events that occurred during the initial months of eruptive activity (September 1 to October 30, 1730). Although the initial vent was associated with Montañas del Fuego, probably because of the presence of surficial thermal anomalies in that location (Bravo 1964), it was finally identified as the Caldera de los Cuervos (CC in Fig. 23.2), a cinder cone located 6 km further to the east, at the end of the volcanic fissure (Carracedo et al. 1990, 1992).

A crucial reference on the location of this vent is provided in the transcription (Buch 1825) of the diary of Andrés Lorenzo Curvelo, the Parish Priest of Yaiza "*On the seventh day of September 1730 a great rock burst upwards with a thunderous sound, and by its pressure forced the lava going northwards to change direction, flowing then to the northwest and west-north west. There it destroyed Maretas and Santa Catalina [...]*". As Fig. 23.3 shows, this rock lies 150 m north of a breach at the flank of the cone, coinciding with the missing part of the volcano detached by the pressure of the lava filling the crater. The block, shifted by the lava, diverted the former flow to the NW towards the village of Santa Catalina. This initial episode (1–19 September 1730) emitted very fluid and extremely SiO_2-undersaturated lavas (melanephelinites) containing abundant ultrabasic (mainly peridotitic) inclusions.

The second eruptive episode of this initial phase started on 10 October 1730 (Real Audiencia de Canarias 1731), with the opening of two new vents: Pico Partido and Caldera de Santa Catalina (PP and SC in Figs. 23.2; a and b in 23.4 and Table 23.1), aligned with the Caldera de Los Cuervos along a NW–SE fissure. These vents emitted large volumes of lava that reached the northern coast and flowed to the SE towards Uga and Yaiza villages, destroying more farmland and villages, and forming an extensive lava field. Some of these vents, particularly Caldera de Santa Catalina, were more explosive and ejected a large volume of lapilli that filled the crater of the Caldera de Los Cuervos and mantled the central sector of the island with a thick lapilli bed (Fig. 23.5).

The Pico Partido lavas were extremely rich in peridotitic and gabbro inclusions, which occasionally constitute 30–40 % of the rock. The magma composition changed progressively during this initial phase from melanephelinites through basanites to alkali basalts. These vents aligned along a NW–SE fissure, at an angle with the main (N80E) volcano–tectonic fracture, which probably had not yet developed.

Second phase: Montañas del Señalo volcanic group

A few months after the end of the first phase, volcanic activity resumed near the Pico Partido volcanic edifice. At least four eruptive episodes seem to have occurred between March and July 1731 (Priest of Yaiza's diary). Successive vents were formed along an N80E direction (Table 23.1), with constant west-to-east progression, indicating the initiation of the main volcanic fracture. Lavas of this phase are more silica-saturated in composition (olivine basalt evolving to olivine tholeiites) and lack peridotitic inclusions (Carracedo et al. 1992); the higher magmatic viscosities and lower effusion rates in this phase led to shorter flows that stopped halfway to the coast (Fig. 23.2).

Third phase: Volcán del Quemado-Montana Rajada-Calderas Quemadas

An abrupt change in the eruption is indicated by the apparently sudden shift of the eruptive activity (by at least 12 km) to the westernmost edge of the main fracture, which propagated westwards into the sea. The initial activity in this phase, at the end of June 1731, formed a submarine (phreatomagmatic) vent on the west coast. According to the Diary of the Parish Priest of Yaiza, there were explosions at the coast, and probably even deeper submarine eruptions, as suggested by the arrival to the shore of deepwater fish species previously unknown in the area. Onshore, the first eruptive centre was El Quemado Volcano, a group of small elongated cinder cones and hornitos located 1 km from the coast (Figs. 23.2 and 23.6a). The eruptive activity migrated

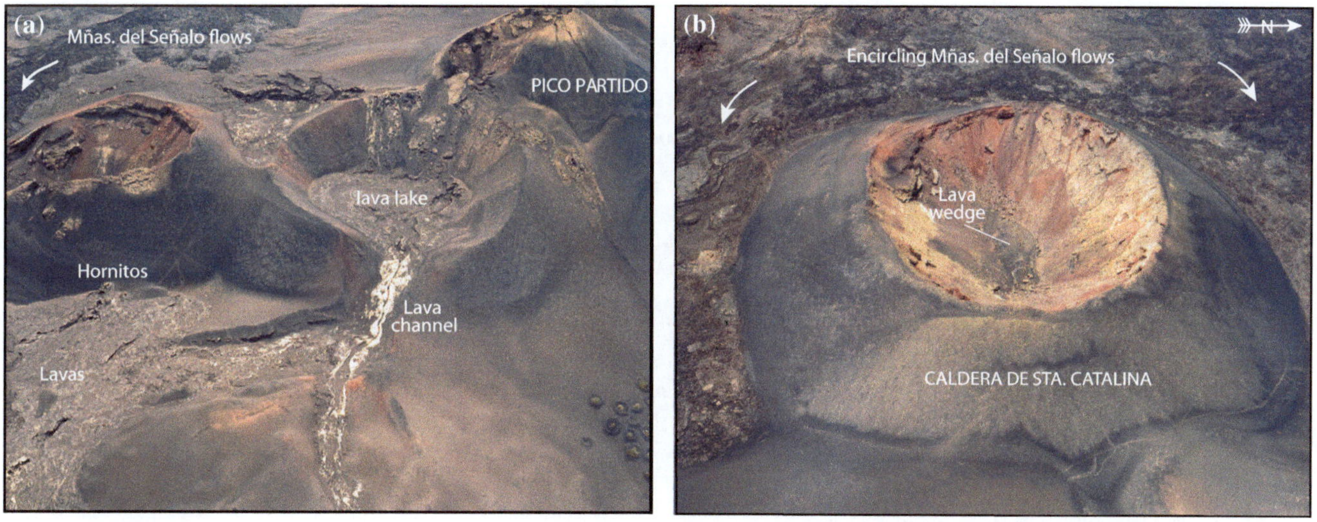

Fig. 23.4 A. Pico Partido group. B. Caldera de Santa Catalina, the explosive vent that emitted the lapilli mantling the centre of the island (Google Earth images)

Table 23.1 Main eruptive phases and centres of the 1730–1736 eruption, Lanzarote

Phase	Eruptive centre	Date and duration	Reference
I	Caldera de los Cuervos	1–13 September 1730	Real Audiencia de Canarias (1731)
	Pico Partido	10 October 1730 to January 1731	Real Audiencia de Canarias (1731)
	Caldera de Sta. Catalina	10–31 October 1730	Real Audiencia de Canarias (1731)
II	Montañas del Señalo	March–June 1731	Real Audiencia de Canarias (1731)
III	Volcán El Quemado	June? 1731	Lorenzo Curvelo (1731)
	Mña. and Caldera Rajada	July? 1731	Lorenzo Curvelo (1731)
	Calderas Quemadas	December 1731 to January? 1732	Lorenzo Curvelo (1731)
IV	Montaña del Fuego	1732? to 1736?	
V	Montaña de las Nueces	Early 1733 to 16 April 1736	De La Hoz (1960), Pallarés (2007)
	Montaña Colorada	2? to 16 April 1736 (end of eruption)	De La Hoz (1960)

continuously to the east along the fracture, forming an alignment of closely spaced cinder cones: Montaña Rajada and the four Calderas Quemadas (Fig. 23.2 and Table 23.1). Lavas of similar composition and evolution to the preceding phase (alkali basalts to olivine tholeiites) continued to flow seawards, spreading over a large area of the western part of the island.

At this stage, Lanzarote had experienced over a year of continued volcanic activity, contrarily to the normal duration of historical eruptions in the Canaries, hitherto never lasting more than 99 days, the time span of the longest historical eruption, Chahorra Volcano in Tenerife, or the 146 days of the 2011–2012 submarine eruption in El Hierro (Carracedo et al. 2012). After devastating most of the farmland and villages, the latest Calderas Quemadas episodes closely threatened Yaiza village. Despaizing of ever seeing an end to the eruption, the villagers and their Parish Priest Lorenzo Curvelo took refuge in Gran Canaria (Buch 1825). The diary ends at this time, and further contemporary information comes from the dossier sent by the Real Audiencia de Canarias to the Crown, whose last entry is a summary of the entire eruption written on 12 April 1731. With a great part of the island blanketed by lapilli, and the best land no longer able to yield food for humans or beast, most of the population fled the island at this stage, four years before the end of the eruption.

Fourth phase. Montañas del Fuego volcanic group

Early in 1732, the eruption became stationary for a long period. Volcanic activity was concentrated in the central part of the main fracture, with an initial explosive period in which piling up cinder cones built the complex volcanic edifice of Montañas del Fuego (Fig. 23.6b). Later on, the eruptive style changed and large volumes of lavas were emitted from clusters of hornitos located around the main volcanic group.

Lava composition is similar to those of phases 2 and 3, ranging from alkali basalts to highly evolved olivine tholeiites. Voluminous lava effusion from the cluster of hornitos northwest of Montañas del Fuego (foreground in Fig. 23.7) concealed the previous topography, described as

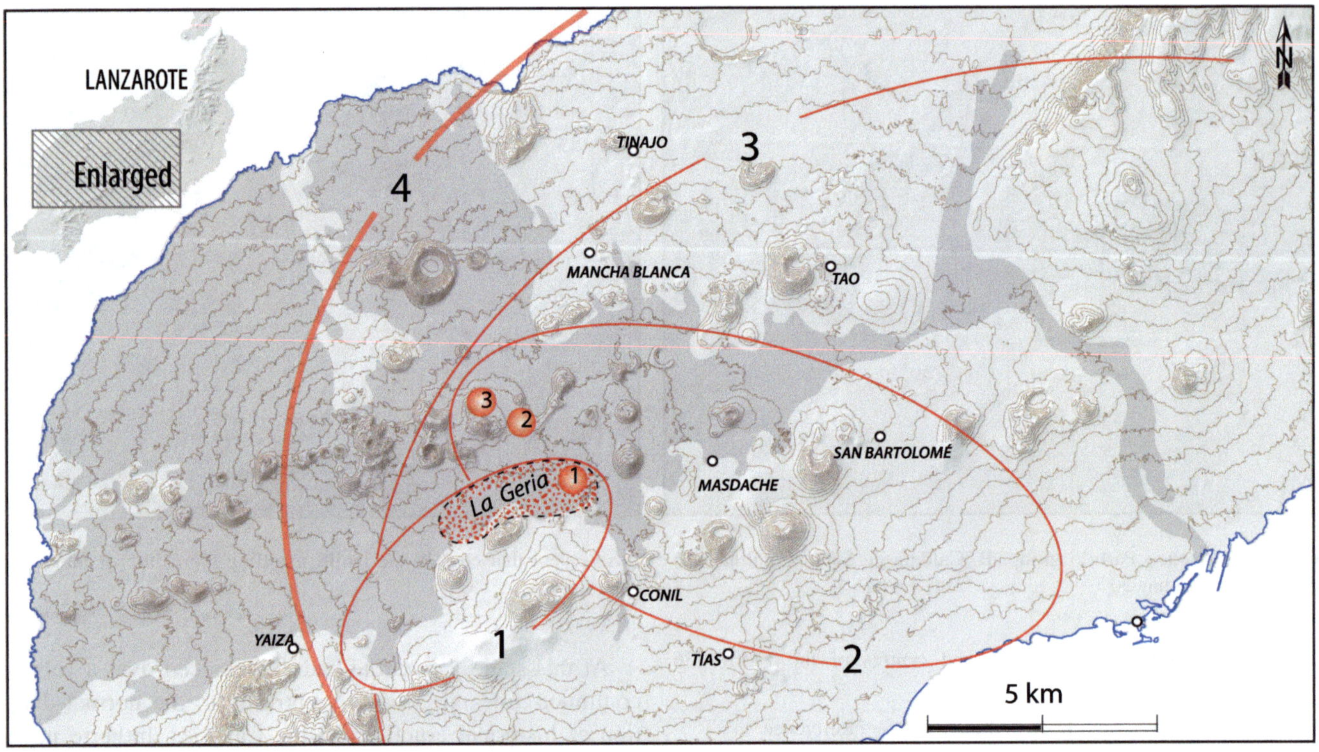

Fig. 23.5 Area covered by lavas (*darker grey*) of the 1730 eruption and approximate distribution of pyroclasts at different stages of the eruption (ellipses) deduced from eyewitness accounts (Real Audiencia de Canarias, 1731): (1) Caldera de Los Cuervos (September 1730); (2) Caldera de Santa Catalina (October 1730); (3) Pico Partido (December 1730); (4) Lapilli accumulated at the end of the first phase, February 1730. Note that La Geria lapilli field coincides with the maximum overlapping and accumulation of lapilli airfall from the successive eruptions. The corresponding eruptive vents and dates are also indicated (modified from Carracedo et al. 1992)

a wide fertile valley in pre-1730 accounts (Bontier and Leverrier 1404).

The duration of this phase and the temporal distribution of its episodes cannot be defined because of the lack of reliable contemporary records for the period 1732–1736. However, based on the size of the volcanic products, it seems plausible that volcanic activity focused in this area for a long time span, probably most of the period of over three years until the onset of the subsequent and last phase of the eruption, in early 1733 (Dávila y Cárdenas 1737), March 1735 (Pallarés 2007), or March 1736 (Buch 1825).

Fifth phase: Montaña de Las Nueces-Montaña Colorada volcanoes

After the long period without information about the eruption, the next account reports the opening of new vents 5 km to the east of Montañas del Fuego, indicating the onset of the fifth and last phase of the eruption. In this phase, only two eruptive centres were active: the Montaña de Las Nueces and Montaña Colorada volcanoes (Fig. 23.8). Both vents are located at the eastern end of the general volcanic fracture, indicating that the eruption resumed its earlier constant west–east progression (see Fig. 23.2).

The timing of the Montaña de Las Nueces eruption is uncertain. The lavas of this vent that evolved very rapidly from alkali basalts to the most differentiated olivine tholeiites produced by the entire 1730 eruption, formed a massive 20-km-long flow that reached the sea near Arrecife, the capital of the island (see Fig. 23.2b). This suggests that the activity in this vent started before February 1733, since Dávila y Cardenas, in their visit to the island in 1733, reported that this flow of Montaña de Las Nueces threatened the harbour of Arrecife.

Similar uncertainties apply to Montaña Colorada, the last vent active during the 1730–1736 eruption. A recent review by Pallarés (2007) ascribes a flow from this volcano to a contemporary account of De La Hoz (1960), who describes an individual flow that travelled north in early April 1736 towards the town of Tinajo.

The lavas issued from this final eruptive centre were, from the initial stages, evolved olivine tholeiites and, as in the first phase of the eruption (Pico Partido Volcano), they had numerous peridotitic inclusions. They flowed northwards but, because of their small volume and low effusion rates, did not reach the sea. This last episode was probably

Fig. 23.6 A. Oblique aerial views of the western part of the 1730 eruptive fissure showing the first vent onshore, Volcán Quemado. B. Montañas del Fuego, with part of Calderas Quemadas to the left

very short, since volcanic activity in the island ceased completely on 16 April 1736 (Buch 1825).

23.3 Volcano-Tectonic Control of the Eruption

The 1730–1736 Lanzarote fissure eruption is in many aspects similar to the Icelandic Laki fissure eruption in 1783, although the former had a higher magnitude (Volcanic Explosivity Index 6, in a scale of 8). Both eruptions aligned along a ca. 15-km-long fissure emitting large amounts of basaltic lavas, lapilli, and gases with global effects, mainly a significant drop in global temperature.

In Lanzarote, the fracture associated with the 1730–1736 eruption has interesting peculiarities. The alignment of the initial phase of the eruption differs significantly from the general trend (N80E). A plausible explanation is that the N80E fracture developed independently 5–6 months after the onset of the eruption. This suggests that the initial phase, a typical Canarian eruption (in terms of extent, duration, number of vents, etc.), was modified by the opening of the N80E fracture, causing the eruption to have a much longer duration and higher magnitude than the average basaltic fissure eruptions in the Canaries, at least in the Holocene.

Although with a slight offset in direction of about 10°, the 1730–1736 eruption is part of the rift-type alignment of recent emission centres of Lanzarote (Series III and IV of

Fig. 23.7 Oblique aerial view of Montañas del Fuego. To the left, the Islote de Hilario, where the geothermal anomalies crop out. In the background, the southern alignment of pre-Holocene cinder cones

Fig. 23.8 Oblique aerial view of Montañas de las Nueces and Montaña Colorada, the final vents of the 1730–1736 eruption (Google Earth image)

Fúster et al. 1968), characterized by a narrow band of closely spaced eruptive vents (Fig. 23.9), typical of rift zones (Walker 1992; Carracedo 1994). As proposed by several authors (e.g. Fiske and Jackson 1972; Swanson et al. 1976), magma injects in these structures through the dykes already formed by hydraulic pressure. A possible explanation is that the magma pressure, acting like a zipper or a blade, forced the fracture to open and keep its direction, generating a fracture that propagates continuously eastwards, from the ocean towards the central part of the island (Fig. 23.10).

Another significant feature of this eruption is the long period of permanence of volcanism focused at the central part of the fracture, giving place to the greater concentration of vents that formed the Montañas del Fuego volcanic group. Probably, the stationary period relaxed the control exerted by the general volcanic fracture on the distribution of the eruptive

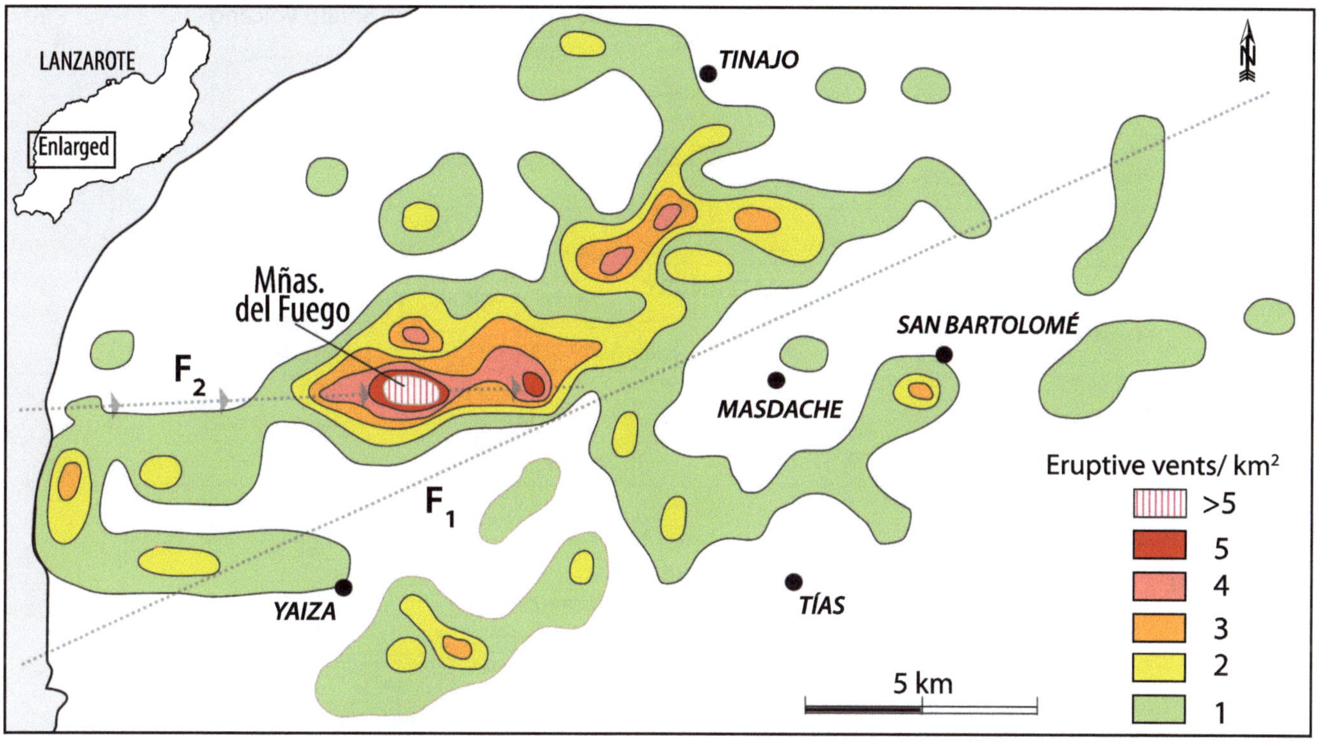

Fig. 23.9 Concentration of eruptive vents in the central part of Lanzarote (Upper Pleistocene–Holocene). F1 shows the general trend and F2 the direction and progression of the 1730–1736 propagating fracture and fissure eruption

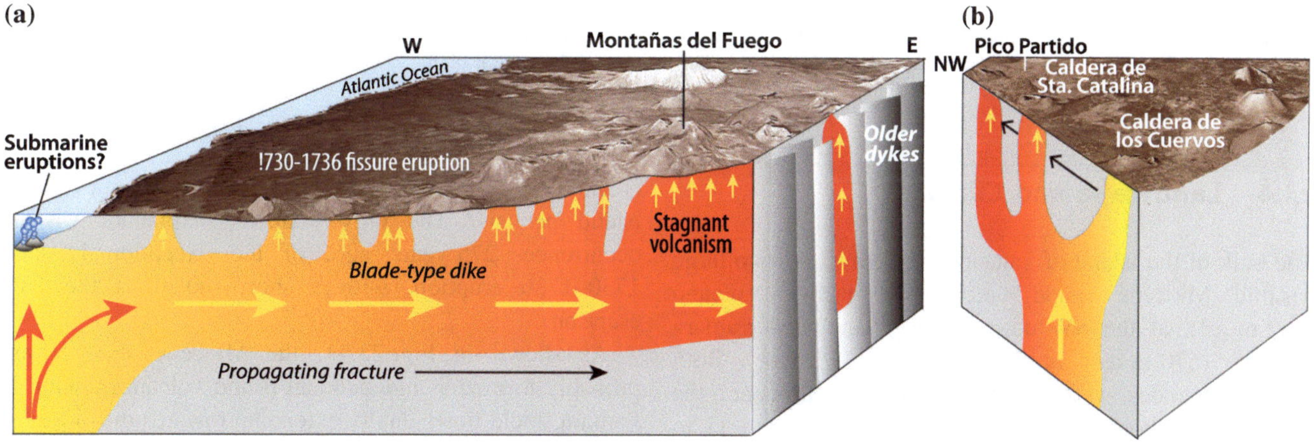

Fig. 23.10 **a** Sketch illustrating the association of the 1730–1736 fissure eruption with a main W–E propagating fracture. **b** The plumbing system of the initial stage of the eruption may have been independent and related to a different (SE–NW trending) eruptive fissure. This two-step development may have lengthened the 1730–1736 eruption well beyond the average duration of the historical eruptions in the Canaries

centres, favouring the progression from a linear array of vents along the main fracture to a more centralized geometry, capturing a greater amount of magma flow, and gaining in mechanical efficiency to raise the magma to the surface (Wadge 1981). This could also explain the exceptional duration of this eruption, over 20 times longer than the lengthiest historical eruption in the Canaries, and 16 times longer than the 2011–2012 submarine eruption of El Hierro Island.

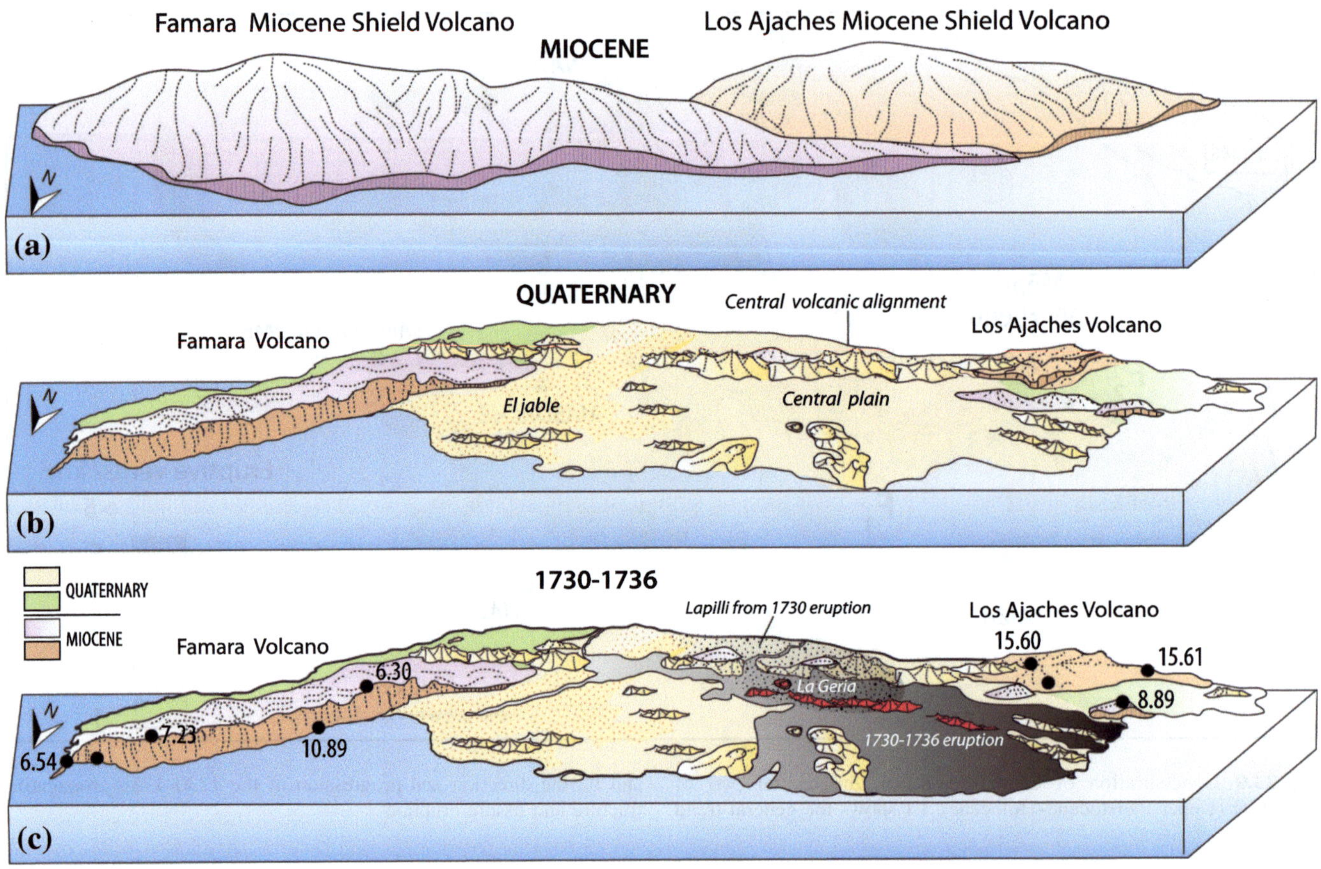

Fig. 23.11 Main stages in the development of Lanzarote: **a** Miocene shield volcanoes, probably disconnected initially as island volcanoes. **b** Restart of volcanism after a long period of repose and erosion. This eruptive rejuvenation connected the Miocene shield volcanoes forming a central shallow lava plain. **c** The 1730–1736 eruption basaltic products (lavas and pyroclasts) mantled and renovated a considerable part of the central plain

23.4 Landscape and Climate

The bulk of the island of Lanzarote is formed by two large basaltic Miocene shield volcanoes: Ajaches Volcano (560 m asl), at the SW edge of the island, and Famara Volcano (670 m asl), at the NE end (Fig. 23.11). Both shield volcanoes were constructed in the Miocene, the former from about 15.5–14 Ma, and the latter from 11 to 6 Ma. At the present time, they are deeply eroded, probably with their north flanks partly mass-wasted by lateral collapses. Post-shield-stage volcanism is restricted to the Quaternary and consists of a relatively thin blanket of basaltic pyroclasts and lava flows erupted in the last 1.8 Ma (Carracedo et al. 1992). Recent volcanism connected the old shields, forming the present-day island of Lanzarote.

Lanzarote and Fuerteventura, the oldest emerged islands of the Canary Volcanic Province (Geldmacher et al. 2005), have evolved to a late post-erosional stage, characterized by their low altitude (maximum elevation 670 m asl) and smooth topography, particularly in the central part of the islands. Why does the idea of Lanzarote as a young volcanic island persists? The only plausible explanation is the occurrence, 275 years ago, of the exceptionally large 1730–1736 eruption, with a short sequel in the 1824 eruption.

As shown in Fig. 23.11, the Miocene shields were probably formed as independent island volcanoes, with the Famara shield flows finally extending towards the older Los Ajaches Volcano (Fig. 23.11a). Just before the 1730–1736 eruption, the landscape in Lanzarote was characterized by a wide central lava flatland connecting the Miocene shields and crossed by NE–SW-trending alignments of cinder cones (Fig. 23.11b). The pre-1730 central plain volcanism, most probably pre-Holocene, was deeply weathered and mantled by abundant aeolian organic sands (*jable,* from the French *sable*), providing extensive areas of fertile soil for cereals and pastures (Bontier and Leverrier 1404), supporting a population of over 5,000 people (Dávila y Cárdenas 1737). The juvenile 1730–1736 black basaltic airfall lapilli and lavas strongly contrast with the light-coloured pre-eruption

Fig. 23.12 Oblique aerial view (from the east) of the area mantled with thick layers of lapilli during the 1730–1736 eruption (area of La Geria). The 1730–1736 lapilli cover, considered initially to be catastrophically damaging to the island's farmland, soon became a traditional mulching technique in Lanzarote

land. In addition, the flat central plain favoured expansion of lava flows over a significant part of the island (Fig. 23.11c); areal extent would have been much lower in an eruption with an equivalent volume of lavas in the steeper western Canaries, where lava flows are usually channelled in deeply incised valleys going into the sea.

Due to the limited altitude of Lanzarote compared with the western Canaries, the island lacks the orographic precipitation from trade winds. Thus, the very low precipitation (100–250 mm) together with the strong winds account for the island's arid climate. The 1730–1736 eruption produced extensive and thick layers of lapilli, which initially had a catastrophic impact on farmland. However, soon after, these highly porous deposits allowed the local people to apply a highly successful traditional mulching technique endemic of Lanzarote, leading to significant savings in irrigation by reducing evaporation losses. A spectacular example of the extensive use of this agricultural practice is found in La Geria area, a peculiar anthropogenic landscape consisting of thousands of small craters carved in the lapilli for growing wine grapes (Fig. 23.12).

23.5 Volcanic Landforms

23.5.1 Cinder Cone Morphology

A characteristic geomorphic feature of the volcanic cones in Lanzarote is the large scatter in morphometric ratios, attributed to eruption dynamics, particularly significant phreatomagmatic activity (Carracedo et al. 1992; Kervyn et al. 2012). High ratios of crater width (Wcr) to cone width (Wco), characteristic of phreatomagmatic vents, are common in Lanzarote (Fig. 23.13), with abundant tuff cones, tuff rings, and maars (Carracedo and Day 2002). A significant proportion of the volcanic vents of the central part of Lanzarote show higher than average Wcr/Wco, explaining the frequent use of the term "caldera" for the volcanic cones of the island.

23.5.2 Lava Flows and Lava Tubes

The long and voluminous lava flows of the 1730–1736 basaltic fissure eruption show a great morphological variety: a'a (predominant) and pahoehoe flows, and the transitional forms between these two main flow types. A particularly interesting lava flow was erupted from the Montaña de las Nueces vent early in 1733 (Dávila y Cárdenas 1737). At the final stage of this eruption, probably lasting for several months (Pallarés 2007), a tholeiite pahoehoe flow about 20 km long reached the SE coast. This is probably the longest flow of the 1730–1736 eruption (see Fig. 23.2). The 10-m-thick tholeiite lava flowed over a nearly horizontal lava plateau apparently as a simple sheet flow, developing abundant tumuli breached by deep cracks that reveal the inner structure of a massive and poorly vesiculated flow. The most spectacular feature of this flow is the Cueva de los Naturalistas lava tube, described by Hernández-Pacheco in 1909 (Fig. 23.14). The tube starts at the NE flank of Montaña de las Nueces and continues towards the east at least for another 7.5 km (Carracedo et al. 2003; Solana et al. 2004; Pallarés 2007).

The most spectacular lava tube of Lanzarote is "Cave of the Greens" (because of the former owner's family name: Los Verdes), part of the 7.5-km-long and up to 35-m-high Corona Volcano lava tube (one of the world's largest). A 40–45-min guided visit shows spectacular views of the complex tube, wide enough at places to host a conference room and a one-time refuge to local people from pirates and slave hunters (Fig. 23.15).

The lava tube has several skylights, the widest—Jameos del Agua—has a 50-m-long, 10-m-wide tidal seawater natural lagoon (Fig. 23.15d), semi-illuminated by the partial

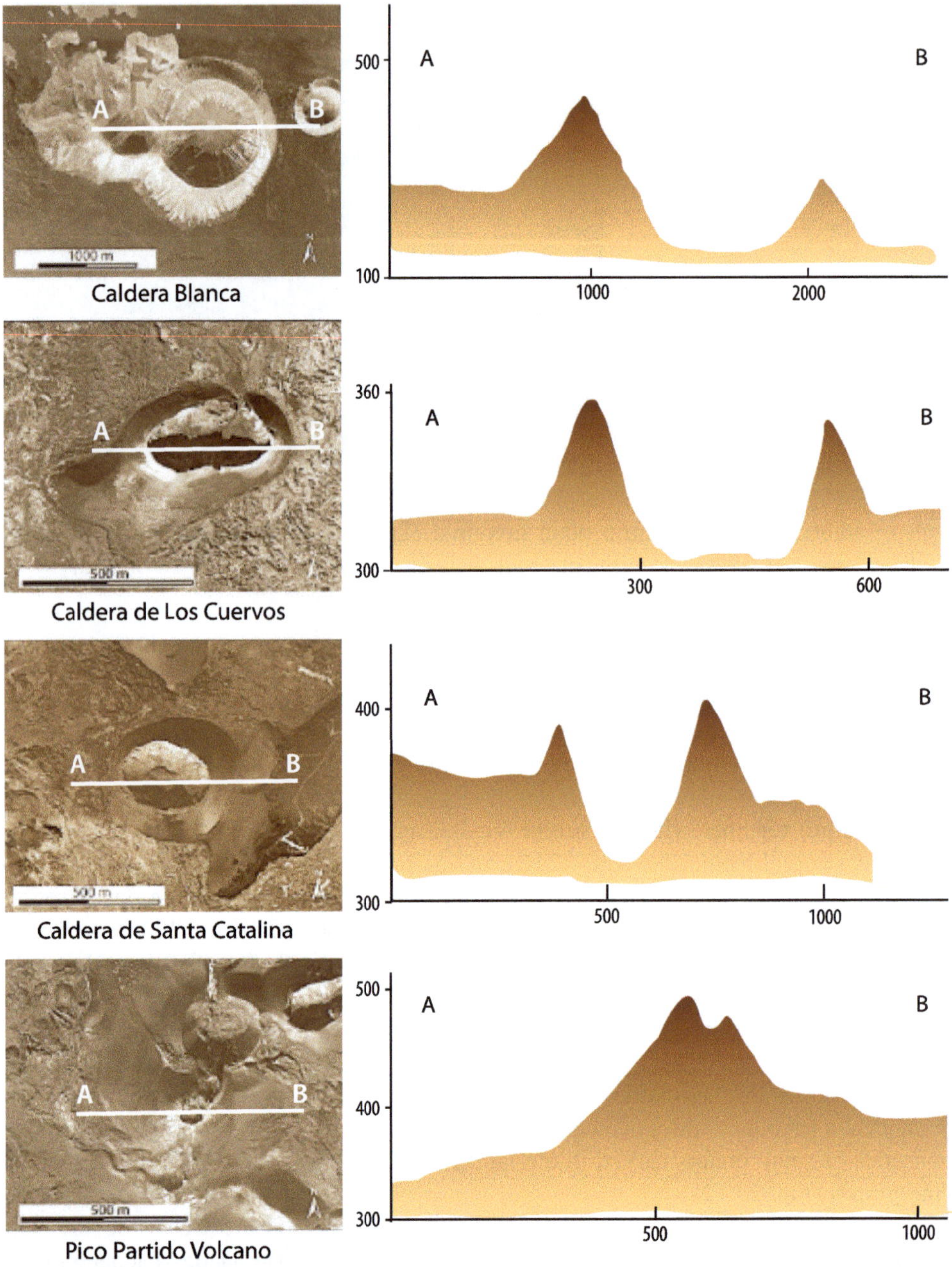

Fig. 23.13 Air photographs and associated topographic profiles along the white lines for specific cones of Lanzarote (modified from Kervyn et al. 2012)

Fig. 23.14 Cueva de los Naturalistas, a 7.5-km-long lava tube in the 20-km tholeite flow of Montaña de las Nueces, developed at the final stages of the 1730–1736 eruption

Fig. 23.15 The Corona lava tube. **a** Skylight in the lava tube. **b** Inside the tube. **c** Swimming pool inside a large skylight in the lava tube. **d** Tidal seawater natural lagoon (Jameos del Agua). The inset shows a specimen of the eyeless and depigmented endemic Munidopsis polymorpha crab that is perfectly adapted to life in this ecological niche

collapse of the ceiling, that has been developed as a tourist attraction. The eyeless and depigmented endemic Munidopsis polymorpha crab lives in this submarine lava tube and can be observed in the lagoon (inset in Fig. 23.15d). Interesting views and features can be observed inside this lava tunnel (enlarged in Fig. 23.15D).

Diving investigations found a 1.6-km-long submerged marine prolongation of the lava tube (the Túnel de la Atlántida), the seaward section ending 64 m below sea level in a cul-de-sac. Although this was initially interpreted as a submarine continuation of the tunnel related to the Corona eruption, the length and depth of the submarine part of the tube is physically unfeasible. Recent $^{40}Ar/^{39}Ar$ dating of the Corona lavas determined the Corona eruption to have occurred 21,000 ± 6,500 years ago, corresponding with the last glacial maximum at 21,000–18,000 years ago. Accordingly, the lava tube can be assumed to have formed under subaerial conditions, only to be flooded during subsequent post-glacial sea-level rise (Fig. 23.16; Carracedo et al. 2003).

23.6 The Geothermal Anomalies

Thermal anomalies have been observed in the area of Montañas del Fuego, particularly in the Islote de Hilario, where surface temperatures reach 100–180 and 300–600 °C at only 5 m depth (Fig. 23.17a).

Differing interpretations on the origin of these anomalies have been put forward by scientists visiting the island virtually since the 1730–1736 eruption: by oxidation of metallic components (Buch 1825); cooling intrusion of magma from that eruption (Hernández-Pacheco 1909); hydrocarbon combustion (Brun 1908); exothermic reaction of calcrete (Bravo 1964).

Calamai and Ceron (1970) carried out studies to assess the potential of these sustained geothermal anomalies for electricity generation. These authors measured 700 °C at 27 m depth at the Islote de Hilario and concluded that the anomalies were caused by a relatively shallow magmatic intrusion and the heat transfer through convection of gases, mainly atmospheric with a small proportion of CO_2 and NH_3. They

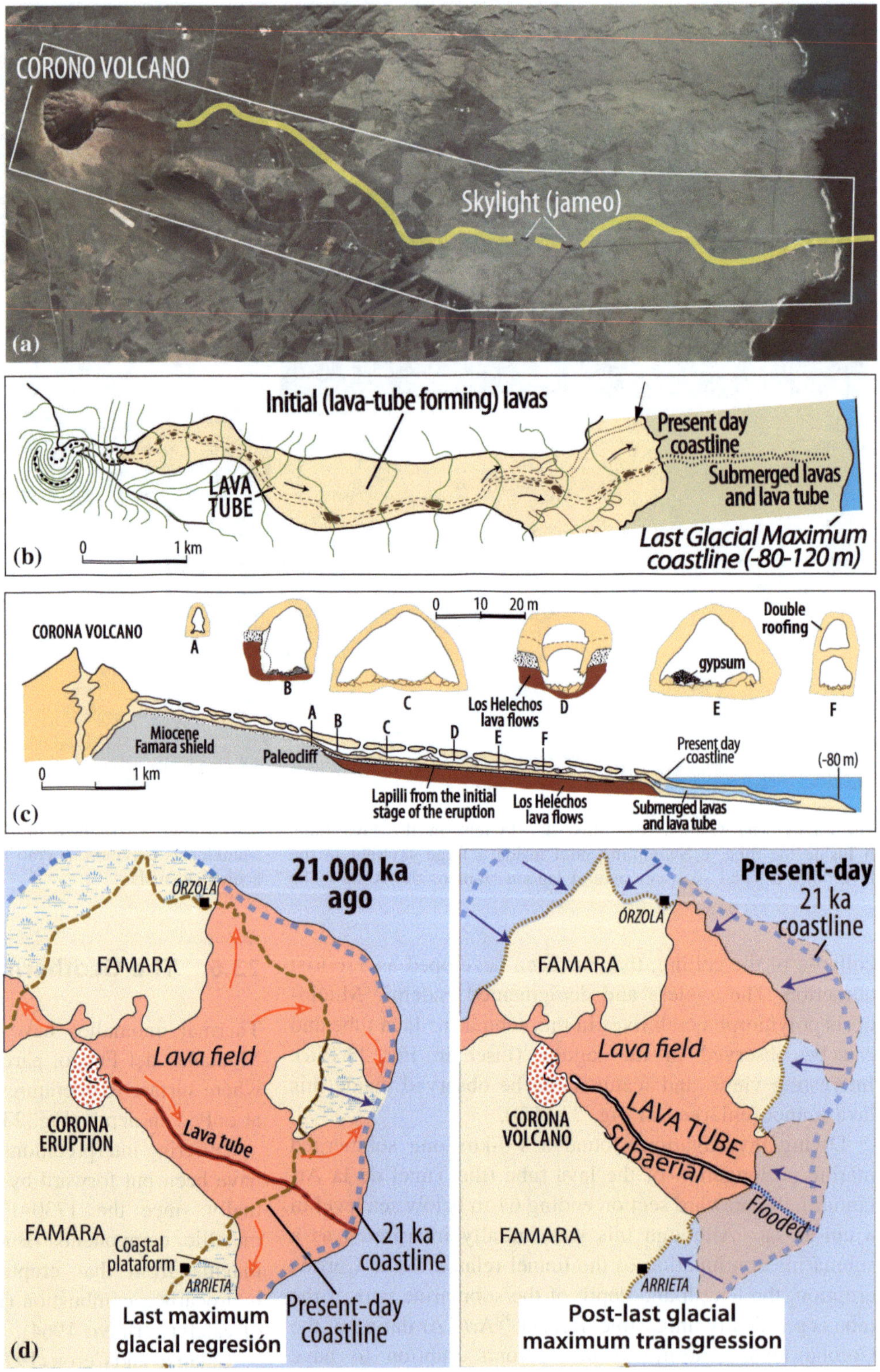

Fig. 23.16 The Corona lava tube. A. Map of the Corona lava tube (Google Earth image). B. Map of the Corona flows and lava tube. C. Cross section of the tube. D. Formation of the lava tube under subaerial conditions during the last glacial maximum at 21,000 to 18,000 years ago (1), and partial inundation of the tube in the subsequent deglaciation and marine transgression (Carracedo et al. 2003)

considered that the temperature of 700 °C was incompatible with the presence of groundwater, necessary for a commercially exploitable geothermal field. Nevertheless, Araña et al. (1973) refuted previous temperature measurements as erroneous, reporting temperatures between 16 and 350 °C in the same area, with a thermal gradient of 0.2 °C/m. These new

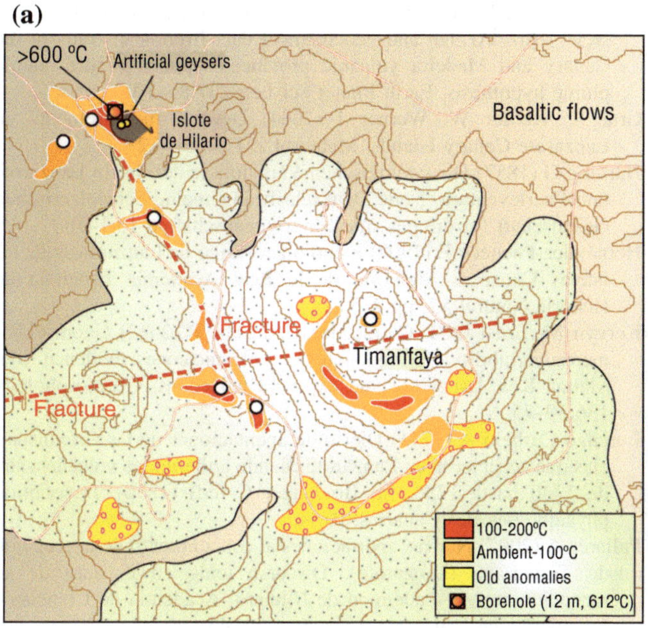

Fig. 23.17 **a** Map of Montañas del Fuego geothermal anomalies. Note the association of these anomalies with the main eruptive fissure of the 1730–1736 eruption (Carracedo and Rodríguez-Badiola 1991). **b** The 2,700-m-deep borehole drilled in 1977 to explore the geothermal anomalies as a potential source for electricity generation

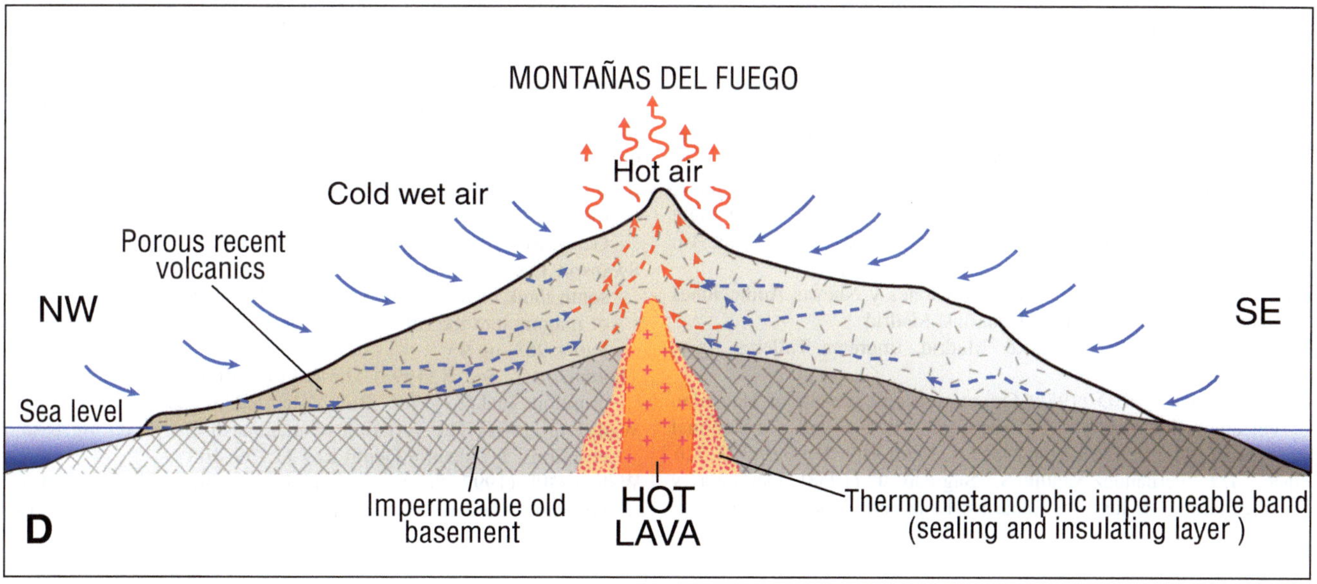

Fig. 23.18 Model depicted by Carracedo and Rodríguez-Badiola (1991) to illustrate the convective system generating the Montañas del Fuego geothermal anomalies by one pass-flow of air and minimal amounts of magmatic gases; a modification of the original idea of Calamai and Ceron (1970)

data apparently fitted with a convective system under impervious layers at great depths, the heat transfer taking place by means of convection in a deep reservoir. Following this assessment, a 2,700-m borehole was drilled in 1977 (the deepest ever carried out in the Canaries, see Fig. 23.15b), with negative results (Sánchez-Guzmán and Abad 1986).

The model put forward by Calamai and Ceron (1970) was in essence correct, the anomalies being caused by a deep intrusion of magma from the 1730–1736 eruption (Fig. 23.18), and the heat transfer by advection of air with minimal amounts of magmatic gases (Carracedo and Rodríguez-Badiola 1991).

References

Araña V, Ortiz R, Yuguero J (1973) Thermal anomalies in Lanzarote (Canary Islands). Geothermics 2:73–75

Bontier P, Leverrier J (1404) Historia del descubrimiento y conquista de Las Canarias. Translation by PM Ramírez of the 1630 Paris edition. Imprenta Isleña, Sta. Cruz de Tenerife, 1847

Bravo T (1964) El volcán y el malpaís de La Corona, La "Cueva de los Verdes" y los "Jameos". Publi. Cabildo Insular de Lanzarote, Arrecife, p 31

Brun A (1908) Quelques recherches sur le volcanisme au Pico de Teyde et au Timanfaya, Genève

Buch L (1825) Physikalische Beschreibung der Canarischen Inseln, Berlin

Calamai A, Ceron P (1970) Air convection within the Montaña de Fuego (Lanzarote Island, Canarian Archipelago). Geothermics 2–1:611–614

Carracedo JC (1994) The Canary Islands: an example of structural control on the growth of large oceanic island volcanoes. J Volcanol Geoth Res 60:225–242

Carracedo JC, Rodriguez-Badiola E, Soler V (1990) Aspectos volcanológicos y estructurales, evolución petrológica e implicaciones en riesgo volcánico de la erupción de 1730 en Lanzarote, Islas Canarias. Estud Geol 46:25–55

Carracedo JC, Rodríguez-Badiola E (1991) La Erupción de Lanzarote de 1730 (con un mapa geológico a color a escala 1/25.000 de la erupción de 1730): Las Palmas de Gran Canaria, pp 184

Carracedo JC, Rodríguez-Badiola E, Soler V (1992) The 1730–1736 eruption of Lanzarote, Canary Islands: a long, high magnitude basaltic fissure eruption. J Volcanol Geoth Res 53:239–250

Carracedo JC, Day SJ (2002) Geological guide of the Canary Islands. Classic geology in Europe. Terra Publishing, London, p 192

Carracedo JC, Singer BS, Jicha B, Guillou H, Rodríguez-Badiola E, Meco J, Pérez-Torrado FJ, Gimeno D, Socorro S, Láinez A (2003) La erupción y el tubo volcánico del volcán Corona (Lanzarote, Islas Canarias). Estudios Geológicos 59: 277–302

Carracedo JC, Pérez-Torrado FJ, Rodríguez-González A, Fernández-Turiel JL, Klügel A, Troll VR, Wiesmaier S (2012) The ongoing volcanic eruption of El Hierro, Canary Islands: Eos (Trans Am Geophys Union) 93: 89–90

Dávila y Cárdenas PM (1737) Constituciones y Nuevas Addicciones Synodales del Obispado de Canarias. Madrid

De la Hoz A (1960) Lanzarote: Madrid, Imprenta Talleres Anro, Madrid

Fiske RS, Jackson ED (1972) Orientation and growth of Hawaiian volcanic rifts: the effect of regional structure and gravitational stresses. Proc R Soc London, Ser A, Math Phys Sci 329:299–326

Fúster JM, Fernández-Santín S, Sagredo J (1968) Geología y volcanología de las Islas Canarias: Lanzarote. Madrid, pp 177

Geldmacher J, Hoernle K, Bogaard P, Duggen S, Werner R (2005) New $^{40}Ar/^{39}Ar$ age and geochemical data from seamounts in the Canary and Madeira volcanic province: support for the mantle plume hypothesis. Earth Planet Sci Lett 237:85–101

Graf A, Kuttler W, Werner J (2004) Dewfall measurements on Lanzarote Canary Islands. Meteorol Z 13:405–412

Hartung G (1857) Die geologischen Verhaltnisse der Inseln Lanzarote und Fuerteventura. Neue Denkschrift allgemeine Schweizerischen Gesellschaft für die gesamte Naturwissenschaften 15:1–168

Hernández-Pacheco E (1909) Estudio geologico de Lanzarote y de las Isletas Canarias. Memoria de la Real Academia Española de Historia Natural VI, pp 235

Kervyn M, Ernst GGJ, Carracedo JC, Jacobs P (2012) Geomorphometric variability of "monogenetic" volcanic cones: evidence from Mauna Kea, Lanzarote and experimental cones. Geomorphology 136:59–75

Lorenzo Curbelo A (1731) Diario de apuntaciones de las Circunstancias que acaecieron en Lanzarote cuando ardieron los volcanes, año de 1730 hasta 1736. In: Boulanger C (ed) (1836). Deseription physique des Iles Canaries. Parls

Pallarés A (2007) Nuevas aportaciones al conocimiento de la erupción de Timanfaya (Lanzarote). Discurso leído en el acto de su recepción como Académico de Número, Academia de Ciencias e Ingenierías de Lanzarote. Lanzarote, pp 43

Real Audiencia De Canarias (1731) Copia de las Ordenes, y providencias dadas para el alivio de los Vezinos de la Isla de Lanzarotte en su dilatado padezer a causa del prodigioso Volcan, queen ella rebentó el primer dia de Septiembre del afio immediato pasado de 1730, y continúa asta el dia de la fecha. Va inserto el Mapa de la Isla, del Volcan, y sus bocas con la descripcion del miserable estado aque tiene reducida la Isla. Canaria y Abril 4 de 1731. Legajo manuscrito de la Real Audiencia de Canarias, Gracia y Justicia, Leg. 89, Archivo General de Simancas:1–56

Sánchez-Guzmán J, Abad J (1986) Sondeo geotérmico Lanzarote-l, significado geológico y geotérmico: Anales de Física 82(1986):102–109

Solana MC, Kilburn CRJ, Rodriguez-Badiola E, Aparicio A (2004) Fast emplacement of extensive pahoehoe flow fields: the case of the 1736 flows from Montaña de las Nueces, Lanzarote. J Volcanol Geoth Res 132:18–207

Swanson DA, Duffield WA, Fiske RS (1976) Displacement of the south flank of the Kilauea volcano: the result of forceful intrusion of magma into the rift zones. U.S. Geological Survey, Professional Paper 963, pp 93

Thordarson T, Self S (1993) The Laki (Skaftar Fires) and Grimsvotn eruptions in 1783–1785. Bull Volc 55:233–263

Wadge G (1981) The variation of magma discharge during basaltic eruptions. J Volcanol Geoth Res 11:139–168

Walker GPL (1992) "Coherent intrusion complexes" in large basaltic volcanoes—a new structural model. J Volcanol Geoth Res 50:41–54

Structural Collapses in the Canary Islands

24

Juan Carlos Carracedo

Abstract

Relatively small landslides of the order of millions of m^3 are frequent geological features, while giant landslides or mega-landslides up to thousands of km^3 are rare and mainly related to the development of oceanic islands, principally in the initial shield stages. They were first documented in the Hawaiian Islands, but are also extraordinarily well represented in the Canary Islands, where they have been comprehensively studied onshore (pre- and post-collapse processes and the evolution of nested volcanism) and offshore (characteristics and extent of the debris avalanche deposits). Mega-landslides are important processes in the development of oceanic islands and their geomorphological features, particularly valleys and calderas, spectacular landscapes, which constitute relevant natural and economic resources.

Keywords

Mega-landslides • Landslide valleys and calderas • Tsunamis • Oceanic islands • Canary Islands

24.1 Introduction

Mountains, cordilleras and valleys in continental settings are formed through complex geological processes (sedimentation, folding, faulting, uplift, erosion, etc.) over very long periods and mainly driven by plate tectonics. However, in oceanic islands, the equivalent features are ultimately generated by volcanic construction and mass-wasting processes, usually in significantly shorter time spans. Mountains in oceanic islands like the Canaries are volcanoes or compound volcanoes (i.e. stratocones, shield volcanoes and rift zones), while broad valleys are generally formed through catastrophic gravitational collapses.

Oceanic islands commonly experience giant landslides as parts of the islands' edifices go through phases of intense volcanism, eventual overgrowth and instability. The only observed lateral collapse corresponds to the 18 May 1980 Mount St. Helens eruption. Its 2.5 km^3 debris avalanche deposit, covering 64 km^2 (Glicken 1996), is insignificant compared with previously reported mega-landslides in the Hawaiian Islands, where similar deposits attained prodigious magnitudes (e.g. the Nuuanu, NE Oahu, 23,000 km^2 and 5,000 km^3; Moore et al. 1989). The huge volume of these landslides, among the largest on Earth, delayed the general acceptance of this interpretation until the Mount St. Helens eruption and lateral collapse were observed. Nevertheless, the concept that the high cliffs along the coast of some of the Hawaiian islands correspond to the headwalls of giant landslides was anticipated by Stearns and Clark (1930) and Stearns and Macdonald (1946), and more explicitly by Moore (1964).

In the Canaries, Bravo (1962) proposed a similar origin for the large, horseshoe-shaped valley of La Orotava and the Caldera de Las Cañadas, after observing debris avalanche

J. C. Carracedo (✉)
Departamento de Física-Geología, Universidad de Las Palmas de Gran Canaria, Las Palmas de Gran Canaria, Canary Islands, Spain
e-mail: jcarracedo@proyinves.ulpgc.es

F. Gutiérrez and M. Gutiérrez (eds.), *Landscapes and Landforms of Spain*, World Geomorphological Landscapes,
DOI: 10.1007/978-94-017-8628-7_24, © Springer Science+Business Media Dordrecht 2014

deposits interbedded within the formations filling these depressions. Regrettably, the model proposed by this author was erroneous, since he interpreted the debris deposits (fanglomerate) as a pre-collapse plastic layer related to previous explosive eruptions that facilitated the development of sliding surfaces, rather than attributing them to the deposits originated by the gravitational collapse.

The first conclusive evidence supporting the landslide origin of Las Cañadas Caldera was obtained by swath bathymetry surveys, showing the continuation of landslide deposits on the sea floor, north of Tenerife, over an area of 5,000 km^2 (Watts and Masson 1995). Multiple offshore (seismic, side scan sonar) and onshore (volcano-stratigraphic, geochronological) studies allowed the identification of abundant landslide scars in the Canaries (Navarro and Coello 1989; Carracedo 1994, 1999; Masson 1996; Stillman 1999; Walter et al. 2005; Carracedo et al. 2007, 2011; Carracedo and Troll 2013). Currently, wide horseshoe-shaped depressions in the Canaries (e.g. Las Cañadas Caldera and La Orotava and Güímar valleys in Tenerife, the Taburiente Caldera in La Palma, and the El Golfo, Julan and Las Playas embayments in El Hierro) are ascribed to giant landslide scars, revealing that large-scale gravity-driven collapses play an instrumental role in the morphological configuration of the islands.

The crucial role of rift zones in the development of lateral collapses was proposed by Carracedo (1994). In his model, intrusion-related extensional stresses build up in rift zones, contributing to the increase in gravitational stresses related to the overgrowth and oversteepening of volcanic edifices, eventually leading to lateral collapses. The Canary Islands provide a representative example of how rift zones act as primary structures that control the constructional and morphological characteristics of oceanic islands and their main mass-wasting events.

24.2 Giant Landslides

Low-volume slope movements are a common morphogenetic process in many regions, particularly slides, debris flows and rockfalls. In the Canaries, these mass movements occur frequently in the walls of barrancos, steep cliffs and roadcuts and constitute the main geological hazard in terms of number of victims (Fig. 24.1). Recurrent slumps play a major role in coastal recession and reduction in island volume, a particularly intense process in the windward coast of the western Canaries (Fig. 24.2). The Rosiana landslide in the Caldera de Tirajana, Gran Canaria, is considered to be the largest on-land historical slope movement in the Canary Islands. About 3 million m^3 of bedrock was mobilized, damaging houses, roads and a bridge, and causing the evacuation of over 300 residents (Linares et al. 2001).

These landslides that may reach millions of m^3 are insignificant compared with the gigantic slides involving tens and even thousands of km^3 and spreading debris over thousands of km^2 on the ocean floor. Giant landslides or mega-landslides occur very infrequently, with a recurrence of the order of several hundred thousand years, and mainly take place in the early constructional stages of the islands. As an important characteristic, the development of mega-landslides generally requires the concurrence of several factors (i.e. overgrowth instability, extensional forces, etc.), thus are commonly associated with volcanic settings, mainly oceanic islands.

Large volcanoes often show an amphitheatre that truncates the edifice and hummocky terrain extends downhill (Fig. 24.3a). These features, originally of a controversial origin, were widely interpreted after the 1980 Mount St. Helens eruption and lateral collapse as debris avalanches resulting from the sudden collapse of large volumes of rock from the flanks of volcanoes.

Oceanic island mega-slides are commonly identified offshore by large fields of blocky landslide deposits transported several hundred kilometres on the submarine flanks of the islands and covering hundreds of km^2 on the seafloor (Fig. 24.3b). Avalanche blocks, which may reach 1.2 km in length, are randomly scattered on the surface of the deposit (see inset in Fig. 24.4, from Masson et al. 2002). The finer fraction usually detaches from the debris avalanche, forming a highly mobile turbulent flow (turbidite) that may run on slopes $<1°$ and accumulate in the deepest ocean basins (e.g. the Madeira Abyssal Plain, Weaver et al. 1989; Gee et al. 2001). Onshore, giant landslides are frequently recognized by arcuate embayments (Fig. 24.4). Mega-landslide scars may be completely filled and concealed by subsequent volcanism. In these cases, the onshore scar may be only identified by geological mapping and geochronological data (e.g. the Tiñor collapse in El Hierro, Guillou et al. 1996, and the Micheque giant landslide in NW Tenerife, Carracedo et al. 2011, as described below).

24.3 Mega-Landslides in the Canaries

A dozen mega-landslides have been documented in the Canary Basin and several additional ones have been inferred. Likely, some others are obliterated by volcanism and marine sedimentation inland and offshore, respectively. As shown in Fig. 24.5, the volumes of the collapses in the Canaries are considerably lower than those of Hawaii, at least by an order of magnitude (Carracedo 1999). Among other factors, the angle and depth of the sliding planes account for this difference. The failure planes of the structural collapses in the Canary Islands are relatively shallow, compared with those of the Hawaiian Islands (see inset in Fig. 24.5a).

Fig. 24.1 Slumps and rockfalls are common in oceanic islands due to their characteristic steep flanks. These slope movements pose the most frequent hazard for human life and property. Rockfall in the north coast of El Hierro triggered by seismicity in March 2013 (photo by Nieves Cortés)

Fig. 24.2 a Regression of the northern coast of La Palma due to erosion enhanced by frequent slumps. b A recent slump and associated earthflows in Playa de la Veta, northwest La Palma

Giant collapses are more frequent in the early stages of development of the islands, when eruptive rates are greater, leading to overgrown edifices flanked by oversteepened slopes. This explains the conspicuous evidence of mega-collapses in the youngest, shield-stage, western Canaries (Urgelés et al. 2001; Masson et al. 2002), and in Tenerife, at present in its early rejuvenation stage (Watts and Masson 1995; Cantagrel et al. 1999; Carracedo et al. 2007, 2011). It is likely that mega-landslides had a similar frequency in Gran Canaria (Schmincke and Sumita 1998) and in the eastern Canaries (Stillman 1999).

24.4 Rifts and Mega-Landslides

The erosional versus tectonic origin of the more prominent "negative" landforms in the Canaries (escarpments, horse-shoe-shaped valleys and calderas) has been a matter of

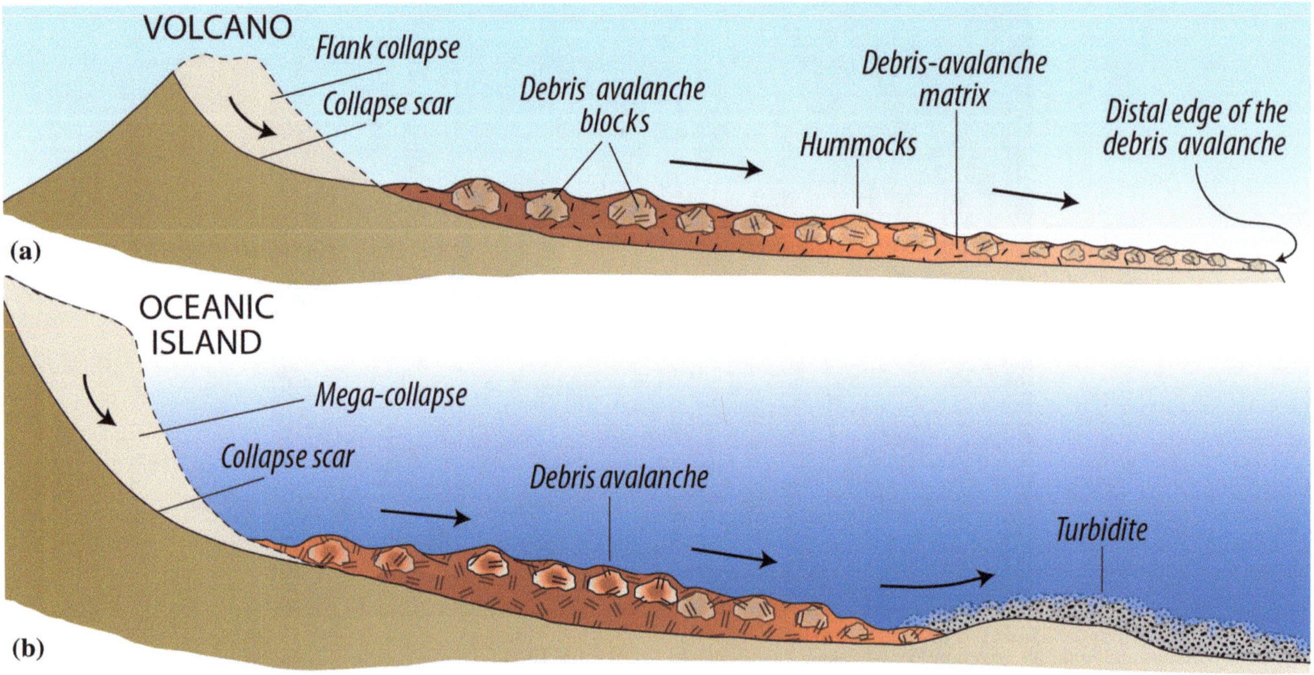

Fig. 24.3 **a** Schematic longitudinal cross section of a debris avalanche related to the lateral collapse of the flank of a volcano with its typical hummocky morphology. **b** Giant landslide in an oceanic island volcano. The debris avalanche generates at its distal end a flow with the finer material forming turbidites

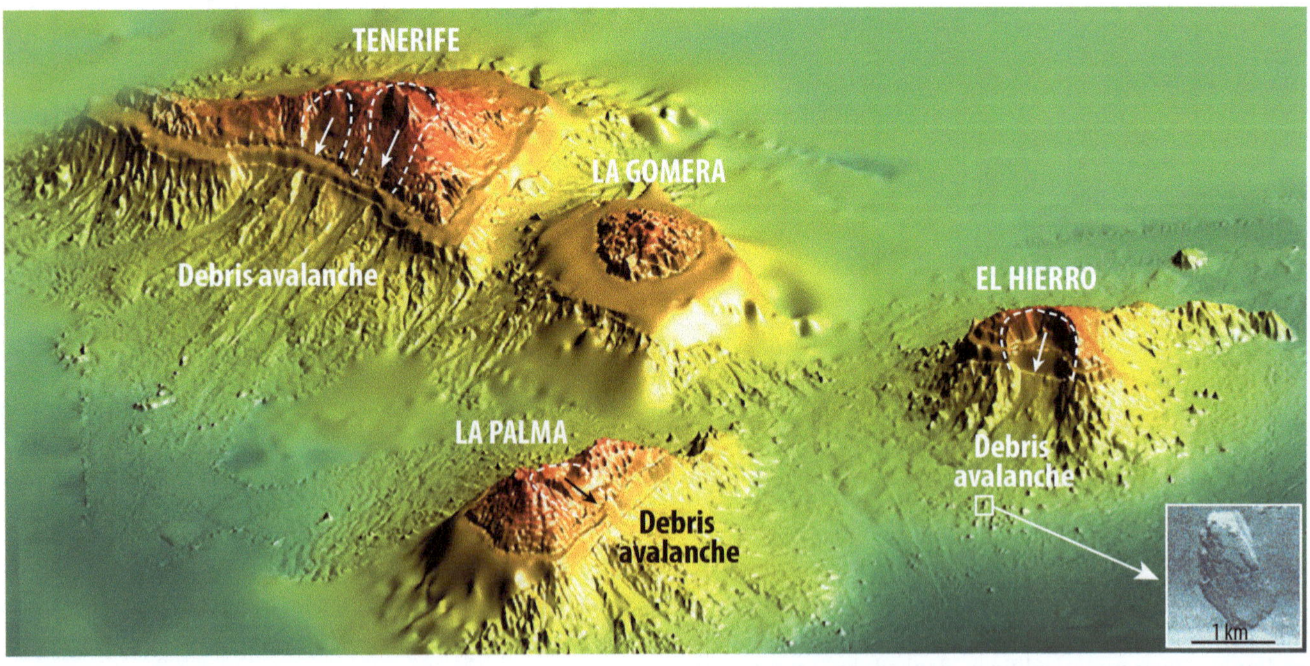

Fig. 24.4 3-D views of some of the islands in the Canaries affected by large-scale landsliding (landslide boundaries shown as dashed lines) and their corresponding debris avalanche deposits. The inset shows a mega-block from the El Golfo lateral collapse. Images courtesy of D. Masson

controversy since Hausen (1961). This author explained the Canaries as emerged remains of tectonic blocks of a faulted subcontinent (the "tableland" formation). In opposition to tectonism, erosion or true collapse-caldera mechanisms were subsequently the preferred interpretations for these outstanding geological features.

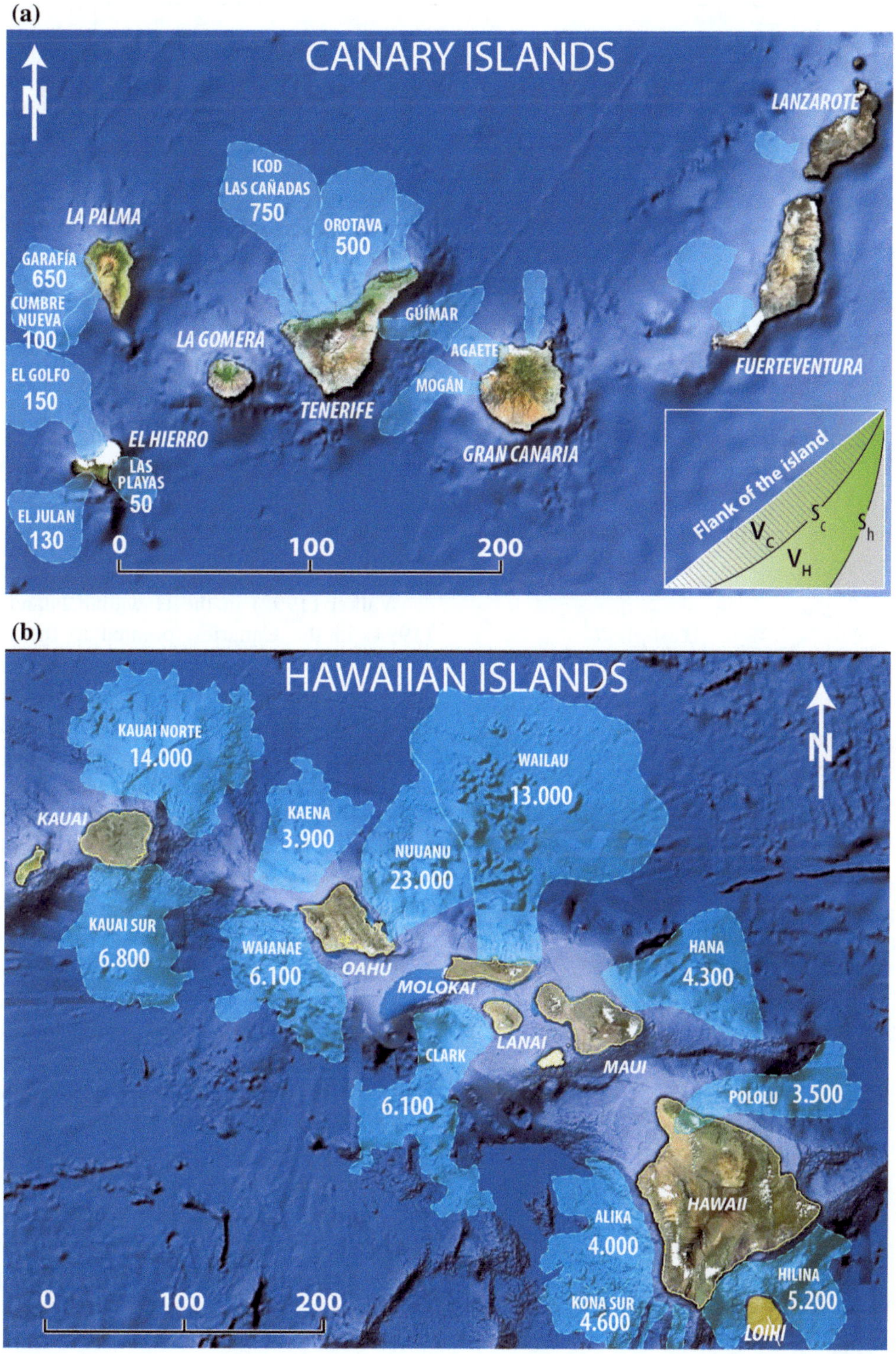

Fig. 24.5 Mega-landslides in the Canaries (**a**) and Hawaii (**b**). Modified from Urgelés et al. (2001) and Moore et al. (1989). Inset in A: Volume versus slide scar angle for the Canaries (Vc and Sc); Vh and Sh, idem for Hawaii (Moore 1964)

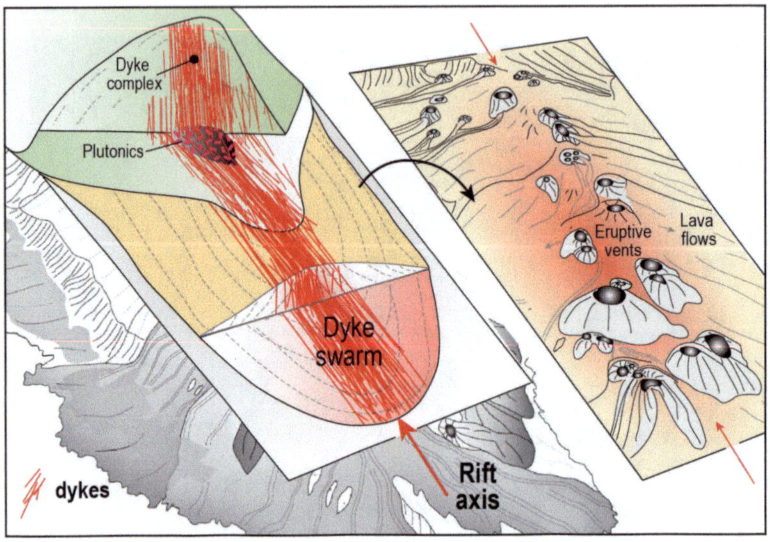

Fig. 24.6 Deep structure of a typical rift zone, the Cumbre Vieja rift, La Palma

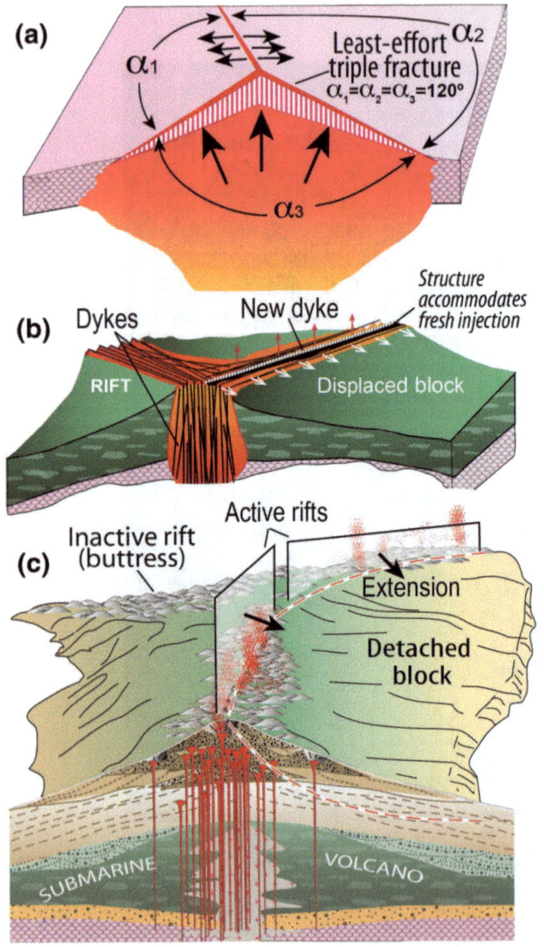

Fig. 24.7 Schematic model for the genesis of a triple-armed rift system, as the oceanic crust is fractured in a "least-effort" configuration by magmatic updoming (modified from Carracedo 1994)

Walker (1992) in the Hawaiian Islands, and Carracedo (1994) in the Canaries, pointed to the presence in these oceanic archipelagos of highly concentrated "coherent intrusion complexes" or "rift zones", consisting of tight clusters of eruptive centres distributed along narrow ridges, known in the Canaries as "*dorsales*". Their deep structure comprises a swarm of feeding dykes, increasing in density with depth and towards the axis of the rift zones (Fig. 24.6).

How do these rift zones form? According to Carracedo (1994), endogenous mechanisms play a major role in establishing axial architectures in the volcanic edifices. This process is analogous to that forcing the initiation of triple junctions (Luongo et al. 1991). A plume under the oceanic crust causes uplift and doming and eventually fractures the crust forming regular triple-armed junctions (Fig. 24.7a). Doming leads to the development of fractures in the brittle crust at 120° angles, the pattern that requires the least effort to form (Carracedo 1994), while aligned dykes (Fig. 24.7b) and eruptive centres (Fig. 24.7c) concentrate in these tri-axial rift zones. Repetitive injection of blade-like dykes progressively increases the anisotropy of the complex, forcing new dykes to wedge their path parallel to the previous intrusions, like a knife between the pages of a book.

Intrusion can progress in a dyke complex if the structure is able to accommodate fresh injections. Since repetitive dyking involves expansion and generates lateral stresses, new injections can only occur if any of the flanks of the rift zone is free to move apart (see Fig. 24.7b). The rapid growth of rifts, causing lateral stresses and instability, may induce the required extensional setting for sustained intrusions and flank displacement. As extension progresses in growing rift zones, slopes at a marginal equilibrium state

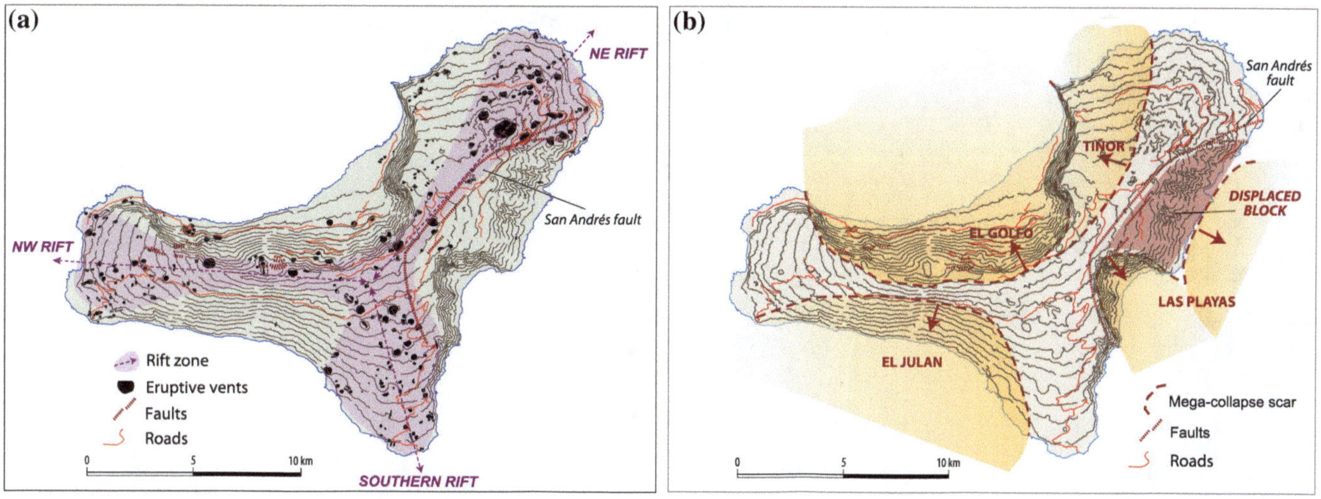

Fig. 24.8 **a** Geometry and disposition of the rifts in El Hierro (Carracedo et al. 2001). **b** Lateral collapses and the resulting debris avalanche deposits in the flanks of El Hierro (Masson et al. 2002)

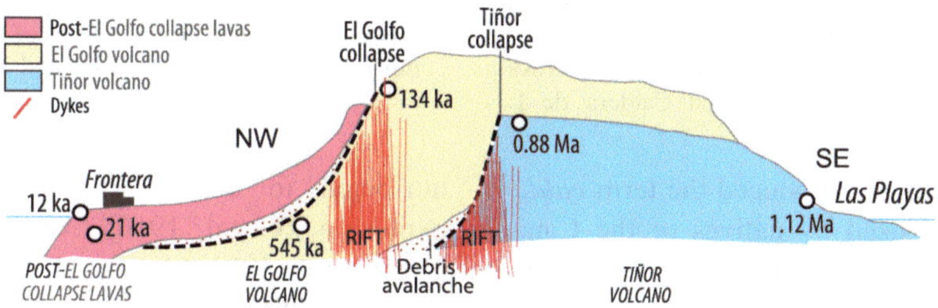

Fig. 24.9 Cross section of El Hierro illustrating the construction of the island by successive cycles of growth and lateral collapses

eventually reach a critical rupture threshold with the initiation of massive landslides.

24.5 El Hierro, an Epitome of Rift-Collapse Dynamic Association

The island of El Hierro, the youngest of the Canaries, has been intensely mass-wasted by successive mega-landslides, exposing parts of its internal structure. This island is an exceptional setting to study the characteristics of rift zones, and the role they play in controlling the growth of the island, the building-up of instability and, eventually, the development of massive collapses.

The main morphological features of El Hierro are alignments of tightly packed volcanic cones (see inset in Fig. 24.8a) forming three narrow ridges resembling the edges of a 1,500-m-high tetrahedron. Wide open embayments with steep headwalls form the arcuate sides of the pyramid (Fig. 24.8).

A cross section of El Hierro illustrates the construction of the island by successive cycles of volcanic growth, instability, lateral collapse and filling of the collapse scar by subsequent volcanism, resulting in a succession of juxtaposed and overlapping volcanoes. The latest landslide-related scar (El Golfo) is currently in the process of filling (Fig. 24.9).

24.6 Examples of Mega-Landslides in the Canaries

Other outstanding examples of mega-collapses are the Las Cañadas and Taburiente Calderas in Tenerife and La Palma, respectively (Fig. 24.10).

24.6.1 Caldera de Taburiente

The Caldera de Taburiente is a 15-km-long, 6-km-wide and 1.5-km-deep horseshoe-shaped depression, whose characteristics fit with the "Oahu type" valley as defined by Cotton (1952). von Buch (1825) referred to the Caldera de Taburiente as an archetypal example of "*Erhebungs crater*" or craters of elevation. His hypothesis attempted to explain the origin of large cavities and the distribution of volcanic rocks that surround them (Lyell 1830). This author, in his

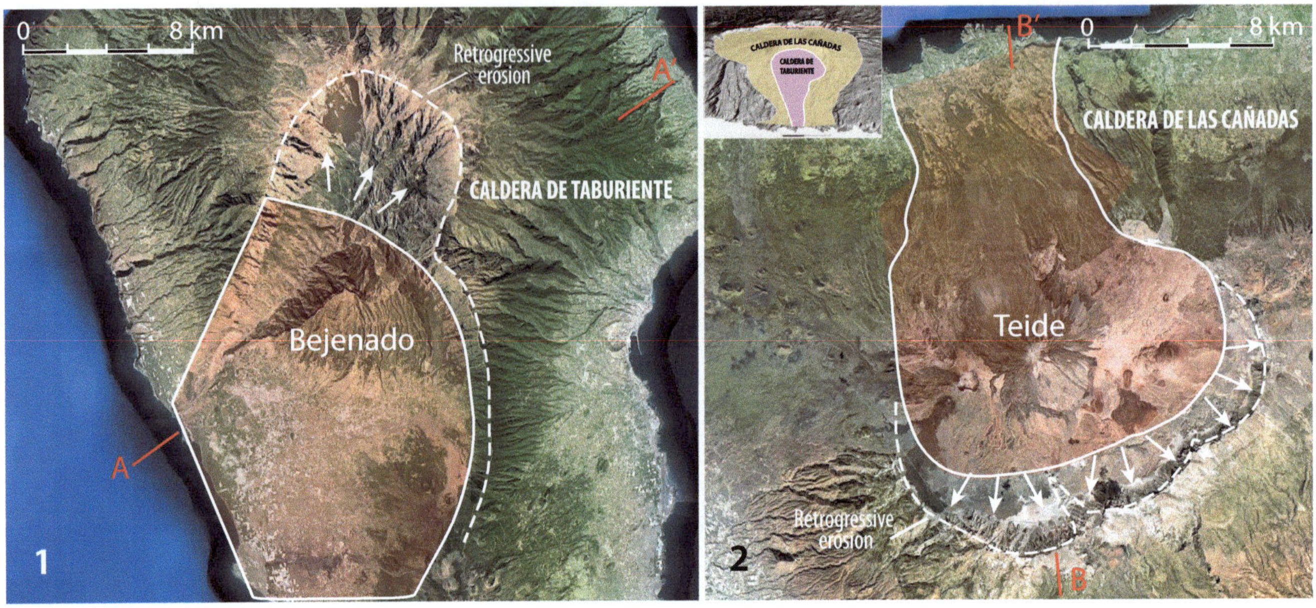

Fig. 24.10 **a** Caldera de Taburiente and Valle de Aridane in La Palma formed by a lateral collapse (solid line) about 560 ky ago. The Taburiente Caldera developed after the collapse by headward erosion of the headwall (dashed line). **b** Partially filled Caldera de Las Cañadas, Tenerife, carved by a mega-collapse about 200 ky ago. A-A' and B-B' indicate the cross sections of Fig. 24.11. The inset shows the comparative size of Las Cañadas and Taburiente calderas

pioneering study of La Palma, introduced the term *caldera* in the geologic literature, used by natives of the Canary Islands for any bowl-shaped depression or crater. Later, it was considered by Lyell (1830) to be a typical erosional feature. Machado (1965) described it as an eroded depression formed originally by a lateral collapse, a recently confirmed hypothesis (Ancochea et al. 1994, 1999; Carracedo 1994; Carracedo et al. 1999a, b, 2001).

The strongest evidence for a lateral collapse as the onset mechanism for the opening of the Caldera de Taburiente came from geochronological and volcano-stratigraphic studies. As shown in Figs. 24.10a and 24.11b, the Bejenado Volcano in the southern wall of the caldera, previously believed to be a part of the pre-caldera Taburiente Volcano, is nested inside the depression and consequently post-dates the caldera (Fig. 24.11a). The youngest age of the Taburiente Volcano (0.56 Ma) and the oldest of the Bejenado Volcano (0.54 Ma) constrain the timing of the lateral collapse at about 0.56 Ma. The collapse removed about 100 km^3 (Carracedo et al. 1999a). The present-day Taburiente Caldera formed by erosion along a drainage system confined between the Bejenado Volcano and the western edge of the collapse scarp, enlarged by headward erosion (Fig. 24.10a).

24.6.2 Caldera de Las Cañadas

This 17 × 12 km ellipsoidal depression (Fig. 24.10b), the largest and most impressive of the Canarian calderas, was attributed to erosion (Lyell 1830), a Krakatoan-type explosion (Gagel 1910), vertical collapse (Friedlander 1915; Araña 1971; Martí et al. 1997), explosion and erosion (Hausen 1956) and erosion enhanced by slides developed over a plastic layer (Bravo 1962).

In recent years, the debate has focused on the alternatives of vertical versus lateral collapse. Although this is still an open discussion, compelling evidence has been obtained supporting the formation of Las Cañadas Caldera by lateral collapse: (1) The detection by swath bathymetry of deposits related to a giant landslide spread over an area of 5,500 km^2 of the seafloor north of Tenerife, extending onshore into Las Cañadas Caldera (Watts and Masson 1995); (2) On-land geological and geochronological evidence indicating successive north-directed flank failures affecting the Las Cañadas Volcano, the latest giving rise to the formation of the Las Cañadas Caldera (Ancochea et al. 1990, Carracedo 1994, 1999; Cantagrel et al. 1999; Carracedo et al. 2007); (3) Studies carried out in *galerías* (horizontal tunnels excavated in the Canaries for groundwater exploitation) revealed a polymictic breccia layer with the characteristic features of debris avalanche deposits underlying the sequence of lava flows from the Teide Volcano, that partially fill Las Cañadas depression (Navarro and Coello 1989; Carracedo et al. 2007). According to Márquez et al. (2008), the debris avalanche breccia extends into Las Cañadas Caldera and underneath the present Teide stratocone (Fig. 24.11b), an observation that strongly supports a lateral collapse for the origin of the Caldera.

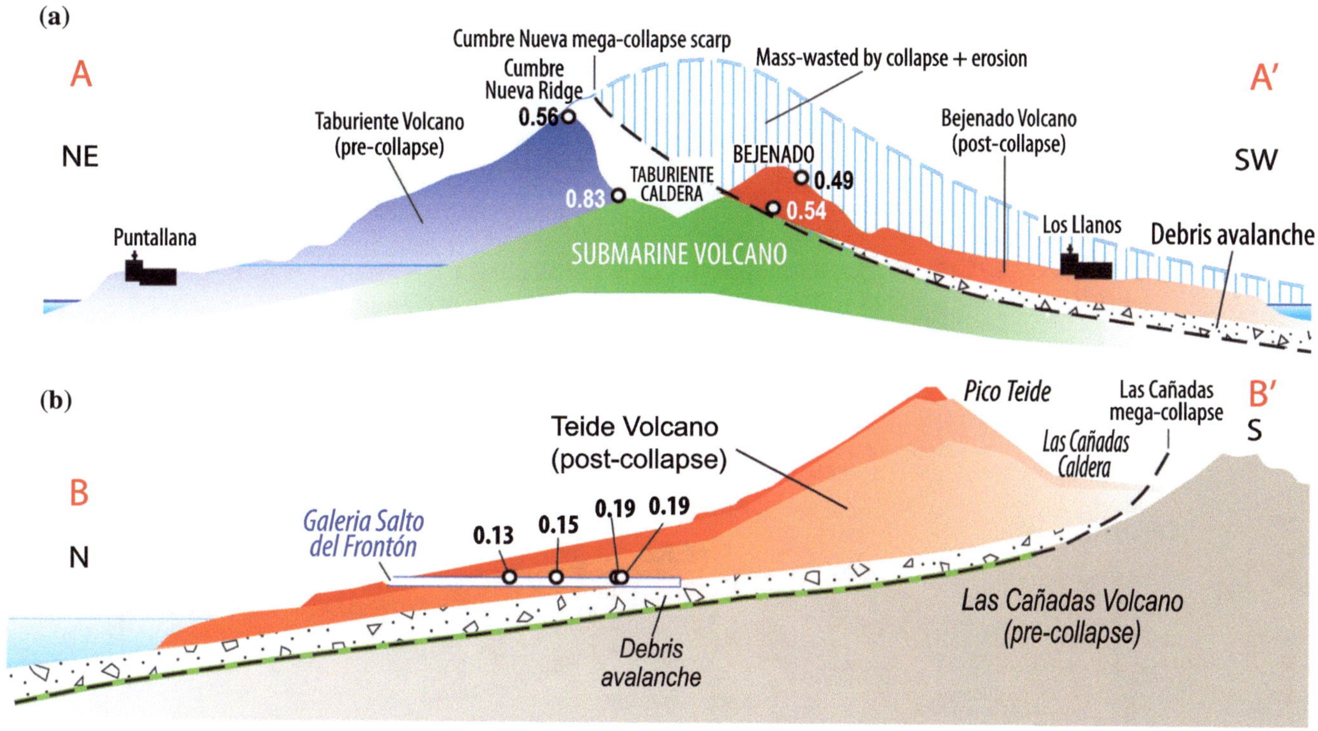

Fig. 24.11 **a** Geologic cross section along the Taburiente collapse scar. The youngest age of Taburiente Volcano (566 ± 8 ka) and the oldest of Bejenado (549 ± 12 ka) constrain the timing of the ~100 km^3 lateral collapse at about 560 ka (Carracedo et al. 1999a). **b** Cross section of the post-collapse Teide Volcanic Complex sequence (Carracedo et al. 2007). The green line indicates the extent of the debris avalanche observed by Márquez et al. (2008). Cross sections traces are indicated in Fig. 24.10

Fig. 24.12 Oblique view of Tenerife from the NE (Google Earth image). The landslide scars of La Orotava and Güímar are clearly visible, unlike the Micheque scar, completely filled and thus only recognizable by geological evidence, particularly through galerías (Carracedo et al. 2011)

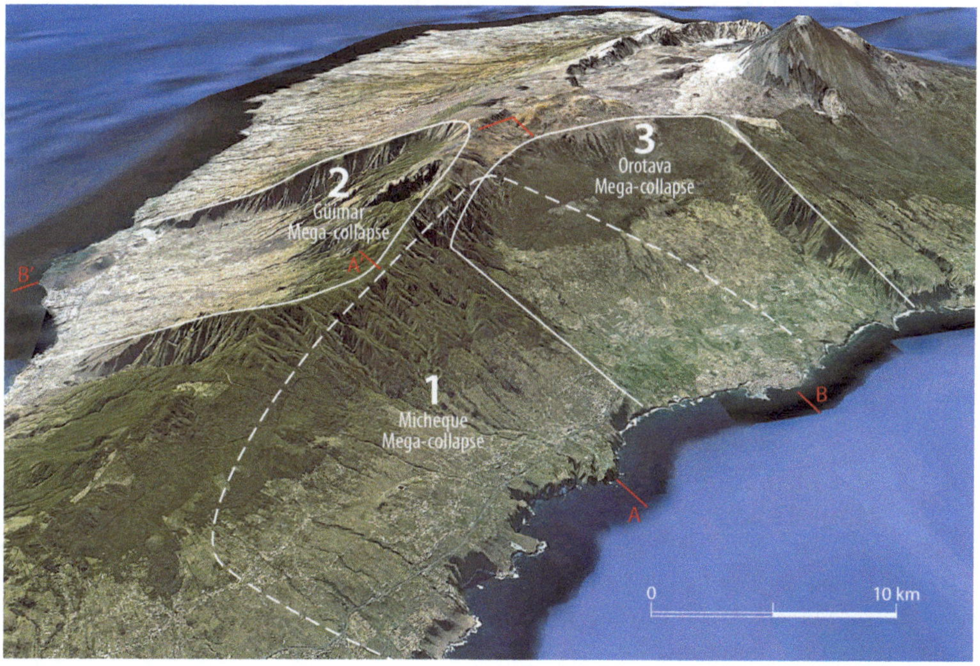

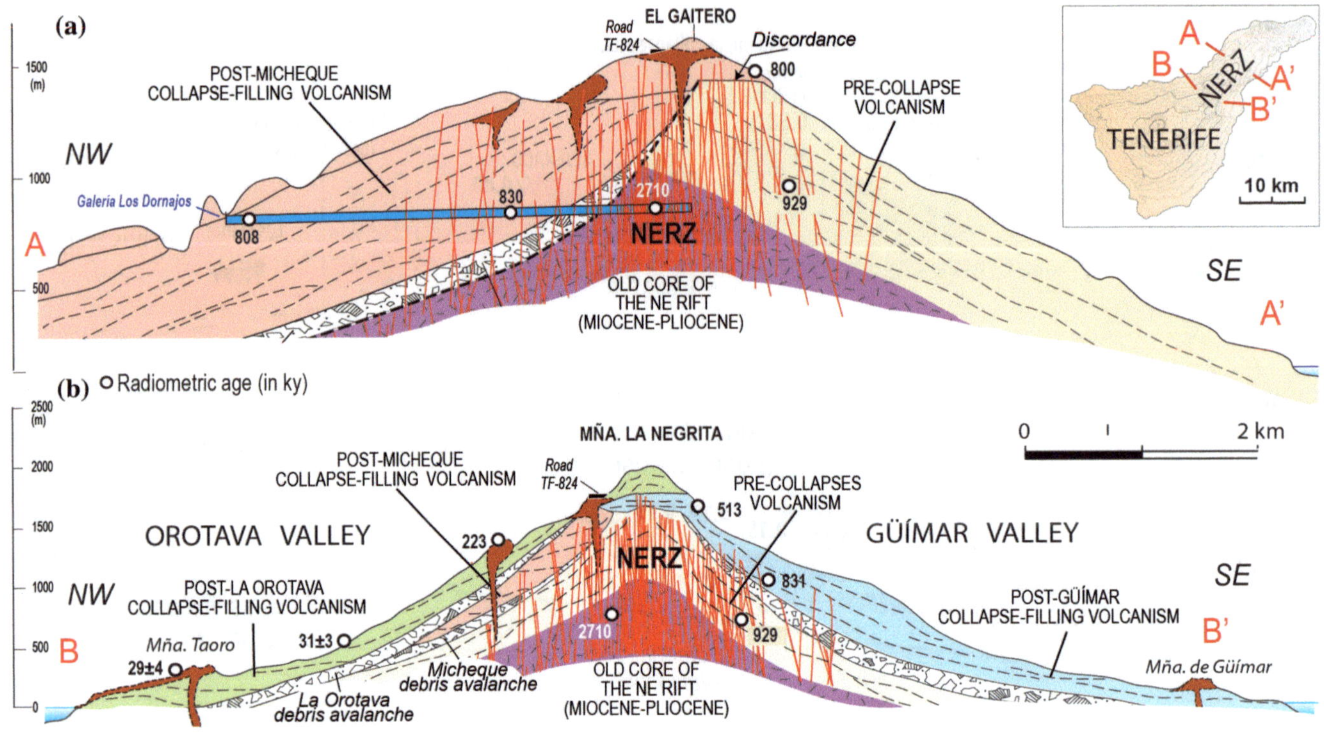

Fig. 24.13 Geological cross sections of the NE rift zone (ages in ka). **a** Section crossing the sequence filling the Micheque collapse embayment. **b** Idem to the southern part of the NERZT, cutting the sequences partially filling the Güímar and La Orotava collapse depressions (from Carracedo et al. 2011)

24.7 Morphological Changes of the NE Rift of Tenerife by Recurrent Landsliding

The NE rift zone of Tenerife (NERZT, Carracedo et al. 2011) is an elongated volcanic edifice that developed very rapidly in the second half of the Quaternary and was successively mass-wasted by recurrent mega-landslides: Micheque (1 in Fig. 24.12), Güímar (2) and La Orotava (3). These gravitational failures drastically reduced the volume of the volcano and changed its morphology, providing a spectacular setting to study the deep structure of a rift zone and the development of lateral collapses on a single rift-related edifice.

24.7.1 The Micheque Landslide, a Lava-Filled Mega-Landslide Depression

The first lateral slide of the NERZT occurred about 830 ky ago (Carracedo et al. 2011). The landslide generated a depression on the northern flank of the rift and extended into the present-day La Orotava valley (see Fig. 24.12). Subsequent volcanism largely filled the basin and concealed the scar and the avalanche breccia, only observable in *galerías* (Fig. 24.13a).

24.7.2 Mega-landslide of Güímar

The Güímar collapse, with an estimated volume of ca. 50 km^3 (Carracedo et al. 2011), formed a conspicuous 10 × 10 km, U-shaped depression bounded by linear escarpments on the east flank of the NERZT (Figs. 24.12 and 24.13b). The Güímar and Micheque collapses apparently occurred at nearly the same time with opposite directions. The Güímar slide was probably the result of gravitational adjustments following the Micheque collapse, possibly involving multiple slides over a period of years, instead of a single catastrophic failure (Carracedo et al. 2011; Delcamp et al. 2012). The Güímar debris avalanche was defined from sonar (GLORIA) data (Krastel et al. 2001).

24.7.3 La Orotava Mega-Collapse

The time of occurrence of this giant landslide, involving the removal of about 60 km^3, is constrained by a lava flow cascading into the Orotava Valley at the eastern wall, dated at 566 ± 13 ka (Carracedo et al. 2011), and the 690 ± 10 ka age obtained from a lava flow at the top of the western wall (Abdel-Monem et al. 1972), thus some hundred thousand years after the Micheque and Güímar landslides (Fig. 24.13b).

Fig. 24.14 The Orotava Valley, with steep rectilinear walls, and the Teide Volcano in the background, is one of the most spectacular scenarios in the Canaries. This dramatic landscape, together with the permanent mild climate at the coast, provided the favourable conditions to attract tourism to Puerto de la Cruz since 1886, the first tourist resort in the Archipelago

Fig. 24.15 **a** The San Andrés fault system bounding the lateral collapse at the eastern flank of the NE rift of El Hierro. After the slump, a large detached block remained attached to the island after a vertical displacement of about 300 m. **b** Fault plane with features that suggest large single-event displacements on the fault (photo J. C. Carracedo)

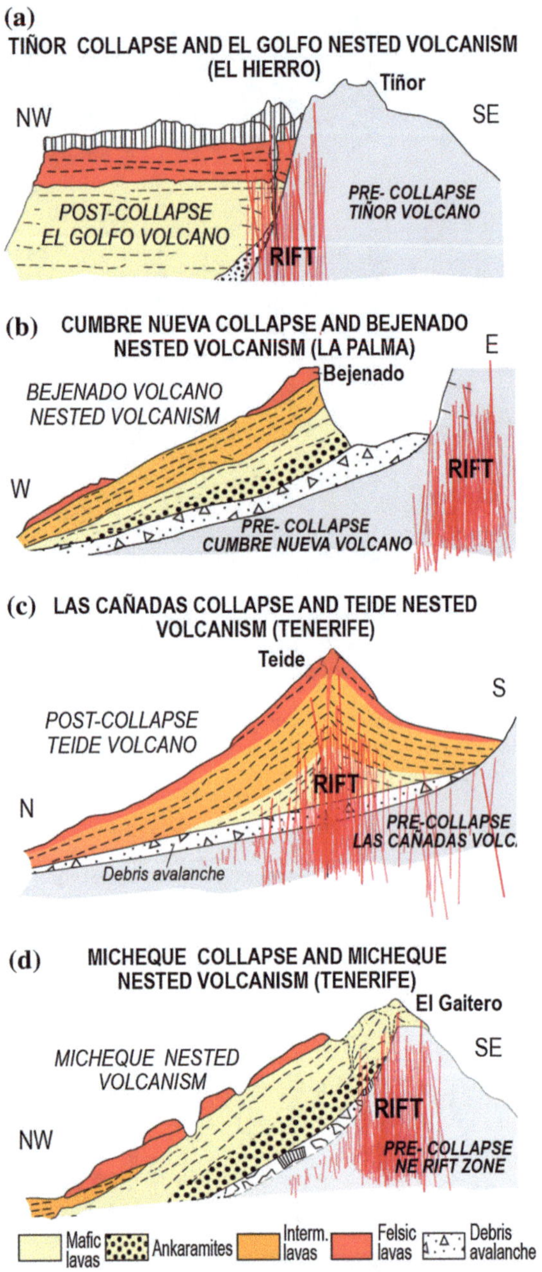

Fig. 24.16 Examples from the Canary Islands of differentiated volcanism associated with giant landslides. Note that progressive magmatic differentiation is recorded in the sequences filling the landslide scars (from Carracedo et al. 2011)

The morphology of the present-day La Orotava Valley, the largest visible landslide amphitheatre in Tenerife, has been modified by lava accumulation from vents of the NERZT located at the head of the Valley, partially filling the collapse scar. Nonetheless, this dramatic depression, with steep and rectilinear walls, displays the characteristic morphology of landslides (Fig. 24.14). The spectacular landscape, with Teide Volcano in the background, and a permanent mild climate at the coast provide favourable conditions for tourism in the Canaries that began in the Orotava Valley (Puerto de la Cruz) in 1886.

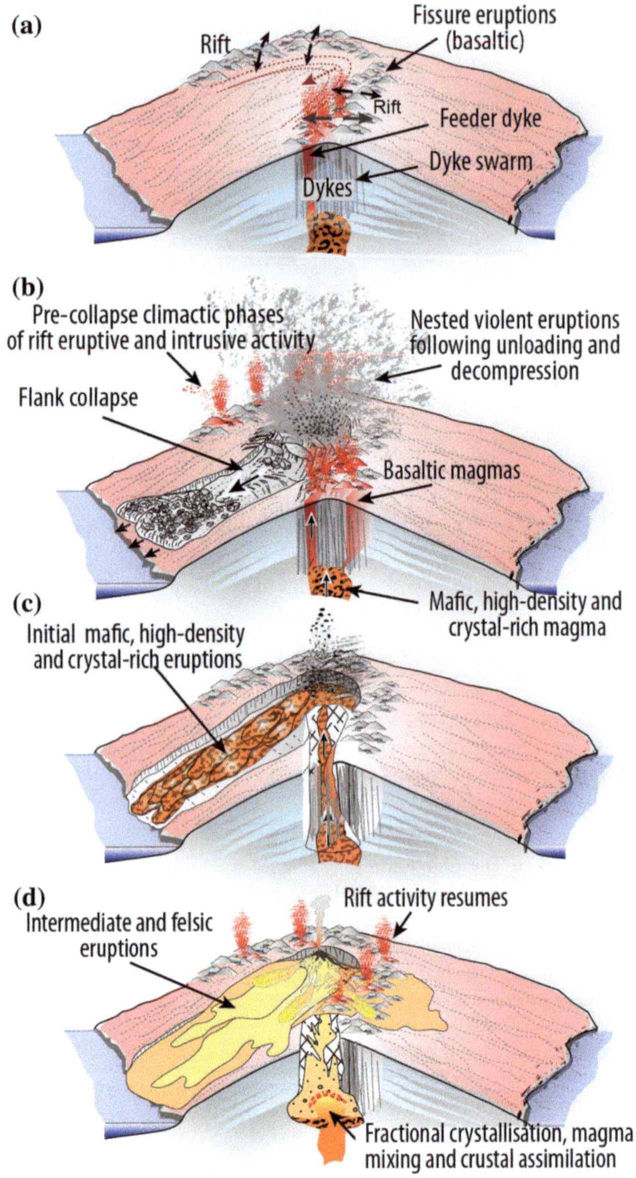

Fig. 24.17 Schematic model illustrating how rift evolution in the Canary Islands can proceed until high intrusive activity causes flank collapses, with the consequent readjustment of the plumbing system and thus leading to petrologic modifications (Carracedo et al. 2011)

24.8 The San Andrés Fault, a Rare Example of Arrested Collapse

The San Andrés fault system runs along the axis of the NE rift zone of El Hierro. These are NE–SW-trending normal faults expressed as escarpments up to 30 m high stretching for 10 km parallel to the coastline and connecting at the

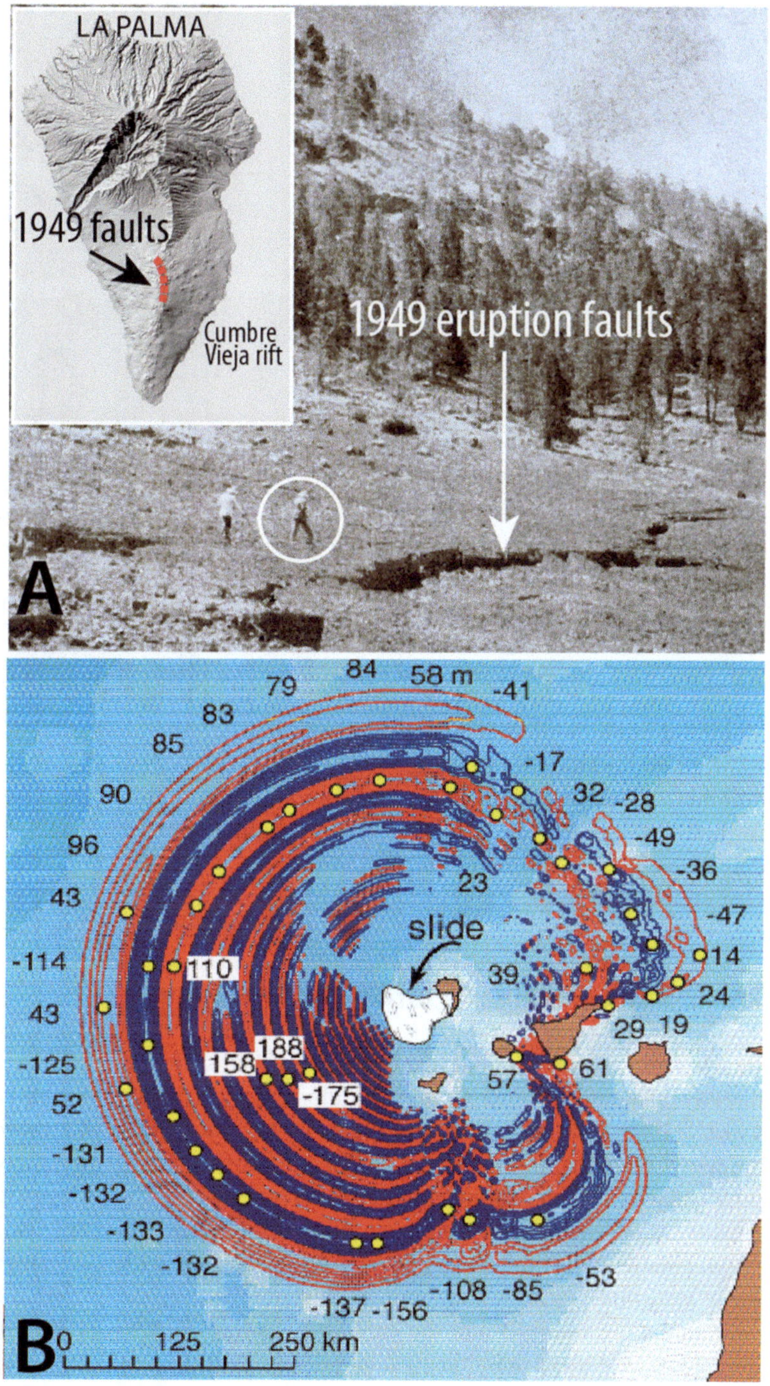

Fig. 24.18 **a** Faults of the 1949 La Palma eruption. **b** Ward and Day (2001) model illustrating the evolution of the La Palma landslide-related tsunami 30 min after the collapse. **c** Idem after 6 h, with 10–25 m high waves reaching the eastern coast of North America

western end with the collapse scar of Las Playas (Fig. 24.15a). Both features have been interpreted as scars of incipient landslides, although with much smaller volumes than the other landslides in El Hierro (Carracedo et al. 1997, 2001; Gee et al. 2001). The faults, dipping 60–70° towards the sea, show in places vertical displacements of approximately 300 m.

The exposed fault planes (Fig. 24.15b) show cataclasites (a metamorphic rock related to shearing on faults) formed close to the palaeosurface, which include zones with partial melting (pseudotachylytes, Day et al. 1997), suggesting large single-event displacements on the fault, probably related to a partially arrested landslide. Sonar and seismic reflection data show evidence of a blocky debris avalanche deposit overlying products of a prior slump extending offshore (Gee et al. 2001). Therefore, a plausible explanation is that the San Andrés fault system may be a slump-type failure similar to those documented on the sea floor of Hawaii (Moore et al. 1989), with two overlapping events: a first large slump (1 in Fig. 24.15a) that left behind the

Fig. 24.18 continued

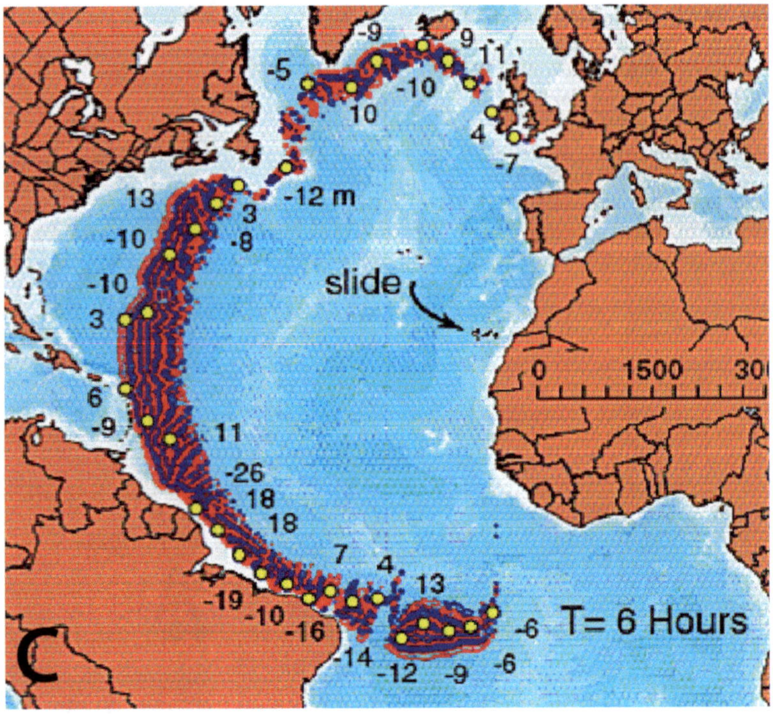

detached block and a minor second lateral collapse that formed the Las Playas embayment and accumulated the debris avalanche deposit (2 in Fig. 24.15b) described by Gee et al. (2001).

24.9 Landslides and Petrological Variation

Besides the crucial role played by massive landslides in directly shaping the morphology of the Canary Islands, these features can indirectly contribute to morphological diversity by inducing variations in magma composition, thus considerably increasing differences in eruptive mechanisms, as well as volcanic products, forms and structures.

Magmatic differentiation appears to be associated with giant landslides in the Canaries; the sequences filling the depressions resulting from the structural collapses frequently evolve from basaltic to trachytic/phonolitic lavas (Fig. 24.16). Mega-landslides may imply disruption of the established feeding system (Fig. 24.17) of a rift, initially erupting the dense mafic magmas that fill the collapse scars (Carracedo et al. 2007, 2011; Delcamp et al. 2010, 2012; Troll et al. 2013). Thus, emplacement of new magma at shallower depths after the collapse may lead to substantial modifications, commonly reaching felsic compositions (trachytes, phonolites). This trend becomes more significant with the progressive increase in height of the volcanoes, as the less evolved denser lavas have more difficulties to reach the summit of a large stratovolcano. A "density filter" is established, promoting later lavas to be more evolved.

24.10 Mega-Landslides and Tsunamis in the Canaries

The Cumbre Vieja rift zone, in La Palma, is a fast-growing, oversteepened and unstable volcanic edifice developed in the last 125 ky (Ancochea et al. 1994; Guillou et al. 1998; Carracedo et al. 1999a, b, 2001). Several lines of evidence were considered to be possible indications of an incipient stage of flank instability, e.g. the fast-growing rate of this volcano, structural changes in the rift suggesting weakening of its western flank and the development of a west-facing normal fault system along the crest of the ridge during the 1949 eruption (Fig. 24.18a), (Day et al. 1999).

Recent audiomagnetotelluric (AMT) surveys confirmed that the western flank of Cumbre Vieja rests over a weak layer of collapse debris material and hyaloclastites (García and Jones 2010). However, apparent deformation data obtained by infrared Electronic Distance Measurement (EDM) and GPS in the period 1994–1997 were within the error margins of the techniques employed (Moss et al. 1999). New interferometric data (InSAR) for the period 1992–2008 indicate movement away from the satellite on the western flank of Cumbre Vieja volcano, interpreted as gravitational aseismic creep. This is considered the dominant deformation

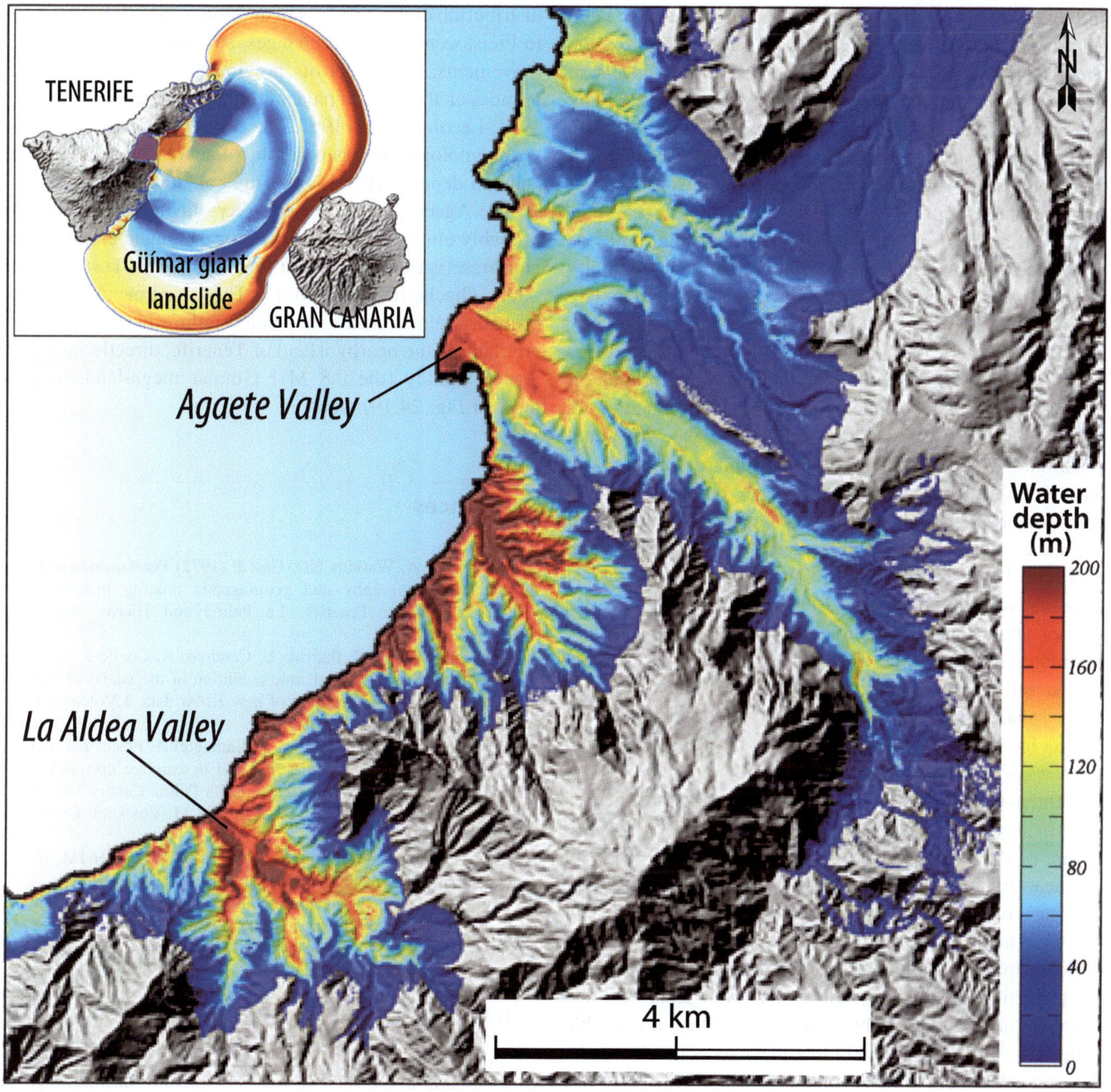

Fig. 19 Maximum water depth recorded inland for one go failure scenario. Inset: extent of the Guímar mega-landslide after 520 s for a one go collapse and a constant retarding stress of 150 kPa (Giachetti et al. 2011)

mechanism at Cumbre Vieja during inter-eruptive periods, probably facilitating stress release and contributing to the stabilization of the edifice (González et al. 2010).

Nevertheless, Ward and Day (2001) attracted worldwide attention by publishing a model for a catastrophic failure of the west flank of La Palma that might occur during a future eruption of Cumbre Vieja Volcano (Fig. 24.18b). According to these authors, the structural collapse up to 500 km^3 could generate waves that would travel across the entire Atlantic and reach within hours the coast of the Americas with a height of 10–25 m (Fig. 24.18c). The lag time to undertake alarm-evacuation measures would be several hours for the Americas, but only a few minutes for the western Canaries. The alleged mega-tsunami would take less than half an hour to reach the coasts of Tenerife, an island with nearly 1 million residents and visited annually by about 5 million people, the greater part concentrated below the tsunami wave run-up; the majority of the visitors and over 40 % of the population occupy areas situated below 100 m a.s.l. According to Ward and Day (2001), the lateral collapse may be triggered by the next eruption of Cumbre Vieja that might occur in the near future, likely with limited warning. Therefore, the only safe measure would require the preventive evacuation of several hundred thousand people in the event of any precursory sign of eruptive reactivation in La Palma.

Recent studies on the same subject rebut this model as overstated and unrealistic, pointing out that media publicity of those predictions has created unnecessary public anxiety among the Atlantic coastal communities (Pararas-Carayannis 2002). Mader (2001) used finite-volume Navier–Stokes modelling to predict that the maximum wave amplitude off the east coast of the United States would be about 1 m, concluding that even with shoaling, the wave would not constitute a significant hazard. Similar findings were reached by Gisler et al. (2006) when calculating the decline with distance of landslide-generated waves, concluding that nonlinear and dispersive effects greatly attenuate tsunami energy at far-field distances in comparison to seismogenic tsunamis (i.e. the Indonesia 2004 and Japan 2011 earthquakes and tsunamis). In relation to the giant landslide postulated for La Palma, these authors affirm that even the largest estimates are considerably smaller than the worrisome values given by Ward and Day (2001). Their calculations for the coasts of Florida, 6,100 km from La Palma, yield wave heights of about 77–120 cm.

Pararas-Carayannis (2002) stressed the fact that there is no sufficient or conclusive geologic evidence that a massive failure of the western flank of Cumbre Vieja will occur in the near future, as postulated by Ward and Day (2001). Similar conclusions were reached by Carracedo et al. (2005), who also suggested that the fault scarps of the 1949 eruption may be a surficial geomorphological feature related to partial collapses of shallow magmatic chambers that supplied lava to the 1949 vents, rather than the visible expression of a deep-seated failure plane associated with the detachment of the Cumbre Vieja western flank, as postulated by Ward and Day (2001).

24.11 The Agaete Tsunami

Although abundant mega-landslides have been documented in the Canaries, evidence for associated tsunamis has been found only in the Agaete Valley, Gran Canaria. Enigmatic marine conglomerates crop out at 41–188 m a.s.l. attached to the valley walls (Fig. 24.19), consisting of polymictic, angular to rounded heterometric volcanic clasts, and fossil rhodolites and marine shells, often broken and never found in growth position (Pérez-Torrado et al. 2006). The altitudinal distribution of the outcrops in the slopes can not be attributed to Pleistocene sea level changes, storm deposits or isostatic movements. These deposits, previously interpreted as marine terraces on the basis of palaeontological criteria (e.g. Denizot 1934; Lecointre et al. 1967; Meco 1989), show stratigraphic, sedimentologic and geomorphic features characteristic of tsunami deposits (Pérez-Torrado et al. 2006).

The Agaete Valley provides very favourable conditions, probably unique in the Canaries for the inland emplacement and preservation of tsunami deposits (Pérez-Torrado et al. 2006; Giachetti et al. 2011). These sediments are most probably related to a catastrophic failure in the north-eastern flank of the nearby island of Tenerife, directly facing the Agaete Valley (the 0.8 Myr Güímar mega-landslide; see inset in Fig. 24.19).

References

Abdel-Monem A, Watkins ND, Gast P (1972) Potassium-argon ages, volcanic stratigraphy and geomagnetic polarity history of the Canary Islands: Tenerife, La Palma and Hierro. Am J Sci 272:805–825

Ancochea E, Fúster JM, Ibarrola E, Cendrero A, Coello J, Hernán F, Cantagrel JM (1990) Volcanic evolution of the island of Tenerife (Canary Islands) in the light of new Kr/Ar data. J Volcanol Geoth Res 44:231–249

Ancochea E, Hernán F, Cendrero A, Cantagrel A, Fúster JM, Ibarrola E, Coello J (1994) Constructive and destructive episodes in the building of a young oceanic island. La Palma Canary Islands and genesis of the Caldera de Taburiente. J Volcanol Geoth Res 60:243–622

Ancochea E, Huertas MJ, Cantagrel JM, Coello J, Fúster JM, Arnaud NM, Ibarrola E (1999) Evolution of the Canadas Edifice and its implications for the origin of the Cañadas Caldera (Tenerife, Canary Islands). J Volcanol Geoth Res 88:177–199

Araña V (1971) Litología y estructura del Edificio Cañadas, Tenerife (Islas Canarias). Estud Geol 27:95–135

Bravo T (1962) El circo de Las Cañadas y su dependencia. Boletín de la Real Sociedad Española de Historia Natural 60:93–108

von Buch L (1825) Physicalische Beschreibung der Canarischen Inseln. Cambridge university press, Berlín

Cantagrel JM, Arnaud NO, Ancochea E, Fúster JM, Huertas MJ (1999) Repeated debris avalanches on Tenerife and genesis of Las Cañadas caldera wall (Canary Islands). Geology 27:739–742

Carracedo JC (1994) The Canary Islands: an example of structural control on the growth of large oceanic-island volcanoes. J Volcanol Geoth Res 60:225–242

Carracedo JC (1999) Growth, structure, instability and collapse of Canarian volcanoes and comparisons with Hawaiian volcanoes. J Volcanol Geoth Res 94:1–19

Carracedo JC, Day S, Guillou H, Torrado FJP (1997) El Hierro geological excursion handbook. Estación Volcanológica de Canarias and the Universidad de Las Palmas, Tenerife, Gran Canaria, pp 43

Carracedo JC, Day SJ, Guillou H, Gravestock P (1999a) The later stages of the volcanic and structural evolution of La Palma, Canary Islands: the Cumbre Nueva giant collapse and the Cumbre Vieja volcano. Geol Soc Am Bull 111:755–768

Carracedo JC, Day S, Guillou H, Pérez-Torrado FJ (1999b) Giant Quaternary landslides in the evolution of La Palma and El Hierro, Canary Islands. J Volcanol Geoth Res 94:169–190

Carracedo JC, Rodríguez-Badiola E, Guillou H, De la Nuez J, Pérez-Torrado FJ (2001) Geology and Volcanology of La Palma and El Hierro (Canary Islands). Estud Geol 57:175–273

Carracedo JC, Pérez-Torrado FJ, Rodríguez-Badiola E, Hansen-Machín A, Paris R, Guillou H, Scaillet S (2005) Análisis de los riesgos geológicos en el Archipiélago Canario: origen, características, probabilidades y tratamiento. Anuario de Estudios Atlánticos 51:513–574

Carracedo JC, Rodríguez-Badiola E, Guillou H, Paterne M, Scaillet S, Pérez-Torrado FJ, Paris R, Fra-Paleo U, Hansen A (2007) Eruptive and structural history of Teide Volcano and rift zones of Tenerife, Canary Islands. Geol Soc Am Bull 119:1027–1051

Carracedo JC, Guillou H, Nomade S, Rodríguez-Badiola E, Pérez-Torrado FJ, Rodríguez-González A, Paris R, Troll VR, Wiesmaier S, Delcamp A, Fernández-Turiel JL (2011) Evolution of ocean-island rifts: the northeast rift zone of Tenerife, Canary Islands. Geol Soc Am Bull 123:562–584

Carracedo JC, Troll V (2013) Structural and geological elements of the Teide volcanic complex: rift zones and gravitational collapses. In: Carracedo JC, Troll V (eds) Teide volcano: geology and eruptions of a highly differentiated oceanic stratovolcano (active volcanoes of the World). Springer, pp 57–74

Cotton CA (1952) Volcanoes as landscape forms. Wiley, New York, p 416

Day SJ, Carracedo JC, Guillou H (1997) Age and geometry of an aborted rift flank collapse: the San Andrés fault, El Hierro, Canary Islands. Geol Mag 143:523–537

Day SJ, Carracedo JC, Guillou H, Gravestock P (1999) Recent structural evolution of the Cumbre Vieja Volcano, La Palma, Canary Islands: volcanic rift zone reconfiguration as a precursor to volcanic flank instability? J Volcanol Geoth Res 94:135–167

Delcamp A, Petronis MS, Troll VR, Carracedo JC, van Wyk de Vries B, Pérez-Torrado FJ (2010) Vertical axis rotation of the upper portions of the north-east rift of Tenerife Island inferred from paleomagnetic data. Tectonophysics 492:40–59

Delcamp A, Troll VR, vanWyk de Vries B, Carracedo JC, Petronis MS, Pérez-Torrado FJ, Deegan FM (2012) Dykes and structures of the NE rift of Tenerife, Canary Islands: a record of stabilisation and destabilisation of ocean island rift zones. Bull Volcanol 74:963–980

Denizot G (1934) Sur la structure des Iles Canaries, considérée dans ses rapports avec le problème de l'Atlantide. Comptes Rendus d'Académie des Sciences 199:372–373

Friedlander I (1915) Uber vulkanische verwerfungstäler. Zeitschr. für Vulkanologie. Band. II, Berlin, Fritsch y Reiss, (1868)

Gagel C (1910) Die mittelatlantischen Vulkaninseln. Handbuch der regionales Geologie 7(10):1–32

García X, Jones AG (2010) Internal structure of the western flank of the Cumbre Vieja volcano (La Palma, Canary Islands) from land magnetotelluric imaging. J Geophys Res 115:B07104

Gee MJR, Watts AB, Masson DG, Mitchell NC (2001) Landslides and the evolution of El Hierro in the Canary Islands. Mar Geol 177:271–293

Giachetti T, Paris R, Kelfoun K, Pérez-Torrado FJ (2011) Numerical modelling of the tsunami triggered by the Güímar debris avalanche, Tenerife (Canary Islands): comparison with field-based data. Mar Geol 284:189–202

Gisler G, Weaver R, Gittings ML (2006) Sage calculations of the tsunami threat from La Palma. Sci Tsunami Hazards 24(4):288–301

Glicken H (1996) Rockslide-debris Avalanche of May 18, 1980, Mount St. Helens Volcano, Washington. U.S. Geological Survey Open-File Report http://vulcan.wr.usgs.gov/Projects/Glicken/OFR96-677.pdf

González P, Kristy F, Tiampo B, Antonio G, Camacho A, Fernandez J (2010) Shallow flank deformation at Cumbre Vieja volcano (Canary Islands): implications on the stability of steep-sided volcano flanks at oceanic islands. Earth Planet Sci Lett 297:545–557

Guillou H, Carracedo JC, Pérez-Torrado F, Rodríguez-Badiola E (1996) K-Ar ages and magnetic stratigraphy of a hotspot-induced, fast grown oceanic island: El Hierro, Canary Islands. J Volcanol Geoth Res 73:141–155

Guillou H, Carracedo JC, Day J (1998) Dating of the Upper Pleistocene-Holocene volcanic activity of La Palma using the unspiked K-Ar technique. J Volcanol Geoth Res 86:137–149

Hausen H (1956) Contributions to the geology of Tenerife (Canary Islands). Soc. Scient. Fenn. Commentationes Physico-mathematicae, 18, l. Helsingfors

Hausen HM (1961) Canarian Calderas: a short review based on personal impressions, 1947–1957. Bulletin de la Commission Geologique de la Finlande 196:179–213

Krastel S, Schmincke HU, Jacobs CL, Rihm R, Le Bas TP, Alibes B (2001) Submarine landslides around the Canary Islands. J Geophys Res 106:3977–3998

Lecointre G, Tinkler KJ, Richards G (1967) The marine Quaternary of the Canary Islands. Proc. Acad. Nat. Sci. Philadelphia 119:331–333

Linares R, Lomoschitz A, Pallí L, Roqué C, Brusí D, Quintana A (2001) Reconocimiento geofísico del deslizamiento de Rosiana (Depresión de Tirajana, Gran Canaria). Scientia Gerundensis 25:35–50

Lyell C (1830) Principles of Geology. London

Luongo G, Cubellis E, Obrizzo F, Petrazzuoli SM (1991) A physical model for the origin of volcanism of the Tyrrhenian margin: the case of Napolitan area. J Volcanol Geoth Res 48:173–185

Machado F (1965) Vulcanismo das ilhas de Cabo Verde e das outras ilhas Atlántidas. Est. Ens. Doc., 17. Junta de Investigaciones do Ultramar, Lisboa, p 83

Mader CL (2001) Modeling the La Palma landslide tsunami. Sci Tsunami Hazards 19:150–170

Márquez A, López I, Herrera R, Martín-González F, Izquierdo T, Carreño F (2008) Spreading and potencial instability of Teide volcano, Tenerife, Canary Islands. Geophys Res Lett 35:L05305

Martí J, Hürlimann M, Ablay GJ, Gudmundsson A (1997) Vertical and lateral collapses on Tenerife (Canary Islands) and other volcanic ocean islands. Geology 25:879–882

Masson DG (1996) Catastrophic collapse of the flank of El Hierro about 15,000 years ago and the history of large flank collapses in the Canary Islands. Geology 24:231–234

Masson DG, Watts AB, Gee MJR, Urgelés R, Mitchell NC, Le Bas TP, Canals M (2002) Slope failures on the flanks of the western Canary Islands. Earth Sci Rev 57:1–35

Meco J (1989) Islas Canarias In: Pérez-González A, Cabra-Gil P, Martín-Serrano A (coords) (eds) Mapa del Cuaternario de España a escala 1:100000. Instituto Tecnologico y Geominero de España

Moore JG (1964) Giant submarine landslides on the Hawaiian Ridge. U.S. Geological Survey Professional Paper 501-D, pp D95–D98

Moore JG, Clague DA, Holcomb RT, Lipman PW, Normark WR, Torresan ME (1989) Prodigious submarine landslides on the Hawaiian Ridge. J Geophys Res 94(17):465–484

Moss J, McGuire WJ, Page D (1999) Ground deformation monitoring of a potential landslide at La Palma, Canary Islands. J Volcanol Geoth Res 94:251–265

Navarro JM, Coello J (1989) Depressions originated by landslide processes in Tenerife. Paper presented at meeting on Canarian Volcanism, European Science Foundation, Strasbourg, France

Pararas-Carayannis G (2002) Evaluation of the threat of mega-tsunamis generation from postulated massive slope failures of island stratovolcanoes on La Palma, Canary Islands, and on the island of Hawaii. Sci Tsunami Hazards 20:251–257

Pérez-Torrado FJ, Paris R, Cabrera MC, Schneider JL, Wassmer P, Carracedo JC, Rodríguez-Santana A, Santana F (2006) Tsunami deposits related to flank collapse in oceanic volcanoes: the Agaete Valley evidence, Gran Canaria, Canary Islands. Marine Geology 227:135–149

Schmincke H-U, Sumita M (1998) Volcanic evolution of Gran Canaria reconstructed from apron sediments: synthesis of VICAP project drilling. In: Weaver PPE, Schmincke H-U, Firth JV, Duffield W (eds) Proceedings of the Ocean Drilling Program, Scientific Results, 57:443–469

Stearns HT, Clark WO (1930) Geology and water resources of the Island of Hawaii. U.S. Geological Survey. Water Supply Paper 616, pp 194

Stearns HT, Macdonald (1946) Geology and ground-water resources. Hawaii Division of Hydrogeology Bulletin 9, pp 363

Stillman CJ (1999) Giant Miocene landslides and the evolution of Fuerteventura, Canary Islands. J Volcanol Geoth Res 94:89–104

Troll V, Deegan FM, Delcamp A, Carracedo JC, Harris C, van Wyk de Vries B, Petronis MS, Pérez-Torrado FJ, Chadwick JP, Barker AK, Wiesmaier S (2013) Pre-Teide Volcanic Activity on the Northeast Volcanic Rift Zone. In: Carracedo JC, Troll V (eds) Teide Volcano: Geology and eruptions of a highly differentiated oceanic stratovolcano (Active Volcanoes of the World). Springer, pp 75–92

Urgelés R, Canals M, Masson DG (2001) Flank stability and processes off the western Canary Islands: a review from El Hierro and La Palma. Sci Marines 65–1:21–31

Walker GPL (1992) Coherent intrusion complexes in large basaltic volcanoes—a new structural model. J Volcanol Geoth Res 50:41–54

Walter TR, Troll VR, Cailleau B, Belousov A, Schmincke H-U, Amelung F, vd Bogaard P (2005) Rift zone reorganisation through flank instability in ocean island volcanoes: an example from Tenerife, Canary Islands. Bull Volcanol 67:281–291

Ward SN, Day SJ (2001) Cumbre Vieja Volcano; potential collapse and tsunami at La Palma, Canary Islands. Geophys Res Lett 28:3397–3400

Watts AB, Masson DG (1995) A giant landslide on the north flank of Tenerife, Canary Islands. J Geophys Res 100:24487–24498

Weaver PPE, Thomson J, Jarvis I (1989) The geology and geochemistry of Madeira abyssal plain sediments: a review. In: Freeman TJ (ed) Advances in underwater technology, Ocean Science and Offshore Engineering, vol 18, London (Graham and Trotman), pp 51–78

25 Geomorphological Heritage and Conservation in Spain

Ángel Salazar, Luis Carcavilla, and Andrés Díez-Herrero

Abstract

Spain is the country with the largest surface of protected natural areas in the European Union (around 27 % of the territory). This is due to some peculiar biogeographic factors, as well as to the high diversity (geodiversity) and uniqueness of the Spanish geomorphological landscapes. This chapter deals with the preservation of this important geomorphological heritage, which must be analyzed in a broader context of nature conservation and within the framework of geoconservation guiding principles. Nature conservation and geoconservation had a promising beginning in Spain during the first decades of the twentieth century, but due to historical ups and downs, this policy was not continued. The environmental movement during the sixties and seventies, together with the democratization of the country, led to a promising shift in nature conservation, but with minor changes regarding geological and geomorphological heritage management. During the last few years, scientific societies, universities, and other related groups have positively influenced the Spanish politicians and the public, resulting in significant advances in geoconservation. Nowadays, the Spanish geomorphological heritage is managed by regional governments through a complicate set of national, regional, and sectoral regulations. Additionally, the participation in international programs of nature conservation, such as those promoted by UNESCO (World Heritage Sites, Biosphere Reserves, and Geoparks), as well as some local and private initiatives, has recently undergone an important development.

Keywords

Geoconservation • Geodiversity • Nature conservation • Protected areas • Geoparks

25.1 Introduction: Geomorphological Heritage in the Context of Nature Conservation

Until the nineteenth century, humans used to perceive wilderness as a dangerous place, or as a storehouse of raw material. The creation of the Yellowstone National Park (NP) in 1872 can be considered the worldwide kickoff of the nature conservation movement, as it was the first natural area protected by a government. Soon after, this idea was welcomed in other countries, where natural areas were protected through the creation of national parks. Although

Á. Salazar (✉)
Instituto Geológico y Minero de España (IGME), C/La Calera n° 1, 28760, Tres Cantos, Madrid, Spain
e-mail: a.salazar@igme.es

L. Carcavilla · A. Díez-Herrero
Instituto Geológico y Minero de España (IGME), C/Ríos Rosas n° 23, 28003, Madrid, Spain
e-mail: l.carcavilla@igme.es

A. Díez-Herrero
e-mail: andres.diez@igme.es

during those early stages, politicians were perhaps more interested in spreading their nationalist ideas than in protecting Nature, important efforts were made with two principal aims. The first one was to safeguard the most representative wild areas of each country from the destructive activities brought about by the industrial revolution, and the second one was to route the first tourists to those iconic locations with high natural values and outstanding scenic landscapes. In this historical context, all the elements of the landscape architecture (geological, geomorphological, and biological) were viewed intuitively as a whole. Also, the American transcendentalist philosophers had strong influence on the early conservation movement, such as the naturalist and writer John Muir (1838–1914), who interpreted the unspoiled wilderness as a manifestation of a unique divinity, probably closer to a pantheistic interpretation of nature than to a true ecological concept (Quick 1999). Hence, four of the five first national parks of the world have important geological and geomorphological values: volcanic and geothermal features in Yellowstone (USA, 1872) and Tongariro (New Zealand, 1886), and structural and glacial landforms in Banff (Canada, 1885) and Glacier National Park (Canada, 1886).

At the end of nineteenth century, the most important scientific disciplines for understanding the principal elements of landscapes, such as Geology, Geomorphology, and Biology, had already achieved a remarkable development, and Forestry emerged as a new discipline aimed at ensuring timber resources and sufficiently large populations of game species. Landforms are the core abiotic component of landscape, and therefore, Geomorphology played an essential role in those early times of the conservation movement. However, the abiotic elements of nature were viewed and valued more as an aesthetic backdrop of the natural environment, rather than for their specific values or their role in the ecological relationships.

During the 1960s and early 1970s, several environmental disasters and the peculiar social and cultural context (Rome 2003) gave rise to the "environmental movement," and forced the administrations to add the concept of sustainable management of resources in the international and national political agendas. Thereby, the conservation movement received a decisive support and reinforced also by the development of modern Ecology, Environmental Sciences and land-use planning tools. As a result of the 1972 UNESCO meeting held in Paris, the term "Natural Heritage" began to be broadly used (UNESCO 1972). The term "heritage" adds a new and important concept to nature conservation; the most valuable elements of nature have been transmitted to us by previous generations, and we must preserve them carefully for the next generations.

Biotic elements of natural heritage have received adequate worldwide attention in environmental policies. However, abiotic elements were commonly regarded as a physicochemical support for life or as a static decoration. Since the early 1970s, many institutions throughout the world (geological surveys and institutes, universities, geoscience societies) have tried to appraise the geological and geomorphological heritage (geoheritage; Sharples 1993) and promote the protection of the most important localities for their scientific, educational, and touristic value. These efforts joined together during the 1st International Symposium on the Conservation of Geological Heritage, held in Digne, France (June 1991), where experts from over 30 countries endorsed the declaration of the "Rights of the Memory of the Earth," an important document that defines the essential needs for the conservation of the geoheritage (Martini and Pagès 1994). Since then, terms such as geosites, geomorphosites, or Geoparks started to be used. A "Geosite" designates locations containing a geological or geomorphological heritage that should be protected (Wimbledon 1996; Panizza 2001). Geopark is used to label a territory that comprises a certain number of geological or geomorphological sites and a sustainable territorial strategy linked to the development of geological tourism (Martini and Zouros 2001).

Biodiversity, a term initially used only in science and environmental policy, became colloquially used in the 1990s due to the influence of the "Convention on Biological Diversity" (informally "Biodiversity Convention"), an international agreement reached at the 1992 Rio de Janeiro Earth Summit. Shortly after, the abiotic equivalent of biodiversity, geodiversity was first used by Sharples (1993) and, nowadays, geodiversity is a guiding principle in the selection of protected areas in several countries (Gray 2008). The most worldwide accepted definition for geodiversity is that provided by Gray (2004): *the natural range (diversity) of geological (rocks, minerals, fossils), geomorphological (landform, processes) and soil features. It includes their assemblages, relationships, properties, interpretations, and systems*. Another key concept in modern nature conservation policy is "ecosystem services": goods and functions of ecosystems that benefit society. Ecosystem services can be classified into: provisioning, regulating, supporting, and cultural (Millennium Ecosystem Assessment 2003). Most of the research efforts address exclusively biotic services, and therefore, the abiotic values and services have not received much attention so far. A comprehensive analysis of nature services must include those that are purely abiotic (geosystem services), and particularly, appreciation services (resources for recreation, cultural, spiritual and historical meanings, and artistic inspiration) and knowledge services derived from geological, paleontological, geomorphological, geochemical, and geophysical research (Webber et al. 2006; Gray 2012). Geomorphological diversity must be considered as a fundamental component of geodiversity because (1) it strongly influences biodiversity (Hopkins 1994; Burnett et al. 1998; Nichols et al. 1998), (2) provides outstanding scientific and education resources, (3) includes major landscape features and unique habitats such as caves and wetlands (Webber et al. 2006), and (4) has a central role on soil diversity, landscape complexity and sensitivity of soil and landscape to environmental change (Thomas 2012).

Spain has played a significant role in the nature conservation movement and currently is the country with the largest surface of protected areas in the European Union; more than 137,000 km^2 representing 27 % of the country (Múgica de la Guerra et al. 2012). This fact is partly due to the biogeographic position of Iberia, linking Africa and Europe, the ascription of the Canary Islands to the Macaronesian biogeographic region, and the diversity and uniqueness of the Spanish geomorphological landscapes (Gutiérrez 1994; Desir et al. 2005; Benito-Calvo et al. 2009). This nature conservation effort is the outcome of a long history and a complex mixture of scientific, political, and social actions, similar to other countries but with some peculiarities.

25.2 Geoconservation in Spain, a Brief History

Nature conservation and geoconservation history in Spain have evolved in parallel with the political fluctuations of the twentieth century. After a promising beginning, with a significant effort in the country to improve culture and science, the civil war (1936–1939) and the policy during the dictatorship of General Franco (1939–1975) represented a backward step in nature conservation. In the 1960s, still under the dictatorship, the environmental movement in Spain had some popular support that, together with the international environmental concern, forced the government to make some minor changes in their nature conservation policy. The democratization of the country, and particularly, the regional management of nature and environment established by the 1978 Constitution, provided a fresh impetus to nature conservation managed by the regional governments, closer to the natural environment and the local population.

25.2.1 Pioneers and Early Efforts

Often referred to as the earliest nature conservation efforts in Spain, regulations for the use and exploitation of natural resources included in medieval territorial laws, such as those from Sepúlveda or Cuenca cities, served as blueprints for subsequent legislation. To avoid the depletion of ore minerals or the rapid disappearance of exploitable granite tors, operating restrictions for their exploitation were established, as well as jurisdiction on the management and the preservation of some elements. For instance, in the early modern period, the decree by King Philip II (mid-sixteenth century) that regulated the use and exploitation of the Segovia Forest (Bosque de Segovia; currently the Montes de Valsaín, Central Spain) to guarantee water, forestry, and hunting resources for the palace and the enjoyment of the court. Perhaps unintentionally, he preserved several geomorphological heritage features of interest, such as granite boulders and slabs, whose quarrying was restricted, and nivation hollows, which were not modified for the construction of artificial ice houses and snow wells.

The true initial steps of the conservation movement in Spain can be placed around the middle of the nineteenth century. Unlike the USA, in Spain, the beginnings were not a reaction to the industrial development and urbanization, which were very scarce. In fact, first steps were an indirect result of the desamortization (disentailment); a long, intermittent, and complex process implemented by the government during the nineteenth century to collect extra income, in order to pay war debts and increase the cultivation of land (Ramos-Gorostiza 2005). The disentailment involved the forced expropriation and subsequent public auction of lands owned by the church, as well as communal lands of the municipalities. From 1859 to 1926, nearly 50,000 km^2 (10 % of the country) of public land became private property (Grupo de Estudios de Historia Rural 1994). In 1859 and 1901, the State forestry engineers cataloged and assessed the quality of the publicly owned lands, considering the protection suitability of their forests and other natural elements such as geomorphological features, in order to label those that should be saved from the disentailment, becoming the pioneers of nature conservation in Spain. The early Spanish conservation movement was also promoted by two other social groups, aristocracy and naturalists, each one with very different interests on Nature (Casado 1996; Ramos-Gorostiza 2005).

Historically, the Spanish aristocracy and monarchy used the wilderness for recreation, mainly through sport hunting and fishing. They were interested in suitable habitats for game species by owning the land to ensure an exclusive use or by implementing restricting hunting rules. The disentailment allowed the aristocracy and the wealthier people to acquire, at bargain prices, wide territories for hunting in the nineteenth century (Vías 2011). The leader of this group was undoubtedly Pedro Pidal (1870-1941), an aristocratic hunter and reputed mountain climber, who as a senator promoted the first laws in order to protect species and areas. To save from extinction the ibex and chamois, Pidal

Fig. 25.1 Due to the occurrence of unique karst features, such as mushroom rocks, the Ciudad Encantada (Enchanted City, Cuenca province) was proposed as a National Geological Park in 1914 and protected in 1929 as a Natural Site of National Interest

promoted the creation of two royal hunting reserves in 1905 (Cotos Reales), Picos de Europa and Gredos, with the endorsement of King Alfonso XIII. Shortly after, in 1916, Pidal also promoted the first legislation on national parks, greatly inspired by the United States' law (Casado 1996; Ramos-Gorostiza 2005). The first national parks, Picos de Europa and Ordesa-Monte Perdido, were established in 1918, both sites with high outstanding geological and geomorphological values (Chaps. 13 and 14).

The Spanish naturalists paid attention to the Nature protection actions developed in other countries, such as the establishment of NPs, because they were deeply concerned about the extinction of some Spanish endemic botanical and zoological species and the degradation of sites of singular geological and geomorphological interest. Such efforts were developed through official institutions (universities, Junta de Ampliación de Estudios, Natural History Museum) and other societies (Royal Society of Natural History, Institución Libre de Enseñanza, mountaineering clubs). One of the Spanish pioneers in geoconservation was the high school Professor Juan Giménez de Aguilar (1876–1947), who suggested to the Royal Society of Natural History in 1914 to promote the establishment of a National Geological Park in the Ciudad Encantada (Enchanted City), Cuenca Province (Real Sociedad de Historia Natural 1914, Casado 1996), with magnificent examples of karst landforms (Fig. 25.1).

The involvement of the Spanish researchers in nature conservation policy became possible through the establishment of the Central Board of National Parks (Junta Nacional de Parques Nacionales, later designated as Comisaria de Parques Nacionales). The geologist Eduardo Hernández-Pacheco (1872–1965), at the University of Madrid, was appointed a member of such Board and had a decisive contribution. Hernández-Pacheco understood that the National Park management approach, imported from the USA, was difficult to implement in Europe and Spain, where there were no longer strictly wild territories, and therefore actively promoted the development of regulations according to the true situation of the Spanish nature, allowing the involvement of private and local actors in conservation proposals (Casado 1996; Ramos-Gorostiza 2005). That new regulation, promulgated in 1927, was a success. Until 1936, sixteen Natural Sites and Natural Monuments were protected in Spain, most of which with high geological and geomorphological values: granite landforms in La Pedriza (see Chap. 5) and Peña del Arcipreste de Hita (Madrid province); structural and karst landforms in Torcal de Antequera (Málaga), Ciudad Encantada (Cuenca, Fig. 25.1), Picacho de la Virgen de la Sierra (Córdoba), and Monte del Valle—Sierra de Espuña (Murcia); glacial landforms and deposits in Peñalara (Madrid, Fig. 25.2) and Moncayo (Zaragoza); coastal landforms in Monte Curotiña, Cabo Villano, and Cabo de Vares (La Coruña). Eduardo Hernández-Pacheco also performed groundbreaking divulgation work through the publication of the first guidebooks of the Spanish protected areas.

25.2.2 Shadowy Times

The conservation policy was interrupted during the Spanish Civil War and the following dictatorship. Although the protected areas continued under the guardianship of the

Fig. 25.2 The Natural Park of Peñalara (Central Range, Madrid province) was protected in 1930 due to the high scientific interest of the glacial features and its outstanding landscape

state administration, the Board of National Parks disappeared, the budget for the management of protected areas was drastically reduced, and new nature conservation laws, focused on timber harvest and recreational hunting and fishing, were approved (Ramos-Gorostiza 2006). Additionally, the geological, geographical, and biological research teams were broken up and the conservation leaders fell into oblivion, went into exile or were persecuted, such as the professor Juan Giménez de Aguilar, who was condemned to death (López-Villaverde 2005).

As a consequence of that lack of concern in nature conservation, only four new sites became protected by the government in the following 33 years, although all of them have remarkable geomorphological features: Sanabria glacial lake (the largest glacial lake of Spain) became protected Natural Site in 1946 and two of the most remarkable volcanic landscapes of the Canary Islands, Teide volcano (see Chap. 22) and Caldera de Taburiente (Fig. 25.3) became National Parks in 1954, and Aigües Tortes—Lago de San Mauricio National Park, with Pyrenean glacial landscapes and lakes, was declared in 1955.

During the early 1960s, forest harvesting and tourism development policy entailed a great impact on the Guadalquivir marshes and dunes (see Chap. 19). Due to the international alarm, in 1963 the World Wildlife Fund (WWF) acquired some lands and created a nature reserve and a research center (Estación Biológica de Doñana), which currently belongs to the National Research Council—CSIC. Finally, in 1969, Doñana National Park was established through an agreement between the WWF and the Spanish government. The environmental movement, closely linked in Spain to the vindication of democracy, emerged in those years and became a way of protest against the dictatorship. Modern ecology began to develop in Spanish universities thanks to researchers such as Ramón Margalef (1919–2004) and Fernando González-Bernáldez (1933–1992). Also the TV series and books of the outstanding communicator Félix Rodríguez de la Fuente (1928–1980) brought the Spanish nature closer to the public. However, these new efforts had not any link to those

Fig. 25.3 These cliffs, spurs, and gullies in Caldera de Taburiente NP, with a local relief greater than 2,000 m, are related to fluvial erosion of the "Las Angustias" ravine (boulder deposits and terraces at bottom) and tributaries, whose headwaters are associated with the headscar of a giant landslide developed over a former stratovolcano (La Palma Island, Canaries)

conducted long time ago by the pioneers (Ramos-Gorostiza 2006) and did not had much appreciation on the geomorphological heritage.

In this scenario, the government made some changes in conservation policy through (1) the creation of a powerful government agency, the Institute for the Conservation of Nature (ICONA) in 1971; (2) the update of the forestry, hunting, and protected areas laws; (3) and the declaration of two new national parks: Tablas de Daimiel in 1973 (continental wetlands, Ciudad Real province) and Timanfaya in 1974 (landscape related to a major historical volcanic eruption, Lanzarote Island). The ICONA (1975) produced a first national inventory of outstanding landscapes, taking into account some geological and geomorphological features, which was revised a few years later (García-Cortés 1996). The first inventory of geological, geomorphological, and paleontological sites was constructed by Professor Emiliano Aguirre and colleagues from Madrid University, as part of the land-use planning of Madrid region and the surrounding areas (COPLACO 1975).

25.2.3 Democracy and New Times

The 1978 Spanish Constitution recognized the right of the citizens to enjoy a suitable environment and the responsibility of the state and public administration to take care of the environment and natural areas. According to the Constitution, these jurisdictions should fall on the regional governments. A new national regulation about wildlife and natural areas conservation (Law 4/1989) was approved in 1989. Although this law relies on a modern and more holistic approach to nature, geoconservation principles were essentially neglected. In accordance with the allocation of powers set by the new Constitution, all the regional governments, except Madrid government, enacted their own nature conservation laws, but all of them were based mainly on the national regulation (Law 4/1989) and also paid little attention on geoconservation (Carcavilla et al. 2009).

The transfer of power to the new regional governments in 1984 was a success for nature conservation, and 121 natural areas were protected only during the year 1987; more than twice the legally protected natural areas since 1918 (Fernández-Sañudo and de Lucio 1994). Nonetheless, the increase in the number and surface of protected areas did not involve a significant improvement in the protection of geological and geomorphological heritage (Gallego-Valcarce and García-Cortés 1996).

The geologist Emilio Elízaga (1945–1992), from the Geological Survey of Spain (Instituto Geológico y Minero de España, IGME), put forward an important initiative in 1978, leading the construction of a national inventory of geological sites, including geomorphosites. Unfortunately, only 20 % of the country had been examined in 1987. Due to the scarcity of financial resources, since 1988 the assessment of the geological heritage was included in the geological mapping projects, but with irregular results (Elízaga 1988; García-Cortés et al. 1992; Carcavilla et al. 2009).

Some regional administrations also produced their own catalogs of geological and geomorphological heritage, such as those of Guipúzcoa province (Portero et al. 1991), Segovia province (Díez 1991), and Madrid (Instituto Tecnológico Geominero de España 1988), and faintly began to consider the geological and geomorphological heritage in nature conservation, land-use planning, and environmental impact assessment (Salazar-Rincón et al. 1996).

25.3 Geomorphological Heritage Conservation in Spain Today

The Geological Society of Spain (SGE) was founded in 1985 and in 1994 created a Geological Heritage Commission. Since then, it has organized meetings on geological heritage periodically. Through its Commission on Geological Heritage, the SGE joined the International Union for Conservation of Nature (IUCN) as a national non-governmental organization (NGO) in 2008, followed by ProGEO (European Association for the Conservation of the Geological Heritage), that joined IUCN as an international NGO. These are the first scientific societies related to geology and geomorphology in the world that became IUCN members The Spanish Society of Geomorphology (SEG) was founded in 1989 and has also developed a major activity in the protection and promotion of the geomorphological heritage.

Both societies (SGE and SEG) and their members, in conjunction with other related NGOs, have advised the Spanish deputies, senators, and politicians on the assessment and management of the geomorphological and geological heritage. The enactment of the new laws 5/2007 (about National Parks) and 42/2007 (about Natural Heritage and Biodiversity) is probably the most significant achievements.

25.3.1 National and Regional Legislation About Geomorphological Heritage Conservation

The National Parks law (Law 5/2007) included as criteria for the declaration of these spaces the uniqueness of the geological and geomorphological elements, and particularly, their ecological, aesthetic, cultural, educational, and scientific values. The National Parks Network should be the best synthesis of the Spanish natural heritage and should integrate the most representative samples of the Spanish natural systems. The biological and ecological issues have been considered exclusively for defining several of those natural systems. However, the geological and geomorphological characteristics are the key issues for defining the following ones (Annex of the Law 5/2007):

- Unique landforms and geological features of the Iberian Massif and Alpine Ranges.
- Unique formations and landforms of mountains and high mountains (such as the granitic landforms described in Chap. 5, in the Guadarrama NP).
- Unique natural systems of glacial and periglacial origin (such as those of Ordesa and Monte Perdido NP, Chap. 14).
- Unique natural systems of karstic origin (as those of the Picos de Europa NP, Chap. 13).
- Natural systems and formations related to the continental and marine Tertiary basins.
- Fluvial canyons on structural reliefs.
- Unique deposits and landforms of fluvial and eolian origin.
- Coasts, cliffs, dunes, and coastal deposits.
- High mountain wetlands and lakes.
- Halophilic ponds, salt pans, and gypsum outcrops.
- Freshwater ponds, reedbeds, bulrushes, and water-table grasslands with temporary flooding.
- Coastal wetlands and coastal marshes (such as the Doñana NP marshes in Chap. 19).
- Unique natural systems of volcanic origin (as those illustrated in Chaps. 22 and 23 referring to Teide and Timanfaya NPs).

The list above includes the principal geomorphological landscapes present in the current NPs. A very important matter concerning another law enacted shortly after (Law 42/2007) is the fact that the most valuable Spanish geomorphological landscapes (such as all the ones described in this volume) are also included in this set of natural systems.

The Law of natural heritage and biodiversity (Law 42/2007) entailed also a major update in geoconservation, with the incorporation of terms such as geodiversity, geological heritage, and Geoparks in the basic guidelines of nature conservation (Díaz-Martínez et al. 2008). The Law 42/2007 also included the requirement of a national inventory of geosites that should be representative of the principal geological units (based on the list of natural systems of law 5/2007 mentioned above) and the geological frameworks of global relevance identified for Spain (see below Geosites Program). It is worth to mention that 5 out of the 20 geological frameworks of global relevance established have a geomorphological origin (Annex VIII, Law 42/2007):

- Fluvial network, piedmonts, and Appalachian-type relief of the Iberian Massif.
- Flat coasts of the Iberian Peninsula (such as the Ebro River Delta or the coastal dunes and marshes in Doñana NP, Chaps. 18 and 19).
- Karst systems developed in carbonates and evaporites of the Iberian Peninsula and Balearic Islands (such as those described in Chaps. 6–14).
- Fossil sites of Plio-Pleistocene vertebrates of Spain (including paleoanthropological sites such as Atapuerca, Chap. 8).
- Volcanic mountains and landforms of the Canary Islands (such as Teide, Timanfaya and giant landslides described in Chaps. 22–24).

Some of the Spanish regional governments have also included the conservation of the geological and geomorphological heritage, as well as the sustainable management of the geodiversity in their political agendas. The regional governments of Andalusia and the Basque Country have developed specific strategies for the management of the geodiversity in their own territories (Consejería de Medio Ambiente 2012; Mendía et al. 2010; Mendía and Monge-Ganuzas 2011; Monge-Ganuzas et al. 2011). The Aragón government is also preparing a decree in order to protect the geological heritage, which is currently under debate.

25.3.2 Geomorphological Heritage Protection Through Other Sectoral Regulations

In addition to specific legislation related to the protection of natural areas, other sectoral and environmental regulations could be useful to preserve and protect our geomorphological heritage. These include as follows:

- Legislation to protect historical and artistic heritage. Some Heritage laws passed by the state (Law 16/1985) and the regional governments to protect artistic or cultural elements, including archeological sites that are associated with features of geomorphological interest, ensure geomorphological heritage conservation. For example, Atapuerca Range and its caves (see Chap. 8) are protected for their archeological and paleontological value as a cultural space (Agreement 199/2007 of the Castile and Leon regional Government).
- Legislation related to spatial planning, land management, and urban planning. At the state, regional, provincial, county, and local levels, different regulations such as land laws, spatial planning guidelines, regional plans, and urban plans may declare geomorphological heritage elements to be non-developable or afford them special protection. Article 12 of the national land law itself (Royal Legislative Decree 2/2008) includes the preservation of nature as one of the reasons for declaring land as rural (non-urban). As an example of geomorphological heritage preservation in local planning regulations, the regional planning

guidelines of Segovia and its Environment (DOTSE in its Spanish acronym), from the Castile and Leon Government, preserved over fifty sites of natural interest (LIN in Spanish), including tors, scree slopes, glacial cirques, etc. (Santos et al. 2006). There are also many local examples, such as the management plan for the metropolitan area of Madrid (COPLACO 1975), which in the 1970s already included some geological and geomorphological heritage elements. Another example is the revised general urban plan of Segovia city, which includes over 80 places of geological and geomorphological interest and grants them the highest level of protection.

- Legislation on sustainable rural development. The State Law 45/2007 on sustainable development of rural environment establishes that rural development strategies must take into consideration the conservation and the sustainable use of the geological heritage as a scientific, cultural, and touristic resource.
- Water resource management and coastline legislation. The Water Law (Royal Decree 1/2001) and other regulations created for its implementation, entrust the administration with the management of the so-called public water domain, which includes the fluvial channel (bed and banks) and the area flooded during events with "ordinary" peak discharge. This legislation also defines two zones 5 and 100-m wide on both sides of the channel, called service and police areas, whose use and exploitation are regulated by the public authority. Since many geomorphological heritage features exist in these zones (waterfalls and rapids, abandoned meanders, potholes, sinter deposits, river bars, and islands), the conservation and restricted use of these sites of geomorphological interest could be assured. As an example, the geomorphological place of interest "Las Calderas del Río Cambrones" (Segovia province) is managed under these regulation, and its conservation is guaranteed. The Coastal Law (Law 22/2008) also establishes a maritime-terrestrial public domain where multiple elements of geomorphological heritage can be found (cliffs, beaches, tidal flats, dune fields, marshes, etc.).
- Environmental impact assessment (EIA) regulation. As in the other countries of the European Community, the EIA is a mandatory procedure in Spain for many projects and could be an indirect protection tool for geomorphological heritage. An example is the fluvial canyon known as "La Risca" (Segovia province), where a dam construction project was rejected thanks to the EIA and the geomorphological uniqueness of the site (Díaz-Martínez and Lozano 2011, Fig. 25.4). Recently, the environment ministry commissioned to the Geological Survey of Spain (IGME) a manual to incorporate geological heritage (including geomorphology) in EIA protocols, while waiting for its definitive incorporation (Vegas et al. 2013).

Fig. 25.4 The projected dam at "La Risca" (Valdeprados, Segovia province) did not pass the environmental impact assessment due to the high geomorphological value of the site, a canyon carved by fluvial erosion in a metamorphic massif, mainly gneisses (Image by Enrique Díaz)

There are many other sectoral and environmental regulations that could be used to protect geomorphological heritage in Spain, although they are more limited or local in nature, such as legislation regulating mining, tourism, mountains, or roads.

25.3.3 Participation in International Programs

Nowadays, there is not any specific international program or European directive related to geomorphological heritage conservation. In 2004, the European Council put in writing recommendations concerning the conservation of geological (including geomorphological) heritage (Rec 2004-3), unfortunately with limited effect on environmental policies in the European Union so far. Because of the lack of legal framework, participation in other international nature conservation programs that indirectly contribute to the preservation of the geomorphological heritage is particularly important. Two types of programs can be distinguished: specific programs on geoconservation, in which geomorphological heritage is a key part, and other more general nature conservation programs in which geomorphological heritage is one of the natural components under consideration.

Perhaps, the most important scientific program on geoconservation is the Global Geosites project, promoted by IUGS and UNESCO. The Geological Survey of Spain developed this program for Spain starting in 1999 and following the standard methodology (Wimbledon 1996). The complete list included 20 geological frameworks and 144 geosites corresponding to 226 areas distributed all over the national territory (Carcavilla et al. 2008b; García-Cortés et al. 2009). Approximately 15 % of these Spanish Geosites include geomorphological heritage, and all the frameworks selected were included in the Annex VIII of Law 42/2007 of Nature heritage and biodiversity as areas to be considered in protection policies.

Regarding the development of the European Geopark Network, Spain has always taken an active role in the definition and declaration of Geoparks. One of the four initial Geoparks defined was Maestrazgo. There are seven more Geoparks approved in Spain (Cabo de Gata in Almería; Sobrarbe in the central Pyrenees; Subbetic mountains and Sierra Norte de Sevilla in the Betic Cordillera, Costa Vasca in Vasque Country, Villuercas-Ibores-Jara in Extremadura, and central Cataluña in Catalonia), and there are at least two more projects in progress (El Hierro in the Canary Islands and Molina-Alto Tajo in Central Spain). In all of them geomorphology plays a key role in their landscapes, including glacial, periglacial, costal, karst, fluvial, or volcanic features.

Fig. 25.5 The unique anthropogenic landscape of "Las Médulas" (León province) was generated by a Roman mining technique known as "ruina montium." It consisted in the injection of water in pits and galleries to cause the collapse of the mountain and to drag the gold-bearing alluvium

There are three important international programs that offer direct proposals and opportunities to protect the geomorphological heritage:

- UNESCO World Heritage Convention: promoted by the UNESCO and adopted in 1972. It includes three categories in connection with geological–geomorphological heritage (Dingwall et al. 2005): (1) human heritage locations with geological–geomorphological aspects as their main feature. There are two of them in Spain with geomorphological features: Pyrenees-Monte Perdido National Park (shared with France) and Peak of the Teide National Park (Canary Islands); the most visited Spanish national park with more than 3.5 million visitors per year; (2) places of geological value that are not listed as such, but included in the list for other, non-geological reasons. There are four of these areas in Spain: Doñana National Park in Andalusia (see Chap. 19), Archeological Site of Atapuerca (see Chap. 8), Cultural Landscape of the Serra de Tramuntana in Mallorca Island (see Chap. 7) and the Roman open-cast gold mining area of Las Médulas in León province (Fig. 25.5); (3) Areas with "minor" or indirect geological value. Two areas out of 60 are in Spain: Ibiza and Garajonay NP, both with interesting geomorphological features.
- Biosphere Reserves: also promoted by the UNESCO through the Man and Biosphere (MaB) Program. This program aims at promoting conservation and improving the relationships between man and the environment. In Spain, there are around 30 reserves and almost all of them hold important geomorphological heritage values. However, only very few of them have been identified and managed properly, but there are excellent initiatives to coordinate geomorphological heritage management in these areas (Mendía and Monge-Ganuzas 2011).
- Wetlands of International Importance (RAMSAR agreement): This program is focused on the conservation and sustainable use of the most important wetlands, whose origin is in most cases related to geomorphic processes and factors. Even though it focuses on biodiversity, especially water birds, the ecological and hydrological aspects are also treated. In Spain, there are 63 wetlands included in the list, the fourth country in the world with the largest number of wetlands included.

Fig. 25.6 Puentedey village (Burgos province), meaning Bridge of God. The village is constructed over a geomorphosite, a natural bridge generated by the partial unroofing of a limestone cave (Image by Julian Cuesta, AGB)

On the other hand, the Natura 2000 network is a program of the European Union and other nearby countries whose main aim is to create a network to stop the loss of biodiversity in Europe, through the protection of habitats and species of interest. In spite of the great biological and especially phytosociological importance, the Directive on Habitats deals with geological and geomorphological aspects, as some of the habitats include geomorphic features and environments such as caves, dunes, glaciers, playa lakes, or limestone formations. Some work has been conducted to analyze the geological component of these habitats and assess their level of conservation and expected future evolution (Carcavilla et al. 2008a).

25.3.4 Other Initiatives and Ideas on Geoconservation

In addition to the official government legislative initiatives, other actions have been initiated aimed at involving society as a whole in geoconservation. An example is the Asociación Geocientífica de Burgos (AGB), a cultural association created in 1996, which has promoted the protection of geological and geomorphological heritage of Burgos province through outreach activities (http://asociaciongeocientificadeburgos.com/). The AGB, with the financial support of the provincial council, has edited a collection of leaflets divulgating geosites and geomorphosites of the province, including locations with karst (Fig. 25.6), glacial, and structural landforms.

A relevant initiative, because of its impact and widespread implementation in Spain, is the "Land Stewardship" (LS, Custodia del Territorio) in its various forms. According to Law 42/2007 on Natural Heritage and Biodiversity, LS is a set of strategies or legal instruments that involve owners and users in the conservation and management of natural and cultural resources and the landscape. To achieve this, it fosters agreements and continuous collaboration between owners, stewardship entities, and other public and private stakeholders (Basora and Sabaté 2006). In Spain, there are about 200 LS entities grouped under the Forum of Networks and Land Stewardship Entities (FRECT in its Spanish acronym) and the LS Platform at the Fundación Biodiversidad (www.custodia-territorio.es), which began in 2003 with the Xarxa de Custodi del Territori de Catalunya. Although most of the LS initiatives are focused on flora and fauna habitats (habitat banks), some initiatives also include the conservation of landscapes and geomorphological features. This is the case of the so-called river stewardship entities that work in the public water domain, like the project currently developed along 40 km of the Tormes River by the Fundación Tormes-EB and the Douro River Water Authority.

Another geoconservation initiative that attempts to involve society is the volunteer project "Adopt a rock" (www.apadrinaunaroca.es, Díez-Herrero et al. 2012), which encourages individuals and associations to sponsor a place of geological interest (including geomorphological elements). The only commitment is to monitor these elements to avoid damage and the presentation of claims in case potential or actual threats are detected. So far, it has been solely implemented in places of geological interest in Segovia province (Central Spain), but there are plans to implement it on the GeoCamp platform and extend it to the rest of the Spanish territory.

25.4 Conclusions

Geomorphology plays a relevant role in geodiversity assessment because it determines landscape diversity and has a strong influence on biodiversity. Geomorphological heritage, including geomorphological elements with conservation significance, is part of natural heritage and geoheritage and must be considered in the context of nature conservation and according to geoconservation guiding principles.

Due to historical ups and downs, nature conservation and geoconservation have a peculiar history in Spain. Geoconservation had a promising onset at the beginning of the twentieth century. The consequence of the Civil War and the dictatorship policy was the lack of concern in nature conservation until the end of the 1960s. The international awareness regarding nature conservation, the protests of the Spanish environmental movement and other social and cultural changes forced the dictatorial administration to update its nature conservation and environmental policy. Democratization and particularly the decentralized management of the natural resources during the 1980s led to an important improvement in nature conservation, but the achievements in geoconservation were far less significant.

Scientific societies, universities, and other organizations have recently broadened the Spanish nature conservation policy and legislation with the consideration of the geoconservation values. The Spanish geomorphological heritage is currently being managed by regional governments through the framework of the national, regional, and sectoral regulations. Spanish participation in international programs of nature conservation (World Heritage, Biosphere Reserves, and Geoparks) has increased considerably during the last years and constitutes an additional support to geoconservation policy. Unofficial initiatives regarding geoconservation have been initiated, but are still scarce.

References

Basora X, Sabaté X (2006) Custodia del territorio en la práctica. Manual de introducción a una nueva estrategia participativa de conservación de la naturaleza y el paisaje. Xarxa de Custodi del Territori y Fundació Territori i Paisatge, Caixa Catalunya

Benito-Calvo A, Pérez-González A, Magri O, Meza P (2009) Assessing regional geodiversity: the Iberian Peninsula. Earth Surf Proc Land 34(10):1433–1445

Burnett MR, August PV, Brown JH Jr, Killingbeck KT (1998) The influence of geomorphological heterogeneity on biodiversity: I A patch-scale perspective. Conserv Biol 12:363–370

Carcavilla L, De la Hera A, Durán JJ, Gracia FJ, Pérez-Alberti A, Robledo PA (2008a) El papel de la geología y la geomorfología en la Directiva Hábitats de la Unión Europea. In: Benavente, J, Gracia, FJ (eds) Comunicaciones de la XII Reunión Nacional de Geomorfología, Cádiz, pp 431–434

Carcavilla L, García-Cortés A, Díaz E (2008b) The Spanish global geosites project and its influence on recent legislation for the conservation of natural heritage. In: Abstracts 33th international geological congress, p 114

Carcavilla L, Durán JJ, García-Cortés A, López-Martínez J (2009) Geological heritage and geoconservation in Spain: past, present, and future. Geoheritage 1(2–4):75–91

Casado S (1996) Los primeros pasos de la ecología en España. Ministerio de Agricultura, Pesca y Alimentación—Publicaciones de la Residencia de Estudiantes, Madrid

Consejería de Medio Ambiente (2012) Estrategia andaluza de gestión integrada de la geodiversidad. http://www.juntadeandalucia.es/medioambiente/portal_web/servicios_generales/doc_tecnicos/2011/estrategia_andaluza_geodiversidad/estrategia_andaluza_geodiversidad.pdf

Desir G, Gutiérrez F, Gutiérrez M (2005) Sixth international conference on geomorphology. Field trip guides, 2 vols, 1007 p

Díaz-Martínez E, Lozano G (2011) La conservación del patrimonio geológico en la garganta de La Risca (Valdeprados, Segovia). In: Fernández-Martínez E, Castaño de Luis R (eds) Avances y retos en la conservación del Patrimonio Geológico en España. Actas de la IX Reunión Nacional de la Comisión de Patrimonio Geológico (Sociedad Geológica de España). Universidad de León, pp 91–96

Díaz-Martínez E, Guillén-Mondéjar F, Mata JM, Muñoz P, Nieto L, Pérez-Lorente F, Santisteban C (2008) Nueva legislación española de protección de la Naturaleza y desarrollo rural: implicaciones para la conservación y gestión del patrimonio geológico y la geodiversidad. Geo-Temas 10:1307–1310

Díez A (1991) Puntos de Interés Geológico de la Provincia de Segovia. Litos 6:31–33

Díez-Herrero A, Gutiérrez-Pérez I, Vegas-Salamanca J (2012) 'Apadrina una roca', una iniciativa de voluntariado popular para la conservación del patrimonio geológico. Geo-Temas 13, 388 (1–4)

Dingwall P, Weighell T, Badman T (2005) Geological world heritage: a global framework. A contribution to the global theme study of world heritage natural sites. Protected Area Programme, IUCN. http://whc.unesco.org/uploads/activities/documents/activity-504-1.pdf

Elízaga E (1988) Georecursos culturales. In: ITGE (ed) Geología Ambiental. ITGE, Madrid, pp 85–100

Fernández-Sañudo P, de Lucio JV (1994) Los Espacios Naturales Protegidos del Estado Español. Sección del Estado Español de la Federación de Parques Naturales y Nacionales de Europa, Soto del Real (Madrid)

Gallego-Valcarce E, García-Cortés A (1996) Patrimonio Geológico y Espacios Naturales Protegidos. Geogaceta 19:202–206

García-Cortés A, Gallego-Valcarce E, Palacio-Suárez J (1992) El Patrimonio Geológico—The geological heritage. ITGE, Madrid

García-Cortés A (1996) Inventario del Patrimonio Geológico. In: El Patrimonio geológico, bases para su valoración, protección, conservación y utilización. Serie monografías. MOPTMA, Madrid, pp 53–60

García-Cortés A, Águeda-Villar J, Palacio-Suárez J, Salvador-González CI (eds) (2009) Spanish geological frameworks: an approach to Spanish geological heritage of internacional relevance. Instituto Geológico y Minero de España, Madrid

Gray M (2004) Geodiversity: valuing and conserving abiotic nature. Wiley, Chichester

Gray M (2008) Geodiversity: a new paradigm for valuing and conserving geoheritage. Geosci Can 35(2):51–59

Gray M (2012) Valuing geodiversity in an 'ecosystem services' context. Scott Geogr J 128(3–4):177–194

Grupo de Estudios de Historia Rural (1994) Más allá de la 'propiedad perfecta'. El proceso de privatización de los montes públicos españoles (1859–1926). Noticiario de Historia Agraria 8:99–152

Gutiérrez M (ed) (1994) Geomorfología de España. Rueda, Madrid

Hopkins J (1994) Geology, geomorphology and biodiversity. Earth Heritage 2:3–5

Instituto Tecnológico Geominero de España (ed) (1988) Atlas Geocientífico de la Comunidad de Madrid. ITGE—Comunidad de Madrid, Madrid

Instituto para la Conservación de la Naturaleza—ICONA (1975) Inventario Nacional de Paisajes Sobresalientes. Monografías n°6, vol 2. Ministerio de Agricultura, Servicio de Publicaciones Agrarias, Madrid

López-Villaverde AL (2005) Juan Giménez Aguilar (1876–1947) Conciencia Crítica de la Sociedad Conquense. Centro de Estudios de Castilla-La Mancha (UCLM)—Almud ediciones, Ciudad Real

Martini G, Pagès J (eds) (1994) Actes du premier symposium international sur la protection du patrimoine géologique, Digne-les-Bains, 11–16 juin 1991. Mémoires de la Société Géologique de France, 165. 276 p

Martini G, Zouros N (2001) European geoparks: geological heritage & European identity—cooperation for a common future. Eur Geoparks Mag 1:4

Mendía M, Aramburu A, Carracedo M, González MJ, Monge-Ganuzas M, Pascual A (2010) Lugares de Interés Geológico de la Reserva de la Bioesfera de Urdaibai. Departamento de Medio Ambiente, Planificación Territorial, Agricultura y Pesca, Gobierno Vasco. http://www.ingurumena.ejgv.euskadi.net/contenidos/informe_estudio/ligs_de_urdaibai_2010/es_doc/adjuntos/INVENTARIO%20de%20LIGs%20de%20URDAIBAI.pdf

Mendía M, Monge-Ganuzas M (2011) Estrategia de Geodiversidad de la Reserva de la Biosfera de Urdaibai. In: Fernández-Martínez E, Castaño de Luis R (eds) Avances y retos en la conservación del Patrimonio Geológico en España. Sociedad Geológica de España & Universidad de León, León, pp 196–200

Monge-Ganuzas M, Mendía M, Hilario A (2011) La estrategia de geodiversidad en el País Vasco. In: Fernández-Martínez E, Castaño de Luis R (eds) Avances y retos en la conservación del Patrimonio Geológico en España. Sociedad Geológica de España & Universidad de León, León, pp 201–206

Millennium Ecosystem Assessment (2003) Ecosystems and human well-being: a framework for assessment. Island Press, Washington, DC

Múgica de la Guerra M, Martínez-Alandi C, Gómez-Limón J, Puertas-Blázquez J, Atauri-Mezquida JA (2012) Anuario 2011 del estado de las áreas protegidas en España. Fundación Fernando González Bernáldez, Madrid. http://www.redeuroparc.org/img/publicaciones/Anuario2011.pdf

Nichols WF, Killingbeck KT, August PV (1998) The influence of geomorphological heterogeneity on biodiversity: II A landscape perspective. Conserv Biol 12:371–379

Quick TD (1999) American transcendentalism and deep ecology in the history of ideas. Thesis degree, Master of Arts in the School of Environmental Studies. University of Victoria. http://dspace.library.uvic.ca:8080/bitstream/handle/1828/728/quick_2004.pdf?sequence=1

Panizza M (2001) Geomorphosites: concepts, methods and examples of geomorphological survey. Chin Sci Bull 46(1):4–5

Portero G, Salazar A, Pascual MH, Ortega L, Olivé A (1991) Puntos de Interés geológico de Gipuzkoa. Diputación foral de Gipuzkoa, Donostia—San Sebastian

Ramos-Gorostiza JL (2005) Concepciones económicas en los inicios de la conservación de la naturaleza en España: nexos y contrastes con el caso estadounidense. Revista de Historia Industrial 28:11–46

Ramos-Gorostiza JL (2006) Gestión ambiental y política de conservación de la naturaleza en la España de Franco. Revista de Historia Ind 32:99–138

Real Sociedad de Historia Natural (1914) Sesión del 3 de Junio de 1914—Comunicaciones. Boletín de la Real Sociedad de Historia Nat 14:285–294

Rome A (2003) 'Give Earth a chance': the environmental movement and the sixties. Journal of American History 90:525–554

Santos L, Martín JF, Díez A (2006) Aspectos geomorfológicos en las Directrices de Ordenación Territorial de Segovia y Entorno (DOTSE). In: Pérez-Alberti A, López-Bedoya J (eds) Geomorfología y territorio: Actas de la IX Reunión Nacional de Geomorfología. Universidade de Santiago de Compostela, Santiago de Compostela (Spain), pp 945–962

Salazar-Rincón A, Ortega-Ruiz I, Portero-García G, Mendiola-Gómez I, Tamés-Urdain P (1996) El Patrimonio Geológico de Gipuzkoa: Inventario, divulgación y gestión. Geogaceta 19:221–223

Sharples C (1993) A methodology for the identification of significant landforms and geological sites for geoconservation purposes. Report to Forestry Commission, Tasmania, Australia

Thomas MF (2012) A geomorphological approach to geodiversity—its applications to geoconservation and geotourism. Quaestiones Geographicae 31(1):81–89

UNESCO (1972) Records of the general conference seventeenth session Paris, 17 October to 21 November 1972, Resolutions, Recommendations, vol. 1. United Nations Educational Scientific and Cultural Organization, Paris, France

Vegas J, Alberruche E, Carcavilla L, Díaz-Martínez E, García-Cortés Á, García de Domingo A, Ponce de León D (2013) Guía metodológica para la integración del patrimonio geológico en la evaluación de impacto ambiental. Dirección General de Calidad y Evaluación Ambiental, Ministerio de Agricultura, Alimentación y Medio Ambiente & IGME, Madrid. Available via http://libros.igme.es/product_info.php?products_id=115.

Vías J (2011) Memorias del Guadarrama, historia del descubrimiento de unas montañas. Ediciones La Librería, Madrid

Webber M, Christie M, Glasser N (2006) The social and economic value of the UK's geodiversity. English Nature Research Report, 709, English Nature, Peterborough

Wimbledon WAP (1996) Geosites, a new conservation initiative. Episodes 19:87–88

26 Geomorphic Hazards in Spain

Jaime Bonachea, Viola M. Bruschi, Gema Fernández-Maroto, Juan Remondo, Alberto González-Díez, José Ramón Díaz de Terán, and Antonio Cendrero

Abstract

An overview of the main geomorphic hazards in Spain is presented. For each one of the processes analysed (floods, landslides, sinkholes, and coastal hazards), a brief description of their distribution, socioeconomic effects, and main causes is given. The main lines of research undertaken in recent times on these hazards, including development of new tools or techniques, are discussed. Finally, legislation and land-use planning measures for mitigation of risks due to such processes are described.

Keywords

Geomorphic hazards • Floods • Landslides • Subsidence • Coastal processes • Spain

26.1 Introduction

Due to its geographical location, geotectonic setting, and climatic diversity, Spain is affected to a greater or lesser extent by all types of geological and geomorphic hazards. A fairly comprehensive review was previously presented by Díaz de Terán et al. (1997). Unlike the former contribution, the present review focuses on four main types of geomorphic processes (floods, slope movements, karst subsidence, coastal hazards), which represent a risk for both lives and material elements. It does not cover geological–geophysical hazards (volcanoes, earthquakes), nor gradual processes that do not imply a significant threat for human life (erosion, shrink–swell processes). It also addresses some issues related to existing legislation, planning, and management measures.

J. Bonachea (✉) · V. M. Bruschi · G. Fernández-Maroto ·
J. Remondo · A. González-Díez ·
J. R. Díaz de Terán · A. Cendrero
DCITIMAC, University of Cantabria, Santander, Spain
e-mail: jaime.bonachea@unican.es

26.2 Floods

26.2.1 Floods and Their Distribution in Spain

Spain has a varied and in some cases extreme climate and has suffered historically from droughts and floods. Since ancient times, floods have been a severe social and economic problem. There are historical records on floods and their consequences since the first century BC (Dirección General de Protección Civil y Emergencias 2011).

Floods in the main fluvial systems usually occur between the end of autumn and the beginning of spring, although flash floods in small catchments can also occur in summer (Fig. 26.1), as were the cases of the catastrophic events of 1983 in the Cantabrian strip, and the 1996 Arás flood, Pyrenees. In these and other mountain regions, floods are often linked to convective storms of high intensity and magnitude.

For the purposes of this description, the Spanish territory can be classified in the following areas: (a) Mediterranean strip; (b) Cantabrian strip, mountain ranges, and Canary Islands; (c) Catchment areas of the big rivers in Spain. Figure 26.2 shows the distribution of those areas as well as the limits of the autonomous communities (equivalent to states in federal countries, such as the USA or Germany) and the number of casualties recorded in them between

F. Gutiérrez and M. Gutiérrez (eds.), *Landscapes and Landforms of Spain*, World Geomorphological Landscapes,
DOI: 10.1007/978-94-017-8628-7_26, © Springer Science+Business Media Dordrecht 2014

Fig. 26.1 Image of the February 2003 Ebro River flood taken from the gypsum escarpment on the northern margin of the valley. The medieval El Castellar castle has been destroyed by active landslides triggered by fluvial undercutting (Photograph by F. Gutiérrez)

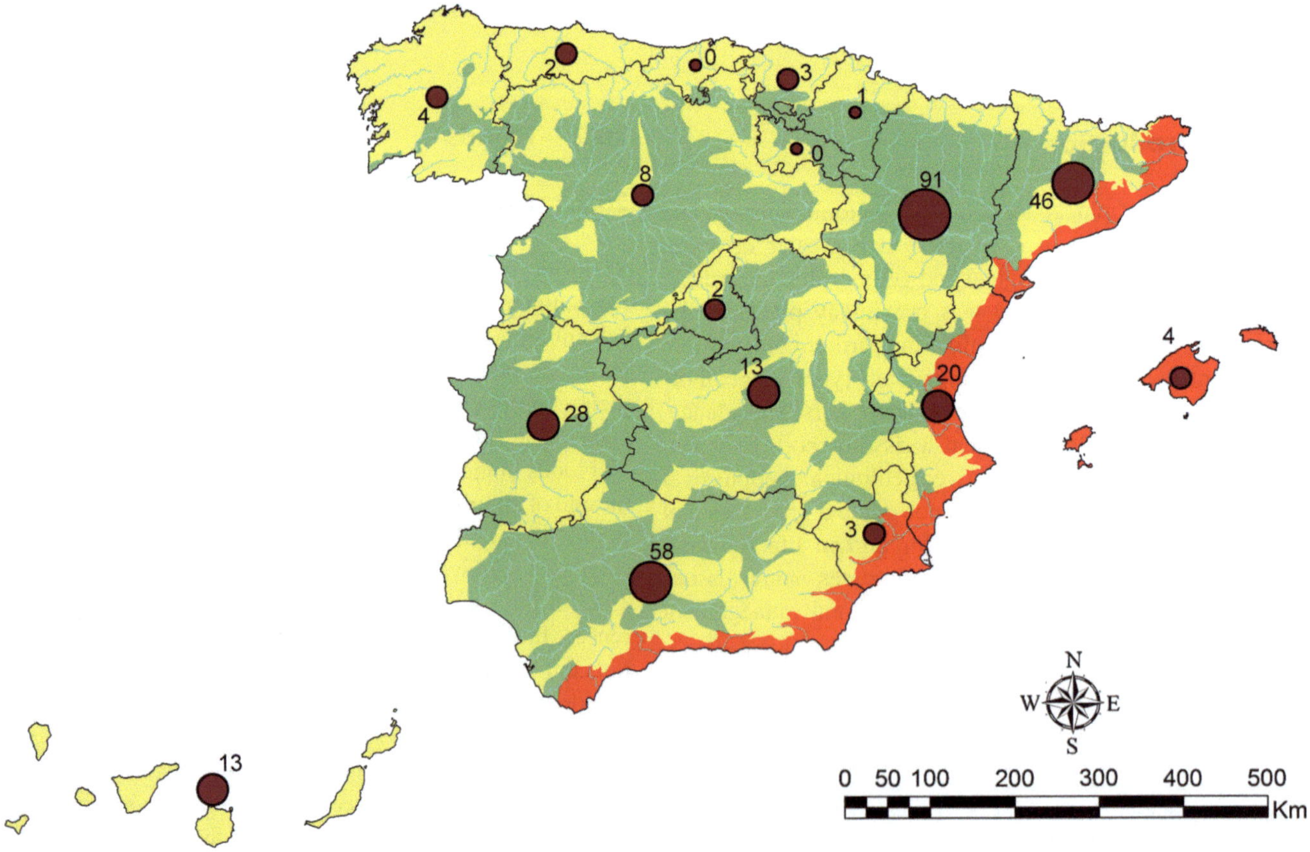

Fig. 26.2 Distribution of flood hazard areas in Spain and number of casualties (297 in total) between 1990 and 2008 in each autonomous community. *Red* Mediterranean strip; *yellow* mountain ranges, Canary Islands and Cantabrian strip; *green* catchment areas of the main peninsular rivers. The high figure in Aragón (91) is related to the 1996 Arás flash flood, which caused 87 fatalities in a camp site built on the active lobe of an alluvial fan fed by a mountain torrent. Adapted from Dirección General de Protección Civil y Emergencias (2011)

1997 and 2008. At least 297 fatalities occurred in that period.

Over 50 % of the floods which occurred during the 1997–2008 period have taken place in the densely populated Mediterranean strip. It is, undoubtedly, the area most affected, especially between September and November. In this region, precipitations of 500–600 mm in 48–72 h, with peaks over 150 mm/h, are relatively common. The floods are normally linked to areas with scarce vegetation and intermittent water courses known as "ramblas", often subject to intensive and poorly regulated use of land.

Around 30 % of the floods recorded in the period 1997–2008 occurred in catchment areas of the main mountain ranges and the Canary Islands. The valleys south of the Pyrenees are the most seriously affected. They have more vegetation and better land-use regulations than the Mediterranean strip, and the water courses are frequently regulated by water reservoirs. However, severe floods happen locally either during the rainy season (autumn–spring) or in relation to the snow-melting period (spring–summer). In Galicia, Asturias, Cantabria, and the Basque Country, floods often occur during the arrival of wet fronts at the end of autumn and the beginning of spring. In these areas, floods are favoured by the existence of low drainage density in catchment areas, short and high-gradient rivers, and narrow valleys. The barrier effect caused by high ocean tides often contributes to the process. Sporadically, floods have been caused by sudden snow melting, frequently induced by rainfall events or by rainstorms, in summer and winter, respectively. In the Canary Islands, floods occur mainly in winter, between December and March, and are caused by south-westerly depressions or by intense cold fronts locally known as *gota fría* (cold drop). These precipitations are often irregular and torrential, and their effects enhanced by the existence of deeply entrenched and steep valleys or canyons.

In the catchment areas of the main rivers (Ebro, Tajo, Duero, Guadiana, Guadalquivir), the floods amount to less than 20 % of the total and happen after long rainy periods affecting a significant part of the peninsula. These floods, characterised by a prolonged lag time, tend to develop after the progressive rise in river discharge due to the contribution of tributaries.

26.2.2 Social and Economic Issues

Floods are considered by the Spanish government as one of the most relevant natural threats affecting the country, with a strategic importance in national security plans (Gobierno de España 2011; González-García 2011). However, until the 1980s, the policies concerning prevention and/or mitigation of this and other natural hazards in the country were limited to engineering measures.

In the early 1980s, cold fronts that affected the Mediterranean in 1982 and the Cantabrian strip in 1983, causing between 30 and 40 casualties each. Consequently, interest on alternative approaches increased. A prognostic assessment on the economic and social impact of geological risks in Spain (Ayala et al. 1987) clearly pointed to floods as the main geohazard for both lives and public or private property and estimated an economic loss of US$28.2 billion for 1986–2016. Later on, Ferrer et al. (2004) estimated economic losses caused by floods during 1987–2001 at around €11.92 billion (2002 euros), with an annual average of €745 million. That is, actual losses were not too different from the former estimation, although comparisons cannot be made directly without taking into account the effect of inflation. The latter authors estimated a total loss of €25.72 billion for 2004–2033, with an annual average of €857 million. This is by far the main contribution to total potential damage.

A complementary assessment for the period 1995–2001, by the Ministry of Food, Agriculture and Environment (MAGRAMA 2013), showed that floods are the main cause of casualties due to natural hazards, 289 fatalities in the time lapse studied, about 28 % of the total number of victims due to all types of natural hazards.

26.2.3 The Contribution of Geomorphology to Flood Prevention Studies in Spain

As a consequence of new legislation on natural hazards prevention and mitigation, a series of methodological manuals have been produced in the last couple of decades dealing with topics such as mapping of flood-prone areas and flood hazard, as well as the use of new cartographic and geomorphic techniques for the assessment of flood hazard and risk, including state-of-the-art reviews (Ministerio de Obras Públicas y Transportes 1988; Ayala 2002; Díez-Herrero 2002; Díez-Herrero et al. 2006, 2008; Olcina-Cantos 2007; MARM 2011a, b).

Interest in natural hazards in general, and flood hazard in particular, has increased considerably among geomorphologists in Spain in the last few years. This is clearly shown in the content of proceedings of recent scientific meetings (Benavente and Gracia 2008; Úbeda et al. 2010; González-Díez et al. 2012). The latter includes a particularly interesting contribution on the calibration and validation of hazard maps (Díez-Herrero et al. 2012a).

A review of the main scientific advances on flood risk assessment during the previous 25 years has been presented by Marco-Segura (2006). The use of GIS techniques for modelling and mapping of flood-prone zones (Díez-Herrero 2003) was an important factor, as was the application of DEMs with high precision and resolution obtained by means of LiDAR techniques for modelling river floodplain geometry (Fernández et al. 2011; Crespo et al. 2012; Vericat et al. 2012). Several authors proposed very innovative techniques for the delimitation of flood-prone areas by using geomorphologic criteria and GIS tools (Díez-Herrero et al. 2006; Marquínez et al. 2006, 2008; Anadón et al. 2008).

The last few years have also seen the publication of a number of papers on the use of dendrogeomorphic techniques for the calibration of hydraulic parameters, estimation of discharges, erosion rates, and reconstruction of peak flows (Bodoque et al. 2005; Rubiales et al. 2008; Ballesteros-Cánovas et al. 2011; Bodoque et al. 2011, 2012; Díez-Herrero et al. 2012b).

The analysis of palaeodischarges and its application to flood hazard analysis has been addressed in different publications (Camarasa and Segura 2001; Benito et al. 2003a, b; Benito and Thorndycraft 2004; Rico 2004; Thorndycraft et al. 2005; Thorndycraft and Benito 2006a, b; Benito et al. 2010). These studies have provided an independent test for the existing climatic models and helped to broaden the information on the time span covered by the flood record and geomorphic controls of flood events (Ortega and Garzón 2003, 2004; Trigo et al. 2003; Benito et al. 2004; Vaquero 2004; Guerrero et al. 2005; Potenciano and Garzón 2008; Machado et al. 2012; García-García et al. 2012), also, on the influence of land-use changes on flood hazards (Benito et al. 2010).

Other works have focused on the role of factors such as suspended sediment load or objects transported by the current (e.g. plant debris of a certain size) and their incorporation into flood hazard models. Vericat and Batalla (2005) and Ruiz-Villanueva et al. (2012) have shown that tree and other plant debris may block the passage of water under bridges affecting the flood-water stage. Martínez-Carreras et al. (2009) and Estrany et al. (2012) have focused on the study of the source areas of suspended load and the role these sediments play in fluvial processes. These authors illustrate the need to take them into account when modelling floods and for erosion assessment in catchment areas.

The effects of river infrastructure, mostly reservoirs, on the hydrodynamic regime of water courses, sediment transport, and river ecology during natural and artificially controlled floods have been studied by several authors (Batalla et al. 2004; Vericat and Batalla 2004, 2005, 2006; Balasch et al. 2007; Ortega and Garzón 2008; Vericat et al. 2008; Ruiz-Bellet et al. 2011; Tena et al. 2011).

26.3 Landslides

26.3.1 Occurrence and Distribution in Spain

Landslides in Spain occur more frequently in the main mountain chains, such as the Pyrenees and the Cantabrian and Betic ranges, where steep topography and/or mechanically weak rocks favour instability. However, they are not restricted to these areas; the valley sides along the big rivers in the Tertiary basins also show mass movements of different magnitudes (Corominas 2005). Landslide activity is enhanced by the climatic characteristics of the country, periods with significant cumulative precipitation, and frequent convective storms, especially in the Mediterranean and mountain areas. To a lower extent, the sparse land cover, as well as land-use practices, also have an influence in many areas. The main triggering factors are, as would be expected, rainstorms, erosion and, to a lesser extent, earthquakes, as well as human actions (Corominas 2005).

Table 26.1 shows some of the most relevant mass movements in Spain since the eighteenth century. Figure 26.3 represents the areas where these phenomena are more frequent (based on the map by M. Ferrer; IGME 1987), and also the location of the movements included in the table. Slope instability frequently affects roads and other infrastructure in mountain areas. Moreover there are numerous villages built next to unstable slopes, and buildings that have been affected by rockfalls (Arcos de La Frontera, Castelfollit de la Roca, Ubrique or Azagra, among others) or slides (Benamejí, Olivares, Inza, Puigcercós).

According to Corominas (2006), unstable areas are located in three distinct settings: (a) Alpine ranges, (b) Neogene depressions, and (c) coastal cliffs and volcanic islands. The first two include the Pyrenees, the Iberian Chain, the Cantabrian (Fig. 26.4) and Betic ranges, and the coastal mountain ranges in Catalonia. In the Pyrenees, steep slopes on Palaeozoic slates have experienced numerous large landslides and flows (Bru et al. 1984a; Guerrero et al. 2012). Other instability-prone rock types in the Pyrenees areas are Keuper (Triassic) marls and evaporites (rotational slides and lateral spreads; Gutiérrez et al. 2012), flysch facies (Gutiérrez et al. 2010), the continental Cretaceous–Eocene clay and calcareous formations (Pinyol et al. 2012), and tills, in which flows and mudslides are common (Brocal 1984; Bru et al. 1984b).

In the Cantabrian Range, Palaeozoic materials intensely folded and faulted (schists, sandstones, siltstones, and limestones) are affected by large deep-seated mass movements (Fig. 26.4), as occurred in Camaleño municipality (Cantabria), in June 2013. Moreover, clays, marls, and siltstones of the Weald and Keuper facies are often affected by rotational and translational slides (González-Díez et al. 1996). In the Sil valley, large translational slides have occurred on lignite layers within the Carboniferous formations. Colluvial deposits and weathered bedrock (regolith) in this region often experience shallow landslides and debris flows (Corominas 2006).

In the Betic ranges, mud flows are frequent in the lower–mid-Cretaceous clays and marls. Mass movements are somewhat different in the Betic and Sub-Betic domains. In the former, the presence of slates favours the development

Table 26.1 Historical and recent slope movements in Spain. Based in part on Bonachea (2006), Corominas (2005, 2006), and Gutiérrez (2013)

No.	Locality	Date	Type	Effects
1	Inza (Navarra)	1714, 1715	Debris flow	Village destroyed
2	Felanitx (Majorca)	31 March 1844	Landslide	414 fatalities, 200 injured
3	Azagra (Navarra)	1856, 1874, 1903, 1946	Rockfalls	11 fatalities 92 fatalities, and 72 houses destroyed 2 fatalities 2 fatalities
4	Puigcercós (Lleida)	13 January 1882	Landslide	Houses destroyed, village abandoned
5	Güevejar (Granada)	1755, 1874	Complex landslide	Village destroyed
6	Albuñuelas (Granada)	25 December 1884	Landslide	102 fatalities, over 500 injured, 463 houses destroyed
7	Bono (Lleida)	26 October 1937	Debris avalanche	River blocked
8	Rocabruna (Girona)	18 October 1940	Debris flow	6 fatalities
9	Alcalá del Júcar (Albacete)	1946	Rockfall	12 fatalities, several houses destroyed
10	Rosiana (Gran Canaria)	17 February 1956	Landslide	Bridge and houses destroyed, 250 evacuated
11	Puebla de Arenoso (Castellón)	October 1957	Earth flow	Cracks in buildings
12	Benamejí (Córdoba)	February 1963, December 1989	Landslide	55 houses destroyed and 50 damaged
13	Senet (Lleida), Benasque (Huesca)	3 August 1963	Debris flow	Obstruction of river, road affected
14	Villanueva de San Juan (Sevilla)	May 1964	Earth flow	Partial obstruction of river. Road blocked
15	Alcoy (Alicante)	December 1964	Rotational landslide	Cracks in buildings
16	Tudela de Veguin (Asturias)	1975	Earth flow	Cracks in buildings
17	Caregue (Lleida)	November 1982	Rotational landslide	Road and bridge destroyed
18	Pont de Bar (Lleida)	7 November 1982	Landslide	Houses destroyed, village abandoned
19	Capdella (Lleida)	7 November 1982	Debris flow	3 fatalities
20	La Guingueta (Lleida)	November 1982	Debris avalanche	Village isolated
21	La Coma (Lleida)	November 1982	Mud flow	Village affected
22	Cantabria and Basque Country	August 1983	Regolith landslides and debris flows	Houses and roads affected, heavy losses
23	Olivares (Granada)	12–25 April 1986	Debris flow	Village affected, obstruction of river. Over 6×10^6 € loss
24	Cabra del Camp (Tarragona)	September 1987	Rockfall	Bus struck, 1 fatality
25	Guixers (Lleida)	October 1987	Rockfall	Car struck, 2 fatalities
26	La Massana (Andorra)	October 1987	Rockfall	Car struck, 2 fatalities
27	Villahermosa del Río (Castellón)	July 1988	Landslide	Road destroyed, river dammed up
28	Camprodón (Girona)	May 1992	Debris flow	2 fatalities
29	Collado Escobal (Asturias)	December 1993	Landslide and debris flow	3 fatalities. House destroyed
30	Eastern Pyrenees and Coastal Ranges	October–November 1994	Landslides and debris flows	Reactivation of landslides, 1 fatality
31	Bens, O Portiño (Coruña)	September 1996	Spoil dump flow	1 fatality
32	Sant Corneli (Barcelona)	17 December 1997	Landslide	1 seriously injured. Road interrupted
33	Ampuero (Cantabria)	10 January 1999	Landslide and earth flow	Several houses destroyed
34	Montserrat (Barcelona)	10 June 2000	Debris flows and falls	Several roads and funicular damaged
35	Tenerife	31 March 2002	Rockfalls	3 roads interrupted
36	Mogán (Gran Canaria)	12 December 2002	Rockfall	1 car struck, 1 fatality

(continued)

Table 26.1 (continued)

No.	Locality	Date	Type	Effects
37	Cala Ramón de Palafrugell (Girona)	25 August 2003	Rockfall	2 fatalities and 2 injured
38	Barruera, Vall de Boí (Lleida)	20 September 2003	Rockfall	2 injured, road interrupted
39	Buscabrero de Salas (Asturias)	16 November 2003	Landslide–debris flow	2 fatalities, house destroyed
40	Bajo Deba Itziar	December 2005	Landslide	Motorway interrupted during 30 h
41	Cantabria Valle río Miera	4 January 2006	Rock avalanche	Road interrupted
42	Playa Valle Gran Rey (La Gomera, Canary Islands)	February 2013	Rockfall	1 fatality
43	Valle Gran Rey (La Gomera, Canary Islands)	4 March 2013	Rockfall	1 fatality

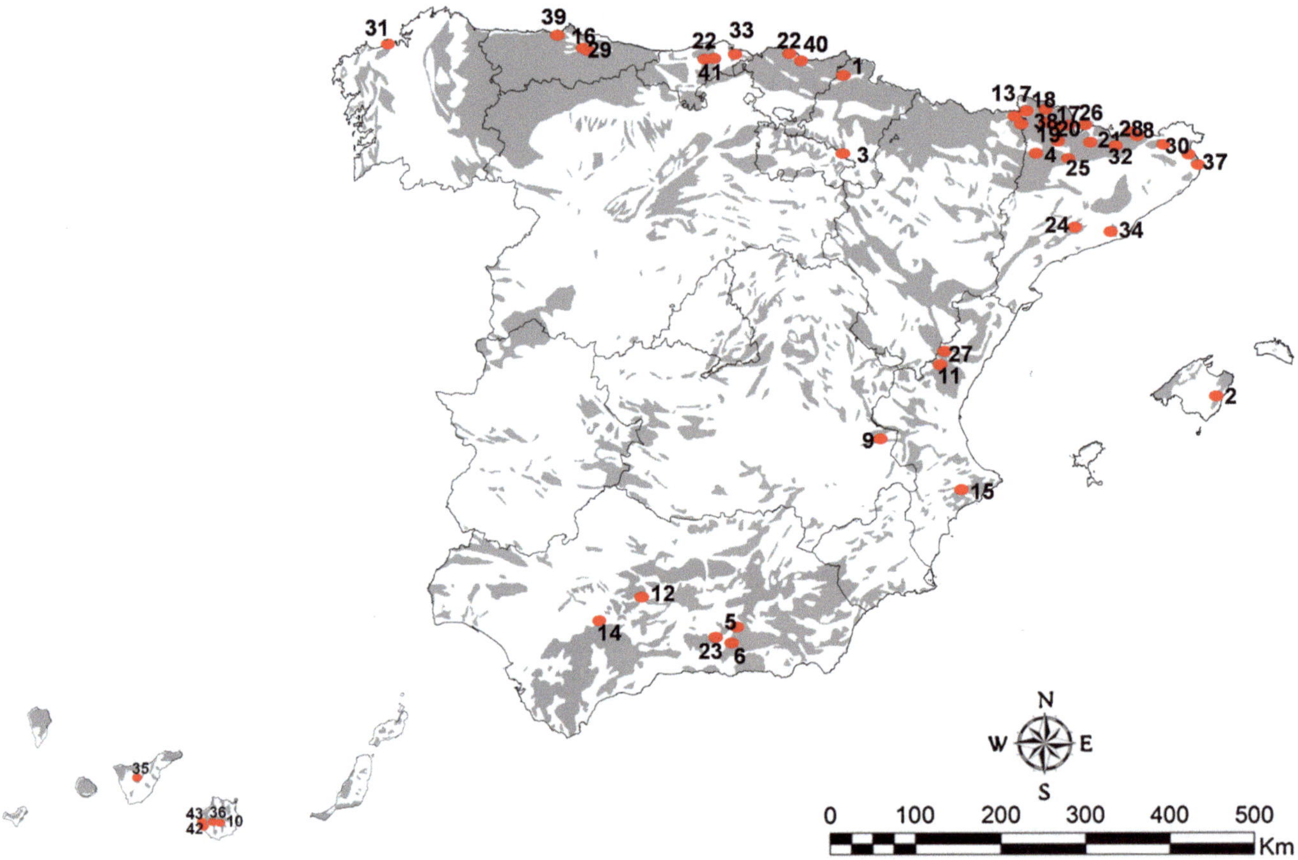

Fig. 26.3 Map depicting the main landslide-prone areas in Spain, based on M. Ferrer (IGME 1987). *Red dots* represent historical and recent mass movements in Table 26.1

of rotational and translational slides (Chacón et al. 2003), as well as debris flows. In the latter, Jurassic and Cretaceous marls give place to earthflows (Irigaray and Chacón 1991).

In the Iberian Chain, rotational slides and earthflows occur in the alternating Aptian blue marls and limestones (Corominas 2005). In the Neogene depressions (Ebro, Duero, Tajo, and Guadalquivir valleys), basal erosion of slopes on interlayered clay, gypsum, and marl formations by fluvial undercutting favours the occurrence of translational and rotational slides (Martínez and García-Yagüe 1988), but topples and falls occur too (Gutiérrez et al. 1994).

The cliffs of the Costa Brava, Cantabrian, Andalusian, and Balearic coasts, commonly produce rockfalls, rock-slides, and topples. Undermining of cliffs formed by

Fig. 26.4 Large deep-seated landslide developed in Palaeozoic rocks affecting the village of Sebrango (Picos de Europa, Cantabria). The landslide occurred in June 2013 and led to the evacuation of the village (Photograph by A. González Díez)

volcanic materials in the Canary Islands induces many falls. There are descriptions of landslides due to basal erosion of Cretaceous–Eocene flysch in the Basque Country (Salazar and Ortega 1990), in the Keuper clays in Asturias (González-Villarías 2001) and on Majorca's north coast (Ferrer et al. 1997).

26.3.2 Causes of Landslides in Spain

Most instability phenomena in Spain are triggered by rainfall. Ferrer and Ayala (1997) studied about twenty major landslides throughout Spain and found that they were related to abnormally high-intensity, short rain events (15–20 % of the annual average). According to Corominas (2005), all major landslides, which occurred in Catalonia during the twentieth century, were triggered by intense precipitations. Corominas (2006) presents a detailed study of the links between climate and landslide activity in Spain.

Shallow slides can be triggered by high-intensity and short duration rains on slopes underlain by regolith. For example, in June 2000, an intense precipitation of 171 mm in 19 h caused many landslides, debris flows, and falls in Montserrat (Marquès et al. 2001). Episodes of shallow landslides caused by heavy rains have been documented in the Basque Country (Remondo et al. 2005b). In the Eastern Pyrenees, precipitations above 185 mm/day normally trigger shallow landslides (Gallart and Clotet 1988; Corominas and Moya 1999), a threshold similar to the one estimated by Marquès et al. (2001) for the slope movements recorded in Montserrat. However, in artificial slopes, the thresholds tend to be lower (Moya and Corominas 1997; Domínguez et al. 1999).

On the contrary, historical records of major landslides show that slope ruptures are normally not caused by weather phenomena, even though they may reactivate or accelerate them (Corominas 2000). Generally, long rainy periods show a direct relationship with the reactivation of major landslides (Corominas et al. 2004). However, in specific geomorphologic contexts that favour water accumulation, landslides can be reactivated by short episodes of heavy rain (Corominas and Alonso 1990).

According to Moya and Corominas (1997) and Corominas et al. (2002), there are three weather situations that cause instability in the Eastern Pyrenees: short duration and high-intensity storms that trigger shallow landslides; moderate- to low-intensity rains extended for several days or weeks that reactivate rotational and translational slides or mudflows; and abnormally damp periods that reactivate major landslides.

Other types of mass movements, such as falls, are often caused by freeze–thaw cycles, but they can also be linked to intense precipitations like those of February and March 2013 in La Gomera (Canary Islands), which caused several rockfall episodes and two deaths. In high mountain areas, particularly the Pyrenees and the Cantabrian Range, periglacial conditions favour solifluction processes and landslides affecting moraine deposits are common.

In the Cantabrian Range, a relationship seems to have existed between high rainfall periods and landslide frequency over the last 100,000 years (González-Díez et al.

1996, 1999). In recent times, episodes of slope instability linked to rainy periods in Spain took place in 1907, 1940, and 1982 in Catalonia, 1983 and 1994 in Cantabria and the Basque Country, and the 1996–1997 winter in Andalusia (Corominas 2006). In the 2012–2013 winter, abnormally wet and with many intense rainstorms in the Cantabrian Range and Andalusia, the frequency of slope instability processes has been particularly high.

According to Corominas (2006), in the coastal ranges in Catalonia, Eastern Pyrenees, and Iberian Range, landslide activity takes place mainly in autumn, whereas in the Cantabrian and Betic ranges, the main Neogene depressions, and the Canary Islands, it happens more frequently in winter and spring. The heavy rainfall of August 1983, which produced the most intense instability period ever recorded in Cantabria and the Basque Country, is an exception.

Although less frequently, fluvial undercutting is also a cause of slope instability. Examples of multiple rotational slides due to lateral migration in the Ebro River have been described by Gutiérrez et al. (1994) in the scarps near Zaragoza. A large planar slide in Vallcebre (Eastern Pyrenees) regularly moves due to erosion undercutting at its foot (Corominas et al. 1999). Basal erosion is also the cause of numerous mass movements in coastal cliffs. Good examples are landslides that almost every year affect the flysch coast of Guipúzcoa (northern Spain) and frequently damage roads.

Seismic activity in Spain is moderate. Earthquakes capable of triggering landslides in the most seismically active areas, such as the Betics and the Pyrenees, have return periods significantly higher than those of heavy rainstorms (Corominas 2006). The 1755 Lisbon earthquake caused many falls and landslides throughout the country, for instance, falls in scarps near Arcos de la Frontera (Jiménez 1992) or the reactivation of a landslide in Güevéjar, Granada (Sanz 1992). Other examples are mass movements recorded in Murcia Province, related to the 11 May 2011 earthquake (Alfaro et al. 2012), or a rotational slide in Bàlitx (Majorca) linked to active faults (Mateos et al. 2013). Zarroca et al. (2013) related the fourteenth century reactivation of the translational landslide that destroyed Montclús village, Pyrenees, to the 1373 Ribagorza earthquake, with an estimated magnitude of M_w 6.2. At a different timescale, González-Díez et al. (1999) identified an important episode of deep-seated landslide activity around 5,500–5,000 BP, likely related to earthquake activity. Additionally, the episodic displacement inferred via trenching in some *sackung* features in the Pyrenees could be linked to earthquake activity (Gutiérrez et al. 2008c).

The instability linked to the rapid building of volcanic edifices in the Canary Islands has caused mass movements of several cubic kilometres, the biggest ever recorded in Spain (Cendrero and Dramis 1996). Some examples of these are the Orotava and Güimar valleys or Las Cañadas in Tenerife (Ancochea et al. 1990), El Golfo in El Hierro (Fúster et al. 1993; Soler 1997), and those of La Palma (Ancochea et al. 1994; Carracedo et al. 1999). These extremely large movements were related to Pleistocene volcanic processes, but other factors should not be discarded such as climate influence (Hürlimann et al. 1999), the lowering of sea level (Ablay and Hürlimann 2000), or the presence of weak stratigraphic levels (Lomoschitz et al. 2002).

The influence of human activity on slope instability processes has been analysed in several studies. Beguería (2006) and García-Ruiz et al. (2010) studied the effects of land-use changes on landslide activity in the Pyrenees. Gutiérrez et al. (1995) found that the slides affecting buildings in Ejea de los Caballeros (Zaragoza) are caused by leaks in the sanitation network. According to Domínguez et al. (1999), around 20 % of mass movements in the Carboniferous terrain in central Asturias were caused by human activity. Remondo et al. (2005b) found that 7 % of the landslides, which occurred in the second half of the twentieth century in the lower Deba valley (Guipúzcoa), were directly triggered by human activity and a further 30 % showed evidence of a possible relationship. Wood logging, changes in slope drainage, excavations, and clearings are among the main causes of instability (Corominas 2005). Failures are frequent in road cuts and anthropogenic fills (deposits). For instance, the rainy winters of 1995–1996 and 1996–1997 in Andalusia caused hundreds of mass movements along the main roads. The 10-km-long Ardales–Campillo road (Málaga) was affected by over 100 failures (González et al. 1997).

As pointed out by Sánchez and Soriano (2001), dams and reservoirs are often threatened by or contribute to mass movements. In Spain, examples are found in Zahara (Cádiz), Arenós (Castellón), Beninar (Granada), Lanuza (Huesca), Giribaile (Jaén), La Viñuela (Málaga), Las Picadas and El Atazar (Madrid), Urdalur, Yesa, and Itoiz (Navarra), Contreras and Cortes de Pallás (Valencia). Costly monitoring and stabilisation actions have been necessary in some of them (Confederación Hidrográfica del Ebro 2007).

A more general, fuzzy relationship between human influence and slope instability may exist at global level (Remondo et al. 2005b; Cendrero et al. 2006). Evidence of such possible relationship in northern Spain has been described (González-Díez et al. 1996, 1999; Cendrero et al. 2006, 2007; Remondo et al. 2008) and appears to be linked to increasing sedimentation rates (Remondo et al. 2005b; Bruschi et al. 2013). The former authors mention that land surface modification by human activities may decrease slope resilience to different triggering factors (rain being the main one). That is, the rainfall threshold necessary to trigger landslides might become lower with time, as a result of growing land transformation. If this is correct, we should expect landslide activity to be greater in the future, even if there are no changes in rainfall regime.

26.3.3 Social and Economic Issues

Several studies have tried to assess the economic and social consequences of landslides. The Spanish Geological Survey (IGME) made a prognostic assessment for the period 1986–2016 (Ayala et al. 1987; González de Vallejo 1988). The authors estimated that losses could be €30–50 billion. However, data obtained for the first 15 years of this period indicated that actual losses were much lower, in the order of €180 million per year (Ferrer 1995). In another study (Ayala et al. 2004), economic damage during 1990–2000 was estimated as €42 million per year. On the other hand, an analysis for the Basque Country (about 1.4 % of the territory and 4.5 % of the population of Spain) estimated that in 1982–1995 areas affected by slope instability suffered losses of about €10 million per year (Del Val et al. 1996).

There are some studies at local scales. Remondo et al. (2005a) estimated damage to the road network in the small Deba municipality (Guipúzcoa Province) (about 50 km^2) for the next 50 years, at 20,000 €/year. Bonachea et al. (2009) for a somewhat larger area in the same zone come to €1.1–3.3 million losses for a 10-year period. Only direct costs were taken into account in both cases. In the Tena valley (central Pyrenees), Herrera et al. (2013) have estimated €15 million landslide damage for the last decade.

Corominas (2006) pointed out that the main damage is caused by mass movements produced by some sort of human influence, land-use changes, excavations, accumulations, alterations of slope drainage, etc. No exact figures are available, but losses amount to several hundred million euros every year. The above-mentioned 1995–1996 and 1996–1997 winter episodes in Andalusia caused very considerable damage in the road network (González et al. 1997). The direct cost and indirect disturbance caused by events like the collapse of motorway A-3 near Madrid, or motorway AP-7, near Girona, in March 2004 are difficult to estimate, but nevertheless they are high.

An additional factor leading to increasing landslide damage is the growing development of mountain areas for ski resorts and recreation in general. Linear infrastructure and buildings built in these areas in the last few decades are often affected by instability problems.

Mass movements have also caused numerous deaths. Ayala (1995) mentioned 17 landslide-related casualties between 1991 and 1993. The Statistical Yearbook of the Ministry of Food, Agriculture and Environment (MAGRAMA 2013) reported 36 deaths in 1995–2011. In 2011, only in Andalusia landslides killed three persons (Ministerio de Interior 2013).

26.3.4 Prediction and Mitigation of Landslides

The study of landslide hazard and risk has been approached through two main research lines. One can be called "basin approach"; it covers fairly large areas and is based on process inventories. Different methods are then used to build models of susceptibility, hazard and, less frequently, risk, and to represent them in the form of maps, with or without a probabilistic expression. The second approach focuses on more restricted areas or individual landslides and is based on the determination of different geotechnical and hydrological parameters, in order to assess quantitatively their degree of stability and to predict their behaviour, often through the use of deterministic models. Numerous research groups and institutions are working with either approach.

Unlike the case of floods, no national standard for landslide hazard mapping or a systematic national landslide data base exists. Díaz de Terán et al. (1997) indicate that in the 1970s the Geological Survey of the Ministry for Public Works published a series of maps at a 1:100,000 scale that showed problematic areas, susceptible geological formations, instability angles, etc. There is a 1:1,000,000 scale map for the whole country by M. Ferrer (IGME 1987), 1:400,000 scale maps like those for Andalusia and Castille and Leon or 1:25,000 scale for Asturias (Marquínez et al. 2003) or Catalonia (Oller et al. 2011). Maps at a provincial level have been elaborated for Madrid (IGME 1986) and Granada (IGME 2007). Landslide hazard maps at the 1:25,000 scale or more detailed have been published for several urban areas. In some cases, 1:5,000 scale hazard and risk maps were made (Cendrero et al. 1987). In general, these types of maps are based on qualitative geomorphologic criteria.

During the last decade, the development of spatial data analysis techniques and the generalised use of GIS have favoured the incorporation of statistical methods in landslide analyses, allowing the estimation of probabilities and making maps more quantitative and easier to validate. Chacón et al. (2006) presented a review of mapping methods for landslide hazard. Irigaray et al. (1999, 2007) and Fernández et al. (2003) used mainly the matrix method in several areas in the Granada Province. Santacana et al. (2003) applied discriminant analysis techniques in La Pobla de Lillet (Catalonia), based on previous works by Baeza and Corominas 2001. Amorim et al. (2009) used logistic regression techniques and artificial neural networks (ANN) to develop and evaluate susceptibility maps in the same area. In the lower Deba (Guipuzcoa), Remondo et al.

(2003), Bonachea et al. (2009), and Felicísimo et al. (2013) applied several techniques to model the susceptibility to shallow slides. In spite of the diversity of methods used, the evaluations carried out show that the results obtained are not very different. The volume edited by Ayala and Corominas (2002) presents several examples of GIS mapping methods in Spain.

Time series on landslides are necessary for estimating probability (hazard). In general, the lack of these data, to a great extent, limits the application of probabilistic methods. Some examples of time series are those obtained through dendrochronology techniques for falls and landslide reactivations (Corominas and Moya 1999; Moya and Corominas 2005). Time series based on air photographs of different periods in the Basque Country show that the landslide frequency and mobilisation rate increased tenfold between 1954 and 1997 (Remondo et al. 2005b) and decreased in the following years (Bonachea et al. 2009). For older movements, there is great uncertainty, but time data obtained in the Cantabrian Range through relative and absolute dating methods, showed different periods of landslide activity increase during the Holocene (González-Díez et al. 1996, 1999). During the last years, radar interferometry techniques (DInSAR) (Castañeda et al. 2009; Fernández et al. 2009; Delgado et al. 2011; Herrera et al. 2011, 2013; García-Davalillo et al. 2013), as well as terrestrial laser (Abellán et al. 2010), have been applied for landslide monitoring.

A few examples of detailed analyses will be mentioned. Based on the analysis of past frequency, Remondo et al. (2005a, 2008) and Bonachea et al. (2009) assessed and mapped landslide hazard (spatio-temporal probability) in the Deba River valley, considering different scenarios of future frequency. They also proposed a method to evaluate vulnerability and generate several direct and indirect risk models for the area. Hürlimann et al. (2010) were trying to develop and test an early warning system for debris flows in the Eastern Pyrenees, based on rain thresholds and the detection of vibrations with geophones.

Other specific studies cover a wide variety of topics such as the analysis of the conditioning and triggering factors that determine the occurrence of landslides, rupture mechanisms, movement typology, dating of movements, local studies for prevention and correction of slope failure, economic analysis of damage, cost of corrective measures, national inventories of movements, etc.

To conclude, it is worth mentioning that in the last 20 years, there have been over one hundred articles published in internationally recognised journals, as well as many other contributions in national periodicals (Cuaternario y Geomorfología, Boletín Geológico y Minero de España, Geogaceta, among others) and in volumes of national congresses, symposia, and meetings. This shows both the relevance of the landslide hazard in Spain and the interest of the Earth science community.

26.4 Sinkhole and Subsidence Hazards in Karst Areas

26.4.1 Dissolution-Induced Subsidence in Spain

Opposite to other widely distributed hazards, such as landslides or floods, karst processes are logically restricted to areas with soluble bedrock, essentially carbonates and evaporites. Karst studies in the country have traditionally focused on carbonate terrains, mostly linked to speleological or groundwater issues (Puch 1998; Andreo and Durán 2004). The latter authors presented a complete review of the research work to date in the Spanish karst areas. The study of evaporite karst has received much less attention due to the limited quality of the groundwater and the scarcity of cave systems. This has changed over the last couple of decades mainly due to an increasing demand of scientific and technical basis to manage sinkhole hazards. Evaporite rocks have much higher solubility and lower mechanical strength than carbonate rocks (Gutiérrez et al. 2008a) and are therefore much more prone to sinkhole development. Subsidence can also occur in relation to other natural or human-induced processes, such as hydrocompaction, collapse of underground mines, or exploitation of aquifers, but these will not be dealt with here.

About 25 % of the world's population lives on karst zones (Ford and Williams 2007). In Spain, carbonates and evaporites cover around 30 % of the country's area (IGME 2013; Fig. 26.5). Carbonate rocks crop out in some 110,000 km^2 ($\sim$23 % of the country; Durán and López-Martínez 1989) and evaporites in 35,000 km^2 ($\sim$7 %; Macau and Riba 1962). These rocks are present in the main Alpine orogens: Betic Cordillera, Pyrenees, and Cantabrian Mountains, Catalan Coastal Chain, Iberian Chain, as well as in the main Cenozoic basins of the rivers Ebro, Duero, Tajo, or Guadalquivir. Examples of landscapes with striking sinkholes in carbonate rocks can be found in the Garraf Massif (Catalonia), central Pyrenees, the Picos de Europa, Iberian Chain, Sierra Morena and the Guadalquivir depression (Andalusia), Majorca, etc.

Subsidence sinkholes may display three different kinematic behaviours: gradual, slow settlement, and sudden collapse. The latter is the one that most frequently causes damage. Although deaths related to karst hazards in Spain are very few, material damage can be considerable in certain parts of the country (Durán 2002; Gutiérrez and Cooper 2002; Gutiérrez et al. 2008b, 2009; Galve et al. 2012). In the last few decades, some notable events have caused increasing social awareness about these processes and their possible consequences.

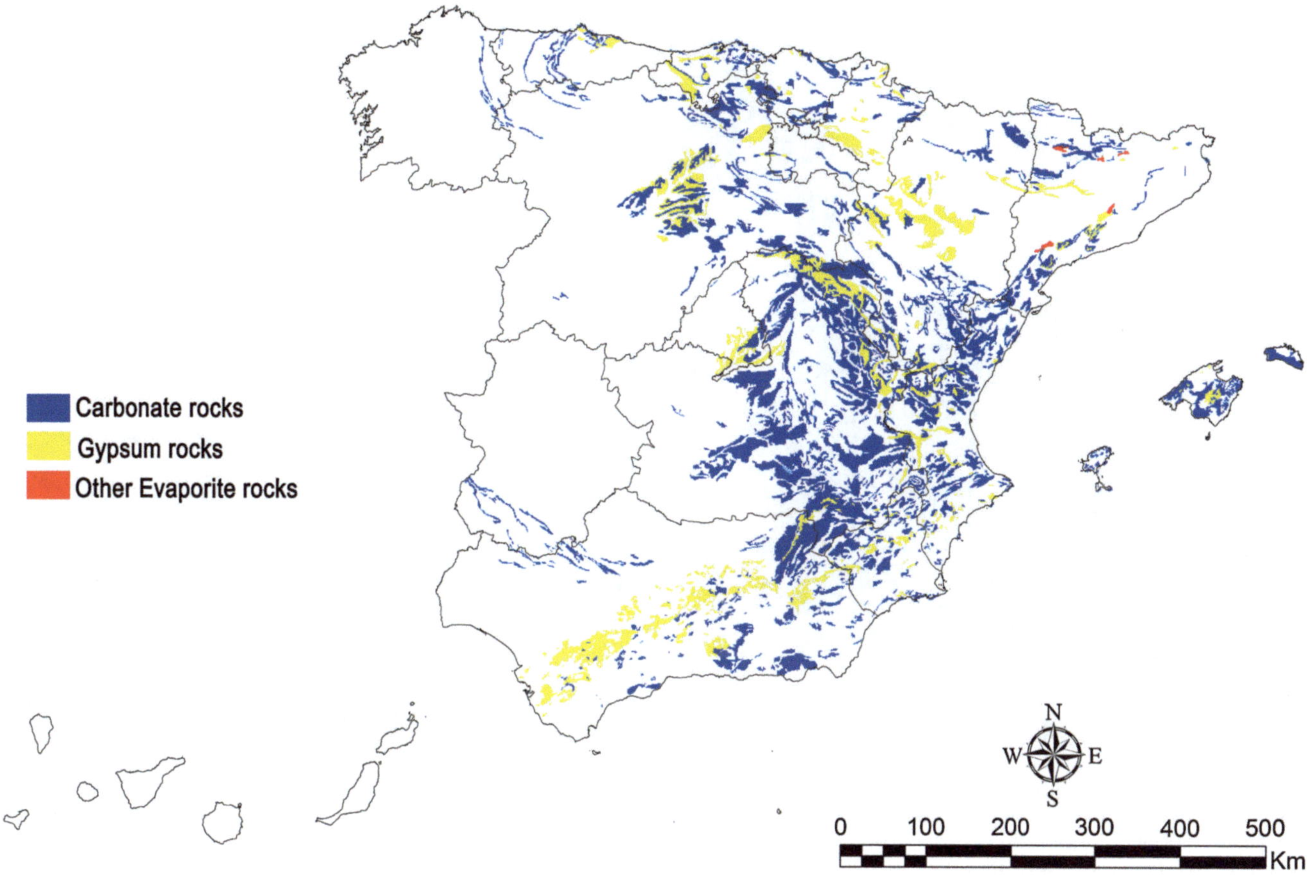

Fig. 26.5 Areas of Spain with bedrock in which karst processes occur. After IGME (2013)

A spectacular example of collapse over a karst aquifer took place in 1982 in Pedreguer (Alicante), where a 14 × 12 m wide and 70 m deep collapse happened overnight in a farming field (Garay 1986). Another, more recent example is a collapse in Majorca (García-Moreno and Mateos 2011). During the last few years, in the municipality of Camargo (Cantabria), built over the solution residue (*terra rosa)* derived from cavernous Aptian limestones, numerous small collapses have formed affecting several houses and other constructions and causing considerable social alarm. Here, the subsidence process is mainly related to groundwater extraction from the local karst aquifer and the consequent decline in the water table.

In the review carried out by Ayala et al. (1987) on the consequences of geological hazards in Spain, the estimation of damage due to karst processes was aggregated to landslides and other mass movements, and therefore, no estimate was provided for this particular hazard. There is no systematic data compilation, nor country-wide estimate of damage due to karst processes, but there are examples that show their importance.

Spain is one of the European countries where subsidence due to dissolution of evaporite rocks has significant economic consequences (Gutiérrez et al. 2008b). This process is very active in relation to Tertiary evaporite rocks covered by unconsolidated alluvial deposits. Some areas in Spain where karst subsidence causes important economic losses are the urban areas in Madrid, Oviedo, Besalú (Gerona), Cardona (Barcelona), Calatayud (Zaragoza), or the surroundings of the city of Zaragoza (Gutiérrez et al. 2008b).

The Ebro basin is the area in which these phenomena have been better documented. Damage to farming fields caused by sinkholes has been frequent for decades. However, these landforms are often refilled by farmers and disappear fairly quickly from the landscape. Expansion of cities, with Zaragoza as the main example, has entailed the construction of infrastructure and many buildings in areas where subsidence or collapse are frequent (Soriano and Simón 2002), and considerable expense on mitigation and repair measures has been necessary. In 1975, the town of Puilatos, near Zaragoza, was evacuated due to instability problems produced by subsidence and was totally demolished 5 years later (Benito et al. 1995). In Calatayud (Zaragoza), a collapse in November 2003 affected a five-storey building (Fig. 26.6a) which had to be demolished.

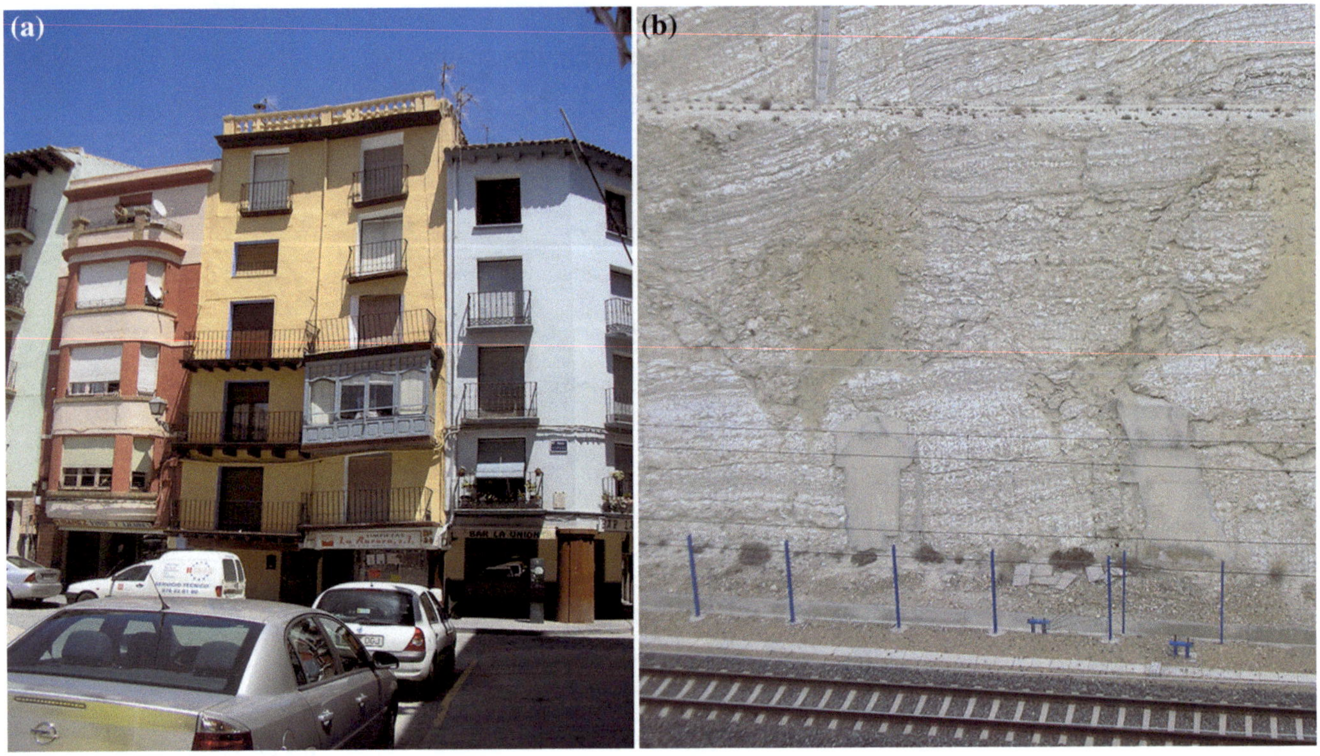

Fig. 26.6 (a) Buildings affected by subsidence damage in Calatayud, built upon evaporite bedrock and collapsible gypsum-bearing silts (Calatayud Neogene Graben, Iberian Chain); (b) cuts along the high-speed Madrid–Zaragoza railway, showing dissolution and subsidence features in gypsum. The concrete patches in the lower part of the exposure correspond to plugged air-filled cavities (Photographs by J. Bonachea)

Direct damage was over €4.8 m (Gutiérrez et al. 2004). In 1991, a collapse caused the derailment of a freight train near El Burgo de Ebro, also in the surroundings of Zaragoza (Gutiérrez et al. 2007).

In March 2003, the high-speed railway between Madrid and Barcelona was affected by a collapse sinkhole (Fig. 26.6b) shortly before its inauguration (Guerrero et al. 2008). Recently, a six-storey building in Zaragoza, with 100 homes, built on a previously known sinkhole, suffered an important deformation (Gutiérrez et al. 2009). The municipality has decided to demolish one-fifth of the building and the owners demand in a law suit a compensation of over €30 m arguing inappropriate permission for construction.

In 1998, in Oviedo, the construction of an underground car park over a gypsum karst cavity resulted in the demolition of 362 apartments and a loss of €18 m (Pando et al. 2013). There are other areas such as Polanco and Cabezón de la Sal (Cantabria), Fuentealbilla (Albacete), or Sallent and Cardona (Barcelona), where subsidence has caused serious problems in relation with dry or solution salt mining. In Sallent, located over a potash mine, a total of 333 homes were affected by subsidence in 2004; the local government spent €14 m million in compensations. In Cardona (Barcelona), the Cardener River flowed through several sinkholes into a salt mine causing massive dissolution and uncontrolled subsidence. The flooding event was induced by the excavation of a shallow gallery meant to be a store for hazardous waste. The mine has been abandoned, and the course of the river has been modified by a cut-off tunnel (Lucha et al. 2008).

26.4.2 Risk Identification, Assessment, and Mapping; Main Research Lines and Results

Several monographs on karst research in Spain, including hazards, have been published (Durán and López-Martínez 1989; Durán 2002; Andreo and Durán 2004). In earlier studies, the techniques used to analyse the distribution of karst processes were limited to the elaboration of geomorphologic maps indicating the distribution of karst landforms, mainly sinkholes (Gutiérrez et al. 1985; Ayala et al. 1987). Durán and Martínez-Goytre (1990) developed a collapse and subsidence hazard map of the Alicante Province at the 1:200,000 scale, taking into account factors like rock type, surface landforms, or degree of karstification. A qualitative rank of karst hazard based on expert judgement was thus established and represented.

More recently and prompted by instability problems in industrial areas of Zaragoza, in which subsidence rates of

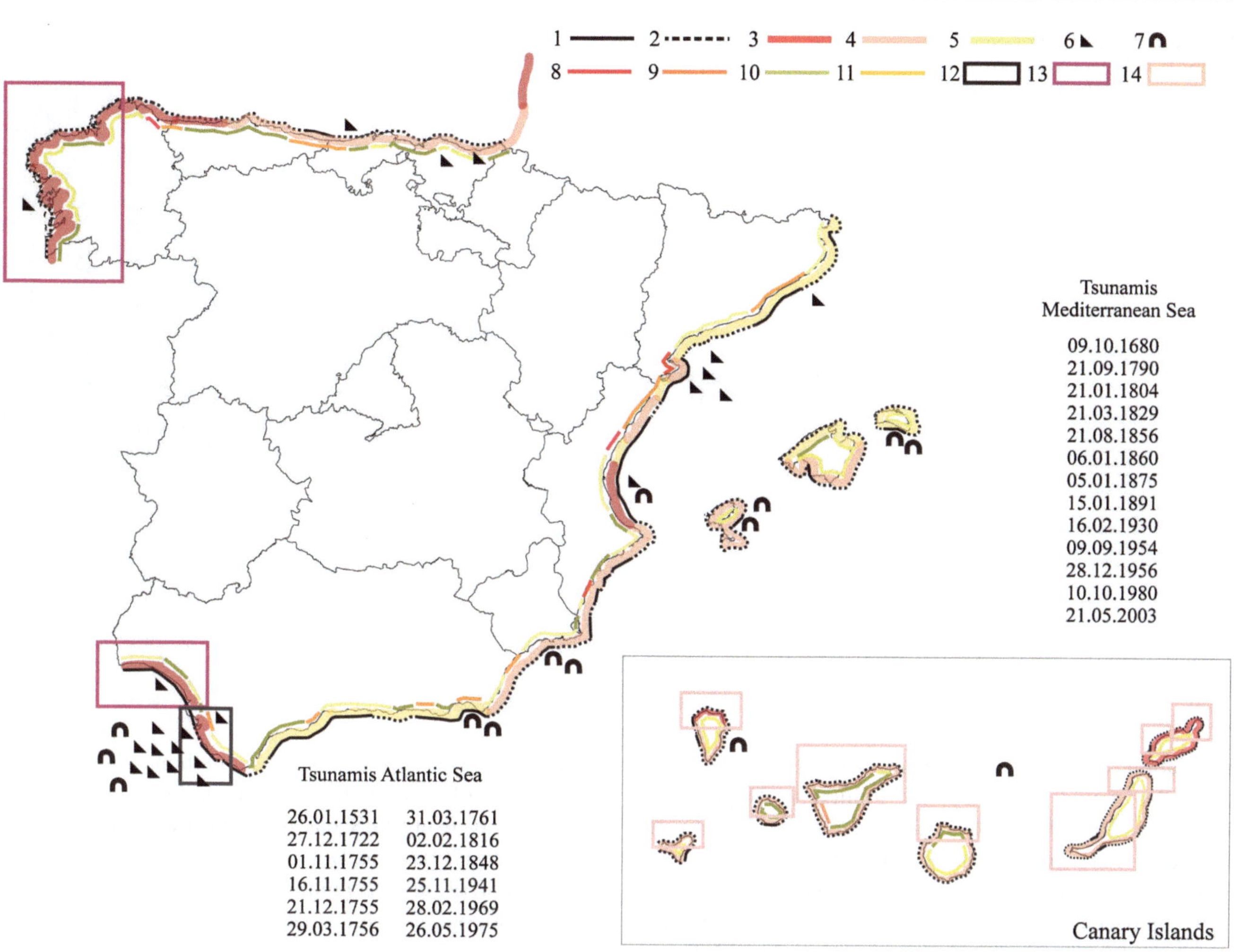

Fig. 26.7 Synthetic map of coastal hazards in Spain. Coast typology: *1* coastal lowlands, *2* cliffs. Annual retreat, up to 2050, *3* >20 cm, *4* 20–15 cm, *5* 15–5 cm. Number of scientific contributions (from SCOPUS) on: *6* sea-level change and coastal erosion, and *7* tsunamis. Coastal erosion hazard: *8* high, *9* moderate, *10* low, *11* nil. Tsunami hazard: *12* high, *13* moderate, *14* low. The main tsunami events recorded in the Spanish coast are listed. Based in part on data from De Andrés and Gracia (1988), Ayala et al. (1987), and IPCC (2007)

the order of centimetres per year have been measured (e.g. Soriano and Simón 1995, 2002), innovative techniques have been applied, normally in conjunction with traditional studies based on field surveys and interpretation of historical maps and aerial photographs.

The usefulness of mapping landforms and making inventories of sinkholes to assess karst hazard and incorporate it into land-use planning has been pointed out by Gutiérrez-Santolalla et al. (2005). The analysis of the origin, typology, and spatial distribution of sinkholes in the proximities of Zaragoza was addressed by Gutiérrez et al. (2007). They found that the spatial pattern of different types of sinkholes reflects different combinations of factors, pointing out to different possible processes involved. For instance, determinant factors for the occurrence of most small collapse sinkholes were geomorphology, geology, and hydrogeology, with human activities (irrigation and other water-related activities) as the main triggering factor. On the basis of this analysis, Galve et al. (2008) used spatial data analysis techniques to develop and validate probabilistic susceptibility models, in which the study area was ranked in terms of relative likelihood of new events. Using data from sinkhole inventories, Galve et al. (2009) were able to transform these models into hazard models, in which probability of occurrence of new collapses was estimated in quantitative terms.

To analyse sinkhole susceptibility along the high-speed railway between Zaragoza and Barcelona, Guerrero et al. (2008) proposed an approach for identifying areas with different degrees of susceptibility (low, medium, high), mainly based on the distribution and typology of dissolution and subsidence features exposed in the adjacent cuttings. Galve et al. (2012), using a more quantitative approach, developed a methodology for assessing the cost-effectiveness of mitigation measures along a recently built road in a sinkhole-prone area. According to their estimates, without

Fig. 26.8 Cliff-foot erosion in Playa de la Vega (Asturias) (Photograph by V. M. Bruschi)

mitigation measures damage in the sector affected by sinkholes would amount to €7 million in 50 years, and casualties would be likely to occur. However, if mitigation measures were implemented total cost could be only €1.5 million, and human lives would be saved.

Geophysical techniques, such as ground penetrating radar (GPR), electrical resistivity tomography (ERT), gravimetry, or magnetometry have been used to detect anomalies related to dissolution processes and investigate sinkholes (e.g. Pueyo-Anchuela et al. 2010; Gómez-Ortiz and Martín-Crespo 2012; Carbonel et al. 2013; Rodríguez et al. 2013).

The trenching technique has been used to map the edges of sinkholes, to characterise the deformation style and identify the subsidence mechanisms, to infer the kinematics of the process (gradual vs. episodic), and, with the aid of numerical dates, to determine the age of the sinkholes and calculate long-term subsidence rates (Gutiérrez et al. 2009, 2011; Carbonel et al. 2013).

Another innovative and non-destructive technique used for the analysis of subsidence and other types of ground movements is radar interferometry (InSAR). Through the interpretation of satellite images, it is possible to measure displacements in the order of millimetres over large areas and with considerable spatial resolution (Castañeda et al. 2009; Gutiérrez et al. 2011). This technique could be very helpful to monitor structures affected by slow subsidence processes, helping to detect precursor movements. Its application depends on the availability of satellite images of adequate spatial resolution and revisit time.

In the case of karst areas in carbonate terrains, the elaboration of sinkhole maps based on digital terrain model (DTM) data using GIS has been illustrated by López-Vicente et al. (2009) and Durán et al. (2012). The successful application of this technique requires the availability of detailed topographic information.

26.5 Coastal Hazards

26.5.1 Distribution of Coastal Hazards in Spain

Spain has 50 provinces, 22 of which border the sea, including the Balearic and Canary islands. These, together with the North African coastal towns of Ceuta and Melilla, have a population of 27,516,000, about 58 % of the total. Around 35 % of the population lives in a strip less than 5 km wide along the coastline (Díaz de Terán and Cendrero 1992). Human pressure is doubled during the summer months, by tourists flocking to the Spanish seaside (INE 2013). The coastal strip has an obvious strategic and economic interest and concentrates a variety of activities (urban and industrial development, tourism, commerce). This has led to a very intense transformation of coastal areas during the last 50 years or so. Presently, some 700 km of coastline have been altered by artificial structures, and in some provinces, over 75 % of the coast is urbanised (De Andrés and Gracia 1988; MARM 2007).

The Spanish coastline is about 7,000 km long, with 58 % cliffs, 30 % beaches, and the remainder being low-lying coasts. Two main coastal environments can be differentiated: Atlantic and Mediterranean (Fig. 26.7). The former is subject to intense processes (strong waves and high tidal range) and consists mainly of rocky cliffs (Fig. 26.8), with some pocket beaches and inlets as well as small estuaries. In the latter, beaches and low-lying coasts form much of the coastline, which has lower-intensity storms and a 30–40-cm tidal range. Meteorological tides and associated strong waves can elevate sea-level about 60 cm during several days (Marcos et al. 2009). As most littoral areas, Spanish coasts are affected to some extent by storms and cliff landslides or rockfalls. However, the main problems are caused by sea-level rise, coupled with sediment supply reduction due to river regulation, as well as subsidence in low-lying coastal areas underlain by unconsolidated sediments. These circumstances lead to coastal erosion and retreat, changes in depositional landforms like sandspits, or coastal flooding. Tsunamis are the main threat for the future (De Andrés and Gracia 1988). In the most affected sector, the Gulf of Cádiz, the return period of events comparable to the one related to the 1755 Lisbon earthquake ($M_w \geq 8$) has been estimated at around 250–400 years by Damaskinidou-Georgiadou et al. (1987). Gracia et al. (2010) and Lario et al. (2011) estimated return periods of 1,200–1,500 years by means of palaeoseismological analyses based on the study of turbidites, and onshore geomorphic and stratigraphic evidence of palaeotsunamis covering the last 7,000 years.

Coastal erosion is a serious problem in Spain, with over 80 % of the country's beaches affected (Nuhfer et al. 1993). Ayala et al. (1987) estimated that losses due to this process would amount to nearly €2 billion for 1986–2016, out of which around 25 % would correspond to Andalusia. The cost for the recovery of 1 km of sandy coast has been estimated at €3 million (Nuhfer et al. 1993). A total of €92 million was spent in 2003–2005 by the central government on the recovery of beaches, dunes, and wetlands, as well as on coastal defence works. The Andalusian government foresaw an expenditure of €47 million in 2004–2010 (IGME 2007).

According to the IPCC (2007), average global sea-level rise was 1.8 mm a^{-1} in 1961–1993 and 3.1 mm a^{-1} in 1993–2003. In the case of the northern Spanish coast, values of 3.3–3.8 mm a^{-1} for the period 1972–1990 were obtained by Gómez-Gallego (1994), whereas Marcos et al. (2005) estimated 2–3 mm a^{-1}. An average estimate of 2.5–3.0 mm a^{-1} for Spain has been proposed by Medina et al. (2004). Leorri et al. (2012) estimated an average of 1.9 ± 0.3 mm a^{-1} for the twentieth century, particularly after 1950. In a report for the Ministry of the Environment (Moreno 2005), a projection of 10–68 cm for the end of the present century was made, with 50 cm as the most likely scenario. This is not very different from the 18–79 cm estimate by the IPCC (2007) (Fig. 26.7). This, together with an expected increase in wave height and storm frequency, would raise the flooding level in 2050 to 35 cm in Galicia, Bay of Biscay, and Canary Islands and 20 cm in the Mediterranean. This increase would imply an estimated retreat of 15 m in the beaches of the Canary Islands, Huelva, and Cádiz. The latter area, where much of the coastline is formed by soft-rock cliffs, with present erosion rates of 1.25–2.2 m a^{-1} (Rodríguez-Ramírez 1998), would be particularly affected. However, data from the Permanent Service for Mean Sea Level (PSMSL 2013) show substantial differences between sectors of the Spanish coast, with relative lowering in some cases, like the Alicante coast.

Sea-level rise, together with reduction in sediment supply due to the construction of reservoirs, would increase erosion retreat in beaches and deltas. According to Cendrero et al. (2005), if there is no increase in sediment supply, a 50-cm sea-level rise would imply a loss of 40 % of the beaches in the Cantabrian Sea (or Bay of Biscay) and 50 % of the Ebro delta. About 20 km^2 of the areas with coastal wetlands, marshlands, or alluvial plains would be flooded in the eastern Cantabrian coast. In the Mediterranean, more vulnerable areas would be the Ebro and Llobregat deltas and the Manga del Mar Menor lagoon (20 km of coastline). The estimates for the Cabo de Gata lagoons and Gulf of Cádiz are 5 and 10 km of coastline affected, respectively. In the Doñana wetlands, about 100 km^2 would be invaded by the sea.

The main tsunami source areas are in the Gulf of Cádiz, the Alborán Sea, and the northern Algerian coast (González-García 2009). The most vulnerable area is between the Gulf of Cádiz and Gibraltar, due to seismic activity along the Azores–Gibraltar fault zone. Tsunamis generated by seismicity in the Algerian coast and Alborán Sea represent a threat for the Balearic Islands (Alasset et al. 2006) and the Almería–Cartagena coast (Amir and Cisternas 2010). Potential damage due to this process for 1987–2016 was estimated at €2.3 billion in a pessimistic scenario and nil in the most likely one (Ayala et al. 1987). The latter has been the case to date.

Since 1531, the Spanish coast has been hit by 31 tsunamis (Martínez-Solares 2005; Rodríguez-Vidal et al. 2011). The most important event was the "Lisbon earthquake" of 1755, with 13-m-high waves that severely affected the coast between Huelva and Gibraltar, causing 1,214 fatalities (Campos 1992; Martínez-Solares 2005; Blanc 2008; Rodríguez-Vidal et al. 2011). The effects were also felt in Galicia and the Bay of Biscay. Carreño and Seller (2005) estimated that a similar event nowadays would produce, only in the province of Huelva, over 112,000 casualties and €2.1 billion damage. Between 1950 and 1990, the Gulf of Cádiz has experienced seven lower magnitude tsunami episodes (Campos 1992; Ministerio de

Fomento 2012). Tsunamis were identified in 1941 and 1980 in Tenerife and Alicante, respectively, but no significant damage was reported. In 2003, about 50 boats were damaged by a tsunami in Mahón (Menorca).The damage related to this type of hazard is expected to increase as a consequence of sea-level rise and growing occupation of low-lying coastal areas, particularly along the Mediterranean and South-West Iberian coasts.

26.5.2 Identification, Mapping, and Assessment Initiatives: Main Research Lines

In the last few years, efforts have been devoted to the analysis of Holocene sea-level changes and estimation of future rise, coastal erosion and setback problems, effects of human activities on coastal dynamics, and storm-generated coastal flooding (e.g. Somoza and Díaz del Río 1992; Zazo et al. 1994; Rodríguez-Ramírez et al. 1996; Soria et al. 1999).

The assessment of the effects of climate change on the Spanish coasts (Medina et al. 2004; Méndez et al. 2004) to design adaptation strategies for 2050–2100 has been approached through the elaboration of a simulator (C3Sim). Different sea-level rise scenarios can be simulated and potential impacts on beaches, estuaries, or coastal infrastructure estimated (http://www.c3sim.ihcantabria.com/).

The Gulf of Cádiz has been the focus of researches on coastal dynamics, erosion, and beach evolution in both natural and human-modified environments (Anfuso and Gracia 2005; Domínguez et al. 2005; Gracia et al. 2005; Anfuso et al. 2007; Del Río et al. 2012). More directly focused on hazards, Plomaritis et al. (2011) developed a methodology for assessing the effects of storms on beach erosion and flooding in the city of Cádiz and proposed a coastal risk index as a support for decision making. Benavente et al. (2006) produced a risk map based on coastal retreat data, geomorphologic mapping, and a detailed digital elevation model. Potential flood areas were estimated for storms with 1–10 years return period. Using a more empirical approach, Rangel-Buitrago and Anfuso (2011) analysed coastal changes, which occurred during different storms, to assess possible future impacts.

The effect of human actions on the coastal environments has also been addressed by several groups. Anfuso et al. (2001) and Rodríguez-Ramírez et al. (2008) studied the effects of human actions and sea-level rise on the acceleration of beach erosion in Huelva. They estimated a 10–15-m retreat for 2050. With a different approach, Domínguez et al. (2005) studied the effect of land-use changes on coastal retreat and assessed vulnerability in a 23-km-long coastal sector. A methodology to assess cliff erosion hazard in the coast of Cádiz has been proposed by Del Río and Gracia (2009), including determination of hazard, impact, and risk indicators, on the basis of cliff characteristics, processes, and human factors.

The study of coastal retreat and its relationship with human influence and sea-level rise has also been addressed in the Mediterranean coast. Sanjaume and Pardo-Pascual (2005) analyse, in the coast of Valencia, the direct and indirect human influence in the reduction of sediment supply, and the consequent coastal retreat and reduction in dune area. The effects of sediment-supply decrease as well as sea-level rise on the evolution of the Ebro Delta have been studied by Guillén et al. (1992), Jiménez et al. (1997), Sánchez-Arcilla et al. (2008), and Alvarado-Aguilar et al. (2012). The vulnerability of beaches to erosion and flooding as a consequence of storms in Catalonia has been assessed by Mendoza and Jiménez (2009), who established hazard and risk ranks.

Some potential effects of sea-level rise on the eastern coast of the Bay of Biscay were assessed by Rivas and Cendrero (1991, 1995). They estimated that a 0.5-m rise would imply the flooding of 23 km^2 of coastal lowlands, most of them resulting from previous reclamation, and the loss of 40 % of the confined beaches. If the rise were 1.0 m, the values would be 79 km^2 and 60 %, respectively. Liria et al. (2011), using a high-resolution DTM (LiDAR), estimated that sea-level rise and increased storminess will lead to the flooding of nearly 3 km^2 in the coast of Guipúzcoa by the end of the century.

Marcos et al. (2012) applied a similar model to the whole of the Bay of Biscay and estimated that by 2040, with a 40-cm rise in sea level, flood hazard in the region would triple. Nevertheless, Losada et al. (2011) point out the need to carefully consider uncertainties when trying to forecast the geomorphic evolution of the coast, especially for long-term periods and impact or hazard assessments.

Analyses of the combined effects of human actions and sea-level rise on erosion processes and coastal retreat in northern Spain have been carried out by González-Villanueva et al. (2007), Lorenzo et al. (2007), or Cardineau and González (2005). The latter authors describe a 300-m retreat in only 9 years in the sandspit of Liencres (Cantabria).

Although, as indicated before, tsunami hazard is not very important in the country, a return period of 250–1500 years for a tsunami similar to the 1755 event has been estimated and its assessment has been addressed by several authors. Possible tsunami sources, data on past events, and wave propagation models have been used to assess risk (López-Marinas and Salord 1990; Zitellini et al. 2009; Amir and

Cisternas 2010; Lima et al. 2010; Duarte et al. 2011; Roger et al. 2011; Periáñez and Abril 2013). Vulnerability and risk of different coastal sectors on the basis of past events have been estimated (Roger and Hébert 2008; Lima et al. 2010). A methodology for a quantitative risk analysis has been developed for the city of Cádiz (Jelínek et al. 2012). Flood and tsunami hazard areas were identified on the basis of water depth and wave velocity, and exposure and vulnerability maps were combined with the former to obtain risk models. Birkmann et al. (2010) developed tsunami vulnerability and risk maps for the area of Cádiz, within the framework of the European project TRANSFER (Tsunami Risk and Strategies for the European Region). They include consideration of qualitative information on societal vulnerability.

A particular case is the one of the Canary Islands, in which large landslides due to the collapse of volcanic edifices might cause large tsunamis (Cendrero and Dramis 1996; Masson et al. 2002; Hürlimann et al. 2004). This process is of particular concern in the island of La Palma (Løvholt et al. 2008; Abadie et al. 2012), where the possible frequency, direction, and height of waves generated by structural collapses of different sizes, as well as their effects in the islands, have been analysed.

In the Mediterranean, Alvárez-Gómez et al. (2011) have studied tsunami risk in Murcia and Almería (with relatively high exposure), as well as Ibiza and Menorca (with lower exposure), based on the analysis of 22 tsunami sources in the Alborán Sea and northern Algeria. Wave elevation maps and tsunami itinerary times were obtained, to identify highest risk areas and provide the basis for the elaboration of alert systems.

As a consequence of the devastating 2004 tsunami in SE Asia, efforts on tsunami risk assessment and mitigation have increased (González-García 2009). For the Gulf of Cádiz, Omira et al. (2009) determined that three tsunamometres and 29 tide gauges would be needed for a tsunami detection network. Within this general context, several European and international programmes have been launched. These are NEAREST (Integrated observations from near shore sources of tsunamis: towards an early warning system), which includes the analysis of tsunami sources and risks, as well as early warning systems for the Gulf of Cádiz, DEWS (Distant Early Warning System), also with Spain's cooperation, for the development of tsunami early alert systems (http://www.dews-online.org/home; Lendholt and Hammitzsch 2012), and TRANSFER (Tsunami Risk And Strategies For the European Region; http://www.utm.csic.es/idi_proyecto.asp?idipryid={D870EC62-2D0B-4DCB-90E1-5FD9D0A1E6BC}); with participation of University of Cantabria (UC), University of Barcelona (UB), National Geographical Institute (IGN), and the *Organismo Público Puertos del Estado* (OPPE; national body which coordinates port authorities). Its main aim is to produce hazard and risk assessments and early warning systems for the Euro-Mediterranean region. A map of tsunami hazard for the South-East coast of Spain has already been generated (UC-IGN 2009).

26.6 Legislation and Land-Use Planning

The full incorporation of geomorphic hazards into the land-use planning process is quite recent. The national Land Act of 1998 (*Ley 6/1998 de 13 de abril, sobre Régimen del Suelo y Valoraciones*) established that it was not allowed to build in areas subject to natural hazards, but these were not specifically dealt with. The new Land Act 8/2007 states that all land-use, master plans, etc., must include a natural risk map to be approved. Although this law is national, its application is the responsibility of the autonomous communities ("states"). With the present legislation, all municipal or provincial master, urban or land-use plans must take into account natural hazards. Natural hazard assessment and mapping is a mandatory part of the Sustainability Report for those plans. Up to that date, some natural hazards (floods, earthquakes) were contemplated in the existing legislation, but others (landslides, subsidence) were not explicitly mentioned.

There is a national office, the *Dirección General de Protección Civil y Emergencias* (Civil Protection and Emergencies Directorate), that establishes the organisation of services and procedures to guarantee an effective response from the public administration in emergency situations caused by natural and other hazards. In addition, the different autonomous communities have their own civil protection plans regarding risk situations.

Damage caused by some natural risks is covered by the *Consorcio de Compensación de Seguros* (a consortium integrated and funded by all insurance companies), which covers the consequences of extraordinary events (Royal Act 2022/1986) and includes flooding, earthquakes, tsunamis, volcanic eruptions, storms of certain intensity, and falls of meteorites or other planetary objects. Landslides are not specifically included, except when they are linked to high-intensity storms, and damage related to sinkholes is not covered.

As would be expected, certain hazards have had more specific regulations. The fact that the Iberian Peninsula is affected by periodic droughts makes water a valued resource. Modern Spanish legislation to deal with water problems goes back to 1879 (Ministerio de Fomento 1879). In 1926, the water authorities (*confederaciones hidrográficas*) responsible for river regulation were created. These bodies, under the jurisdiction of the Central Government, continue to exist nowadays and regulate the administration of the 10 main river basins or catchment areas in Spain.

However, when a basin is entirely within one autonomous community, the river authority depends on the autonomous government of that community.

After the flood events of 1982 in the Mediterranean coastal area and 1983 in the Cantabrian strip, the Spanish administration revised its flood prevention policies. Among the actions undertaken, it is worth mentioning the meteorological and hydrological monitoring in the main catchment areas. In 1984, the first Automatic System for Meteorological Information—SAIH in its Spanish acronym—was installed in the Júcar basin, on the Mediterranean coastal strip, and is being gradually extended to all main river basins. The SAIH is a network of sensors located in different points of each basin, which collect data on precipitation, water flow, and reservoir level (among other factors) and transmit them to a SAIH control point or to the main SAIH Processing Centre. These data are used to monitor floods, predict their future development, and manage risk scenarios through several actions, like flood lamination using reservoirs or the declaration of emergency state.

The new Water Act of 1985 makes it compulsory for river authorities to draft hydrological regulation plans for their catchment areas, which must include "the criteria on studies, interventions and works to foresee and avoid the damage due to floods and other hydraulic events".

Spain's adhesion to the International Decade for the Reduction of Natural Disasters 1990–1999, declared by the UN's General Assembly in 1989, represented an official acknowledgement of the importance of flood risks in the country. Thus, in 1995, a regulation known in Spain as Basic Guideline for Civil Protection Planning for Flood Risk was issued. This guideline establishes the need to develop, at the national level, an analysis of the types of floods and an estimation of flood hazard, including a classification and zoning according to risk level. The National Catalogue of Historical Floods—CNIH in its Spanish acronym—was created. It includes information on floods in fluvial systems and other environments with effects on life or property since the first century BC until present. More than 3,000 episodes have been recorded, with a mean of 10 new ones per year; it is an open catalogue that keeps being updated with time. All the information in the CNIH is georeferenced to municipal level.

The 2000/60/CE Directive, approved by the European Parliament and Council on 23 October 2000, established a common setting for water policies in Europe and created two structures for basin management. One of these is equivalent to the already existing Spanish *confederaciones hidrográficas*. The other, the newly created Hydrographical Demarcation, encloses the land and water areas between two or more neighbouring catchment areas and the transition waters, both underground and coastal, associated with those catchments.

Flood risk assessment and management, as well as risk reduction through structural and non-structural measures, are included in the 23 October 2007, 2007/60/CE Directive, which was incorporated into Spanish legislation in 9 July 2010, through a legal ruling known as *Real Decreto* 903/2010, which was later reinforced by the Regulations of the Hydraulic Public Domain. The 2007/60/CE Directive and corresponding Spanish legislation take into account, among other issues, the assessment of flood risk through the application of different tools, such as the analysis of historic floods with significant negative impacts and the elaboration of flood hazard and risk maps. An important consequence of its implementation in Spain was the National Flood-Prone Areas Cartographic System (SNCZI in its Spanish acronym), a valuable support tool for risk assessment, land-use planning, and management in fluvial areas (MARM 2011a). The SNCZI is strengthened by the elaboration of a cartographic viewer for flood-prone areas (http://sig.magrama.es/snczi/visor.html?herramienta=DPHZI), which allows citizens to consult studies on the limits of the Hydraulic Public Domain (DPH in its Spanish acronym) and the maps of flood risk areas produced by bodies of both national and autonomous governments. Flooding in coastal areas and those related to coastal-fluvial transition zones are regulated by the Coastal Act of 1988 (Ley de Costas de 1988) and are also incorporated in the SNCZI.

The already mentioned *Dirección General de Protección Civil y Emergencias* is responsible for the elaboration of a National Plan for Civil Protection against Flood Risks (*Plan Estatal de Protección Civil ante el Riesgo de Inundaciones*). The autonomous communities have similar plans. Finally, the Nuclear Security Council—(CSN in its Spanish acronym), the national body responsible for the security of the country's nuclear power plants, created in 2005 a protocol on Protection against Severe Meteorological Conditions and Floods that introduced a number of compulsory regulations for power plants in these situations.

The national legislation does not make explicit or specific reference to landslide hazard. But specific regulations have been laid down by some autonomous governments, such as the 153/1997 decree-law in the Basque Country, which describes natural risks in the community including slope movements (Bonachea 2006) and incorporates civil protection plans and emergency strategies. The plans establish rules and protocols to face a natural disaster and establish that every municipality with more than 20,000 inhabitants must have a specific emergency plan.

Cases of an explicit incorporation of sinkhole hazard assessment into regulations area are even fewer. A good example is urban planning in some sectors of Zaragoza city. Simón-Gómez et al. (1998), based on a cartographic sinkhole inventory, proposed a 15-m radius restriction area around mapped sinkholes (possible "evolution area") and an additional 50 m security buffer in which construction should be avoided. At present, the city of Zaragoza demands developers to carry out a sinkhole hazard study in some sinkhole-prone areas prior to allowing any building activity. Another example of regulations concerning this hazard is from Catalonia, where the Cartographic Institute is responsible for mapping and assessment of subsidence and collapse, in view of incorporating it into municipal planning. Finally, an interesting example is the one of Asturias, where the filling of sinkhole areas is prohibited.

In the case of coastal hazards, the main framework is based on two legal documents, the Coastal Act (*Ley de Costas de 1988*) and the Spanish Integrated Strategy for Coastal Zone Management (*Estrategia Española de Gestión Integrada de Zonas Costeras 2002*). A new law has been approved by the Parliament on 9 March 2013, to substitute the former (*Ley de Protección y Uso Sostenible del Litoral y de Modificación de la Ley de Costas*). The integrated strategy aims at coordinating policies affecting the coastal area and its economic development, through sustainable management of resources, conservation, and hazard protection actions. Within this strategy, several bodies were created, such as a Master Plan for Coastal Sustainability, an Observatory on Sustainability of the Spanish Coast, or the National Coastal Council. Actions such as purchase of land along the coast, for protection and conservation purposes, are contemplated. The strategy establishes the basis for cooperation agreements with the autonomous communities (states) for conservation, defence, or rehabilitation of coastal areas (MAGRAMA 2013). The main official institutions responsible for the observation, monitoring, and assessment of coastal hazards are as follows: *Instituto Español de Oceanografía*, IEO (Spanish Oceanographic Institute), *Instituto Geográfico Nacional*, IGN (National Geographical Institute), *Real Observatorio de la Armada*, ROA (Royal Navy Observatory), *Organismo Público Puertos del Estado*, OPPE.

26.7 Final Comments

Geomorphic hazards affect a large part of Spain and cause casualties or important damage every year. However, with some exceptions, they have not been explicitly contemplated in legislation and land-use regulations until recently. Geomorphologic research in Spain has significantly contributed to a better understanding of hazardous geomorphic processes in the country and to the development of tools and methods for their assessment, prediction, and mitigation, which can be applied elsewhere. Advance of knowledge on geomorphic hazards has been particularly important during the last couple of decades, in parallel with the development experienced by Spanish geomorphology in general (Gutiérrez et al. 2013). The review by the former authors provides a good sample for assessing the relative weight of hazards within geomorphologic research in Spain in recent times. Out of the 129 references listed by Gutiérrez et al. (2013), 32 correspond to topics related to geomorphic hazards analysed here. That is, about 25 % of the total contributions. Although research on these topics has very much improved in quantity and quality in recent years, its future does not look promising, due to drastic government cuts on research funds since 2010. The continuity of research groups and the training of new scientists are seriously jeopardised and interruption of progress or even a setback is not unlikely (Moro-Martín 2012; Pain 2012).

References

Abadie SM, Harris JC, Grilli ST, Fabre R (2012) Numerical modelling of tsunami waves generated by the flank collapse of the Cumbre Vieja Volcano (La Palma, Canary Islands): tsunami source and near field effects. J Geophys Res: Oceans 117(5):C05030

Abellán A, Calvet J, Vilaplana JM, Blanchard J (2010) Detection and spatial prediction of rockfalls by means of terrestrial laser scanner monitoring. Geomorphology 119(3–4):162–171

Ablay GJ, Hürlimann M (2000) Evolution of the north flank of Tenerife by recurrent giant landslides. J Volcanol Geoth Res 103:135–159

Alfaro P, Delgado J, García-Tortosa FJ, Giner JJ, Lenti L, López-Casado C, Martino S, Scarascia Mugnozza G (2012) The role of near-field interaction between seismic waves and slope on the triggering of a rockslide at Lorca (SE Spain). Nat Hazards Earth Syst Sci 12(12):3631–3643

Alasset PJ, Hebert H, Maouche S, Calbini V, Meghraoui M (2006) The tsunami induced by the 2003 Zemmouri earthquake (MW = 6.9, Algeria): modelling and results. Geophys J Int 166:213–226

Alvarado-Aguilar D, Jiménez JA, Nicholls RJ (2012) Flood hazard and damage assessment in the Ebro Delta (NW Mediterranean) to relative sea level rise. Nat Hazards 62:1301–1321

Alvárez-Gómez JA, Aniel-Quiroga I, González M, Otero L (2011) Tsunami hazard at the Western Mediterranean Spanish Coast from seismic Sources. Natural Hazards Earth Syst Sci 11(1):227–240

Amir L, Cisternas A (2010) Appraisal of the 1790 Alboran tsunami source in the west Mediterranean Sea as inferred from numerical modelling: insights for the tsunami hazard in Algeria. In: 9th US National and 10th Canadian conference on earthquake engineering, including papers from the 4th international tsunami symposium 6, pp 4937–4945

Amorim S, Corominas J, Lantada N, Angulo C, Baeza C (2009) Comparative analysis of three data treatments techniques for landslide susceptibility assessment in the Eastern Pyrenees, Spain EGU General Assembly 2009, Vienna, p 13058

Anadón S, Fernández E, Domínguez R, Fernández Alonso M, Fernández García M (2008) Metodología para la actualización de la cartografía de peligrosidad de inundación fluvial en Asturias (NO España). In: Benavente J, Gracia FJ (eds) Trabajos de Geomorfología en España, 2006–2008. Actas de la X Reunión Nacional de Geomorfología, Cádiz, pp 369–372

Ancochea E, Fuster JM, Ibarrola E, Cendrero A, Coello J, Hernán F, Cantagrel J, Jamond C (1990) Volcanic evolution of the island of Tenerife (Canary Islands) in the light of new K–Ar data. J Volcanol Geoth Res 44:231–399

Ancochea E, Hernán E, Cendrero A, Cantagrel JM, Fúster JM, Ibarrola E, Coello J (1994) Constructive and destructive episodes in the building of a young oceanic island, La Palma, Canary Islands, and the genesis of the Caldera de Taburiente. J Volcanol Geoth Res 60:243–262

Andreo B, Durán JJ (eds) (2004) Investigaciones en sistemas kársticos españoles. IGME, Madrid

Anfuso G, Gracia FJ (2005) Morphodynamic characteristics and short-term evolution of a coastal sector in SW Spain: implications for coastal erosion management. J Coastal Res 21(6):1139–1153

Anfuso G, Benavente J, Gracia FJ (2001) Morphodynamic responses of nourished beaches in SW Spain. J Coast Conserv 7(1):71–80

Anfuso G, Domínguez L, Gracia FJ (2007) Short and medium-term evolution of a coastal sector in Cadiz, SW Spain. Catena 70:229–242

Ayala FJ (1995) Probabilidad y vulnerabilidad en movimientos de ladera. In: Reducción de riesgos geológicos en España, ITGE, Madrid, pp 95–113

Ayala FJ (2002) Estrategias y medidas de mitigación del riesgo de inundaciones. Gestión de zonas inundables. In: Ayala FJ, Olcina Cantos J (eds) Riesgos Naturales. Ariel Ciencia, Barcelona, pp 977–995

Ayala FJ, Corominas J (eds) (2002) Mapas de susceptibilidad a los movimientos de ladera con técnicas SIG. IGME, Serie Medio Ambiente No 4, Madrid

Ayala FJ, Elízaga E, González de Vallejo LI, Durán JJ, Beltrán de Heredia F, Oliveros MA, Carbó A, Guillamont ML, Capote R (1987) Impacto económico y social de los riesgos geológicos en España. IGME, Madrid

Ayala FJ, Olcina J, Vilaplana JM (2004) Impacto social de los riesgos naturales en España en el periodo 1990–2000 (II). Gerencia de riesgos XXI:17–29

Baeza C, Corominas J (2001) Assessment of shallow landslide susceptibility by means of multivariate statistical techniques. Earth Surf Proc Land 26:1251–1263

Balasch JC, Vericat D, Batalla RJ (2007) Deposición de sedimentos finos durante una crecida en un tramo del Ebro medio. Cuaternario y Geomorfología 21(1–2):41–55

Ballesteros-Cánovas JA, Bodoque JM, Díez-Herrero A, Sánchez-Silva M, Stoffel M (2011) Calibration of floodplain roughness and estimation of flood discharge based on tree-ring evidence and hydraulic modeling. J Hydrol 403(1–2):103–115

Batalla RJ, Gómez CM, Kondolf GM (2004) Reservoir induced hydrological changes in the Ebro River basin (Northeastern Spain). J Hydrol 290(1–2):117–136

Beguería S (2006) Changes in land cover, shallow landslide activity: a case study in the Spanish Pyrenees. Geomorphology 74:196–206

Benavente J, Gracia FJ (eds) (2008) Trabajos de geomorfología en España 2006–2008. Actas de la X Reunión Nacional de Geomorfología, Cádiz

Benavente J, Del Río L, Gracia FJ, Martínez del Pozo JA (2006) Coastal flooding hazard related to storms and coastal evolution in Valdelagrana spit (Cádiz Bay Natural Park, SW Spain). Cont Shelf Res 26(9):1061–1076

Benito G, Thorndycraft VR (2004) Systematic, paleoflood and historical data for the improvement of flood risk estimation, methodological guidelines. CSIC, Madrid

Benito G, Pérez del Campo P, Gutiérrez M, Sancho C (1995) Natural and human-induced sinkholes in gypsum terrain and associated environmental problems in NE Spain. Environ Geol 25:156–164

Benito G, Sopeña A, Sánchez-Moya Y, Machado MJ, Pérez-González A (2003a) Palaeo-flood record of the Tagus River (Central Spain) during the Late Pleistocene and Holocene. Quatern Sci Rev 22:1737–1756

Benito G, Díez-Herrero A, Fernandez de Villalta M (2003b) Magnitude and frequency of flooding in the Tagus Basin (Central Spain) over the last millennium. Clim Change 58:171–192

Benito G, Díez-Herrero A, Fernández de Villalta M (2004) Flood response to NAO and solar activity in the Tagus Basin (Central Spain) over the last millennium. Clim Change 66:27–28

Benito G, Rico M, Sánchez-Moya Y, Sopeña A, Thorndycraft VR, Barriendos M (2010) The impact of late Holocene climatic variability and land use change on the flood hydrology of the Guadalentín River, southeast Spain. Global Planet Change 70:53–63

Birkmann J, Teichman KV, Welle T, González M, Olabarrieta M (2010) The unperceived risk to Europe's coasts: tsunamis and the vulnerability of Cadiz, Spain. Nat Hazards Earth Syst Sci 10(12):2659–2675

Blanc PL (2008) The tsunami in Cádiz on 1 November 1755: a critical analysis of reports by Antonio de Ulloa and Louis Godin. C R Geosci 340:251–261

Bodoque JM, Díez-Herrero A, Martín-Duque FJ, Rubiales JM, Godfrey A, Pedraza J, Carrasco RM, Sanz MA (2005) Sheet erosion rates determined by using dendrogeomorphological analysis of exposed tree roots: two examples from Central Spain. Catena 64(1):81–102

Bodoque JM, Lucia A, Ballesteros JA, Martín-Duque JF, Rubiales JM, Genova M (2011) Measuring medium-term sheet erosion in gullies from trees: a case study using dendrogeomorphological analysis of exposed pine roots in central Iberia. Geomorphology 134(3–4):417–425

Bodoque JM, Ballesteros JA, Díez-Herrero A, Ruiz-Villanueva V, Nadal-Romero ME, Génova M, Rubiales JM, Baeza-Chico E, Hernández-Ruiz M (2012) Nueva propuesta para la mejora de la estimación de tasas de erosión hídrica mediante técnicas dendrogeomorfológicas. In: González-Díez A, et al (eds) Avances de la Geomorfología en España 2010–2012. Actas de la XII Reunión Nacional de Geomorfología, Publican, Santander, pp 143–146

Bonachea J (2006) Desarrollo, aplicación y validación de procedimientos y modelos para la evaluación de amenazas, vulnerabilidad y riesgo debidos a procesos geomorfológicos. Tesis Doctoral. Universidad de Cantabria, Santander

Bonachea J, Remondo J, González-Díez A, Díaz de Terán JR, Cendrero A (2009) Landslide risk models for decision making. Risk Anal 29(11):1629–1643

Brocal J (1984) Obras hidráulicas realizadas en el valle de Arán y Alto Ribagorzana. Jornadas sobre inestabilidad de laderas en el Pirineo, Barcelona

Bru J, Julià R, Marquès MA (1984a) El movimiento de masa de Pont de Bar. Dinámica geomorfológica. Jornadas sobre inestabilidad de laderas en el Pirineo. Barcelona, pp I.2.1–I.2.10

Bru J, Serrat D, Vilaplana JM (1984b) La dinámica de la cuenca del Torrent de Jou-La Guingueta. Jornadas sobre inestabilidad de laderas en el Pirineo. Barcelona, pp I.2.1–I.2.10

Bruschi VM, Bonachea J, Remondo J, Gómez-Arozamena J, Rivas V, Méndez G, Naredo JM, Cendrero A (2013) Analysis of geomorphic systems' response to natural and human drivers in northern Spain; implications for global geomorphic change. Geomorphology 196:267–279

Camarasa AM, Segura F (2001) Flood events in Mediterranean ephemeral streams (ramblas) in Valencia region, Spain. Catena 45:229–249

Campos ML (1992) El riesgo de Tsunamis en España. Análisis y valoración geográfica. Monografía 9. Instituto Geográfico Nacional, Madrid

Carbonel D, Rodríguez V, Gutiérrez F, McCalpin JP, Linares R, Roqué C, Zarroca M, Guerrero J, Sasowsky I (2013) Evaluation of trenching, ground penetrating radar (GPR) and electrical resistivity tomography (ERT) for sinkhole characterization. Earth Surf Proc Land. doi:10.1002/esp.3440

Cardineau CA, González JA (2005) Natural and human erosive factors in Liencres beach spit and dunes (Cantabria, Spain). J Coastal Res 49:70–75

Carracedo JC, Day SJ, Guillou H, Pérez-Torrado FJ (1999) Giant Quaternary landslides in the evolution of La Palma and El Hierro, Canary Islands. J Volcanol Geoth Res 94:169–190
Carreño E, Seller A (2005) Simulación de daños en la costa occidental de Huelva como consecuencia de tsunami. Fundación Mapfre GR 90:23–40
Castañeda C, Gutiérrez F, Manunta M, Galve JP (2009) DInSAR measurements of ground deformation by sinkholes, mining subsidence, and landslides, Ebro River, Spain. Earth Surf Proc Land 34(11):1562–1574
Cendrero A, Dramis F (1996) The contribution of landslides to landscape evolution in Europe. Geomorphology 15(3–4):191–212
Cendrero A, Díaz de Terán, JR, Fernández O, Garrote R, González Lastra JR, Inoriza I, Lütting G, Otamendi J, Pérez M, Serrano A, Grupo Ikerlana (1987) Detailed geomorphological hazard mapping for urban and rural planning in Vizcaya (Northern Spain). In: Wolf FC (ed) Geology for environmental planning. Geological Survey of Norway, Special Publ. 2, Trondheim, pp 25–41
Cendrero A, Sánchez-Arcilla A, Zazo C (2005) Impactos sobre las zonas costeras. Evaluación preliminar de los impactos en España por efecto del cambio climático. In: Moreno JM (coord) Proyecto ECCE - Informe Final, pp 469–524
Cendrero A, Remondo J, Bonachea J, Rivas V, Soto J (2006) Sensitivity of landscape evolution and geomorphic processes to direct and indirect human influence. Geografia Fisica e Dinamica Quaternaria 29(2):125–137
Cendrero A, Remondo J, Bonachea J, Rivas V, Soto J (2007) Global change, global geomorphic change and natural hazards; a new scenario? In: Dinâmicas Geomorfológicas, Metodologias, Aplicaçao Associaçao Portuguesa de Geomorfólogos, Lisboa, pp 19–38
Chacón J, Irigaray C, Fernández T, El Hamdouni R (2003) Susceptibilidad a los movimientos de ladera del Sector Central de la Cordillera Bética. In: Ayala FJ, Corominas J (eds) Mapas de susceptibilidad a los movimientos de ladera con técnicas SIG. Fundamentos y aplicaciones en España. IGME, Madrid, pp 83–96
Chacón J, Irigaray C, Fernández T, El Hamdouni R (2006) Engineering geology maps: landslides and geographical information systems (GIS). Bull Eng Geol Environ 65:341–411
Confederación Hidrográfica del Ebro (ed) (2007) Jornadas técnicas sobre estabilidad de laderas en embalses. CHE. Zaragoza, 597 p
Corominas J (2000) Landslides and climate. In: Bromhead EN (ed) VIII International symposium on landslides cardiff UK keynote lectures CD-ROM 2000
Corominas J (2005) Impactos sobre los riesgos naturales de origen climático. B. Riesgos de inestabilidad de laderas. In: Moreno JM (coord) Evaluación preliminar de los impactos en España por efecto del cambio climático. Ministerio de Medio Ambiente, Madrid, pp 549–579
Corominas J (2006) El clima y sus consecuencias sobre la actividad de los movimientos de ladera en España. Cuaternario y Geomorfología 20(3–4):89–113
Corominas J, Alonso EE (1990) Geomorphological effects of extreme floods (November, 1982) in the southern Pyrenees. In: Hydrology in mountainous regions. IAHS 194:295–302
Corominas J, Moya J (1999) Reconstructing recent landslide activity in relation to rainfall in the Llobregat River basin, Eastern Pyrenees, Spain. Geomorphology 30:79–93
Corominas J, Moya J, Ledesma A, Rius J, Gili JA, Lloret A (1999) Monitoring of the Vallcebre landslide, Eastern Pyrenees, Spain. In: Proceedings international symposium on slope stability engineering: IS-Shikoku'99. Matsuyama, pp 1239–1244
Corominas J, Moya J, Hürlimann M (2002) Landslide rainfall triggers in the Spanish Eastern Pyrenees. In: Proceedings 4th EGS conference on Mediterranean storms. Mallorca. CD-ROM
Corominas J, Moya J, Masachs I, Baeza C, Hürlimann M (2004) Identificación de episodios de reactivación en grandes deslizamientos pirenaicos mediante técnicas dendrocronológicas., In: Benito G, Díez-Herrero A (eds) Riesgos Naturales y Antrópicos en Geomorfología. Actas de la VIII Reunión Nacional de Geomorfología, Toledo, pp 403–411
Crespo J, Menéndez R, Fernández E (2012) Aplicación de modelos digitales del terreno generados a partir de tecnología LIDAR en la caracterización de zonas inundables In: González-Díez A, et al (eds) Avances de la Geomorfología en España 2010–2012. Actas de la XII Reunión Nacional de Geomorfología, Publican, Santander, pp 267–270
Damaskinidou-Georgiadou A, Asce AM, Johnson WJ (1987) Evaluation of tsunami risk at the southwest coast of the Iberian Peninsula. In: Magoon OT (ed) Coastal zone '87. In: Proceedings of the 5th symposium on coastal and ocean management, Seattle, American Society of Civil Engineers
De Andrés JR, Gracia FJ (1988) La franja litoral: riesgos y protección. In: IGME (ed) Riesgos Geológicos. Madrid, pp 1022–1057
Delgado J, Vicente F, García-Tortosa F, Alfaro P, Estévez A, López-Sánchez JM, Tomás R, Mallorquí JJ (2011) A deep seated compound rotational rock slide and rock spread in SE Spain: structural control and DInSAR monitoring. Geomorphology 129(3–4):252–262
Del Río L, Gracia J (2009) Erosion risk assessment of active coastal cliffs in temperate environments. Geomorphology 112:82–95
Del Río L, Gracia FJ, Benavente J (2012) Shoreline change patterns in sandy coasts. A case study in SW Spain. Geomorphology 196:252–266
Del Val J, Delgado J, Eguiguren E, García I, Garrote R (1996) Ejemplo de análisis de inestabilidades del terreno para la atención de emergencias, el caso del País Vasco. Actas del VI Congreso Nacional y Conferencia Internacional de Geología Ambiental y Ordenación del Territorio, Granada 1
Díaz de Terán JR, Cendrero A (1992) Riesgos geológicos en España. XXVI Curso de Geología Práctica. Universidad de Verano, Teruel, pp 179–231
Díaz de Terán JR, Cendrero C, Araña V, Badiola ER, Berga L, Carracedo JC, Chacón J, Coello J, Corominas J, Dabrio CJ, Durán JJ, Elízaga E, Ferrer M, García-Fernández M, Garzón MG, González-Díez A, Goy JL, López-Martínez J, Martínez-Goytre J, Mezcua J, de la Nuez J, Salinas JL, Soler V, del Val J, Zazo C (1997) Natural hazards in Spain. In: Embleton C, Embleton-Hamann C (eds) Geomorphological hazards of Europe. Developments in Earth surface processes 5. Elsevier, The Netherlands, pp 429–456
Díez-Herrero A (2002) Aplicaciones de los sistemas de información geográfica al análisis del riesgo de inundaciones fluviales. In: Laín-Huerta L (ed) Los Sistemas de Información Geográfica en la Gestión de los Riesgos Geológicos y el Medio Ambiente. Serie: Medio Ambiente. Riesgos Geológicos, no 3, Publicaciones del Instituto Geológico y Minero de España, Ministerio de Ciencia y Tecnología, Madrid, pp 85–112
Díez-Herrero A (2003) Geomorfología e Hidrología fluvial del río Alberche. Modelos y SIG para la gestión de riberas. Serie Tesis Doctorales 2. Publicaciones del Instituto Geológico y Minero de España, Madrid
Díez-Herrero A, Laín-Huerta L, Llorente-Isidro M (eds) (2006) Mapas de peligrosidad de avenidas e inundaciones. Métodos, experiencias y aplicación. Publicaciones del Instituto Geológico y Minero de España. Serie Medio Ambiente, Riesgos Geológicos No 7, Madrid
Díez-Herrero A, Laín-Huerta L, Llorente-Isidro M (2008) Mapas de peligrosidad de avenidas e inundaciones. Guía metodológica para su elaboración. Publicaciones del Instituto Geológico y Minero de España. Serie Riesgos Geológicos/Geotecnia No 1, IGME, Madrid

Díez-Herrero A, Baillo-Calvo R, Garrote-Revilla J, Gutiérrez-Pérez I, Laín-Huerta L, Mancebo-Mancebo MJ, Martín-Duque JF, Pérez-Cerdán FL (2012a) Propuesta para la calibración y validación de los análisis y cartografías de peligrosidad y riesgos naturales mediante el estudio de eventos recientes. In: González-Díez A, et al (eds) Avances de la Geomorfología en España 2010–2012. Actas de la XII Reunión Nacional de Geomorfología, Publican, Santander, pp 155–158

Díez-Herrero A, Génova-Fúster M, Mayer-Suárez P, Ballesteros-Cánovas JA, Becerril-Carretero L, Rubiales-Jiménez JM, Hernández-Ruiz M, Saz-Sánchez MA, Bodoque del Pozo JM, Ruiz-Villanueva V (2012b) Investigaciones dendrogeomorfológicas aplicadas al estudio de la peligrosidad por avenidas e inundaciones en el Parque Nacional de la Caldera de Taburiente (La Palma, Islas Canarias). In: González-Díez A, et al (eds) Avances de la Geomorfología en España 2010–2012. Actas de la XII Reunión Nacional de Geomorfología, Publican, Santander, pp 159–162

Dirección General de Protección Civil y Emergencias (2011) Plan Estatal de Protección Civil ante el Riesgo de Inundaciones. BOE 210 (de 1 de septiembre de 2011), Madrid

Domínguez MJ, Jiménez M, Rodríguez A (1999) Press archives as temporal records of landslides in the North of Spain: relationships between rainfall and instability slope events. Geomorphology 30:125–132

Domínguez L, Anfuso G, Gracia FJ (2005) Vulnerability assessment of a retreating coast in SW Spain. Environ Geol 47:1037–1044

Duarte JC, Rosas FM, Terrinha P, Gutscher MA, Malavieille J, Silva S, Matias L (2011) Thrust-wrench interference tectonics in the Gulf of Cadiz (Africa-Iberia plate boundary in the North-East Atlantic): insights from analog models. Mar Geol 289:135–149

Durán JJ (2002) Riesgos en terrenos kársticos. In: Ayala FJ, Olcina Cantos J (eds) Riesgos naturales. Ariel Ciencia, Barcelona, pp 365–378

Durán JJ, López-Martínez J (1989) El karst en España. Monografía No 4 Sociedad Española de Geomorfología, Madrid

Durán JJ, Martínez-Goytre J (1990) Riesgos geológicos asociados a procesos kársticos en la provincia de Alicante. Actas del IV Congreso Español de Espeleología, Santander, pp 426–435

Durán JJ, Pardo-Igúzquiza E, Robledo PA (2012) Automatic detection of depressions using digital terrain models and their application in geomorphological mapping. In: González-Díez A, et al (eds) Avances de la Geomorfología en España 2010–2012. Actas de la XII Reunión Nacional de Geomorfología, Publican, Santander, pp 259–262

Estrany J, Garcia C, Martínez-Carreras N (2012) Determinación del origen del sedimento en suspensión en una pequeña cuenca agrícola mediterránea: Can Revull, Mallorca. In: González-Díez A, et al (eds) Avances de la Geomorfología en España 2010–2012. Actas de la XII Reunión Nacional de Geomorfología, Publican, Santander, pp 159–162

Felicísimo A, Cuartero A, Remondo J, Quirós E (2013) Mapping landslide susceptibility with MLR, MARS, CART, and MAXENT methods: a comparative study. Landslides 10:175–189

Fernández T, Irigaray C, El Hamdouni R, Chacón J (2003) Methodology for landslide susceptibility mapping by means of a GIS. Application to the Contraviesa area (Granada, Spain). Nat Hazards 30(3):297–308

Fernández P, Irigaray C, Jiménez J, El Hamdouni R, Crosetto M, Monserrat O, Chacón J (2009) First delimitation of areas affected by ground deformations in the Guadalfeo River Valley and Granada metropolitan area (Spain) using the DInSAR technique. Eng Geol 105(1–2):84–101

Fernández E, Anadón S, Fernández M, Vázquez D, Crespo J, Moro V, Menéndez R (2011) Aportaciones geomorfológicas de los modelos derivados del LIDAR en el estudio de la Zona de Flujo Preferente. XIV Congreso de la Asociación Española de Teledetección (AET)

Ferrer M (1995) Los movimientos de ladera en España. In: Reducción de Riesgos Geológicos en España. ITGE, Madrid, pp 69–82

Ferrer M, Ayala FJ (1997) Relaciones entre desencadenamiento de movimientos y condiciones meteorológicas para algunos deslizamientos en España. IV Simposio Nacional sobre Taludes y Laderas Inestables, Granada 1, pp 185–197

Ferrer M, López JM, Mateos R, Morales R, Rodríguez A (1997) Análisis y estimación del riesgo de desprendimientos rocosos en la cala de Banyalbufar (Mallorca). IV Simposio Nacional sobre Taludes y Laderas Inestables, Granada 1, pp 365–376

Ferrer M, González de Vallejo LI, García López-Davalillo JC, Rodríguez-Franco JA (2004) Pérdidas por terremotos e inundaciones en España durante el periodo 1987–2001 y su estimación para los próximos 30 años (2004–2033). IGME-Consorcio de Compensación de Seguros, Madrid

Ford DC, Williams P (2007) Karst hydrogeology and geomorphology. Wiley, Chichester

Fúster JM, Hernán F, Cendrero A, Coello J, Cantagrel JM, Ancochea E, Ibarrola E (1993) Geocronología de la isla de El Hierro (Islas Canarias). Boletín de la Real Sociedad Española de Historia Natural (Geología) 88:86–97

Gallart F, Clotet N (1988) Some aspects of the geomorphic process triggered by an extreme rainfall event: the November 1982 flood in the Eastern Pyrenees. Catena Supplement 13:75–95

Galve JP, Bonachea J, Remondo J, Gutiérrez F, Guerrero J, Lucha P, Cendrero A, Gutiérrez M, Sánchez JA (2008) Development and validation of sinkhole susceptibility models in mantled karst settings. A case study from the Ebro valley evaporite karst (NE Spain). Eng Geol 99:185–197

Galve JP, Gutiérrez F, Lucha P, Guerrero J, Bonachea J, Remondo J, Cendrero A (2009) Probabilistic sinkhole hazard modelling. Earth Surf Proc Land 34:437–452

Galve JP, Gutiérrez F, Guerrero J, Alonso J, Diego I (2012) Application of risk, cost-benefit and acceptability analyses to identify the most appropriate geosynthetic solution to mitigate sinkhole damage on roads. Eng Geol 145–146:65–77

Garay P (1986) Informe geológico sobre la sima de hundimiento de Pedreguer (Provincia de Alicante). Jornadas sobre el Karst en Euskadi 1:323–334

García-Davalillo JC, Herrera G, Notti D, Strozzi T, Álvarez-Fernández I (2013). DInSAR analysis of ALOS PALSAR images for the assessment of very slow landslides: the Tena Valley case study. Landslides. doi:10.1007/s10346-012-0379-8

García-García F, Bohorquez P, Martínez-Sánchez C, Pérez-Valera F, Pérez-Valera LA (2012) Régimen fluvial, frecuencia e hidrología de paleoinundaciones en el Río Segura (Holoceno superior, SE de España). In: González-Díez A, et al (eds) Avances de la Geomorfología en España 2010–2012. Actas de la XII Reunión Nacional de Geomorfología, Publican, Santander, pp 441–444

García-Moreno I, Mateos RM (2011) Sinkholes related to discontinuous pumping: susceptibility mapping based on geophysical studies. The case of Crestatx (Majorca, Spain). Environ Earth Sci 64:523–537

García-Ruiz JM, Beguería S, Alatorre LC, Constante A, Puigdefábregas J (2010) Land cover changes, shallow landsliding in the Flysch sector of the Spanish Pyrenees. Geomorphology 124:250–259

Gobierno de España (2011) Estrategia Española de Seguridad. Una responsabilidad de todos (Spanish security strategy. Everyone's responsability), Madrid. (www.lamoncloa.gob.es/NR/rdonlyres/EF784340-AB29-4DFC-8A4B-206339A29BED/0/SpanishSecurityStrategy.pdf)

Gómez-Gallego J (1994) Estudio de las variaciones del nivel del mar: anuales estacionales y mensuales en el puerto de Santander. Actas

IV Coloquio Internacional sobre Oceanografía del Golfo de Vizcaya, Santander, pp 83–97
Gómez-Ortiz D, Martín-Crespo T (2012) Assessing the risk of subsidence of a sinkhole collapse using ground penetrating radar and electrical resistivity tomography. Eng Geol 149–150:1–12
González B, Recchi C, Oliveros MA, Ortuño L (1997) Adaptación de dos proyectos de carreteras a las condiciones geológico-geotécnicas reales deducidas de la experiencia en tramos cercanos. IV Simposio Nacional sobre Taludes y Laderas Inestables, Granada 1, pp 157–169
González-Díez A, Salas L, Díaz de Terán JR, Cendrero A (1996) Late quaternary climate changes and mass movement frequency and magnitude in the Cantabrian region, Spain. Geomorphology 15:291–309
González-Díez A, Remondo J, Díaz de Terán JR, Cendrero A (1999) A methodological approach for the analysis of the temporal occurrence and triggering factors of landslides. Geomorphology 30:95–113
González-Díez A, Fernández-Maroto G, Doughty MW, Remondo J, Bruschi VM, Bonachea J, Díaz de Terán JR, Martínez-Cedrún P, Rivas V, Flor G, Serrano E, Gutiérrez F, Benito G, Desir G, López J, Romero MA, Batalla R, Gracia FJ, Flor-Blanco G, Garzón G, Cardenal J, Vilaplana JM, Corominas J, Díaz-Martínez E, Jiménez-Sánchez M, Guerra JC (eds) (2012) Avances de la geomorfología en España 2010–2012. Actas de la XII Reunión Nacional de Geomorfología, Publican, Santander
González-García JL (2009) Mapas de riesgos naturales en la ordenación territorial y urbanística. Ilustre Colegio Oficial de Geólogos, Madrid
González-García JL (2011) Natural hazards in the Spanish security strategy. Eur Geol 32:15–16
González de Vallejo L (1988) La importancia socioeconómica de los riesgos geológicos en España. Riesgos Geológicos. IGME, Serie Geología Ambiental, Madrid, pp 21–36
González-Villanueva R, Costas S, Alejo I, Pérez-Arlucea M (2007) Morphological changes forced by the tidal cycle in a low energy estuarine beach. J Coastal Res 50:1010–1014
González-Villarías FJ (2001) Estudio de estabilidad del acantilado de Rebolleres y Sequiro (Candas-Asturias). IV Simposio Nacional sobre Taludes y Laderas Inestables, Granada 1, pp 267–284
Gracia FJ, Anfuso G, Benavente J, Del Río L, Domínguez L, Martínez JA (2005) Monitoring coastal erosion at different temporal scales on sandy beaches: application to the Spanish Gulf of Cadiz coast. J Coastal Res 49:22–27
Gracia E, Vizcaíno A, Escutia C, Asioli A, Rodes A, Pallàs R, García-Orellana J, Lebreiro S, Goldfinger C (2010) Holocene earthquake record offshore Portugal (SW Iberia): testing turbidite paleoseismology in a slow-convergence margin. Quatern Sci Rev 29(9–10):1156–1172
Guerrero J, Gutiérrez F, Lucha P (2005) Riesgos geológicos en Cadrete (Zaragoza). Cuaternario y Geomorfología 19(1–2):63–82
Guerrero J, Gutiérrez F, Bonachea J, Lucha P (2008) A sinkhole susceptibility zonation based on paleokarst analysis along a stretch of the Madrid-Barcelona high-speed railway built over gypsum- and salt-bearing evaporites (NE Spain). Eng Geol 102:62–73
Guerrero J, Gutiérrez F, Carbonel D, Bonachea J, García-Ruiz JM, Galve JP, Lucha P (2012) 1:5000 Landslide map of the upper Gállego Valley (central Spanish Pyrenees). J Maps 8(4):484–491
Guillén J, Díaz JI, Palanques A (1992) Cuantificación y evolución durante el siglo XX de los aportes de sedimento transportado como carga de fondo por el río Ebro al medio marino. Revista de la Sociedad Geológica de España 5:27–37
Gutiérrez F, Cooper AH (2002) Evaporite dissolution subsidence in the historical city of Calatayud, Spain: damage appraisal and prevention. Nat Hazards 25:259–288
Gutiérrez F, Arauzo T, Desir G (1994) Deslizamientos en el escarpe en yesos de Alfajarín (Zaragoza). Cuaternario y Geomorfología 8:57–68
Gutiérrez F, Arauzo T, Sancho C, Gutiérrez M (1995) Factores desencadenantes de los deslizamientos de la cantera del Barrio de La Corona (Ejea de los Caballeros, Depresión del Ebro). Cuaternario y Geomorfología 9:33–50
Gutiérrez F, Lucha P, Guerrero J (2004) La dolina de colapso de la casa azul de Calatayud (noviembre de 2003). Origen, efectos y pronóstico. In: Benito G, Díez-Herrero A (eds) Riesgos naturales y antrópicos en Geomorfología. Actas de la VIII Reunión Nacional de Geomorfología, Toledo, pp 477–488
Gutiérrez-Santolalla F, Gutiérrez-Elorza M, Marín C, Maldonado C, Younger PL (2005) Subsidence hazard avoidance based on geomorphological mapping in the Ebro River valley mantled evaporite karst terrain (NE Spain). Environ Geol 48:370–383
Gutiérrez F, Galve JP, Guerrero J, Lucha P, Cendrero A, Remondo J, Bonachea J, Gutiérrez M, Sánchez JA (2007) The origin, typology, spatial distribution, and detrimental effects of the sinkholes developed in the alluvial evaporite karst of the Ebro River valley downstream Zaragoza city (NE Spain). Earth Surf Proc Land 32:912–928
Gutiérrez F, Cooper AH, Johnson KS (2008a) Identification, prediction and mitigation of sinkhole hazards in evaporite karst areas. Environ Geol 53:1007–1022
Gutiérrez F, Calaforra JM, Cardona F, Ortí F, Durán JJ, Garay P (2008b) Geological and environmental implications of evaporite karst in Spain. Environ Geol 53:951–965
Gutiérrez F, Ortuño M, Lucha P, Guerrero J, Acosta E, Coratza P, Piacentini D, Soldati M (2008c) Late Quaternary episodic displacement on a sackung scarp in the central Spanish Pyrenees. Secondary paleoseismic evidence? Geodin Acta 21:187–202
Gutiérrez F, Galve JP, Lucha P, Bonachea J, Jordá L, Jordá R (2009) Investigation of a large collapse sinkhole affecting a multi-storey building by means of geophysics and the trenching technique (Zaragoza city, NE Spain). Environ Geol 58:1107–1122
Gutiérrez F, Lucha P, Galve JP (2010) Reconstructing the geochronological evolution of large landslides by means of the trenching technique in the Yesa Reservoir (Spanish Pyrenees). Geomorphology 124:124–136
Gutiérrez F, Galve JP, Lucha P, Castañeda C, Bonachea J, Guerrero J (2011) Integrating geomorphological mapping, trenching, InSAR and GPR for the identification and characterization of sinkholes: a review and application in the mantled evaporite karst of the Ebro Valley (NE Spain). Geomorphology 134(1–2):144–156
Gutiérrez F, Linares R, Roqué C, Zarroca M, Rosell J, Galve JP, Carbonell D (2012) Investigating gravitational grabens related to lateral spreading and evaporite dissolution subsidence by means of detailed zapping, trenching, and electrical resistivity tomography (Spanish Pyrenees). Lithosphere 4:331–353
Gutiérrez F, Harvey A, García-Ruiz JM, Silva P, Cendrero A (2013) Geomorphological research in Spain. Geomorphology 196:1–12
Gutiérrez M (2013) Geomorphology. CRCPress/Balkema, The Netherlands (1017 p)
Gutiérrez M, Ibáñez MJ, Peña JL, Rodríguez J, Soriano A (1985) Quelques exemples de karst sur gypse dans la dépression de l'Ebre. Karstologia 6:29–36
Herrera G, Notti D, García-Davalillo JC, Mora O, Cooksley G, Sánchez M, Arnaud A, Crosetto M (2011) Analysis with C-and X-band satellite SAR data of the Portalet landslide area. Landslides 8(2):195–206
Herrera G, Gutiérrez F, García-Davalillo JC, Guerrero J, Notti D, Galve JP, Fernández-Merodo JA, Cooksley G (2013) Multi-sensor advanced DInSAR monitoring of very slow landslides: the Tena

Valley case study (Central Spanish Pyrenees). Remote Sens Environ 128:31–43
Hürlimann M, Ledesma A, Martí J (1999) Conditions favouring catastrophic landslides on Tenerife (Canary Islands). Terra Nova 11:106–111
Hürlimann M, Martí J, Ledesma A (2004) Morphological and geological aspects related to large slope failures on oceanic islands. The huge La Orotova landslides on Tenerife, Canary Islands. Geomorphology 62:143–158
Hürlimann M, Abancó C, Moya J (2010) Debris-flow initiation affected by snowmelt. Case study of the Seneset monitoring site, Eastern Pyrenees. Mountain Risks: Bringing Science to Society, Florence
IGME (1986) Mapa de estabilidad de taludes en las formaciones blandas de la comunidad de Madrid. Instituto Geológico y Minero de España, Madrid
IGME (1987) Mapa de movimientos del terreno en España. Escala 1:1,000,000. Instituto Geológico y Minero de España, Madrid
IGME (2007) Atlas de riesgos naturales en la provincia de Granada. Instituto Geológico y Minero de España y Diputación de Granada, Madrid
IGME (2013) Mapa del Karst de España a escala 1/1.000.000. http://mapas.igme.es/gis/services/Cartografia_Tematica/IGME_Karst_1M/MapServer/WMSServer
Irigaray C, Chacón J (1991) Los movimientos de ladera en el sector de Colmenar (Málaga). Revista de la Sociedad Geológica de España 4:203–214
Irigaray C, Fernández T, El Hamdouni R, Chacón J (1999) Verification of landslide susceptibility mapping. A case study. Earth Surf Proc Land 24(6):537–544
Irigaray C, Fernández T, El Hamdouni R, Chacón J (2007) Evaluation and validation of landslide-susceptibility maps obtained by a GIS matrix method: Example from the Betic Cordillera (Southern Spain). Nat Hazards 41(1):61–79
INE (2013) Instituto Nacional de Estadística. http://www.ine.es/
IPCC—Intergovernmental Panel on Climate Change (2007) Fourth assessment report. Climate change 2007: synthesis report. Summary for policy makers, Valencia
Jelínek R, Krausmann E, González M, Álvarez-Gómez JA, Birkmann J, Welle T (2012) Approaches for tsunami risk assessment and application to the city of Cádiz, Spain. Nat Hazards 60:273–293
Jiménez JA (1992) Estabilidad de las ciudades colgadas. III Simposio Nacional sobre Taludes y Laderas Inestables, La Coruña 3:881–893
Jiménez JA, Sánchez-Arcilla A, Valdemoro HI, Gracia V, Nieto F (1997) Processes reshaping the Ebro delta. Mar Geol 144:59–79
Lario J, Zazo C, Goy JL, Silva PG, Bardají T, Cabero A, Dabrio CJ (2011) Holocene palaeotsunami catalogue of SW Iberia. Quatern Int 242:196–200
Lendholt M, Hammitzsch M (2012) Towards an integrated information logistics for multi hazard early warning systems. Open Environ Eng J 5:27–43
Leorri E, Cearreta A, Milne G (2012) Field observations and modelling of Holocene sea-level changes in the southern Bay of Biscay: implication for understanding current rates of relative sea-level change and vertical land motion along the Atlantic coast of SW Europe. Quatern Sci Rev 42:59–73
Lima VV, Miranda JM, Baptista MA, Catalão J, Gonzalez M, Otero L, Olabarrieta M, Álvarez-Gómez JA, Carreño E (2010) Impact of a 1755-like tsunami in Huelva, Spain. Nat Hazards Earth Syst Sci 10(1):139–148
Liria P, Chust G, Epelde I, Caballero A (2011) Extreme wave flood-risk mapping within the Basque coast. J Coastal Res 64:225–229
Lomoschitz A, Meco J, Corominas J (2002) The Barranco de Tirajana basin, Gran Canaria (Spain). A major erosive landform caused by large landslides. Geomorphology 42:117–130
López-Marinas JM, Salord R (1990) El periodo sísmico oranés de 1790 a la luz de la documentación de los archivos españoles. Instituto Geográfico Nacional, Monografía 6, Madrid
López-Vicente M, Navas A, Machín J (2009) Geomorphic mapping in endorheic catchments in the Spanish Pyrenees: an integrated GIS analysis of karstic features. Geomorphology 111:38–47
Lorenzo F, Alonso A, Pagés JL (2007) Erosion and accretion of beach and spit systems in northwest Spain: a response to human activity. J Coastal Res 23(4):834–845
Losada MA, Baquerizo A, Ortega-Sánchez M, Ávila A (2011) Coastal evolution, sea level, and assessment of intrinsic uncertainty. J Coastal Res 59:218–228
Løvholt F, Pedersen G, Gisler G (2008) Oceanic propagation of a potential tsunami from the La Palma Island. J Geophys Res: Oceans 113(9):C09026
Lucha P, Cardona F, Gutiérrez F, Guerrero J (2008) Natural and human-induced dissolution and subsidence processes in the salt outcrop of the Cardona Diapir (NE Spain). Environ Geol 53:1023–1035
Macau F, Riba O (1962) Situación, características y extensión de los terrenos yesíferos en España. I Coloquio Internacional sobre las Obras Públicas en los terrenos yesíferos, Madrid
Machado MJ, Benito G, Rico MT, Sánchez-Moya Y, Sopeña A (2012) Controles climáticos y ambientales en los registros de paleoavenidas de la Rambla de la Viuda (Castellón). In: González-Díez A, et al (eds) Avances de la Geomorfología en España 2010–2012. Actas de la XII Reunión Nacional de Geomorfología, Publican, Santander, pp 437–440
MAGRAMA (2013) Ministerio de Agricultura, Alimentación y Medio Ambiente. http://www.magrama.gob.es/es/
Marco-Segura JB (2006) Los mapas de riesgo de crecidas en España. Algunas reflexiones sobre 25 años cruciales. In: Díez-Herrero A, Laín-Huerta L, Llorente-Isidro M (eds) Mapas de peligrosidad de avenidas e inundaciones. Métodos, experiencias y aplicación. Publicaciones del Instituto Geológico y Minero de España. Serie Medio Ambiente. Riesgos Naturales No 7. Madrid
Marcos M, Gomis D, Monserrat S, Álvarez-Fanjul E, Pérez B, García-Lafuente J (2005) Consistency of long sea-level time series in the northern coast of Spain. J Geophys Res 110:C03008
Marcos M, Tsimplis MN, Shaw GP (2009) Sea level extremes in southern Europe. J Geophys Res: Oceans 114(C1):C01007
Marcos M, Chust G, Jordà G, Caballero A (2012) Effect of sea level extremes on the western Basque coast during the 21st century. Clim Res 51(3):237–248
MARM-Ministerio de Medio Ambiente y Medio Rural y Marino (2007) Estrategia para la sostenibilidad de la costa. Documento de inicio, Madrid
MARM-Ministerio de Medio Ambiente y Medio Rural y Marino (2011a) Guía Metodológica para el Desarrollo del Sistema Nacional de Cartografía de Zonas Inundables. MAGRAMA, Madrid
MARM-Ministerio de Medio Ambiente y Medio Rural y Marino (2011b) Sistema Nacional de Cartografía de Zonas Inundables en la Demarcación Hidrográfica del Cantábrico Occidental. Confederación Hidrográfica del Cantábrico, Madrid
Marquès MA, Martín E, Gascón M (2001) Episodio extremo y distribución espacial de movimientos de laderas en Montserrat (Barcelona, España). V Simposio Nacional sobre Taludes y Laderas Inestables, Madrid 3:1315–1326
Marquínez J, Menéndez-Duarte R, Fernández-Menéndez S, Fernández-Iglesias E, Jiménez B, Wozniak E, Lastra J, Roces J, Adrados L (2003) Riesgos Naturales en Asturias. KRK Ediciones, Oviedo
Marquínez J, Lastra J, Fernández E (2006) Metodología utilizada para cartografiar la peligrosidad de inundaciones en las cuencas del Norte. In: Díez-Herrero A, Laín-Huerta, L, Llorente-Isidro M (eds)

Mapas de peligrosidad de avenidas e inundaciones. Métodos, experiencias y aplicación. Publicaciones del IGME. Serie: Medio Ambiente. Riesgos Geológicos No 7. Madrid, pp 125–141

Marquínez J, Díez A, Fernández E, Lastra J, Llorente M (2008) Aspectos geomorfológicos en la modificación del Reglamento del Dominio Público Hidráulico y el Sistema Nacional de Cartografía de Zonas Inundables. In: Benavente J, Gracia FJ (eds) Trabajos de Geomorfología en España, 2006–2008. Actas de la X Reunión Nacional de Geomorfología, Cádiz, pp 377–380

Martínez JM, García-Yagüe A (1988) Importancia de los fenómenos de laderas en el Mioceno continental: estudio del río Matayeguas. II Simposio sobre Taludes y Laderas Inestables, Andorra la Vella, pp 79–88

Martínez-Carreras N, Gallart F, Iffly JF, Pfister L, Walling DE, Krein A (2009) Uncertainty assessment in suspended sediment fingerprinting based on tracer mixing models: a case study from Luxembourg. In: Schmidt J, Cochrane T, Phillips C, Elliott T, Basher L (eds) Sediment dynamics in changing environments. IAHS 325:94–105

Martínez-Solares JM (2005) Tsunamis en el contexto de la península Ibérica y del Mediterráneo. Enseñanza de las Ciencias de la Tierra 31(1):52–59

Masson DG, Watts AB, Gee MJR, Urgeles R, Mitchell NC, Le Bas TC, Canals M (2002) Slope failures on the flanks of the western Canary Islands. Earth-Sci Rev 57:1–35

Mateos RM, Rodríguez-Peces MJ, Azañón JM, Rodríguez-Fernández J, Roldán FJ, García-Moreno I, Gelabert B, García-Mayordomo J (2013) El deslizamiento de Bàlitx (Mallorca) y su posible origen sísmico. Procesos activos desde el Pleistoceno superior. Boletín Geológico y Minero 124(1):41–61

Medina R, Losada IJ, Méndez FJ, Olabarrieta M, Liste M, Menéndez M, Tomás A, Abascala J, Agudelo P, Guanche R, Luceño A (2004) Impacto en la Costa Española por Efecto del Cambio Climático. Ministerio de Medio Ambiente, Oficina Española de Cambio Climático-Dirección General de Costas, Madrid

Méndez FJ, Medina R, Losada IJ, Olabarrieta M, Tomás A, Liste M, Menéndez M, Abascal AJ, Agudelo P, Castanedo S (2004) Estudio de los impactos en la costa española por efecto del cambio climático. In: García-Codrón JC, Diego-Liaño C, Fdez. de Arróyabe-Hernáez P, Garmendia-Pedraja C, Rasilla-Álvarez D (eds) El Clima entre el Mar y la Montaña. Asociación Española de Climatología y Universidad de Cantabria, Serie A, no 4, Santander

Mendoza ET, Jiménez JA (2009) Regional vulnerability analysis of Catalan beaches to storms. Proc Inst Civil Eng: Marit Eng 162(3):127–135

Ministerio de Fomento (1879) Ley de Aguas, Gaceta de Madrid. Madrid

Ministerio de Fomento (2012) Catálogo de tsunamis en las costas españolas. http://www.fomento.gob.es/MFOM/LANG_CASTELLANO/DIRECCIONES_GENERALES/INSTITUTO_GEOGRAFICO/Geofisica/sismologia/

Ministerio de Interior (2013) http://www.interior.gob.es/anuarios-estadisticos-12/anuario-estadistico-de-2011-1892

Ministerio de Obras Públicas y Transportes (1988) Las Inundaciones en la España Peninsular. Informe de Síntesis. Dirección General de Obras Hidráulicas, Dirección General de Protección Civil, Madrid

Moreno JM (Coord.) (2005) Evaluación Preliminar de los Impactos en España por Efecto del Cambio Climático. Proyecto ECCE - INFORME FINAL. Ministerio de Medio Ambiente, Madrid

Moro-Martín A (2012) Spanish changes are scientific suicide. Nature 482:277

Moya J, Corominas J (1997) Condiciones pluviométricas desencadenantes de deslizamientos en el Pirineo Oriental. IV Simposio Nacional sobre Taludes y Laderas Inestables, Granada 1, pp 199–212

Moya J, Corominas J (2005) Cálculo de la frecuencia de desprendimientos mediante la dendrocronología en el Solà d'Andorra, Andorra la Vella. VI Simposio Nacional sobre Taludes y Laderas Inestables, Valencia 2, pp 616–627

Nuhfer EB, Proctor RJ, Moser PH (1993) The citizens' guide to geologic hazards. Am Inst Prof Geol, p 167

Olcina-Cantos J (2007) Riesgo de inundaciones y ordenación del territorio en España. Fundación Instituto Euromediterráneo del Agua, Murcia

Oller P, González M, Pinyol J, Barberà M, Martínez P (2011) The geological hazard prevention map of Catalonia 1:25,000. A tool for geohazard mitigation. Proceedings of the Second World Landslide Forum, Rome

Omira R, Baptista MA, Matías L, Miranda JM, Catita C, Carrilho F, Toto E (2009) Design of a sea-level tsunami detection network for the Gulf of Cádiz. Nat Hazards Earth Syst Sci 9(4):1327–1338

Ortega JA, Garzón G (2003) Palaeohydrology of the lower Guadiana river basin. In: Thorndycraft VR, Benito G (eds) Palaeofloods, historical floods and climatic variability: applications in flood risk assessment. Proceedings of the PHEFRA workshop. CSIC, Madrid, pp 33–38

Ortega JA, Garzón G (2004) Influencia de la Oscilación del Atlántico Norte en las inundaciones del río Guadiana. In: Benito G, Díez-Herrero A (eds) Riesgos naturales y antrópicos en Geomorfología. Actas de la VIII Reunión Nacional de Geomorfología, Toledo, pp 117–126

Ortega JA, Garzón G (2008) La influencia de los embalses en la regulación de crecidas de alta frecuencia en el río Guadiana. In: Benavente J, Gracia FJ (eds) Trabajos de Geomorfología en España, 2006–2008. Actas de la X Reunión Nacional de Geomorfología, Cádiz, pp 385–388

Pain E (2012) Research cuts will cause exodus from Spain. Science 336:139–140

Pando L, Pulgar JA, Gutiérrez-Claverol M (2013) A case of man-induced ground subsidence and building settlement related to karstified gypsum (Oviedo, NW Spain). Environ Earth Sci 68(2):507–519

Periáñez R, Abril JM (2013) Modeling tsunami propagation in the Iberia-Africa plate boundary: historical events, regional exposure and the case-study of the former Gulf of Tartessos. J Mar Syst 111–112:223–234

Pinyol NM, Alonso EE, Corominas J, Moya J (2012) Canelles landslide: modelling rapid drawdown and fast potential sliding. Landslides 9:33–51

Plomaritis TA, Del Río L, Benavente J (2011) Validating storm threshold using a single environmental parameter: the case of Cadiz coast. J Coastal Res 64:1876–1880

Potenciano A, Garzón G (2008) Condicionantes geomorfológicos e hidrológicos en la generación de avenidas. El caso de las cuencas del río Córcoles (Guadiana) y Escabas (Tajo). In: Benavente J, Gracia FJ (eds) Trabajos de Geomorfología en España, 2006–2008. Actas de la X Reunión Nacional de Geomorfología, Cádiz, pp 389–392

PSMSL (2013) http://www.psmsl.org/products/trends/, (visited, May 20, 2013)

Puch C (1998) Grandes cuevas y simas de España. Federación Española de Espeleología, Barcelona

Pueyo-Anchuela O, Casas-Saínz AM, Soriano MA, Pocoví-Juan A (2010) A Geophysical survey routine for the detection of doline areas in the surroundings of Zaragoza (NE Spain). Eng Geol 114(3–4):382–396

Rangel-Buitrago N, Anfuso G (2011) Morphological changes at Levante Beach (Cádiz, SW Spain) associated with storm events during the 2009–2010 winter season. J Coastal Res 64:1886–1890

Remondo J, González-Díez A, Díaz de Terán JR, Cendrero A (2003) Quantitative landslide susceptibility models by means of spatial data analysis techniques; a case study in the lower Deva valley, Guipúzcoa (Spain). Nat Hazards 30(3):267–279
Remondo J, Bonachea J, Cendrero A (2005a) Probabilistic landslide hazard and risk mapping on the basis of occurrence and damages in the recent past. In: Lacerda W, Ehrlich M, Fontoura S, Sayão A (eds) Landslides: evaluation and stabilization, vol 1, pp 125–130
Remondo J, Soto J, González-Díez A, Díaz de Terán JR, Cendrero A (2005b) Human impact on geomorphic processes and hazards in mountain areas in Northern Spain. Geomorphology 66(1–4):69–84
Remondo J, Bonachea J, Cendrero A (2008) Quantitative landslide risk assessment and mapping on the basis of recent occurrences. Geomorphology 94:496–507
Rico M (2004) Las paleocrecidas en la cuenca media del Río Segre durante el Pleistoceno superior-Holoceno: registros morfosedimentarios y análisis hidrológico. Ph. D. Thesis. Universidad de Zaragoza, Zaragoza
Rivas V, Cendrero A (1991) Use of natural and artificial accretion in the north coast of Spain; historical trends and assessment of some environmental and economic consequences. J Coastal Res 7(2):491–507
Rivas V, Cendrero A (1995) Human influence in a low-hazard coastal area; an approach to risk assessment and proposal of mitigation strategies. J Coastal Res 12:289–298
Rodríguez V, Gutiérrez F, Green AG, Carbonel D, Horstmeyer H, Schmelzbach C (2013) Characterizing sagging and collapse sinkholes in a mantled karst by means of ground penetrating radar (GPR). Environ Eng Geosci (in press)
Rodríguez-Ramírez A (1998) Geomorfología del Parque Nacional de Doñana y su entorno. Ministerio de Medio Ambiente
Rodríguez-Ramírez A, Rodríguez-Vidal J, Cáceres L, Clemente L, Belluomini G, Manfra L, Improta S, Ramón de Andrés J (1996) Recent coastal evolution of the Doñana National Park (SW Spain). Quatern Sci Rev 15:803–809
Rodríguez-Ramírez A, Morales JA, Delgado I, Cantano M (2008) The impact of man on the morphodynamics of the Huelva coast (SW Spain). J Iberian Geol 34(2):73–108
Rodríguez-Vidal J, Ruiz F, Cáceres LM, Abad M, González-Regalado ML, Pozo M, Carretero MI, Monge-Soares AM (2011) Geomarkers of the 218–209 BC Atlantic tsunami in the Roman Lacus Ligustinus (SW Spain): a palaeogeographical approach. Quatern Int 242:201–212
Roger J, Hébert H (2008) The 1856 Djijelli (Algeria) earthquake and tsunami: source parameters and implications for tsunami hazard in the Balearic Islands. Nat Hazards Earth Syst Sci 8(4):721–731
Roger J, Hebert H, Ruegg JC, Briole P (2011) The El Asnam 1980 October 10 inland earthquake: a new hypothesis of tsunami generation. Geophys J Int 185:1135–1146
Rubiales JM, Bodoque JM, Ballesteros JA, Díez-Herrero A (2008) Response of Pinus sylvestris roots to sheet-erosion exposure: an anatomical approach. Nat Hazards Earth Syst Sci 8(2):223–231
Ruiz-Bellet JL, Balasch JC, Tena A (2011) Transporte de sedimento en suspensión durante una crecida en un río regulado: el río Segre (NE Península Ibérica). Cuaternario y Geomorfología 25(3–4):87–101
Ruiz-Villanueva R, Bodoque JM, Díez-Herrero A, Eguibar MA (2012) La influencia del transporte de detritos leñosos (woody debris) en la peligrosidad por avenidas torrenciales. In: González-Díez A, et al (eds) Avances de la Geomorfología en España 2010–2012. Actas de la XII Reunión Nacional de Geomorfología, Publican, Santander, pp 143–146
Salazar A, Ortega LI (1990) Los deslizamientos en la alineación costera guipuzcoana (Sector Menditorrotz-Igueldo). Actas de la I Reunión Nacional de Geomorfología, Teruel, pp 551–560
Sánchez-Arcilla A, Jiménez JA, Valdemoro HI, Gracia V (2008) Implications of climatic change on Spanish Mediterranean low-lying coasts: the Ebro delta case. J Coastal Res 24(2):306–316
Sánchez FJ, Soriano A (2001) Metodología para el estudio de estabilidad de laderas en embalses. V Simposio Nacional sobre Taludes y Laderas Inestables, Madrid 1, pp 87–98
Sanjaume E, Pardo-Pascual JE (2005) Erosion by human impact on the Valencian coastline (E of Spain). J Coastal Res 49:76–82
Santacana N, De Paz A, Baeza C, Corominas J, Marturià J (2003) A GIS-based multivariate statistical analysis for shallow landslide susceptibility mapping in La Pobla de Lillet area (Eastern Pyrenees, Spain). Nat Hazards 30(3):281–295
Sanz E (1992) El deslizamiento de ladera de Güevejar (Granada) durante los terremotos de Lisboa (1755) y Andalucía (1884). III Simposio Nacional sobre Taludes y Laderas Inestables, La Coruña 1, pp 195–203
Simón-Gómez JL, Martínez-Gil J, Soriano-Jiménez MA, Arlegui-Crespo L, Caballero-Burbano J (1998) Estudio de riesgos naturales de los terrenos de la orla sudoeste del suelo urbanizable. Informe público Ayuntamiento de Zaragoza (Anexo 3.1. PGOU). http://www.zaragoza.es/contenidos/urbanismo/pgouz/memoria/anejos/anejo03/anejo031.pdf
Soler C (1997) Gigantescos deslizamientos en islas volcánicas y su repercusión en la hidrogeología insular. IV Simposio Taludes y Laderas Inestables, Granada 1, pp 325–337
Somoza L, Díaz del Río V (1992) Expected morphological impact of sea level rise in a beach-barrier complex: Mar Menor (Spain). In: Suárez de Vivero JL (ed) The Ocean Change: management patterns and the environment. IGU Commission on Marine Geography, Sevilla, pp 13–18
Soria JM, Alfaro P, Estévez A, Delgado J, Durán JJ (1999) The Holocene sedimentation rates in the Lower Segura Basin (eastern Betic Cordillera, Spain): eustatic implications. Bulletin de la Société Géologique de France 170(3):349–354
Soriano MA, Simón JL (1995) Alluvial dolines in the central Ebro basin, Spain: a spatial and developmental hazard analysis. Geomorphology 11:295–309
Soriano MA, Simón JL (2002) Subsidence rates and urban damages in alluvial dolines of the Central Ebro basin (NE Spain). Environ Geol 42:476–484
Tena A, Batalla RJ, Vericat D, Lopez-Tarazon JA (2011) Suspended sediment dynamics in a large regulated river over a 10-year period (the lower Ebro, NE Iberian Peninsula). Geomorphology 125:73–84
Thorndycraft VR, Benito G (2006a) The Holocene fluvial chronology of Spain: evidence from a newly compiled radiocarbon database. Quatern Sci Rev 25:223–234
Thorndycraft VR, Benito G (2006b) Late Holocene fluvial chronology of Spain: the role of climatic variability and human impact. Catena 66(1–2):34–41
Thorndycraft V, Benito G, Rico M, Sopeña A, Sánchez Y, Casas M (2005) Palaeoflood hydrology of the Llobregat River, NE Spain: a 3000 year record of extreme floods. J Hydrol 313(1–2):16–31
Trigo RM, Pozo-Vázquez D, Castro-Díez Y, Osborn T, Gamis-Fortis S, Esteban-Parra MJ (2003) NAO influence on precipitation, river flow regimes and hydroelectric power generation in the Iberian Peninsula. Geophys Res Abstr 5:54–94
Úbeda X, Vericat D, Batalla RJ (eds) (2010) Avances de la geomorfología en España 2008–2010. Actas de la XI Reunión Nacional de Geomorfología, Lérida
UC-IGN (2009) Tsunami inundation maps along the south west Spanish coast. Deliverable d7.1 of the 6th European framework programme project TRANSFER. http://www.transferproject.eu/
Vaquero JM (2004) Solar signal in the number of floods recorded for the Tagus River Basin over the last millennium. Comment on

"Magnitude and frequency in the Tagus Basin (Central Spain) over the last millennium" by G. Benito et al. (2003). Clim Change 66:23–26

Vericat D, Batalla RJ (2004) Efectos de las presas en la dinámica fluvial del curso bajo del río Ebro. Cuaternario y Geomorfología 18(1–2):37–50

Vericat D, Batalla RJ (2005) Sediment transport in a highly regulated fluvial system during two consecutive floods (lower Ebro River, NE Iberian Peninsula). Earth Surf Proc Land 30:385–402

Vericat D, Batalla RJ (2006) Sediment transport in a large impounded river: the lower Ebro, NE Iberian Peninsula. Geomorphology 79:72–92

Vericat D, Batalla RJ, García C (2008) Bed-material mobility in a large river below dams. Geodin Acta 21:3–10

Vericat D, Brasington J, Wheaton J, William R, Rychkov I (2012) Monitoring channel change in gravel bed rivers: hyperscale surface modelling by means of terrestrial laser scanning. In: González-Díez A, et al (eds) Avances de la Geomorfología en España 2010–2012. Actas de la XII Reunión Nacional de Geomorfología, Publican. Santander, pp 465–468

Zarroca M, Linares R, Roqué C, Rosell J, Gutiérrez F (2013) Integrated geophysical and morphostratigraphic approach to investigate a coseismic? Translational slide responsible for the destruction of Montclús village (Spanish Pyrenees). Landslides (in press)

Zazo C, Goy JL, Somoza L, Dabrio C, Belluomini G, Improta S, Lario J, Bardají T, Silva PG (1994) Holocene sequence of relative sea level fluctuations in relation to the climatic trends in the Atlantic-Mediterranean linkage coast. J Coastal Res 10(4):933–945

Zitellini N, Grácia E, Matias L, Terrinha P, Abreu MA, De Alteriis G, Henriet JP, Dañobeitia J, Masson DG, Mulder T, Ramella R, Somoza L, Díez S (2009) The quest for the Africa-Eurasia plate boundary west of the strait of Gibraltar. Earth Planet Sci Lett 280(1–4):13–50

Index

F. Gutiérrez and M. Gutiérrez (eds.), *Landscapes and Landforms of Spain*, World Geomorphological Landscapes,
DOI: 10.1007/978-94-017-8628-7,

MIX
Papier aus verantwortungsvollen Quellen
Paper from responsible sources
FSC® C105338

If you have any concerns about our products,
you can contact us on
ProductSafety@springernature.com

In case Publisher is established outside the EU,
the EU authorized representative is:
Springer Nature Customer Service Center GmbH
Europaplatz 3, 69115 Heidelberg, Germany

Printed by Libri Plureos GmbH
in Hamburg, Germany